OXIDANTS, ANTIOXIDANTS, AND FREE RADICALS

OXIDANTS, ANTIOXIDANTS, AND FREE RADICALS

Edited by

Steven I. Baskin, Pharm.D., Ph.D.
Pharmacology Division,
U.S. Army Medical Research Institute of Chemical Defense
Aberdeen Proving Ground
Edgewood Area, Maryland

and

Harry Salem, Ph.D.
U.S. Army Edgewood Research Development and Engineering Center
Aberdeen Proving Ground
Edgewood Area, Maryland

Taylor & Francis
Publishers since 1798

USA	Publishing Office:	Taylor & Francis 1101 Vermont Avenue, NW, Suite 200 Washington, DC 20005-3521 Tel: (202) 289-2174 Fax: (202) 289-3665
	Distribution Center:	Taylor & Francis 1900 Frost Road, Suite 101 Bristol, PA 19007-1598 Tel: (215) 785-5800 Fax: (215) 785-5515
UK		Taylor & Francis Ltd. 1 Gunpowder Square London EC4A 3DE Tel: 0171 583 0490 Fax: 0171 583 0581

OXIDANTS, ANTIOXIDANTS, AND FREE RADICALS

The author and editors have made every effort to ensure the accuracy of the information herein, particularly with regard to drug selection and dose. However, appropriate information sources should be consulted, especially for new or unfamiliar procedures. It is the responsibility of every practitioner to evaluate the appropriateness of a particular opinion in the context of actual clinical situations and with due considerations of new developments. The author, editors, and the publisher cannot be held responsible for errors or for any consequences arising from the use of the information contained herein. The opinions in this book are personal to the authors and do not represent those of the U.S. Army or Department of Defense.

1 2 3 4 5 6 7 8 9 0   EBEB   9 8 7

This book was set in Times Roman. The copy and production editors were Elizabeth Dugger and Heather Worley respectively. Cover design by Ed Atkeson; Berg Design.

A CIP catalog record for this book is available from the British Library.
⊗ The paper in this publication meets the requirements of the ANSI Standard Z39.48-1984 (Permanence of Paper)

Library of Congress Cataloging-in-Publication Data

Oxidants, antioxidants, and free radicals/edited by Steven I. Baskin
  and Harry Salem.
     p. cm.
     Includes bibliographical references and index.

     1. Oxidation, Physiological.   2. Antioxidants—Physiological
  effect.   3. Free radicals (Chemistry)—Pathophysiology.   4. Lipids—
  Peroxidation.   I. Baskin, Steven I.   II. Salem, Harry, 1929–   .
     [DNLM: 1. Antioxidants—metabolism—congresses.   2. Oxidants—
  metabolism—congresses.   3. Oxidative Stress—drug effects—
  congresses.   4. Free Radicals—congresses.   5. Lipid Peroxidation—
  physiology—congresses.   6. Vitamins—therapeutic use—congresses.
  QV 800 O98 1997]
  RB170.O945   1997
  615.9′07—dc21
  DNLM/DLC
  for Library of Congress   97-13108paCIP

ISBN 1-56032-644-1 (case) ✓

For the tender moments, the unbridled joy, and the time
For the love, devotion, and help we receive, we thank

our Wives: Susan and Flo
our Children: Miriam, Tzvi, Lloyd and Kendall and Jerry and Amy and
our Children's children: Shifra, Yehuda, Joel, and Marshall

Your touch made us remember who we are
Your gift made this book happen

Steven and Harry

CONTENTS

CHAPTER 3
Vitamin E   79
Sharon Landvik

CHAPTER 4
Vitamin E and Polyunsaturated Fats: Antioxidant and Pro-oxidant Relationship   95
Melanie A. Banks

CHAPTER 5
α-Tocopherol, β-Carotene, and Oxidative Modification of Human
   Low-Density Lipoprotein   113
Hazel T. Bowen and Stanley T. Omaye

Contents   xi

CHAPTER 13

Toxicity of Oxygen and Ozone   227
Harry Salem and Steven I. Baskin

Oxygen   227
Oxidative Stress   228
Oxygen Injury   229
Ozone Injury   229
Comparison of Models   233
References   233

CHAPTER 14

Peroxidation of Lipids and Liver Damage   237
Pablo Muriel

From Free Radicals to Lipid Peroxidation   237
  Oxygen Toxicity   237
  Properties of Free Radicals   237
  Peroxidation of Lipids   237
Peroxidation of Lipids and Alcohol-Induced Liver Damage   239
  Alcohol and Liver Diseases   239
  Role of Peroxidation of Lipids   239
  Promotion of Lipid Peroxidation through Interactions with Cysteine, Glutathione,
    Vitamin E, and Iron   240
Lipid Peroxidation in Other Types of Liver Injury   241
  Paracetamol-Induced Liver Necrosis and Lipid Peroxidation   241
  Liver Damage Induced by Carbon Tetrachloride   244
  Biliary Obstruction-Induced Liver Damage   247
  Naproxen-Induced Lipid Peroxidation   247
  Halothane and Sevoflurane   250
Conclusions   251
References   252

CHAPTER 15

Role of Free Radicals in Alcohol-Induced Tissue Injury   259
Carol A. Colton and Samir Zakhari

ROI and RNI   259
  Cellular Sources of ROI and RNI   259
  Reactions Involving ROI and RNI   260
Protective Mechanisms   262
Ethanol-Induced ROI   262
  Increased Production of ROI   262
  Possible Sources of Alcohol-Induced ROI   263
Role of ROI in Alcohol-Induced Tissue Injury   264
  Liver Damage   264
  Macrophages   265
  Fetal Alcohol Syndrome   265
  Brain Damage   266
Mechanisms of ROI-Induced Tissue Injury   266
  Increased Extracellular LPS   266
  Altered Intracellular Signal Transduction   266
  Fatty Acid Oxygenases and Eicosanoids   266

FOREWORD

Oxidants, Antioxidants, and Free Radicals serves as an authoritative source of information on current topics of research in the area of oxidative stress and antioxidant defense systems. The book is designed to assist researchers and educators to keep abreast of developments in fields other than their own and to provide a source of reference materials for use in research and in graduate teaching. It was compiled with the intention of stimulating cross-fertilization of ideas and encouraging synergistic interactions among readers.

Dr. Steven I. Baskin and Dr. Harry Salem have accomplished a heroic feat in the coordination into a monograph of 20 excellent articles from investigators at the cutting edge of antioxidant, oxidant, and free radical research. I believe that this book will stimulate much more interest in research into the role of oxidation, free radicals, and antioxidants in health and diseases, particularly with respect to developing more understanding about the many interactions between oxidants and antioxidants and how such substances may act as natural protectants and/or natural toxicants.

The overprotection of oxygen free radical species has been linked to a very large number of human diseases. Thus, by-products of such reactions may be directly or indirectly responsible for biochemical lesions, such as lipid peroxidation, oxidation of proteins, nuclear material, and other biologic material. Fortunately, protection from or modulation of such oxidative-induced damage can be afforded by antioxidants and antioxidant enzymes. Many of the antioxidants can be found as part of our natural diet. Other antioxidants, either man-made or modified from other compounds, are gradually making their way into therapeutic regimens.

Because of the potential of such wide implications of the link between oxidative by-products and human diseases and the modulation of such by-products by various antioxidant-like compounds, readers from many disciplines such as toxicology, nutrition, food science, pathology, pharmacology, and various medical fields will find this monograph a useful resource.

Stanley T. Omaye

PREFACE

Recollections from fundamental chemistry and biology aided us to take early notice of oxidation and reduction reactions that form some of the bases of antioxidant action. As we became more aware of biochemical processes in our environment, we came to realize that oxidation, reduction, and free radical reactions are continually taking place in our surroundings to carry out the everyday biology of life.

Because of the timeliness of this topic, a symposium was cosponsored by the NCACSOT, AGT, and Sigma Xi on this subject. As a result, some of the significant presentations were provided as chapters for this book along with others from investigators active in the field.

The chapters of this book provide the present state of knowledge on important aspects of antioxidants, oxidants, and free radicals that are thought to occur in biological functions. Because it is recognized that it would be a Herculean task to cover all areas robustly, certain key areas were highlighted to allow the reader to gain insight into current trends in research and a glimmer of a vision into the future in this exciting area.

The editors have tried to provide a balance of topics to serve the questioning student, the researcher, and the ardent follower of the literature. This mix, we feel, is unique, and it is anticipated that the blend of a wide variety of areas tied by the common thread of antioxidant, oxidant, and free radical papers will provide some stories never told before. The chapters in this book paint a picture of how the family of antioxidants may all function by multiple mechanisms to carry out the fabric of life. It is hoped that more efficient and judicious use of some of these factors will improve and extend it.

Steven Baskin and Harry Salem
Bel Air, Maryland
April 1996

CONTRIBUTORS

Ian N. Acworth, D.Phil.
ESA, Inc.
22 Alpha Road
Chelmsford, MA 01824-4171
and Massachusetts College of Pharmacy
  and Allied Health Sciences
179 Longwood Avenue
Boston, MA 02115

Carmen M. Arroyo, Ph.D.
U.S. Army Medical Research
  Institute of Chemical Defense
ATTN: SGRD-UV-DA
3100 Ricketts Point Road
Aberdeen Proving Ground, MD 21010-5425

Melanie A. Banks, Ph.D.
Nutrition Svcs/M17
Cleveland Clinic Foundation
Cleveland, OH 44195

Steven I. Baskin, Pharm.D., Ph.D.
Commander
U.S. Army Medical Research
  Institute of Chemical Defense
ATTN: MCMR-UV-PB/Dr. Baskin
3100 Ricketts Point Road
Aberdeen Proving Ground, MD 21010-5425

Hazel T. Bowen, M.S.
Department of Nutrition
University of Nevada
Reno, NV 89557

Carol A. Colton, Ph.D.
Department of Physiology and Biophysics
Georgetown University School of Medicine
Washington, DC 20007

Nabil M. Elsayed, Ph.D.
Department of Respiratory Research
Division of Medicine
Walter Reed Army Institute of Research
Washington, DC 20307-5100

Sidney Green, Ph.D.
Department of Toxicology
Corning Hazleton, Inc.
9200 Leesburg Pike
Vienna, VA 22182

Madeline M. Hall, Ph.D.
Department of Biology
Cleveland State University
Cleveland, OH 44115

Kazim Husain, Ph.D.
Department of Pharmacology
Southern Illinois University School
  of Medicine
P.O. Box 19230
Springfield, IL 62794-1222

Michael Iatropoulos, M.D., Ph.D.
American Health Foundation
One Dana Road
Valhalla, NY 10595

Jill R. Keeler, Ph.D., CRNA
U.S. Army Medical Command
  Center School
Anesthesia Branch
Ft. Sam Houston, TX 78234-6140

Sharon Landvik, M.S., R.D.
Veris Research Information Service
5325 South Ninth Avenue
La Grange, IL 60525

Timothy J. Maher, Ph.D.
Massachusetts College of Pharmacy and
  Allied Health Sciences
179 Longwood Avenue
Boston, MA 02115

Douglas R. McCabe, M.S.
ESA, Inc.
22 Alpha Road
Chelmsford, MA 01824-4171

James J. McGrath, Ph.D.
Department of Physiology
Texas Tech University Health
  Sciences Center
Lubbock, TX 79430

Pablo Muriel, Ph.D.
Centro de Investigación y de
  Estudios Avanzados del I.P.N.
Departamento de Farmacología y Toxicología
Apdo. Postal 14-740
México 07000, D.F.
México

Stanley T. Omaye, Ph.D.
Department of Nutrition
University of Nevada
Reno, NV 89557

Dale W. Porter, Ph.D.
Health Effects Laboratory Division
National Institute for Occupational
  Safety and Health
1095 Willowdale Road
Morgantown, WV 26505

Saul R. Powell, Ph.D.
Department of Surgery
North Shore University Hospital
350 Community Drive
Manhasset, NY 11030

Saura C. Sahu, Ph.D.
Division of Toxicological Research
Center for Food Safety and
  Applied Nutrition
U.S. Food and Drug Administration
8301 Muirkirk Road
Laurel, MD 20708

Harry Salem, Ph.D.
Edgwood Research Development and
  Engineering Center
Aberdeen Proving Ground, MD 21010-5423

Eric C. Schlorff, B.S.
Department of Pharmacology
Southern Illinois University School
  of Medicine
P.O. Box 19230
Springfield, IL 62794-1222

Alfred Mario Sciuto, Ph.D.
Pharmacology Division
Neurotoxicology Branch
U.S. Army Medical Research
  Institute of Chemical Defense
3100 Ricketts Point Road
Aberdeen Proving Ground, MD 21010-5425

Robert D. Short, Pharm.D., Ph.D.
Wellington Environmental Toxicology
  Department
St. Louis, MO 63144

Satu M. Somani, Ph.D.
Department of Pharmacology
Southern Illinois University School
  of Medicine
P.O. Box 19230
Springfield, IL 62794-1222

Viktor Sorokin, Ph.D.
Department of Chemistry and
  Biochemistry
Center for Medicinal Chemistry Research
University of Texas at Arlington
Arlington, TX 76019

Andrew L. Ternay, Jr., Ph.D.
Department of Chemistry and
  Biochemistry
Center for Medicinal Chemistry Research
University of Texas at Arlington
Arlington, TX 76019

Harry Van Keulen, Ph.D.
Department of Biology
Cleveland State University
Cleveland, OH 44115

Kokiku Wakayama, M.D.
Department of Pharmacology
Medical College of Philadelphia
Philadelphia, PA 19110

James M. Willard, Ph.D.
Department of Biology
Cleveland State University
Cleveland, OH 44115-2403

Gary M. Williams, M.D.
American Health Foundation
One Dana Road
Valhalla, NY 10595

Samir Zakhari, Ph.D.
Biomedical Research Branch
National Institute on Alcohol
 Abuse and Alcoholism
6000 Executive Blvd
Bethesda, MD 20892-7003

ABOUT THE EDITORS

DR. STEVEN I. BASKIN

Dr. Steven I. Baskin was born in Los Angeles, CA, on November 14, 1942. He received his Pharm.D. with honors from the University of Southern California in 1966. He served as a pharmacy intern at Los Angeles County General Hospital, 1965–1966. He received his Ph.D. in pharmacology, funded as a National Institutes of Health trainee, from Ohio State University in 1971. He completed his postdoctorate fellowship in 1973 from Michigan State University. In addition to being a registered pharmacist, Dr. Baskin was elected a Diplomat of the American Board of Toxicology in 1987 (re-certified 1993) and a Fellow of the Academy of Toxicological Sciences in 1994.

Dr. Baskin joined the staff of the U.S. Army Medical Research Institute of Chemical Defense in December 1981. While his research has focused primarily on mechanisms of cyanide intoxication and therapies for the treatment of this poison, he also devoted studies to the cardiac actions of oximes and organophosphates, as well as muscarinic and other compounds that would antagonize a cholinergic-induced cardiomyopathy.

Dr. Baskin is a member of the American Chemical Society, American Society of Pharmacology and Experimental Therapeutics, Society of Toxicology, Association of Government Toxicologists, and Cardiac Muscle Society, the American Association of Forensic Science, Sigma Xi, and a fellow of the American College of Cardiology. He is editor or coeditor of 5 books, and author or coauthor of 78 scientific articles, and 36 chapters in books. He is presently STO B (Cyanide) program coordinator and a member of the Drug Assessment Technical Evaluation Committee.

DR. HARRY SALEM

Dr. Harry Salem, Chief Scientist for Life Sciences, U.S. Army Edgewood Research, Development and Engineering Center, was previously Chief of the Life Sciences Department and the Toxicology Division, Chemical Research, Development and Engineering Center, Aberdeen Proving Ground, MD. His research interests and experience are in inhalation and general pharmacology and toxicology. He has served on the editorial boards of several professional journals and is currently a member of the Editorial Board of the *Journal of Inhalation Toxicology* and editor-in-chief of the *Journal of Applied Toxicology*. Dr. Salem has served as associate professor of pharmacology at the University of Pennsylvania, adjunct professor of environmental health at Temple University, and is currently adjunct professor of chemical toxicology at Drexel University.

Dr. Salem was also employed by small and large industry in pharmacology and toxicology. He serves on national and international committees and is Chairman of the Technical Committee for the Inhalation Specialty Section of the Society of Toxicology. His professional affiliations include the American Association for the Advancement of Science, American Chemical Society, American College of Toxicology (charter member), American Conference of Governmental Industrial Hygienists, Inc., American Society of Clinical Pharmacology and Therapeutics, American Society for Pharmacology and Experimental Therapeutics, Association of Governmental Toxicologists (where he served as president), Chemical Corps Association, Inc., European Society of Toxicology, International Regulatory Pharmacology and Toxicology, International Society on

Toxicology, National Capitol Area Chapter Society of Toxicology (where he currently serves as president), Society of Toxicology, and Sigma Xi (where he also served as president).

He serves on the steering committee for the Society of Comparative Ophthalmology, and is a fellow of the New York Academy of Sciences, the American College of Clinical Pharmacology, and the Academy of Toxicological Sciences, where he served on the Professional Standards Evaluation Board and currently serves on the Board of Directors.

In 1989, Dr. Salem was awarded the Decoration for Meritorious Civilian Service for his contributions to the field of toxicology both nationally and internationally.

He has contributed many scientific papers and is the coauthor and coeditor of three volumes of the *International Encyclopedia of Pharmacology and Therapeutics* on antitussive agents, the coauthor of a book, *The Biological and Environmental Chemistry of Chromium*, the editor of the textbook *Inhalation Toxicology*, and coeditor of the book *New Technologies and Concepts for Reducing Drug Toxicities* with Dr. Baskin. Dr. Salem received a B.A. degree (1950) from the University of Western Ontario in Canada, B.Sc. degree (1953) in pharmacy from the University of Michigan, and M.A. (1955) and Ph.D. (1958) degrees in pharmacology from the University of Toronto in Canada.

Chapter One

REDOX, RADICALS, AND ANTIOXIDANTS

Andrew L. Ternay, Jr. and Viktor Sorokin

This chapter presents a synopsis of some radical-mediated processes, focusing on phenomena related to biochemistry and medicine. The importance of radicals in biological processes was recognized over a half-century ago (1). More recently, Willson has provided a convenient summary of the sequence of events involved in free-radical induced cell damage (Figure 1) (2). Since biological damage created by radicals often is associated with "oxidation", the literature places more emphasis upon oxidations than reductions. Yet bear in mind that oxidation reactions are one-half of oxidation–reduction (redox) couples and that every oxidation is accompanied by a reduction.

After a summary in this chapter of terminology dealing with oxidation and reduction, attention shifts to radicals since they are arguably the most important intermediates in medicine as well as in biological oxidation–reduction systems. Since oxygen-containing radicals play a substantial role in initiating tissue damage, these species are introduced in the next section. This is followed by an examination of some of the more widely investigated naturally occurring antioxidants, that is, L-ascorbic acid, β-carotene, vitamin E, and coenzyme Q. The chapter ends with a presentation of radical chain reactions and mechanisms whereby antioxidants can interfere with these processes. Strategies for antioxidant defense in living systems have been reviewed (3–5) and comparisons between natural and synthetic antioxidants made (6).

Given the vastness of the topic, the authors apologize to the many whose excellent work has been omitted because of limitations on chapter size.

THE LANGUAGE OF REDOX

Inorganic chemists frequently classify atoms within molecules by their oxidation number (ON) (7). An element with a given ON is described as having the "oxidation state" corresponding to that value and, while not strictly identical, these terms often are used interchangeably. An

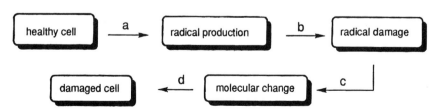

Figure 1. Diminishing radical-induced cell damage: a = radical formation prevention; b = radical scavenging; c = radical repair; d = biochemical repair.

1

oxidation reaction can be described as one in which the oxidation number of an atom becomes more positive while reduction corresponds to it becoming less positive (or more negative).

Although a number of metals have more than one ON, one useful definition of a metal's ON is *the positive number equal to the charge on that metal's oxide*. Thus, the ON of zinc in zinc oxide (ZnO) is +2. Potent inorganic oxidants often contain metals with large, positive oxidation numbers. For example, the ON of manganese in the permanganate ion (MnO_4^-) is +7 while that of chromium in both chromate (CrO_4^{2-}) and dichromate ions ($Cr_2O_7^{2-}$) is +6. After functioning as an oxidant, the ON of the metal becomes less positive, such as MnO_4^- becoming MnO_2 or Mn^{2+} (ON of Mn = +4 or +2, respectively). Particularly large values (e.g., +7) do not represent the actual charge on a metal, with the true value being smaller due to "back bonding" to atoms associated with the metal. Consequently, these values sometimes are termed "formal" oxidation numbers. While metals often have an ON equal to their group number in the traditional periodic table, nonmetals have ONs equal to eight minus their group number. Like metals, nonmetals may exist with several ONs. For example, the ON for oxygen in water or alcohol is −2 but is −1 in molecules containing an −O−O− fragment, such as hydrogen peroxide, and −1/2 in superoxides. Hydrogen can have positive (+1 in water) and negative (−1 in metal hydrides) ONs. Regardless of an individual atom's ON, the sum of the ONs of all elements in any neutral species must equal zero.

The application of ONs to organic molecules is more difficult. With alkanes, for example, the ON of carbon is obtained by dividing the total number of hydrogens by the number of carbons and assigning the value a negative sign. Thus the ON for methane's carbon is −4 while for either carbon of ethane it is −3. It is −2.67 for the average carbon of propane, −2.5 for the average carbon of butane, and so on, although organic chemists do not commonly think of methane, ethane, propane and butane as having carbons in different states of oxidation. It may be more useful to consider the oxidation level of carbon as varying with the number of electronegative substituents to which it is bonded. Since the usual ON for any halogen is −1, the chlorination of methane, to yield chloromethane and hydrogen chloride, is an oxidation of methane since carbon's ON goes from −4 to −3. However, while the conversion of methyl chloride to methanol creates a C−O bond, it is not considered an oxidation since the ON of carbon in both methanol and methyl chloride is −3.

$$CH_3Cl + OH^- \rightarrow CH_3OH + Cl^-$$

Rather than ONs, organic chemists think in terms of *sets* of functional groups arranged in order of increasing extent of oxidation (8). The oxidation reactions shown in Figure 2 involve an obvious increase in the amount of oxygen, or bonds to a given oxygen, in the compound undergoing oxidation.

Oxidation also may be recognized by the loss of hydrogen from a molecule. Such dehydrogena-

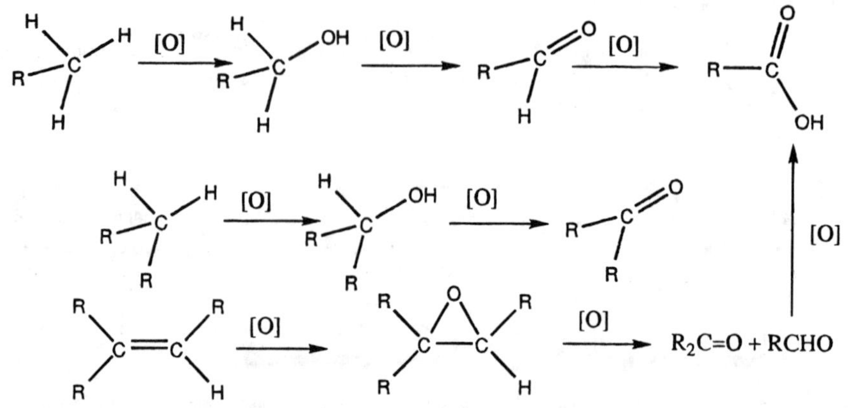

Figure 2.   Oxidation reactions in selected sets of functional groups.

tions may either be intramolecular [e.g., the conversion of ethanol (C_2H_6O) to acetaldehyde (C_2H_4O)] or intermolecular (e.g., conversion of a thiol to a disulfide).

$$2\text{ R}-\text{S}-\text{H} \xrightarrow{\text{[O]}} \text{R}-\text{S}-\text{S}-\text{R}$$

Reductions may be thought of as the reverse of oxidation and defined in terms of hydrogens becoming bonded to a molecule. The conversion of epoxide to alcohol (as follows) and the hydrogenation of a disulfide both involve reduction. Note that all of these definitions are based upon descriptive chemistry and not upon the mechanism of any given reaction.

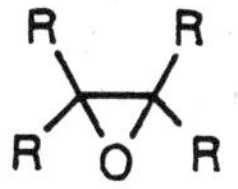

$$\underset{R \quad O \quad R}{\overset{R \qquad R}{\diagdown\!/\!\diagdown\!/}} \xrightarrow{\text{[H]}} \text{HR}_2\text{C}-\text{C(OH)R}_2 \qquad \text{R}-\text{S}-\text{S}-\text{R} \xrightarrow{\text{[H]}} \text{R}-\text{S}-\text{H} + \text{H}-\text{S}-\text{R}$$

In an electrochemical cell, oxidation corresponds to loss of electrons from the anode while reduction refers to a gain of electrons by the cathode. The same is true of a galvanic cell (9, 10).

RADICALS

A radical ("free radical") is a species that possesses one or more unpaired ("odd" or "single") electrons (11). While many radicals have a net charge of zero, those that carry both a charge and an odd electron are "radical ions" and may either be radical cations or radical anions (12). A molecule may lose or gain electrons singly or in pairs. One-electron transfer processes involve radicals (13). Two-electron transfers may involve either a simultaneous transfer of two electrons or two sequential one-electron transfers. Both one- and two-electron transfers occur in vitro (14) and in vivo (15). Indeed, it is likely that all living cells contain some odd-electron species.

Most organic radicals have very short lifetimes. Without stabilizing features, such as steric hindrance at the odd-electron site and/or extensive delocalization of the odd electron, they decompose rapidly—even in the absence of external agents. Two routes that lead to this decomposition are *dimerization* and *disproportionation*. Disproportionation entails the simultaneous oxidation of one radical and reduction of another, frequently through the involvement of a hydrogen β to the carbon bearing the odd electron (Figure 3).

Two other reactions common to many radicals are (a) abstraction of a hydrogen atom from a nearby molecule and (b) addition to molecular oxygen to form a peroxyl radical (Figure 4). Unlike reactions in Figure 3, those in Figure 4 require agents in addition to the radicals themselves. Processes shown in Figures 3 and 4 have in vivo analogs that play an important role in, for example, the peroxidation of lipids.

The structure of radicals usually is determined by EPR (electron paramagnetic resonance spectroscopy) (16). EPR (also termed ESR, electron spin resonance), is similar in a number of

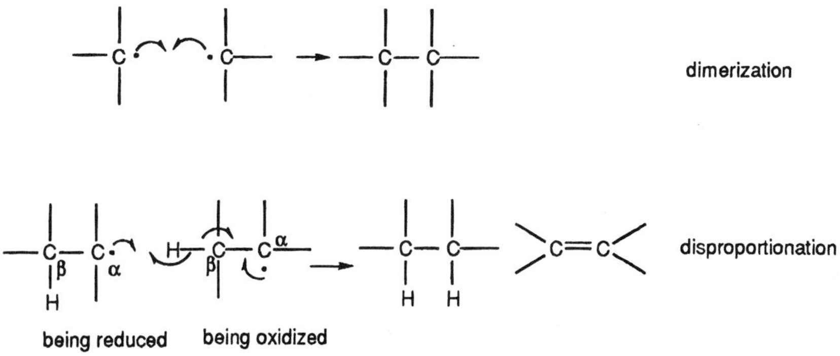

dimerization

disproportionation

being reduced   being oxidized

Figure 3. Usual decomposition pathways of radicals. Single-headed ("fishhook") arrows depict movement of one electron. Their presence always indicates a radical reaction.

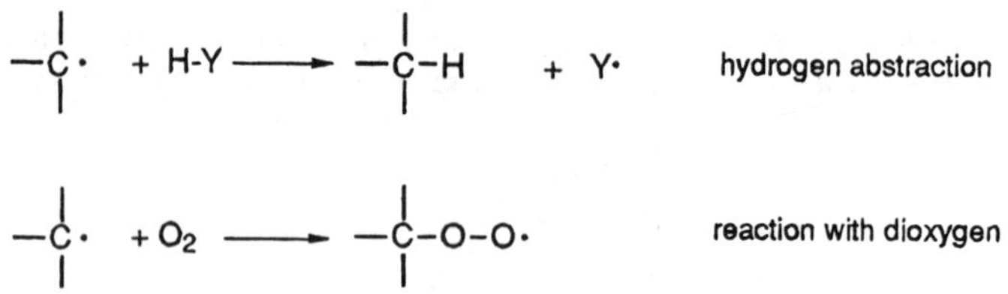

$$-\overset{|}{\underset{|}{C}}\cdot \; + \; H\text{-}Y \longrightarrow -\overset{|}{\underset{|}{C}}\text{-}H \; + \; Y\cdot \qquad \text{hydrogen abstraction}$$

$$-\overset{|}{\underset{|}{C}}\cdot \; + \; O_2 \longrightarrow -\overset{|}{\underset{|}{C}}\text{-}O\text{-}O\cdot \qquad \text{reaction with dioxygen}$$

Figure 4. Two additional routes to the destruction of radicals. Both of these require the presence of a second reactant.

ways to nuclear magnetic resonance (NMR) but relies upon electron spin rather than nuclear spin. Since the technique detects only unpaired electrons, it is specific to the study of radicals. Sometimes very short-lived radicals are difficult to detect and must be converted to less reactive, rather long-lived radical species. This can be done using the technique of spin trapping (17), in which a short-lived, highly reactive radical is converted to a spin adduct by reaction with a spin trapping agent. This conversion often is energetically favored and essentially irreversible (18). The most important spin trapping reagents are nitroso compounds, $R-N=O$, which react with a radical as shown:

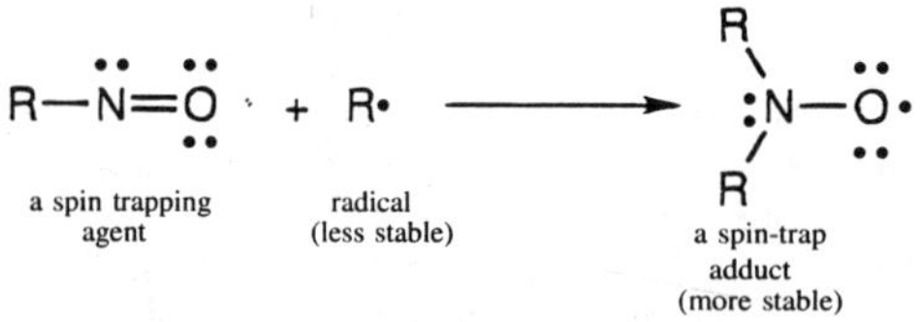

$$R-\overset{..}{N}=\overset{..}{\underset{..}{O}} \; + \; R\cdot \longrightarrow \overset{R}{\underset{R}{\diagdown}}\!\!\!\diagup\!\!\overset{..}{N}\!-\!\overset{..}{\underset{..}{O}}\cdot$$

a spin trapping agent   radical (less stable)   a spin-trap adduct (more stable)

Typical trapping agents include PBN (α-phenyl-*N*-*t*-butylnitrone), DMPO (5,5-dimethyl-1-pyrroline-*N*-oxide) and MNP (2-methyl-2-nitrosopropane). Spin trapping experiments have been used to detect a variety of radicals including both alkyl and aryl radicals, superoxide, alkoxyl and alkylperoxyl radicals, and thiyl radicals (19).

Among the successes in correlating EPR signals with biological activity have been studies of flavin coenzymes and coenzyme Q. Several workers have shown that relatively stable radicals are formed in these systems and that radicals are directly involved as biological intermediates (20). Adriamycin (21) (doxorubicin) is a quinone (22) with important applications to the treatment of breast cancer. Unfortunately, its use is limited by its cumulative cardiotoxicity. Spin trapping experiments have helped correlate this toxicity to the formation of superoxide and hydroxyl radicals (23).

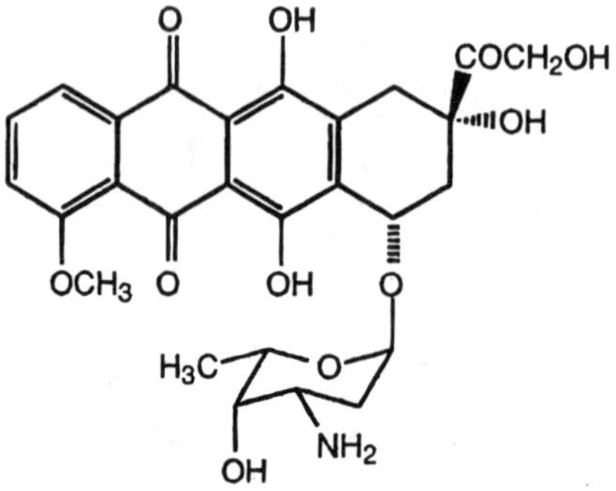

doxorubicin

CIDNP (chemically induced dynamic nuclear polarization) is an NMR-based technique that is of value in studying radical reactions (24). However, it is more limited than is EPR since not all radicals are expected to exhibit the phenomenon (25). A similar technique applied to EPR is termed CIDEP, chemically induced dynamic electron polarization. Finally, a caveat: Since these techniques are extremely sensitive, capable of detecting very small amount of radicals, one must be certain that the signals being observed arise from the process under investigation.

OXYGEN AND ITS DERIVATIVES

Oxygen is a colorless, odorless, and tasteless gas that dissolves in water to the extent of 3.08 ml (at standard temperature and pressure [STP]) in 100 ml of water at 20°C. Solubilities in organic systems (e.g., carbon tetrachloride) are approximately an order of magnitude higher. At low temperatures oxygen condenses to a blue liquid, boiling point $-183°C$. Molecular oxygen (also called dioxygen, O_2) is paramagnetic (26). It has a triplet ground state with its two least strongly held electrons existing in degenerate pi-star orbitals with parallel spins and is, therefore, a diradical (27). When these electrons, which are critical in many biological processes, are excited they remain unpaired and in separate orbitals but develop opposite spins. This state, abbreviated $^1\Delta_g$, is referred to as singlet oxygen. A second excited singlet state, abbreviated $^1\Sigma_g^+$, is known to possess two electrons with opposite spins in a common pi-star orbital. Singlet oxygen, which can be prepared by several processes including the irradiation of "normal" oxygen in a presence of a photosensitizer, can lead to lipid peroxidation (28). The nature of the sensitizer determines which singlet state is produced. The more stable singlet state ($^1\Delta_g$) also results from several chemical reactions, including that of hypochlorite ion with hydrogen peroxide.

While it is essential for human life, there is ample evidence that oxygen also can be toxic. One illustration of this is found in the central nervous system (CNS) damage that often accompanies exposure to cyanide. While cyanide is considered to be primarily a metabolic poison that blocks cytochrome oxidase, exposure leads to peroxidation of lipids and, therefore, damage to membranes. Studies (29) indicate increased levels of conjugated dienes (a result of peroxidation) in the brains of mice after a cyanide treatment. The effect is both dose and time dependent. In some ways, cyanide poisoning represents a useful set of conditions for studying oxidative damage since oxygen is available because its usage is restricted. Beyond this, the activity of enzymes that might limit such damage is restricted by cyanide (30). Unfortunately, the brain appears to be especially sensitive to lipid peroxide formation after treatment with cyanide, although not all tissues exhibit equal sensitivity.

In addition to oxygen itself, several oxygen-rich species are significant in biological processes (31). Among the more important of these is superoxide. It is common to find superoxide written as an anion (O_2^-) rather than as a radical anion ($O_2^{\bullet-}$), although the latter is the more accurate representation. Superoxide is produced by the reduction (i.e., addition of one electron) of dioxygen, has been studied by both conventional EPR and spin trapping techniques (32), has been observed in biological systems (33), and has been implicated in disease states (34). Superoxide is described as a "weak" radical and its ability to abstract hydrogen from unsaturated lipids is modest at best (35). In spite of this, it is considered toxic and its presence is associated with increased pathology (36).

Although superoxide is in equilibrium with the hydroperoxyl radical ($HO_2^\bullet$) in aqueous media, at physiological pH superoxide concentration is significant.

$$\underset{\text{superoxide}}{O_2^-} + H_3O^+ \rightleftarrows H_2O + \underset{\substack{\text{hydroperoxyl}\\\text{radical}}}{HO_2^\bullet}$$

A total of four electrons is required for the complete reduction of dioxygen to water. The three remaining conversions are shown next. Note that the overall process affords both hydrogen peroxide and hydroxyl radicals.

$$HO_2^\bullet + H_3O^+ \xrightarrow{\text{electron}} \underset{\substack{\text{hydrogen}\\\text{peroxide}}}{H_2O_2} + H_2O$$

$$H_2O_2 \xrightarrow{\text{electron}} \underset{\substack{\text{hydroxide}\\\text{ion}}}{OH^-} + \underset{\substack{\text{hydroxide}\\\text{radical}}}{OH^\bullet}$$

$$OH^\bullet + H_3O^+ \xrightarrow{\text{electron}} 2\,H_2O$$

While superoxide may cause pathology directly, it is considered likely that the more reactive hydroxyl radical plays a larger role in producing molecular/biological damage. One source of hydroxyl radical involves a sequence known as the superoxide-driven Fenton reaction (37, 38). (Note that the conversion of superoxide to hydrogen peroxide and oxygen frequently is described as a disproportionation [dismutation] reaction.) As seen in Figure 5, iron plays a critical role in this process. Superoxide easily converts iron(III) to iron(II). It also is able to release iron from ferritin. Iron(II) (ferrous) is more effective than is iron(III) (ferric) in decomposing hydrogen peroxide and lipid hydroperoxides to oxygen-containing radicals. This ability of superoxide to accelerate the formation of oxygen-containing radicals by liberating iron, and converting iron(III) to iron(II), may contribute to its toxicity.

Although it is not a radical, another potential oxidant is hydrogen peroxide. Its oxidation reactions with organic substrates generally are slow, and this may help to explain its relatively low toxicity. However, it is quite reasonable to imagine that hydrogen peroxide, when combined with metals that can exist in more than one oxidation state, could be quite hazardous due to hydroxyl radical formation. This is germane since hydrogen peroxide can diffuse into membrane bilayers and bulk lipid depots. Enzyme systems, including catalase and glutathione peroxidase, play critical roles in controlling intracellular peroxide levels (39).

ANTIOXIDANTS

Historically, the term antioxidant referred to any substance that hindered the reaction of a substance with dioxygen. Since such reactions frequently involve radicals, "antioxidant" has taken on a more mechanistic flavor and now generally refers to any substance which inhibits a free radical reaction. In living systems this often involves oxygen in one manner or another.

Unwanted or excessive oxidation (radical) reactions need to be limited in order for organisms to survive. Moreover, control of radical reactions now is viewed as a potential route for prevention/ intervention in certain diseased states, such as inflammation (40). Antioxidants also have potential in prevention of atherosclerosis (41), as well as in the treatment of viral diseases (42). One method to achieve control of biological oxidation is via antioxidant enzymes (e.g., SOD [superoxide dismutase], catalase, and glutathione peroxidase) while another is via small, nonproteinaceous antioxidant molecules (43). In this section we examine four naturally occurring antioxidants: L-ascorbic acid (vitamin C), β-carotene, α-tocopherol (vitamin E), and ubiquinone (coenzyme Q). However, it is reasonable that other natural products undoubtedly also will prove to be valuable radical inhibitors (44).

L-Ascorbic Acid (Vitamin C)

Vitamin C (45) is a water-soluble (~0.3 g/ml) carbohydrate derivative (lactone) that owes its acidic properties to the presence of the enediol group.

It is a white solid, melting point 190–191°C, with $[\alpha]_D^{25} = +21°$ at $C = 1$, and is stable to air oxidation when dry (46).

Because of storage limitations, research has been conducted to relate the chemistry and the shelf-life of various antioxidants (47). It is known, for example, that in the presence of air, vitamin C's dianion is oxidized much faster than is either the neutral species or the monoanion. However, in the presence of copper(II) ions, the rate of oxidation is proportional to the concentration of the monoanion. The effect of pH on the oxidation of aqueous vitamin C was a part of

$$2\,O_2^- + 2\,H^+ \longrightarrow H_2O_2 + O_2$$

$$O_2^- + Fe^{+3} \longrightarrow O_2 + Fe^{+2}$$

$$Fe^{+2} + H_2O_2 \longrightarrow Fe^{+3} + OH^- + OH^\cdot$$

Figure 5. Superoxide-driven Fenton reaction.

this study (48). Hydroperoxide intermediates in the rose bengal-sensitized photooxygenation of ascorbic acid have been described (49).

Its crystal structure (50) and absolute configuration (51) have been determined and there are numerous studies directed toward its synthesis (52). Because of its hydrophilic substituents, vitamin C is essentially insoluble in lipids and lipid-like solvents (e.g., chloroform). In this way it differs from the other antioxidants discussed in this section, all of which are more soluble in lipids than in water. (Other water-soluble antioxidants include glutathione (53), uric acid, anserine and carnosine.) It also differs from these in being able to exist in tautomeric (prototropic) forms because of its conjugated enediol moiety.

L-ascorbic acid

It would be wrong to think of vitamin C as "just another antioxidant." For example, prolyl hydroxylase requires vitamin C to convert proline to hydroxyproline as part of the biosynthesis of collagen. Since L-ascorbic acid is not synthesized by primates, the absence of vitamin C in their diet produces the collagen disorder known as scurvy. It also has been examined for its role in immunology, cancer prevention, and the common cold, to mention but a few areas (54). The pharmacology of vitamin C has been reviewed (55).

β-Carotene and Carotenoids

A large number of compounds found in nature arise from the combining of isoprene (C_5H_8) units, usually in a head-to-tail manner (*isoprene rule*). Many of these compounds are termed "terpenoids," including hydrocarbons such as limonene and α-pinene as well as oxygenated molecules such as geraniol. The designations monoterpene ($C_{10}H_{16}$), sesquiterpene ($C_{15}H_{14}$), diterpene ($C_{20}H_{32}$), triterpene ($C_{30}H_{48}$), and tetraterpene ($C_{40}H_{64}$) reflect the number of isoprene units in the structure. The stereochemistry of a number of carotenoids has been summarized (56).

isoprene   limonene   α-pinene   geraniol

Carotenoids, which include α-, β-, γ-, and δ-carotene, are tetraterpenes. They are lipid-soluble substances, a category that also includes vitamins E, K and coenzyme Q. All of the carotenes are intensely colored, and while each has a characteristics absorption spectrum, only β-carotene has a fully conjugated π-electron system (57, 58).

The α- and β-isomers of carotene contain two cyclohexenyl rings while the less common γ- and δ-forms contain only one. According to Fieser and Fieser (59), carotene from carrots is ~15% α, 85% β, and a trace of others. All of these compounds are quite sensitive to both light and oxidation.

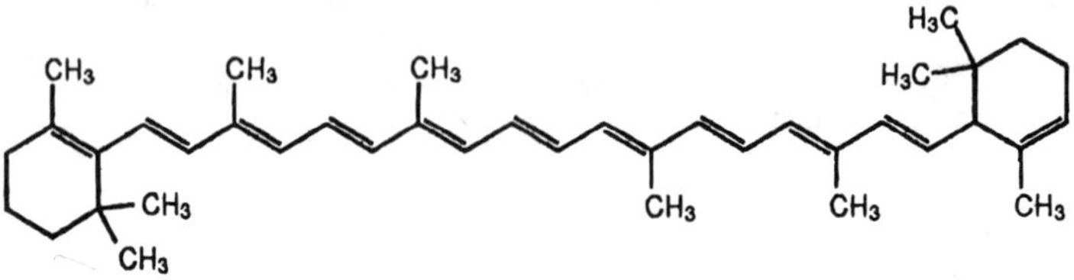

α–carotene

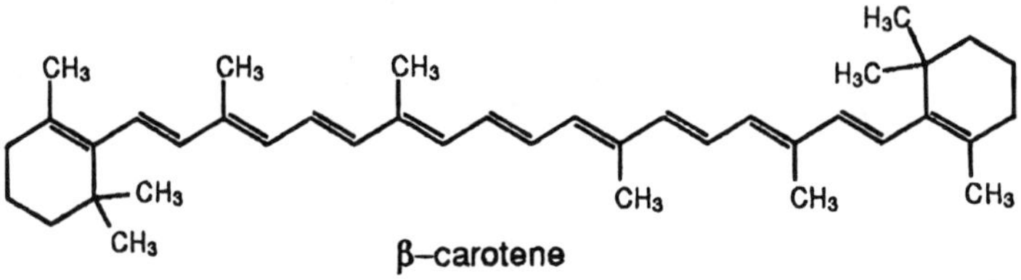

β–carotene

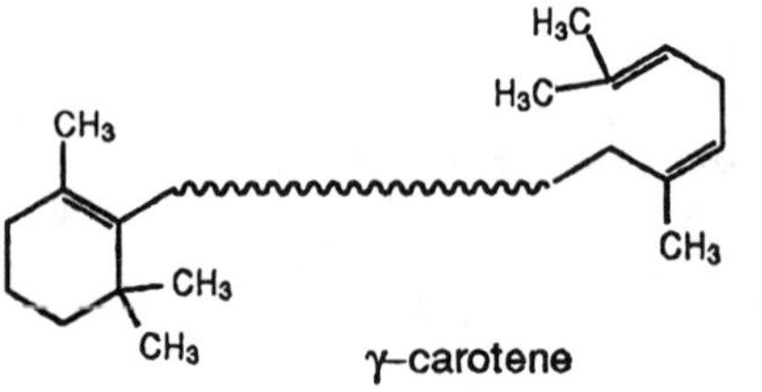

γ–carotene

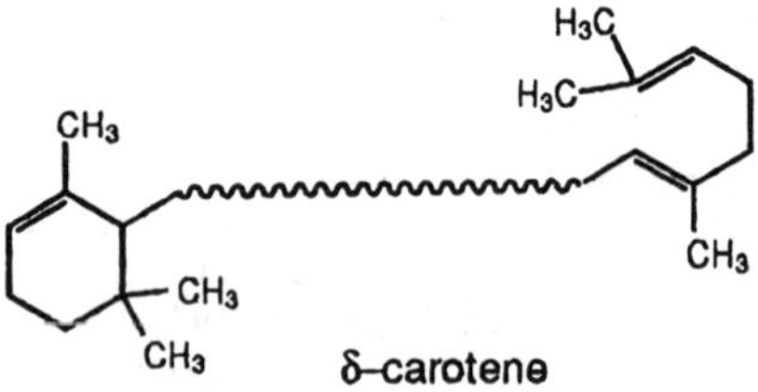

δ–carotene

Both α- and β-forms are rather purple, while γ- and δ-forms are red or red-orange. The electronic spectra of carotenes have been studied extensively (60) and helped to establish many of the rules used to calculate ultraviolet and visible spectra of conjugated systems (61). The mass spectra of carotenoids have been reviewed (62) and studies conducted on crystal structures (63) and syntheses (64).

A related, lipophilic hydrocarbon that may ultimately have significant impact on this area is lycopene. This deep red [λ_{max} at 476nm in hexane] compound, mp 174°C, is found largely in ripe fruit (e.g., ~20 mg/kg ripe tomatoes). There are suggestions that lycopene may be an even more potent antioxidant than is β-carotene (65).

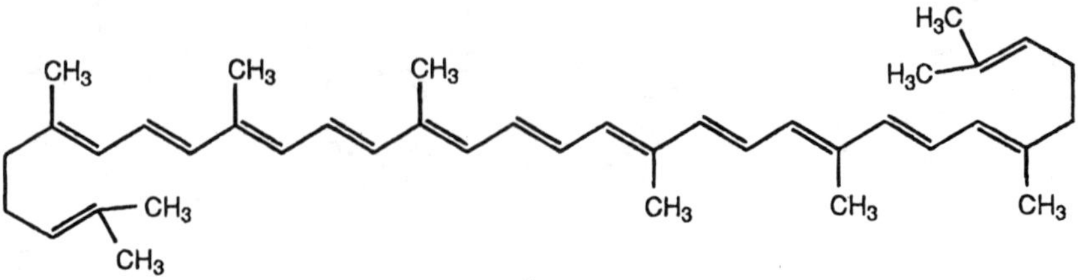

lycopene

The term "vitamin A" actually represents several compounds including vitamin A_1 and vitamin A_2 (retinol and dehydroretinol, respectively). The molecule is synthesized in vivo by the oxidative cleavage of carotene. Several reviews of vitamin A and carotenoids as physiological antioxidants have appeared (66).

vitamin A₁   vitamin A₂

Appropriately substituted phenols may be oxidized, with various reagents, to a number of compounds including quinones, quinonemethides (C=O replaced by C=C), or quinonimines (C=O replaced by C=N).

$$\text{aromatic}-O-H \xrightarrow{[O]} \text{aromatic} - \ddot{\underset{}{O}}\cdot \longrightarrow \text{aromatic}-O-O-\text{aromatic}$$

a phenol   a phenoxyl radical   a peroxide dimer

Quinones can have their carbonyl groups either adjacent to one another (1,2-quinones or *ortho*-quinones) or at opposite ends of a ring (1,4-quinones or *para*-quinones). The latter are generally more important. Two examples follow:

catechol   o-benzoquinone

a quinonemethide-based
aryloxyl radical

Quinones can be reduced to a dianion in two stages (Figure 6). The intermediate radical anion, known as a semiquinone, possesses both the phenoxyl radical and the phenoxide anion fragments. The final product is the dianion of a hydroquinone.

The redox chemistry of phenols and quinones is further complicated by the potential formation of a 1:1 charge-transfer complex between a quinone and hydroquinone, forming a quinhydrone (71).

The tocopherols may be viewed as derivatives of tocol, the principal changes being in the nature of the bicyclic ring system. Vitamin E (α-tocopherol) is the most extensively studied member of the family of tocopherols (72). All chiral centers of α-tocopherol have the *R* absolute configuration. However, with three chiral centers, seven stereoisomers of α-tocopherol are possible. Relative biopotency of these isomers does not appear to vary dramatically, although

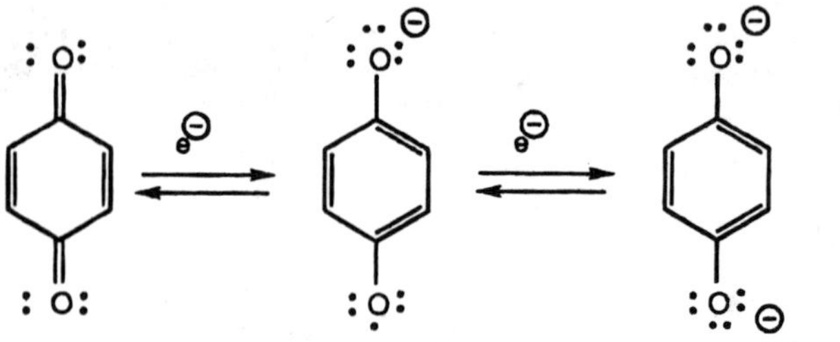

a quinone　　a semiquinone radical anion　　a hydroquinone dianion

Figure 6. Two-electron reduction of a quinone. At appropriate pH, the oxyanions would exist as hydroxyl groups. Note that reaction with one electron and one proton is equivalent to reaction with a hydrogen radical (H·). The intermediate radical anion can be thought of as a benzene analog of the superoxide radical anion.

the alpha isomer generally is most active. Since it is more stable to oxidation during preparation and storage, but biologically equivalent to the free phenol, vitamin E generally is marketed as its acetate ester.

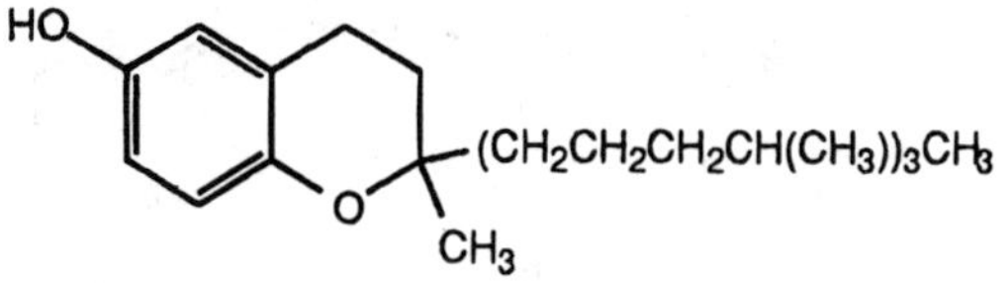

tocol

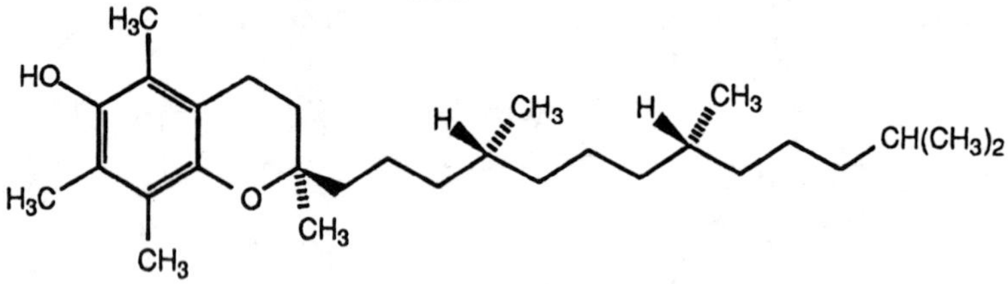

vitamin E (α-tocopherol)

The synthesis of isomeric tocopherols has been described (73). The relationships between molecular structure and antioxidant activity for some vitamin E derivatives has been studied (74). In a related vein, SAR (structure–activity relationship) studies on vitamin E analogs have been reviewed and various qualities (e.g., σ, σ*, heats of formation and molecular orbital parameters) correlated with antioxidant activity (75). Nitric oxide's reactions with α-tocopherol (and model phenolic antioxidants) in sodium dodecyl sulfate (SDS) micelles have been shown to lead to covalent adducts (76). Infrared spectra have been analyzed and used to identify tocopherols in mixtures (77). High-performance liquid chromatography (HPLC) analysis of tocopherols has been reported and detection schemes compared (78).

Ubiquinone–Ubiquinol

While named in the singular, ubiquinone (coenzyme Q; CoQ) actually is a family of benzoquinones which differ in the length of the lipophilic side chain. The existence of dual nomenclature systems is noteworthy. When named as a ubiquinone the appropriate name is "ubiquinone (*x*)," where x represents the total number of carbons in the side chain and is a multiple of five. Named as a coenzyme Q, the appropriate name is coenzyme Q_n, where n represents the number of five-carbon fragments shown in the structure here. The most abundant CoQ in mammals is CoQ_{10}.

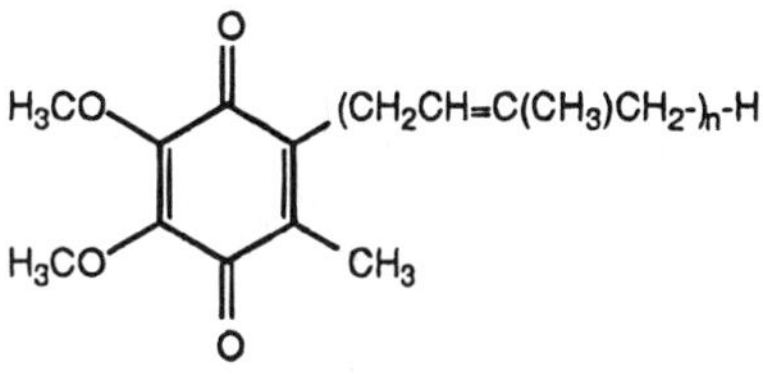

coenzyme Q
ubiquinone

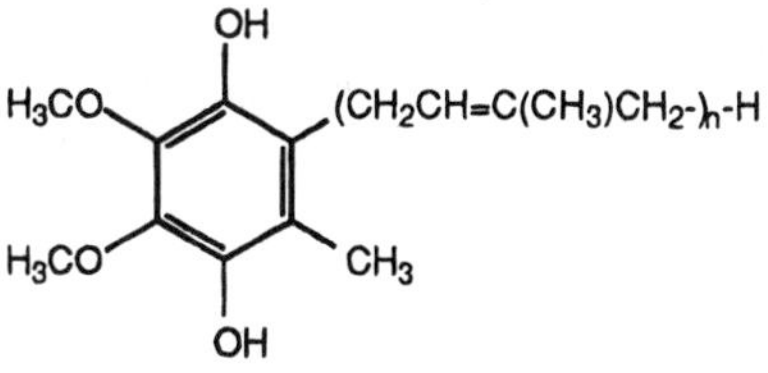

coenzyme Q-H$_2$
ubiquinol

Its structure is similar to that of the vitamin K_2 family, quinoidal compounds of value because of their antihemorrhagic activity. Indeed, a recent report suggests that the vitamin E metabolite vitamin E quinone inhibits the vitamin K-dependent enzyme that controls the clotting of blood. This, in turn, may help to explain its beneficial effects in preventing heart attacks and strokes (79).

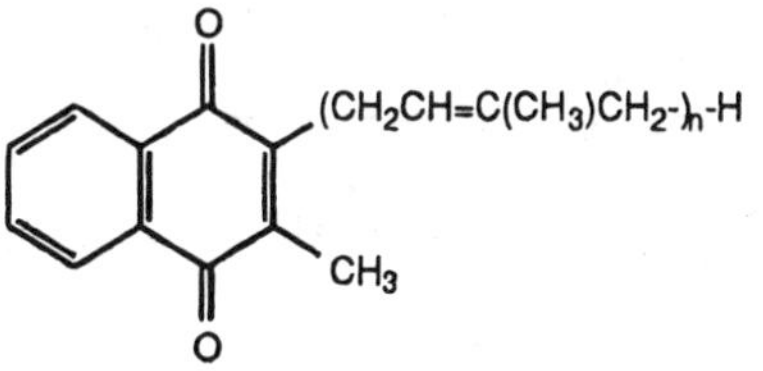

vitamin(s) K$_2$

The reduced form of ubiquinone (80), ubiquinol (CoQ·H$_2$), is a hydroquinone that results from the reaction of two protons and two electrons with CoQ. The intermediate in this reduction is a stable radical, ubisemiquinone (Figure 7).

Ubiquinone acts as both an electron and proton carrier in the electron-transport chain in bacteria and mitochondria (81). It also is a lipid-soluble antioxidant. For example, data suggest that ubiquinol(10) may protect human LDL from peroxidation more effectively than does vitamin E (82).

CHAIN REACTIONS AND THE ROLE OF ANTIOXIDANTS

Many, though certainly not all, radical reactions are chain processes, meaning that one "initiating" event will produce many product molecules (83). The process normally involves three stages: initiation, propagation, and termination. Any process affording the necessary starting radicals

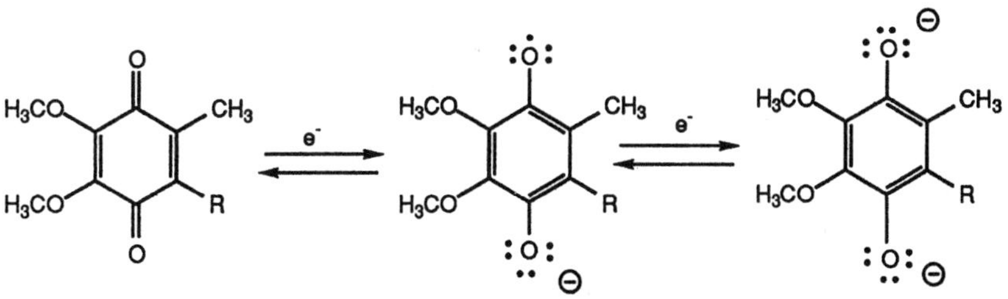

Figure 7. Two-electron reduction of CoQ_{10}. R = $(CH_2CH=C(CH_3)CH_2)_{10}H$. At appropriate pH, the oxyanions would exist as hydroxyl groups.

can be an initiating step, while any process destroying the species in the chain can be a terminating step.

As noted earlier, antioxidants may be defined as compounds that interfere with a radical chain reactions (84). Lewin and Popov have considered antioxidant homeostasis and categorized antioxidants as compartmentalization, detoxification, repair, and utilization. Furthermore, they have described a parameter for characterizing the overall state of the antioxidant system in humans and for detecting its changes under pathological conditions (85).

Among the most extensively studied systems is the oxidation of lipids. Peroxidation of membrane lipids appears to have first been noted in 1959 (86). This chain-reaction can be summarized as shown in Figure 8.

Steps other than those shown in Figure 8 are possible, including the conversion of the hydroperoxide LO_2H to an alkoxyl radical, $LO^{\bullet}$, and its subsequent abstraction of a hydrogen atom from $L-H$ to form an alcohol (LOH) and a new $L^{\bullet}$ radical. This particular sequence illustrates "chain branching" and ultimately leads to a new radical sequence and a molecule of alcohol. The loss of a hydrogen atom from $L-O-O-H$ can reinitiate peroxidation by its conversion to $L-O-O^{\bullet}$. While chain termination may involve coupling of various radicals to form several stable species, such as $L-L$, $L-O-O-L$, and $L-O-L$, it is common to show the initiating radical (e.g., $L^{\bullet}$) undergoing dimerization.

Interfering with lipid peroxidation (87), and other radical chain reactions, can, in principle, be accomplished in several ways (88). One would involve suppressing the initiation step that produces the radical $L^{\bullet}$. A second would involve accelerating the chain termination sequence by, for example, introducing a new termination process. The chain also would be interrupted when an antioxidant functions as a hydrogen atom donor and replaces $L-H$ in the second step of the propagation sequence. Even though such processes produce a radical derived from the antioxidant, this radical may be sufficiently nonreactive that it is, for practical purposes, removed from the system. This appears to be especially significant with thiol-based antioxidants.

Antioxidants such as SOD (superoxide dismutase), catalase, and glutathione peroxidase appear to function by reducing the general rate of radical formation and can be thought of as "preventive" antioxidants. This group of antioxidants acts by destroying/deactivating active species and/or their precursors. For example, catalase decomposes hydrogen peroxide without the formation of radicals. Antioxidants such as vitamins C and E and ubiquinol act by breaking the radical chain process. They do this by destroying radicals involved in the chain carrying sequence (89). However, it probably is simplistic to believe that any given antioxidant must act in only one fashion. Overviews of the antioxidant role of vitamins are available (90–92).

Chain Initiation:

$$L{-}H \xrightarrow{\text{radical}^{\bullet}} L^{\bullet} \ + \ \text{radical}{-}H$$

Chain propagation:

$$L^{\bullet} \xrightarrow{O_2} L{-}O{-}O^{\bullet}$$

$$L{-}O{-}O^{\bullet} \xrightarrow{\text{L-H}} L{-}O{-}O{-}H \ + L^{\bullet}$$

Chain termination:

$$L^{\bullet} \ + L^{\bullet} \longrightarrow L{-}L$$

Figure 8. Sequence describing lipid peroxidation. L represents a lipid and the hydrogen being lost generally is allylic, i.e., $C=C-C-H$.

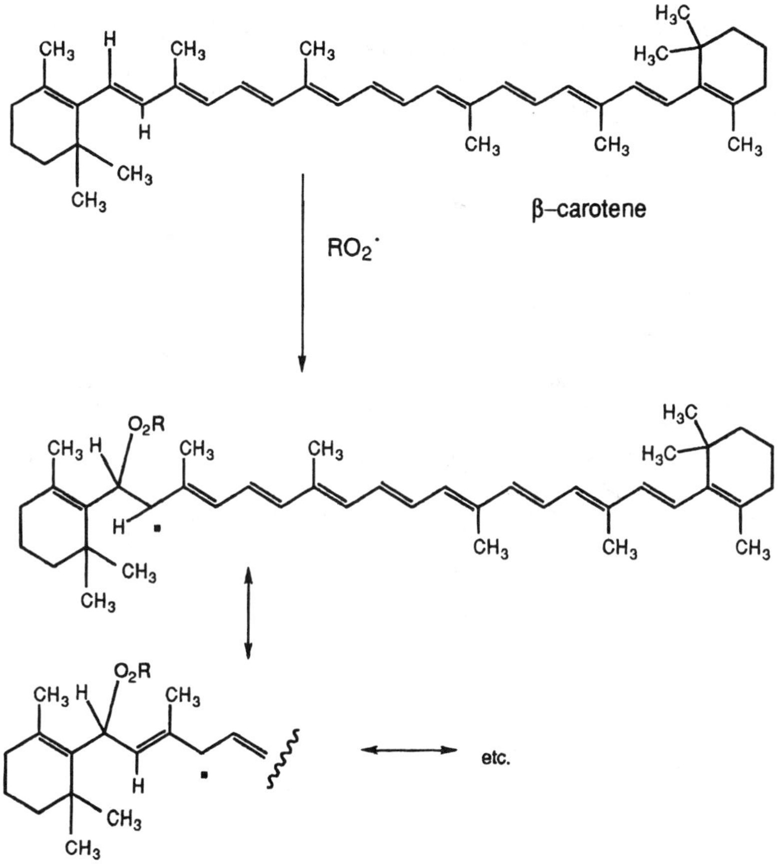

Figure 9. Trapping of a radical by β-carotene.

Antioxidant inhibition of the oxidation of low-density lipoproteins (LDL) has been reviewed. Free radical oxidation of LDL involves a chain mechanism that affords phosphatidylcholine hydroperoxide and cholesteryl ester hydroperoxides as major primary products. The efficacy of antioxidants depends upon radical chain inhibiting activity of antioxidant as well as their location and mobility in the LDL microenvironment (93).

β-Carotene

One of the ways in which β-carotene, and other conjugated polyenes, may act (94–97) is by trapping alkylperoxyl radicals ($R-O-O\cdot$) (98). The result is a highly conjugated/delocalized, and thus less reactive, radical (Figure 9).

L-Ascorbic Acid

While much attention has been given to both the in vivo and in vitro redox activity of L-ascorbic acid, the literature addressing the structures of species involved as intermediates has been somewhat unsettled. And while various suggestions about the nature of the intermediate(s) exist, they are not detailed here. This multiplicity stems from several factors including the nature of the structural studies undertaken, difficulties in interpreting data, and the complexity of the ascorbic acid molecule and of its reactions. This highly functionalized lactone, at pH $\approx$ 7, exists predominately as its monoanion. This anion could be drawn in a variety of isomeric/tautomeric forms, but only one of these, a monocyclic structure, is shown:

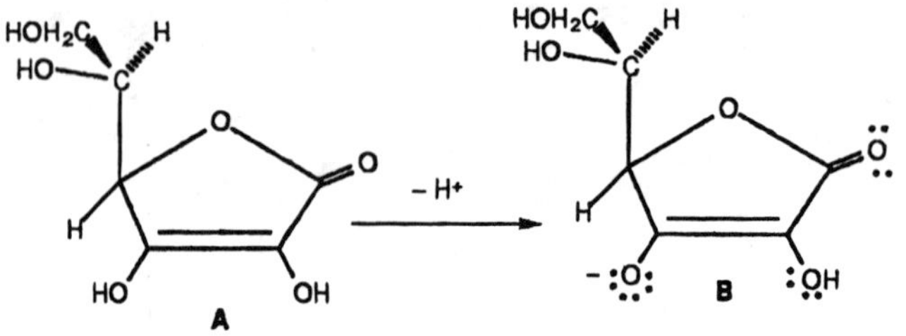

Loss of an electron from this anion then may produce a (neutral) radical, which, by loss of a second proton, affords a radical anion. The loss of both a proton and an electron, equivalent to the loss of a hydrogen atom, from this anion affords a 1,2-semidione (99) radical anion that is resonance stabilized. The resulting species may be drawn in several ways (100), including monocyclic (101) and bicyclic structures. Which species predominates in vitro may be a result, at least in part, of experimental conditions. Recent work (102) at elevated pressures, using volume analysis, may be interpreted to favor the bicyclic structure, one resonance form of which is shown:

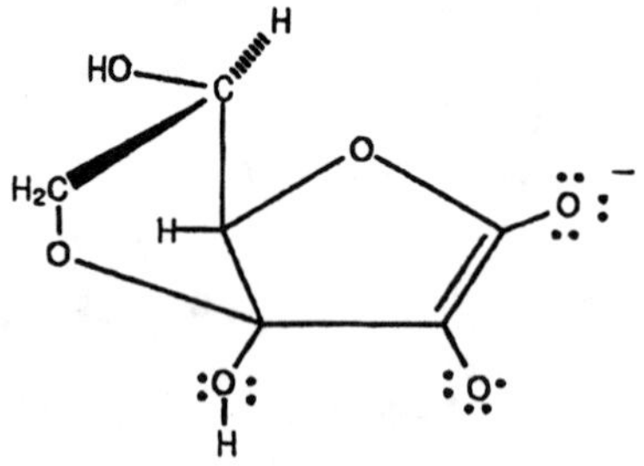

Production of this delocalized, stable intermediate tends to inhibit radical processes by removing reactive radicals from the system and replacing them with less reactive radicals. Ultimately, this anion radical may couple with other radicals or disproportionate, with the assistance of a proton, to yield dehydro-L-ascorbic acid and ascorbic acid. Loss of a second electron from the radical anion also produces the neutral dehydro-L-ascorbic acid, whose bicyclic structure is shown:

dehydro-L-ascorbic acid   dehydro-L-ascorbic acid hydrate

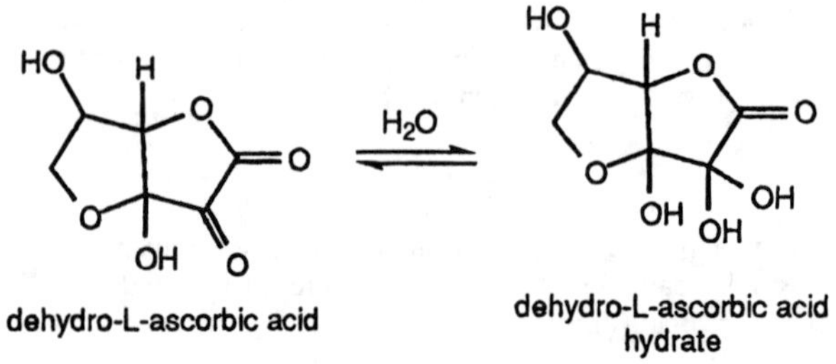

It is noteworthy that the semidione radical anion of vitamin C can be viewed as a vinylog of the superoxide radical anion.

Vitamin E

Vitamin E may be the most studied antioxidant for the lipid phase. It functions by converting chain-carrying alkylperoxyl radicals to hydroperoxides and, simultaneously, being converted to tocopheroxyl radicals.

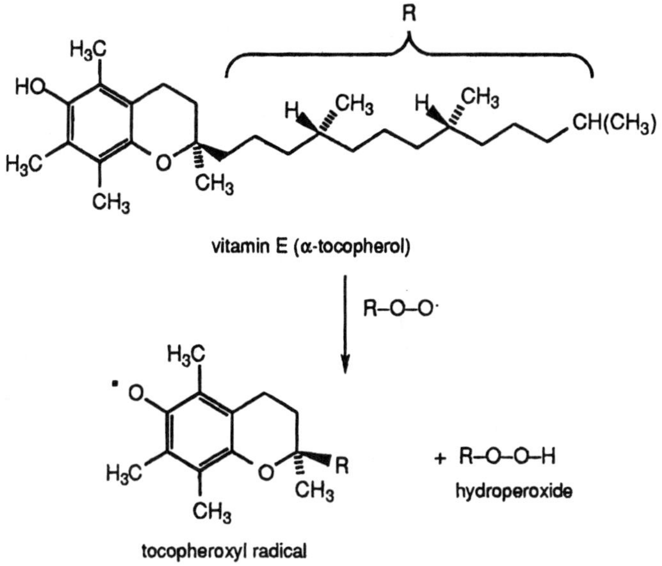

vitamin E (α-tocopherol)

R–O–O·

+ R–O–O–H
hydroperoxide

tocopheroxyl radical

The resulting tocopheroxyl radicals are, at least in part, reduced by ascorbic acid back to the antioxidant tocopherol. The role of the vitamin E antioxidant cycle in normal and diseased states has been reviewed (103). The role of metabolism in the antioxidant activity of vitamin E, and its synergism with vitamin C, reduced glutathione, NADPH, and cellular electron transport proteins, also has been reviewed (104).

The EPR spectra of tocopheroxyl radicals have been studied (105), as has their formation by reaction with peroxyl radicals (106). The antioxidant effectiveness of several tocopherols has been compared. d-γ-Tocopherol was found to be more effective then either the d-α- or d,l-α-forms (107). The relative reactivities of α- and γ-tocotrienol, α-tocopherol, and other lipid soluble antioxidants toward peroxyl radicals were investigated. In homogeneous systems the sequence of relative reactivities was ubiquinol 10 > α-tocopherol ≈ α-tocotrienol > β-carotene ≈ lycopene > γ-tocopherol ≈ γ-tocotrienol. Similar results were found in studies involving human LDL (108,109). The antioxidant effect of vitamin E upon fish-oil diets has been investigated (110). Complex formation between α-tocopherol and fatty acids or their hydroperoxides has been studied using fluorescence (111). The work of Mukai and co-workers (112) suggests that the role of phenolic antioxidants (of which tocopherol is an example) does not depend upon whether the reactive species is singlet oxygen, peroxyl radicals, or phenolic radicals.

Ubiquinone

The mode(s) of action of ubiquinone–ubiquinol as an antioxidant remains a matter of discussion. It has been suggested by Kagan and co-workers (113), for example, that ubiquinol works with vitamin E in protecting cell membranes against oxidation by functioning as a recycling agent for vitamin E. A scheme for accomplishing this, offered by these workers, involves the reduction of ubisemiquinone to ubiquinol and concomitant oxidation of α-tocopheroxyl radical to α-tocopherol. In this scheme, NADH, NADPH, and succinate reduce ubiquinone to ubisemiquinone. (Note that the structure given for ubisemiquinone on p. 243 of reference 113 is incorrect.) This suggestion has, however, been questioned: "It appears that ubiquinol prevents both the initiation and propagation of lipid peroxidation, whereas vitamin E acts exclusively as a chain-breaking antioxidant, inhibiting propagation" (114).

SUMMARY

With the advantages of hindsight, it becomes clear that while the details may not have been on the mark, Michaelis was correct in his assertion that radicals play a significant role in biological redox reactions. And while research on the role of antioxidants in hindering unwanted in vivo oxidations has been ongoing for a number of years, the potential for using antioxidants to prevent or limit illness is in its infancy. Obviously, inhibition of necessary electron-transfer processes must be avoided. Only relatively recently has the interplay ("recycling") between various radical inhibitors gotten the attention that it warrants (115). It is now known, for example, that a major role of vitamin C is to protect hydrophobic compartments by regenerating the antioxidant vitamin E (116). A study of vitamins C and E in plasma subjected to free-radical oxidative stress has led to the suggestion that vitamin C may be the ultimate small-molecule antioxidant in living systems (117). The glutathione–vitamin C antioxidant system in animals has been reviewed (118). And while it may not be the cure-all including for the common cold, one should not forget that Linus Pauling's interest in vitamin C heightened public awareness and stimulated many a discussion within the scientific and funding communities.

As the role and scope of established antioxidants (e.g., coenzyme Q) are expanded, as new natural antioxidants and antioxidant systems are identified and examined, as new synthetic antioxidants are created, and so forth, a challenge to chemists, biochemists, health-related scientists, and, perhaps most importantly, funding agencies is to improve the human condition by developing an even clearer understanding of the area of in vivo free radical chemistry.

REFERENCES

1. Michaelis, L. and Schubert, M. P., The Theory of Reversible Two-Step Oxidation Involving Free Radicals, *Chem. Rev.*, 1938, 22, 437–470.
2. Willson, R. L., Free Radical Repair Mechanisms and the Interactions of Glutathione and Vitamins C and E, *Radioprotectors and Anticarcinogens*, O. F. Nygaard and M. G. Simic, eds., Academic Press, New York, 1983, pp. 1–22.
3. Muggli, R., Free Radical Tissue Damage; the Protective Role of Antioxidant Nutrients, *Free Radicals Antioxid. Nutr.*, F. Corongiu ed., Richelieu Press, London, 1993, pp. 189–204.
4. Diplock, A. T. Antioxidants and Free Radical Scavengers, *New Compr. Biochem.*, 1994, 28 (Free Radical Damage and Its Control), pp. 113–130.
5. Bonorden, W. R. and Pariza, M. W., Antioxidant Nutrients and Protection from Free Radicals, *Nutr. Toxicol.*, 1994, pp. 19–48.
6. Sies, H., Strategies of Antioxidant Defense, *Eur. J. Biochem.*, 1993, 215, 213–219.
7. Shriver, D. F., Atkins, P. and Langford, C. H., *Inorganic Chemistry*, 2nd ed., Freeman, New York, 1994, p. 54.
8. (a) Stewart, R., *Oxidation Mechanisms*, W. A. Benjamin, New York, 1964; (b) Rinehart, Jr., K. L., *Oxidation and Reduction of Organic Compounds*, Prentice-Hall, 1973.
9. Skoog, D. A. and West, D. M., *Principles of Instrumental Analysis*, 2nd. ed., Saunders, Philadelphia, 1980, Ch. 8.
10. Yoshida, K., *Electrooxidation in Organic Chemistry*, Wiley, New York, 1984.
11. (a) Pryor, W. A., *Free Radicals*, McGraw-Hill, New York, 1966; (b) Lowry, T. H. and Richardson, K. S., *Mechanism and Theory in Organic Chemistry*, 3rd ed., Harper and Row, New York, 1987, Ch. 9; (c) March, J., *Advanced Organic Chemistry*, 4th ed., Wiley-Interscience, New York, 1992, pp. 186–195.
12. (a) Kaiser, E. T. and Kevan, L., *Radical Ions*, Interscience, New York, 1968; (b) Szwarc, M., Chemistry of Radical-Ions, in *Progress in Physical Organic Chemistry*, Vol. 6, A. Streitwieser, Jr. and R. W. Taft, eds., Wiley-Interscience, New York, 1968.
13. (a) Pryor, W. A., Free Radicals In Biological Systems, *Sci. Amer.*, August, 1970; reprinted in *Organic Chemistry of Life*, (Readings From *Sci. Amer.*), W. H. Freeman, San Francisco; (b) *Free Radicals*, Vols. I and II, J. K. Kochi, ed., Wiley, New York, 1973; (c) For a view of sulfur-containing radicals see Kaplan, L., *Bridged Free Radicals*, Dekker, New York, 1972, Ch. 8.
14. Bowser, J. R., *Inorganic Chemistry*, Brooks/Cole Publishing, Pacific Grove, 1993, Ch. 9.
15. Isied, S. S., Long-Range Electron Transfer in Peptides and Proteins, *Prog. Inorg. Chem.*, 1984, 32, 443–517.
16. (a) Bersohn, M. and Baird, J. C., *An Introduction to Electron Paramagnetic Resonance*, W. A. Benjamin, New York, 1966; (b) Norman, R. O. C. and Gilbert, B. C., Electron-spin Resonance Studies of Short-lived Organic Radical, in *Advances in Physical Organic Chemistry*, Vol. 5, V. Gold, ed., Academic Press, New York, 1967; (c) Symons, M. C. R., The Identification of Organic Free Radicals by Electron

Spin Resonance, in *Advances in Physical Organic Chemistry*, Vol. 1, V. Gold, ed., Academic Press, New York, 1963.

17. (a) Janzen, E. G., Spin Trapping, *Acc. Chem. Res.*, 1971, 4, 31–40; (b) Hamilton, C. L. and McConnell, H. M., Spin Labels, in *Structural Chemistry and Molecular Biology*, A. Rich and N. Davidson, eds., W. H. Freeman, New York, 1968, pp. 115–149.

18. For a useful discussion of equilibria involving radicals see Hine, J., *Structural Effects on Equilibria in Organic Chemistry*, Krieger Pub. Co., Huntington, 1981, Ch. 10. (A corrected reprint of a monograph published by Wiley, New York, in 1975.)

19. (a) *Free Radicals in Biology*, W. A. Pryor, ed., Vols. I–V, Academic Press, New York, 1976–1978; (b) *EPR and Advanced EPR Studies of Biological Systems*, Dalton, L. R., ed., CRC Press, Boca Raton, 1985; (c) Rosantsev, E. C., *Free Nitroxyl Radicals*, Plenum Press, New York, 1970; Beuttner, G., *Free Radical Biol. Med.*, 1987, 3, 259–303; (d) Thornalley, P. J., Electron Spin Resonance and Spin Trapping, in *Atmospheric Oxidation and Antioxidants*, Vol. III, G. Scott, ed., Elsevier, Amsterdam, 1993; (d) Mason, R. P. and Ramakrishna Rao, D. N., Electron Spin Resonance Investigation of the Thiyl Free Radical Metabolites of Cysteine, Glutathione, and Drugs, in *Sulfur-Centered Reactive Intermediates in Chemistry and Biology*, C. Chatgilialoglu and K-D. Asmus, eds., Plenum Press, New York, 1990.

20. (a) Massey, V. and Hemmerich, P., in *Enzymes*, 3rd ed., P. D. Boyer ed., Vol. 12, Academic Press, New York, 1975, pp. 191–252; (b) Michaelis, L., Schubert, M. P. and Smythe, C. V., *Science*, 1936, 84, 138–139; (c) Massey, V. and Palmer, R., *Biochemistry*, 1966, 5, 3181–3189.

21. Larsen, I. K., DNA and Cell Growth in Chemotherapy of Cancer, in *A Textbook of Drug Design and Development*, P. Krogsgaard-Larsen and H. Bundgaard, eds., Harwood Academic Publishers, Philadelphia, 1991, pp. 218–223.

22. Fieser, L. F. and Fieser, M., *Advanced Organic Chemistry*, Reinhold, 1961, Ch. 26.

23. (a) Doroshow, J. H. and Davies, K. J. A., Redox Cycling of Anthracyclines by Cardiac Mitochondria. II. Formation of Superoxide Anion, Hydrogen Peroxide, and Hydroxyl Radical, *J. Biol. Chem.*, 1986, 261, 3068–3074; (b) Davies, K. J. A. and Doroshow, J. H., Redox Cycling of Anthracyclines by Cardiac Mitochondria. I. Anthracycline Radical Formation by NADH Dehydrogenase, *J. Biol. Chem.*, 1986, 261, 3060–3067; (c) Thornalley, P. J., Bannister, W. H. and Bannister, J. V., Reduction of Oxygen by NADH/NADH Dehydrogenase in the Presence of Adriamycin, *Free Radic. Res. Commun.*, 1986, 2, 163–171.

24. (a) James, T. L., *Nuclear Magnetic Resonance in Biochemistry*, Academic Press, New York, 1975; (b) Hore, P. J. and Kaptein, R., Photochemically Induced Dynamic Nuclear Polarization (Photo-CIDNP) Using Continuous Wave and Time Resolved Methods, in *NMR Spectroscopy: New Methods and Applications*, G. C. Levy, ed., American Chemical Society Symposium Series #191, ACS, Washington, D.C., 1982.

25. (a) Lowry, T. H. and Richardson, K. S., *Mechanism and Theory in Organic Chemistry*, 3rd ed., Harper and Row, New York, 1987, pp. 812–837; (b) Closs, G. L., *Adv. Magn. Reson.*, 1974, 7, 157; (c) Closs, G. L., Miller, R. J. and Redwine, O. D., Time-Resolved CIDNP: Applications to Radical and Biradical Chemistry Acc. *Chem. Res.* 1985, 18, 196–202; (d) Lepley, A. R. and Closs, G. L., eds., *Chemically Induced Magnetic Polarization*, Wiley, New York, 1973.

26. Sanderson, R. T., *Chemical Bonds and Bond Energy*, 2nd ed., Academic Press, New York, 1976, Ch. 8.

27. Bowser, J. R., *Inorganic Chemistry*, Brooks/Cole Publishing, Pacific Grove, 1993, p. 87.

28. Nakano, M., Determination of Active Oxygen Species by Chemiluminescence, in *Active Oxygens, Lipid Peroxides, and Antioxidants*. K. Yogi, ed., CRC Press, Boca Raton, 1993, pp. 97–109.

29. Johnson, J., Conroy, W., Burris, K. and Isom, G. E, Peroxidation of Brain Lipids Following Cyanide Intoxication in Mice, *Toxicology*, 1987, 46, 21–28.

30. Ardelt, B. K., Borowitz, J. L. and Isom, G. E., Brain Lipid Peroxidation and Antioxidant Protectant Mechanisms Following Acute Cyanide Intoxication, *Toxicology*, 1989, 56, 147–154.

31. Erenel, E., Erbas, D. and Aricioglu, A., Free Radicals and Antioxidant Systems, *Mater. Med. Pol.* (English edition), 1993, 25, 37–43.

32. (a) Finkelstein, E., Rosen, G. M., Rauckman, E. J. and Paxton, J., Spin Trapping of Superoxide, *Mol. Pharmacol.*, 1979, 16, 676–685; (b) Briere, R. and Rossat, A., Nitroiydes. LXVIII. Synthèse et Etude Cinetique de la Decomposition de t-Butyl Isopropyl Nitroxide. Effet Isotopique, *Tetrahedron*, 1976, 32, 2891–2898; (c) Buettner, G. R. and Oberley, L. W., Considerations in the Spin Trapping of Superoxide and Hydroxyl Radical in Aqueous Systems Using 5,5-Dimethyl-1-pyrroline-1-Oxide, *Biochem. Biophys. Res. Commun.*, 1978, 83, 69–74.

33. (a) Buettner, G. R., Spin Trapping: ESR Parameters of Spin Adducts, *Free Radic. Biol. Med.*, 1987, 3, 259–303; (b) Arroyo, C. M., Kramer, J. H., Dickens, B. F. and Weglicki, W. B., Identification of Free Radicals in Myocardial Ischemia/Reperfusion by Spin Trapping with Nitrone DMPO, *FEBS Lett.*, 1987, 221, 101–104.

34. To, T., Ito, M., Senoo, K. and Takeuchi, T., Superoxides, *Shindan to Chiryo*, 1993, 81, 1621–1623.

35. Sawyer, D. T. and Valentine, J. S., How Super is Superoxide?, *Acc. Chem. Res.*, 1981, 14, 393–400.

36. (a) Fridovich, I., Superoxide Radical: An Endogenous Toxicant, *Ann. Rev. Pharmacol. Toxicol.*, 1983, 23, 239–257; (b) McCord, J. M., Oxygen-Derived Free Radicals in Postischemic Tissue Injury, *N. Engl. J. Med.*, 1985, 312, 159–163.
37. (a) Fenton, H. J. H., Oxidation of Tartaric Acid in the Presence of Iron, *J. Chem. Soc.*, 1894, 65, 899–910; (b) Fenton, H. J. H. and Jones, H. O., The Oxidation of Organic Acids in Presence of Ferrous Iron, *J. Chem. Soc.*, 1900, 77, 69–83; (c) Waters, W. A., *Chemistry of Free Radicals*, Oxford, London, 1948, p. 247.
38. Thomas, C. E. and Aust, S. D., in *CRC Handbook of Free Radicals and Antioxidants in Biomedicine*, Vol. 1, J. Miguel, A. T. Quintanilha and H. Weber, eds., CRC Press, Boca Raton, 1989, pp. 37–48.
39. (a) Borg, D. C. and Schaich, K. M., Cytotoxicity from Coupled Redox Cycling of Autooxidizing Xenobiotics and Metals. A Selective Critical Review and Commentary on Work-in-Progress, *Isr. J. Chem.*, 1984, 24, 38–53; (b) Link, E. M. and Riley, P. A., Role of Hydrogen Peroxide in the Cytotoxicity of the Xanthine/Xanthine Oxidase System, *Biochem. J.*, 1988, 249, 391–399; (c) Cerutti, P. A., Prooxi-dant States and Tumor Promotion, *Science*, 1985, 227, 375–381; (d) Chance, B., Sies, H. and Boveris, A., Hydroperoxide Metabolism in Mammalian Organs, *Physiol. Rev.*, 1979, 59, 527 605.
40. Yoshikawa, T., Yoshida, N., and Koodo, M., Role of Oxygen-Derived Free Radicals in Inflammation, *Ensho*, 1993, 13, 413–421.
41. Illingworth, D. R., The Potential Role of Antioxidants in the Prevention of Athersclerosis, *J. Nutr. Sci. Vitaminol.*, 1993, 39, S43–S47.
42. Peterhaus, E., Reactive Oxygen, Antioxidants and Autotoxicity in Viral Diseases, in *Oxid. Stress, Cell Act. Viral Infect.*, C. Pasquier, ed., Birkhaeuser, Basel, 1994, pp. 203–215.
43. Keniya, M. V., Lukash, A. L. and Gus'kov, E. P., Role of Low-Molecular Weight Antioxidants in Oxidative Stress, *Usp. Sourem. Biol.*, 1993, 113, 456–470.
44. Colegate, S. M. and Molyneux, R. J., *Bioactive Natural Products, Detection, Isolation and Structural Determination*, CRC Press, Boca Raton, 1993.
45. (a) Szent-Györgi, A., Observations on the Function of Peroxidase Systems and the Chemistry of the Adrenal Cortex. Description of New Carbohydrate Derivatives, *Biochem. J.*, 1928, 22, 1387–1409; (b) Szent-Györgi, A. and Haworth, W. N., Hexuronic Acid [Ascorbic Acid] as the Antiscorbutic Factor, *Nature*, 1933, 131, 24; (c) L-*Ascorbic Acid: Chemistry, Metabolism and Uses; ACS Advances in Chemistry Series No. 200*, P. A. Seib and B. M. Tolbert, eds., American Chemical Society, Washington, D.C., 1982.
46. Al-Meshal, I. A. and Hassan, M. M. A., in *Analytical Profiles of Drug Substances*, Vol. 11, K. Florey, ed., Academic Press, New York, 1982, pp. 45–78.
47. King, D. L., Hahm, T. S. and Min, D. B., Chemistry of Antioxidants in Relation to Shelf Life of Food, *Dev. Food Sci.*, 1993, 33, 629–705.
48. Weissberger, A. and LuValle, J. E., Oxidation Processes. XVII. The Autooxidation of Ascorbic Acid in the Presence of Copper, *J. Am. Chem. Soc.*, 1944, 66, 700–705.
49 Kwon, B-M. and Foote, C. S., Chemistry of Singlet Oxygen. 50. Hydroperoxide Intermediates in the Photooxygenation of Ascorbic Acid, *J. Am. Chem. Soc.*, 1988, 110, 6582–6583.
50. Hvoslef, J., The Crystal Structure of L-Ascorbic Acid, Vitamin C, *Acta Chem. Scand.*, 1964, 18, 841–842.
51. Buckingham, J. and Hill, R. A., Atlas of Stereochemistry, 2nd. ed, supplement, Chapman and Hall, 1986, p. 55.
52. Crawford, T. C. and Crawford, S. A., Synthesis of L-Ascorbic Acid, *Adv. Carbohydr. Chem.*, 1980, 37, 79–155.
53. Chatgilialoglu, C. and Asmus, K-D., eds., *Sulfur-Centered Reactive Intermediates in Chemistry and Biology*, Plenum Press, New York, 1990.
54. (a) Nakazawa, H., Ichinori, K., Shinozaki, Y., Okino, H. and Hori, S., Is Superoxide Demonstration by Electron-Spin Resonance Spectroscopy Really Superoxide?, *Am. J. Physiol.*, 1988, 255, H213–H215; (b) Haase, G. and Dunkley, W. L., Ascorbic Acid and Copper in Linoleate Oxidation. I. Measurement of Oxidation by Ultraviolet Spectrophotometry and the Thiobarbituric Acid Test, *J. Lipid. Res.*, 1969, 10, 555–560.
55. Sauberlich, H. E., Pharmacology of Vitamin C, *Annu. Rev. Nutr.*, 1994, 14, 371–391.
56. Buckingham, J. and Hill, R. A., *Atlas of Stereochemistry*, 2nd. ed, supplement, Chapman and Hall, 1986, and previous volumes in this set.
57. (a) Salem, L., *Molecular Orbital Theory of Conjugated Systems*, W. A. Benjamin, London, 1974; (b) Murrell, J. N., *The Theory of the Electronic Spectra of Organic Molecules*, Methuen, London, 1963.
58. Hubbard, R. and Wald, G., Pauling and Carotenoid Stereochemistry, in *Structural Chemistry and Molecular Biology*, A. Rich and N. Davidson, eds., W. H. Freeman, New York, 1968, pp. 545–554.
59. Fieser, L. and Fieser, M., *Topics in Organic Chemistry*, Reinhold, New York, 1963, p. 205.
60. Jaffe, H. H. and Orchin, M., *Theory and Applications of Ultraviolet Spectroscopy*, Wiley, New York, 1962, Ch. 11, and references cited therein.

61. Silverstein, R. M., Bassler, G. C. and Morrill, T. C., *Spectrometric Identification of Organic Compounds*, Wiley, New York, 1991, pp. 296–299.
62. Budzikiewicz, H., Selected Reviews on Mass Spectrometric Topics, in *Mass. Spectrom. Rev.*, 1984, 3, 317–318.
63. Sterling, *Acta Cryst.*, 1964, 17, 1224.
64. (a) Suzmatis, J. D. and Ofner, A., A New Synthesis of trans-β-Carotene and Decapreno-β-Carotene, *J. Org. Chem.*, 1961, 26, 1171–1173; (b) Fischli, A. and Mayer, H., Carotinoidsynthesen über Sulfone; Synthese von β-Carotin, *Helv. Chim. Acta*, 1975, 58, 1584–1590.
65. DiMascio, P., Devasagayam, T. P. A., Kaiser, S. and Sies, H., Carotenoids, Tocopherols and Thiols as Biological Singlet Molecular Oxygen Quenchers, *Biochem. Soc. Trans.*, 1990, 18, 1054–1056.
66. Olson, J. A., Vitamin A and Carotenoids as Antioxidants in a Physiological Context, *J. Nutr. Sci. Vitaminol.*, 1993, 39 (Suppl.), S57–S65.
67. Shenoy, V. R., Vitamin E—An Appraisal, *J. Sci. Ind. Res.*, 1995, 54, 357–382.
68. Wattenberg, L. W. and Lam, L. K. T., Phenolic Antioxidants as Protective Agents in Chemical Carcinogenesis, in *Radioprotectors and Anticarcinogens*, O. F. Nygaard and M. G. Simic, eds., Academic Press, New York, 1983, pp. 461–469.
69. *Chem. & Eng. News*, 1995, 73, 39.
70. Ternay, Jr., A. L., *Contemporary Organic Chemistry*, W. B. Saunders, Philadelphia, 1979, Ch. 23.
71. Foster, R., *Organic Charge-Transfer Complexes*, Academic Press, New York, 1969.
72. L. Machlin, ed., *Vitamin E: A Comprehensive Treatise*, Dekker, New York, 1980.
73. (a) Cohen, N., Scott, C. G., Neukom, C., Lopresti, R. J., Weber, G. and Saucy, G., Total Synthesis of All Eight Stereoisomers of α-Tocopheryl Acetate. Determination of their Diastereomeric and Enantiomeric Purity by Gas Chromatography, *Helv. Chim. Acta*, 1981, 64, 1158; (b) U.S. pat. 2,723,278 (1955); (c) Scott, J. W., Readily Available Chiral Carbon Fragments and Their Use in Synthesis, in *Asymmetric Synthesis*, Vol. 4., J. D. Morrison and J. W. Scott, eds., Academic Press, New York, 1984, Ch. 1.
74. Mukai, K., Ohbayashi, S., Nagoaka, S., Ozawa, T. and Azuma, N., X-ray Crystallographic Studies of Vitamin E Derivatives. Relationship Between Antioxidant Activity and Molecular Structure, *Bull. Chem. Soc. Jpn.*, 1993, 66, 3808–3810.
75. van Acker, S. A. B. E., Koymans, L. M. H. and Bast, A., Molecular Pharmacology of Vitamin E: Structural Aspects of Antioxidant Activity, *Free Radic. Biol. Med.*, 1993, 15, 311–328.
76. Wilcox, A. L. and Janzen, E. G., Nitric Oxide Reactions with Antioxidants in Model Systems: Sterically Hindered Phenols and α-Tocopherols in Sodium Dodecyl Sulfate (SDS) Micelles, *J. Chem. Soc., Commun.*, 1993, 1377–1379.
77. Rosenkrantz, H., Milhorat, A. T. and Farber, M., Countercurrent Distribution in Identification of Tocopherol Compounds in Feces, *J. Biol. Chem.*, 1951, 192, 9–15, and references cited therein.
78. Chase, Jr., G. W., Akoh, C. C. and Eitenmiller, R. R., Analysis of Tocopherols in Vegetable Oils by High-Performance Liquid Chromatography: Comparison of Fluorescence and Evaporative Light-Scattering Detection, *J. Am. Oil Chem. Soc.*, 1994, 71, 877–880.
79. *Proc. Natl. Acad. Sci. USA*, 92, 8171 (1995).
80. Takeshige, K., Takayanagi, R. and Minakami, S., Lipid Peroxidation and the Reduction of ADP-iron (III) Chelate by NADH-Ubiquinone Reductase Preparation from Bovine Heart Mitochondria, *Biochem. J.*, 1980, 192, 861–866.
81. Voet, D. and Voet, J. G., *Biochemistry*, Wiley, New York, 1990, Ch. 20.
82. Stocker, R., Bowry, V. W. and Frei, B., Ubiquinol-10 Protects Human Low Density Lipoprotein More Efficiently Against Lipid Peroxidation Than Does α-Tocopherol, *Proc. Natl. Acad. Sci. U.S.A.*, 1991, 88, 1646–1650.
83. Huyser, E. S., *Free Radical Chain Reactions*, Wiley-Interscience, New York, 1970.
84. (a) Men'schikova, E. B., and Zenkov, N. K., Antioxidants and Inhibitors of Free Radical Oxidative Processes, *Usp. Sourem. Biol.*, 1993, 113, 442–455; (b) Simic, M. G., Interaction of Free Radicals and Antioxidants in *Radioprotectors and Anticarcinogens*, O. F. Nygaard and M. G. Simic, eds., Academic Press, New York, 1983, pp. 449–460.
85. Lewin, G. and Popov, I., The Antioxidant System of the Organism. Theoretical Basis and Practical Consequences, *Med. Hypotheses*, 1994, 42, 269–275; (b) Pryor, W. A., Oxidative Stress Status Measurements in Humans and Their Use in Clinical Trials, in *Active Oxygen, Lipid Peroxides, and Antioxidants*, K. Yogi, ed., CRC Press, Boca Raton, 1993, pp. 117–126.
86. Ottolenghi, A., Interaction of Ascorbic Acid and Mitochondrial Lipids, *Arch. Biochem. Biophys.*, 1959, 79, 355–363.
87. (a) K. Yogi, ed., *Active Oxygens, Lipid Peroxides and Antioxidants*, CRC Press, Boca Raton, 1993; (b) McCay, P. B., King, M. M., Fong, K-L., Poyer, J. L., Brueggemann, G., Lai, E., Leemaster, L., Olson, L. and Wilkins, T., Nutritional Factors in Vivo; The Effect of Antioxidants on the Formation of Trichloromethyl Radical from CCl_4 in Liver of Intact Rats, in *Radioprotectors and Anticarcinogens*, O. F. Nygaard and M. G. Simic, eds., Academic Press, New York, 1983, pp. 585–604.

88. Niki, E., Noguchi, N. and Gotoh, N., Dynamics of Lipid Peroxidation and Its Inhition by Antioxidants, *Biochem. Soc. Trans.*, 1993, 21, 313–317.
89. Scott, G., Antioxidants: Chain Breaking Mechanisms, in *Atmospheric Oxidation and Antioxidants*, G. Scott, ed. Elsevier, Amsterdam, 1993, pp. 121–160.
90. Terao, J., Antioxidant Vitamins and Their Function, *Kassei Sanso, Furi Rajikaru*, 1993, 4, 288–306; CA 120:53474k, 1994.
91. Nordmann, R., Free Radicals, Oxidative Stress and Antioxidant Vitamins, *C. R. Seances Soc. Biol. Ses Fil.*, 1993, 187, 277–285; CA 120: 70940b, 1994.
92. (a) Maestro Duran, R., and Borja Padilla, R., Antioxidant Activity of Vitamins C, and E, and Provitamin A, *Grasas Aceites (Seville)*, 1993, 44, 107–111; CA 120: 75714r (1994); (b) Maestro Duran, R. and Borja Padilla, R., Antioxidant Activity of Phenolic Compounds, *Grasas Aceites (Seville)*, 1993, 44, 101–106; CA 120: 75713q (1994).
93. Etsuo, E., Noguchi, N. and Gotoh, N., Inhibition of Oxidative Modification of Low Density Lipoprotein by Antioxidants, *J. Nutr. Sci. Vitaminol.*, 1993, 39, S1–S8.
94. Burton, G. W. and Ingold, K. U., β-Carotene: An Unusual Type of Lipid Antioxidant, *Science*, 1984, 224, 569–573.
95. Tamas, V. and Borza, A., *Stud. Cercet. Biochim.*, 1993, 36, 129–138; CA 121: 148156t (1994).
96. Terao, J., Reactive Oxygen Species, Free Radicals and Antioxidant Functions of β-Carotene, *Fragrance J.*, 1993, 21, 49–54.
97. Sies, H. and Sundquist, A. R., Role of Carotenoids in Antioxidant Defense, *Antioxid. Health Dis.*, 1994, 1, 275–287.
98. Tsuchiya, M., Scita, G., Thompson, D. F. T., Packer, L., Kagan, V. E. and Livcrea, M. A., Retinoids and Carotenoids Are Peroxyl Radical Scavengers, *Clin. Dermatol.*, 1993, 525–536.
99. Russell, G. A, Semidione Radical Anions, in *Radical Ions*, E. T. Kaiser and L. Kevan, eds., Interscience Publishers, New York, 1968, Ch. 3.
100. Creutz, C., The Complexities of Ascorbate as a Reducing Agent, *Inorg. Chem.*, 1981, 20, 4449–4452.
101. Lohmann, W. and Holz, D., Structure of Ascorbic Acid and its Biological Function. I. ESR Determination of the Ascorbyl Radical in Biological Samples and in Model Systems, *Biophys. Struct. Mech.*, 1984, 10, 197–204.
102. Kagayama, N., Sekiguchi, M., Inada, Y., Takagi, H. D. and Funahashi, S., Mechanistic Studies of the Reaction of L-Ascorbic Acid with Hexacyanometalate (III) Ions of Iron (III), Ruthenium (III), and Osmium (III) in Aqueous Acidic Solution at Elevated Pressures, *Inorg. Chem.*, 1994, 33, 1881–1885.
103. Packer, L., The Vitamin E Antioxidant Cycle in Health and Disease, *NATO ASI Ser., Ser. A*, 1993, 246, 297–308.
104. Liebler, D. C., The Role of Metabolism in the Antioxidant Function of Vitamin E, *Crit. Rev. Toxicol.*, 1993, 23, 147–169.
105. Lambelet, P. and Loliger, J., The Fate of Antioxidant Radicals During Lipid Autooxidation. I. The Tocopheryl Radicals, *Chem. Phys. Lipids*, 1984, 35, 185–198.
106. Burton, G. W. and Ingold, K. U., Autoxidation of Biological Molecules. 1. The Antioxidant Activity of Vitamin E and Related Chain-Breaking Phenolic Antioxidants in Vitro, *J. Am. Chem. Soc.*, 1981, 103, 6472–6477.
107. Dutie, G. G., McPhail, D. B., Morrice, P. C. and Arthur, J. R., Antioxidant Effectiveness of Tocopherol Isomers, in *Lipid-Soluble Antioxidants*, A. S. H. Ong and L. Packer, eds., Birkhaeuser, 1992, pp. 76–84.
108. Suarna, C., Dean, R. T. and Stocker, R., The Reactivity of Tocotrienols and Other Lipid-Soluble Antioxidants Towards Peroxyl Radicals, in *Lipid-Soluble Antioxidants*, A. S. H. Ong and L. Packer, eds., Birkhaeuser, 1992, pp. 17–26.
109. Puhl, H., Waeg, G. and Esterbauer, H., Inhibition of LDL Oxidation by Vitamin E and Other Antioxidants, *Atheroscler. Rev.*, 1993, 25, 277–285.
110. Cho, S. H. and Kyung, Y. S., Lipid Peroxidation and Antioxidant Status is Affected by Different Vitamin E levels When Feeding Fish Oils, *Lipids*, 1994, 29, 47–52.
111. Chudinova, V. V., Zakharova, E. I., Alekseev, S. M. and Evstigneeva, R. P., Mechanism of the Vitamin E Activity. Study of the Interaction of Vitamin E (α-Tocopherol) and Its Analogs with Fatty Acids and Their Derivatives by the Fluorescence Method, *Bioorg. Khim.*, 1992, 18, 1528–1534.
112. Mukai, K., Daifuku, K., Okabe, K., Tanagaki, T. and Inoue, K., Structure-Activity Relationships in the Quenching Reaction of Singlet Oxygen by Tocopherol Derivatives and Related Phenols. Finding Linear Correlation Between the Rates of Quenching of Singlet Oxygen and Scavenging of Peroxyl and Phenoxyl Radicals in Solution, *Int. Congr. Ser.—Excerpta Med.*, 1992, 998, 625–628.
113. Kagan, V. E., Serbinova, E., Khwaja, S., Catudioc, J., Maguire, J. J. and Packer, L., Ubiquinones and Vitamin E: Partners or Competitors in Antioxidation?, in *Active Oxygen, Lipid Peroxides, and Antioxidants*, K. Yogi, ed., CRC Press, Boca Raton, 1993, pp. 237–245.
114. Ernster, L., Lipid Peroxidation in Biological Membranes: Mechanisms and Implications, in *Active Oxygen, Lipid Peroxides, and Antioxidants*, K. Yogi, ed., CRC Press, Boca Raton, 1993, pp. 1–38.

115. Freisleben, H. J. and Packer, L., Free-Radical Scavenging Activities, Interactions and Recycling of Antioxidants, *Biochem. Soc. Trans.*, 1993, 21, 325–330.
116. Beyer, R. E., The Role of Ascorbate in Antioxidant Protection of Biomembranes: Interactions with Vitamin E and Coenzyme Q, *J. Bioenerg. Biomembr.*, 1994, 26, 349–358.
117. Sharma, M. K. and Buettner, G. R., Interaction of Vitamin C and Vitamin E During Free Radical Stress in Plasma: an ESR Study, *Free Radic. Biol. Med.*, 1993, 14, 649–653.
118. Meister, A., Glutathione-Ascorbic Acid Antioxidant System in Animals, *J. Biol. Chem.*, 1994, 269, 9397–9400.

Chapter Two

THE ANALYSIS OF FREE RADICALS, THEIR REACTION PRODUCTS, AND ANTIOXIDANTS

Ian N. Acworth, Douglas R. McCabe, and Timothy J. Maher

During evolution, as photosynthesizing organisms increased the amount of oxygen in the atmosphere, anaerobic creatures had to make an important choice: to remain anaerobic[1] or to adapt and become aerobic. Aerobic organisms have made good use of oxygen. Besides using it in metabolic transformations (hydroxylases, oxidases, etc.), cells also use oxygen in "metabolic combustion," that is, respiration. There is considerable release of energy ($\Delta G^{0'} = -52.6$ kcal/mol) (see Table 1 for reaction rate constants used throughout this text) when electrons are transferred from the highly reducing NADH (or $FADH_2$) (both of which are formed during catabolism of fuel molecules), through the respiratory chains of the inner mitochondrial membrane, to the terminal electron acceptor, molecular oxygen (a strong oxidizer). This energy, a consequence of the difference in reduction potential between NADH and oxygen ($\Delta E^{0'} = 1140$mV; Table 2), is efficiently harnessed within the cell to the production of ATP molecules, the cell's metabolic fuel. Aerobic respiration is much more effective in producing fuel molecules than anaerobic respiration. For example, aerobic respiration of glucose produces 34 molecules of ATP, whereas anaerobic respiration only produces 2 molecules of ATP. The enormous importance of aerobic respiration becomes all too apparent as death is very rapid following exposure to inhibitors of the respiratory chain such as cyanide or carbon monoxide. However, living in an atmosphere abundant in oxygen (21% by volume) is not without problems. Although oxygen is essential for aerobic organisms, exposure to oxygen at concentrations of >21% can lead to injury, such as neuronal and alveolar damage, and even cellular death (1–3). However, the susceptibility of an organism to the damaging effects of oxygen is complex, and is found to be species and tissue specific and dependent on many variables including age, health, and diet (3).

Initially, oxygen toxicity was thought to be due to inactivation of enzymes (1). However, as few targets could be identified, researchers turned their attention to a number of partially reduced oxygen free radical species (4) and in particular, following the discovery of superoxide dismutase, to the superoxide free radical anion (O_2^-) (5). Currently a number of free radical and nonradical species derived from molecular oxygen (reactive oxygen species, ROS), nitrogen (reactive nitrogen species, RNS), and other atoms are known to exist in biological systems (Table 3). Some are produced as part of normal metabolism (e.g., O_2^- is formed when xanthine is converted to uric acid by xanthine oxidase). Others are even considered to be beneficial and play important

---

[1] Anaerobic organisms still exist but they have restricted themselves to reducing environments where oxygen does not penetrate. Many of these organisms are highly sensitive to oxygen which can inhibit their growth or even kill them. They use electron acceptors other than oxygen. Bacteria have evolved to use ferric iron, sulfate or carbon dioxide which become reduced when energy is released from NADH (producing ferrous iron, hydrogen sulfide and methane, respectively). Most bacteria that reduce nitrate (producing nitrite, nitrous oxide or nitrogen) are called facultative anaerobes as they are not affected by exposure to oxygen and in fact will preferentially transfer electrons to oxygen if it is present or to nitrate if oxygen is absent.

Table 1.   Rate constants for the reaction of ROS, RNS, and other radicals with other molecules

Reaction	Rate constant $(M^{-1}\ s^{-1})^a$	References
Hydroxyl radical reactions		
$HO^• + $ General metabolite	10^9 to 10^{10}	375
$HO^• + $ Albumin	$>10^{10}$	375
$HO^• + $ Ascorbyl	1.0×10^{10}	378
$HO^• + N$-Acetylcysteine	1.4×10^{10}	376
$HO^• + $ Carnosine	2.5×10^9	377
$HO^• + \beta$-Carotene	$<1.0 \times 10^{11}$	421
$HO^• + $ Cysteamine	5.9×10^9	379
$HO^• + $ Cysteic acid	5.3×10^7–1.6×10^8	379
$HO^• + $ Cysteinesulfinic acid	3.2×10^9	379
$HO^• + $ Deoxyguanosine	1.0×10^9	383
$HO^• + $ Deoxyribose	3.1×10^9	375
$HO^• + $ Deferrioxamine	1.3×10^{10}	380
$HO^• + $ DMSO	6.6–7.1×10^9	118, 380
$HO^• + $ "Double bond"	$\geq 10^{10}$	381
$HO^• + $ Ethanol	1.9×10^9	118
$HO^• + $ Glucose	1.0×10^9	375
$HO^• + $ Glutathione (reduced)	8.8×10^9	376
$HO^• + $ Hypotaurine	5×10^9–1.2×10^{10}	379
$HO^• + $ Mannitol	1.7×10^9	118
$HO^• + NO^• \rightarrow HNO_2$	1.0×10^{11}	80
$HO^• + NO_2^• \rightarrow ONO_2H$	4.5×10^9	378
$HO^• + $ Phenylalanine	1.9×10^9	19, 37
$HO^• + $ Phosphate (e.g. DNA)	$<10^7$	381
$HO^• + R-H \rightarrow R^• + H_2O$	1.0×10^9	381
$HO^• + $ Salicylic acid	5.0×10^9–1.0×10^{10}	382
$HO^• + $ Taurine	2.4×10^6–1.4×10^7	379
$HO^• + \alpha$-Tocopherol	1.0×10^{10}	421
$HO^• + $ Trolox	8.0×10^{10}	118
$HO^• + $ Unsaturated fatty acids	1.0×10^9	122
Superoxide radical anion and hydroperoxyl radical reactions		
$O_2^- + N$-Acetylcysteine	1.0×10^3–2.7×10^6	393
$O_2^- + $ Ascorbate	1.0×10^4–2.7×10^5	3, 122
$O_2^- + $ Carnosine	1.0×10^3	377
$O_2^- + $ Catechols	1.0×10^9	3
$O_2^- + Fe^{3+}$	1.0×10^6	122
$O_2^- + H^+ \rightarrow \frac{1}{2}H_2O_2 + \frac{1}{2}O_2$	5.0×10^5–2.4×10^9	380, 384
$O_2^- + HO_2^• + H^+ \rightarrow H_2O_2 + O_2$	2.4×10^5	385
$O_2^- + HOCl \rightarrow HO^• + Cl^- + O_2$	7.5×10^6	376
$O_2^- + NO^• \rightarrow ONO_2^-$	3.4×10^7–7×10^9	118, 122
$O_2^- + NO_2 + O_2$	1.0×10^8	387
$O_2^- + HO_2^• + H^+ \rightarrow H_2O_2 + O_2$	8.0×10^7	3
$O_2^- + O_2^- + 2H^+ \rightarrow H_2O_2 + O_2$	<0.3–5×10^5	3
$O_2^- + $ Quinones	1.0×10^9	3
$O_2^- + $ Taurine	$<1 \times 10^3$	379
$HO_2^• + \alpha$-Tocopherol (TO)	2.5×10^6	80
$HO_2^• + HO_2^• \rightarrow H_2O_2 + O_2$	8.0×10^5	3
Peroxide reactions		
$H_2O_2 + $ Cysteine	1.3×10^1	119
$H_2O_2 + Cu^+ \rightarrow HO^• + HO^-Cu^{2+}$	4.7×10^3	384
$H_2O_2 + Fe^{2+} \rightarrow HO^• + HO^-Fe^{3+}$	7.6×10^1	384
$H_2O_2 + Fe^{2+}$-ADP $\rightarrow HO^• + HO^-Fe^{3+}$-ADP	8.0×10^2	384
$H_2O_2 + Fe^{2+}$-EDTA $\rightarrow HO^• + Fe^{3+}$-EDTA	5.0×10^3	384
$H_2O_2 + ONO_2H \rightarrow O_2^- + NO_2 + H^+ + H_2O$	1.0×10^5	387

Table 1. Rate constants for the reaction of ROS, RNS, and other radicals with other molecules—Continued

Reaction	Rate constant $(M^{-1}\ s^{-1})^a$	References
Reactions involving reactive nitrogen species		
$NO^{\bullet} + Fe^{2+}$(heme)	1×10^7	118
$NO^{\bullet} + Fe^{3+}$(heme)	1×10^2–1×10^7	118
$NO^{\bullet} + Heme$	1×10^3–1×10^4	122
$NO^{\bullet} + \frac{1}{2}O_2 \rightarrow NO_2^{\bullet}$	$3.5 \times 10^6\ (M^{-2}\ s^{-1})$	90
$NO^{\bullet} + Tryptophan^{\bullet}$	1–2×10^9	83
$NO^{\bullet} + Tyrosine^{\bullet}$	1–2×10^9	83
$ONO_2^- + SOD \rightarrow SOD-Cu^+-O^{-\cdots\cdots}NO_2^+$	1.0×10^5	384
$ONO_2^- + Albumin$ (single thiol)	2.6×10^3	119
$ONO_2H + Ascorbate$	2.4×10^1	378
$ONO_2^- + Cysteine$ (or glutathione)	2.0–6.0×10^3	119, 378
$ONO_2^- + CO_2 \rightarrow ONO_2CO_2^-$	3.0×10^4	132
$ONO_2CO_2^- + Tyrosine \rightarrow Tyr^{\bullet} + NO_2^{\bullet} + HCO_3^-$	$>2.0 \times 10^5$	133
$NO_2^{\bullet} + Fatty\ acid \rightarrow NO_2^- + Fatty\ acid^{\bullet} + H^{\bullet}$	1.0×10^5	122
$NO_2^{\bullet} + Tyrosine \rightarrow 3\text{-Nitrotyrosine}$	3.0×10^9	388
$NO_2^+ + Tyrosine \rightarrow 3\text{-Nitrotyrosine}$	1.0	389
Miscellaneous reactions		
$AscH^- + TO^{\bullet} \rightarrow Asc^{\bullet -} + TOH$	2.0×10^5	390
$2Asc^{\bullet -} + H^+ \rightarrow AscH^- + dehydroascorbate$	2.0×10^5	383
$CoQH_2 + TO^{\bullet} \rightarrow CoQ^{\bullet -} + TOH$	2.0×10^5	391
$DNA^{\bullet} + GSH$	$1.0 \times 10^7 - 10^8$	375
$e_{aq} \rightarrow H^{\bullet} + H^-$	1.6×10^1	381
$Fe^{3+} + e^- \rightarrow Fe^{2+}$	1.0×10^6	384
$GS^{\bullet} + GS^- \rightarrow GSSG^{\bullet -}$	8.0×10^8	386
$GS^{\bullet} + O_2 \rightarrow GSO_2^{\bullet}$	2.0×10^9	386
$GSO_2^{\bullet} \rightarrow GS^{\bullet} + O_2$	2.0×10^9	386
$GSSG^{\bullet -} \rightarrow GS^{\bullet} + GS^-$	2.4×10^5	386
$GSSG^{\bullet -} + O_2 \rightarrow O_2^- + GSSG$	1.6×10^8	386
$L^{\bullet} + O_2 \rightarrow LO_2^{\bullet}$	3.0×10^8	386
$L^*O^{\bullet}(L^*O_2^{\bullet}) + NO^{\bullet}$	1.3×10^9	81
$LO_2^{\bullet} + LH \rightarrow L^{\bullet} + LO_2H$	1–5×10^1	386
$2LO_2^{\bullet} \rightarrow Nonradical\ products$	$1.0 \times 10^{7-8}$	386
$L^*O_2H + Fe^{2+} \rightarrow LO_2^{\bullet} + Fe^{3+}$	1.0×10^3	384
$RCH_2^{\bullet} + O_2 \rightarrow RCH_2O_2^{\bullet}$	3.0×10^9	392
$TO + PUFA-O_2^{\bullet} \rightarrow PUFA-O_2H + TO^{\bullet}$	8.0×10^4	386
$Tyrosine^{\bullet} + tyrosine^{\bullet} \rightarrow dityrosine$	4.0×10^8	388

^{a} See reference for reaction conditions, concentrations, and comments.

biological roles when their formation is controlled (e.g., O_2^-, hydrogen peroxide, hypochlorous acid, and nitric oxide are the major "microbicides" of circulating phagocytic leucocytes and stationary cells) (3, 6, 7). However, the major production of ROS and RNS appears to be uncontrolled, and results from what Halliwell describes as "accidents of chemistry" such as "leaking" electron transport chains and from catecholamine redox cycling (8).

Control of ROS and RNS levels poses a major challenge to aerobic organisms. Normally, their rate of production does not exceed the capacity of the body to dispose of them. Indeed, the body is well equipped with a battery of mechanisms that serve to keep the levels of most ROS and RNS within well-defined limits. As discussed later, these mechanisms include prevention of their production by metal chelation (e.g., ferritin, transferrin, ceruloplasmin) and enzymatic destruction once ROS and RNS are formed (e.g., superoxide dismutase, catalase, and peroxidases). Furthermore, various antioxidants (e.g., vitamins A, C, E, and K, ubiquinone, uric acid, glutathione, and polyphenols) are capable of reacting with ROS and RNS or with their reaction products, thereby preventing cellular damage. These are discussed in greater detail later. However, under certain circumstances when the body's natural defenses are compromised (e.g., following exposure to sunlight, smoking, or in individuals with a genetic predisposition) a condition known as

Table 2. Reduction potentials of various redox couples

Redox couple	$E^{\circ\prime}$ (mV)	Reference
Single-electron reactions		
$HO^{\bullet}$, H^+/H_2O	+2310	394
$C_2H_5^{\bullet}$, H^+/C_2H_6	+1900	395
$RO^{\bullet}$, H^+/ROH (aliphatic)	+1600	395
$HO_2^{\bullet}$, H^+/H_2O_2	+1060	394
$RO_2^{\bullet}$, H^+/RO_2H	+770 to +1400	395
O_2^{-2}, $2H^+/H_2O_2$	+940	394
$RS^{\bullet}/RS^-$ (cysteine)	+920	396, 397
$^1\Delta O_2/O_2^{\bullet-}$	+650	398
$PUFA^{\bullet}$, $H^+/PUFA-H$	+600	395
$HU^{\bullet-}$, H^+/UH_2^-	+590	399
$CA-O^{\bullet}$, $H^+/CA-OH$	+530	400
$\alpha\text{-}TO^{\bullet}$, $H^+/\alpha\text{-}TOH$	+500	401–404
H_2O_2, H^+/H_2O, $HO^{\bullet}$	+320	406
$NAD^{\bullet} + e^- + H^+/NADH$	+300	407
$Asc^{\bullet-}$, $H^+/AscH^-$	+282	408
Fe^{3+}/Fe^{2+} (cytochrome)	+260	409
$CoQ^{\bullet-}$, $H^+/CoQH_2$	+200	410
Fe^{3+}/Fe^{2+}(aqueous, pH 7.0)	+110	411
$CoQ/CoQ^{\bullet-}$	−36	410
Dehydroasc/asc$^{\bullet-}$	−174	408
Fe^{3+}/Fe^{2+} (ferritin)	−190	413
$FADH^{\bullet} + e^- + H^+/FADH_2$	−240	414
Riboflavin/riboflavin$^{\bullet-}$	−317	415
$O_2/O_2^{\bullet-}$	−330	416
Fe^{3+}/Fe^{2+} (transferrin)	−400 (pH 7.3)	417
O_2, $H^+/HO_2^{\bullet}$	−460	406
$RSSR/RSSR^{\bullet-}$ (e.g., cystine or glutathione disulfide)	−1500	418, 419
$CO_2/CO_2^{\bullet-}$	−1800	406
H_2O/e_{aq}^-	−2870	420
Multiple-electron reactions[b]		
$O_2 + 2e^- + 2H^+/H_2O_2$	+330	405
$FAD^+ + 2e^- + 2H^+/FADH_2$	−180	412
$NAD^+ + 2e^- + H^+/NADH$	−320	412
$O_2 + 4e^- + 4H^+/2H_2O$	+820	394

Source: Based, in part, on ref. 386. Reproduced with permission from Academic Press, Inc.

[a] We have used $O_2^{\bullet-}$ to represent the superoxide radical anion (see ref. 3 for reasons).

[b] Note that $E^{\circ\prime}$ values should only be compared within a group and for reactions involving the same number of electrons.

oxidative stress results. Consequently, overproduction of ROS and RNS can lead to premature aging and a variety of diseases including cancer, ischemic damage following stroke, arthritis, atherosclerosis, and a host of neurodegenerative disorders (3, 9–11).

This article is divided into three sections. The first section reviews the current methodological approaches used to measure ROS and RNS levels. The second section covers measurement of reaction products formed by the reaction between ROS and RNS with biological molecules. The third section reviews the measurement of antioxidants. In each case particular emphasis is placed on state-of-the-art high-performance liquid chromatography (HPLC)-based techniques.

REVIEW OF CHEMISTRY AND MEASUREMENT OF REACTIVE OXYGEN AND NITROGEN SPECIES

Reactive Oxygen Species

Successive addition of electrons to molecular oxygen can give rise to several intermediate ROS, culminating in the formation of water (Figure 1). Cytochrome oxidase, the terminal enzyme in

Table 3.   ROS, RNS, and other free radicals sometimes found in biological systems

Reactive oxygen species		Reactive nitrogen species		Miscellaneous	
		Free radicals			
Hydroxyl	HO$^•$	Nitric oxide (monoxide)	NO$^•$	Thiyl	RS$^•$
Superoxide	O$_2^-$	Nitrogen dioxide	NO$_2^•$	Hydrogen atom	H$^•$
Alkoxyl	LO$^•$			Carbon-centered radicals	e.g., CCl$_3^•$
Hydroperoxyl	HO$_2^•$				
Peroxyl	LO$_2^•$				
		Nonradicals			
Hydrogen peroxide	H$_2$O$_2$	Dinitrogen trioxide	N$_2$O$_3$	Thiol	RSH
Singlet oxygen	$^1\Delta$GO$_2$	Dinitrogen tetroxide	N$_2$O$_4$		
Lipid peroxides	LO$_2$H	Dinitrogen pentoxide	N$_2$O$_5$		
Ozone	O$_3$	Peroxynitrite	ONO$_2^-$		
		Alkyl peroxynitrites	LO$_2$NO$^-$		
		Nitrocarbonate	O$_2$NOCO$_2^-$		
		Nitrosoperoxycarbonate	ONO$_2$CO$_2^-$		

the respiratory chain, adds four electrons to oxygen fairly efficiently to form water during energy generation in the mitochondria. However, during this process ROS are also generated. Some have suggested that as much as 1–3% of oxygen may be converted to O$_2^-$ in the mitochondria (12). Taking all O$_2^-$ formation together, a 70-kg man would be expected to produce about 2 kg of O$_2^-$ per year. This is a considerable amount but, judging from survival of the human race, not beyond our antioxidant protective capacity. Not all ROS are equal, however. If 2 kg of the more energetic hydroxyl (HO$^•$) free radicals was formed, depending upon their site of production, the resulting damage to macromolecules would probably overpower our protective mechanisms and lead to serious disease.

The Hydroxyl Free Radical

The hydroxyl free radical (HO$^•$) is one of the most aggressive radicals found in the body reacting at a diffusion-controlled rate with almost every molecule in the living cell (Table 1) including DNA, lipids, proteins and carbohydrates (Figure 2). HO$^•$s are usually generated by two principal

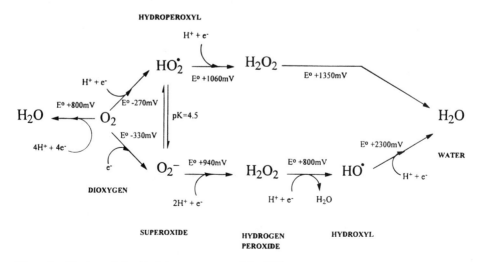

Figure 1.   The interrelationship between oxygen and the ROS.

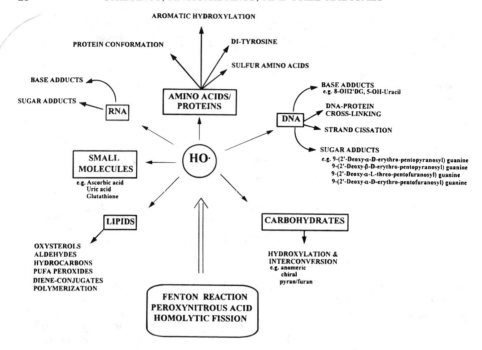

Figure 2. The attack of HO· on a variety of biologically important molecules produces a great diversity of reaction products.

mechanisms: homolytic fission of water molecules by ionizing radiation (ultraviolet, gamma, microwave, x-rays, etc.) or the breakdown of hydrogen peroxide with metals:

$$\text{Metal}^{n+} + H_2O_2 \longrightarrow \text{Metal}^{(n+1)+} + HO\cdot + HO^- \qquad (1)$$

Several metals can react in this way (iron, copper, chromium, vanadium, etc.), but by far the most common is Fe^{2+}, and if this is the case the reaction is then referred to as the Fenton reaction. In actuality the Fenton reaction is far more complicated than suggested in Eq. 1 (10, 13). In the Fenton reaction, Fe^{2+} is likely regenerated from Fe^{3+} by cellular antioxidants (reducing agents) such as ascorbate (14). The resulting ascorbyl free radical is much less reactive and probably dissipates by dismutation into ascorbic acid and dehydroascorbic acid (15). A second similar reaction, the Haber-Weiss reaction (or O_2^- Fenton reaction), has O_2^- reacting with Fe^{3+} to regenerate Fe^{2+} (and O_2), thereby permitting the Fenton reaction to proceed more effectively (10). As discussed later, HO· can also be produced from peroxynitrous acid (16) and from the reaction between O_2^- and hypochlorous acid (17).

The HO· radical is so reactive that it normally diffuses only 5–10 molecular diameters before it reacts, so that HO· damage is very site specific (18). Almost every molecule in the cell will react with the electrophilic HO· with second-order rate constants of 10^9 to 10^{10} M^{-1} s^{-1} (Table 1) such that any molecule encountering HO· will be attacked. The body's major defenses against HO· are twofold: (1) to attempt to try to prevent its generation by the chelation of divalent metal ions (thereby preventing their participation in the Fenton reaction), and (2) to repair or eliminate molecules damaged by HO· attack.

Several assays have been developed for the detection of HO· radicals in biological systems, including the production of ethene from methional and related compounds, formation of degradation products from tryptophan, production of radiolabeled CO_2 from [carboxy-^{14}C]benzoic acid, and production of methanesulfinic acid from dimethylsulfoxide and the deoxyribose assay (19). The deoxyribose assay involves the heating of the products of HO· attack on deoxyribose with thiobarbituric acid. Under acidic conditions pink chromagens are produced which can be determined at 532 nm (reviewed in ref. 10). The chromagen is indistinguishable from the thiobarbituric acid-malondialdehyde reaction product, and the exact species undergoing reaction is unknown as is the reaction mechanism. Although this is a simple test for HO˙ radicals, suitable control experiments must always be performed. By far the most promising techniques for trapping hydroxyl free radicals are considered to be electron spin resonance (ESR) and aromatic hydroxylation.

Electron Spin Resonance.

ESR measures the presence of unpaired electrons (reviewed in ref. 3) and can determine levels as low as 10^{-10} M but only if the free radical survives long enough to be measured. The use of spin-traps prolongs the life of even the most reactive free radicals, thereby extending the usefulness of ESR techniques. In fact ESR has been used successfully to measure HO˙ radicals, the superoxide radical anion, and lipid peroxidation products. A variety of spin traps are currently in use and include nitroso (e.g., nitrosobutane), nitrone (e.g., α-phenyl-*tert*-butylnitrone) and *N*-oxide (e.g., 5,5-dimethylpyrroline-*N*-oxide) derivatives. However, there are several issues to be aware of when using this approach in vivo, including the ability to distinguish between reaction products formed from the spin-trap and different free radicals; production of other free radicals by decomposition of the spin-trap/free radical product; quenching of the spin-trap/free radical product by cellular reducing agents (e.g., ascorbic acid); and tissue availability of the spin-trap and the toxicity and solubility of the spin-trap agent (3, 20, 21).

Aromatic Hydroxylation Reactions.

The reaction of HO˙ with an aromatic compound such as benzene leads to the formation of the hydroxycyclohexadienyl radical, which either can dimerize, forming biphenyl (after the loss of water), or can oxidize, forming phenol. Attack of HO˙ on substituted aromatic compounds is more complex, leading to a variety of products depending upon the pH of the reaction and the presence or absence of metal ions. For example, salicylic acid (SAL) can undergo decarboxylation at low pH, forming phenol, or hydroxylation at neutral pH, forming various isomers of dihydroxybenzoic acid. This ability of HO˙ to attack both endogenous and exogenous aromatic compounds is just a variation on the spin trap approach discussed with ESR earlier, and has been successfully used by researchers to examine HO˙ production in vivo (19). Although colorimetric (22) and gas-liquid chromatography coupled to electron-capture or mass spectrometry (23) have previously been used with some degree of success, there are inadequacies in these procedures. The reaction products are now typically measured using HPLC with a number of different detection systems including those based on ultraviolet (UV) absorbance (21, 24–26), fluorescence (27), electrochemical detection (ECD) (28, 29), ECD with UV (30,31) and ECD with fluorescence (32). Furthermore, a variety of compounds have been investigated as spin trap agents including phenol (28), phenylalanine (27), 2-deoxyguanosine (24), and SAL (25, 26, 28–32). The success of this approach depends upon the measurement of a unique compound that is not normally present in biological tissue but is formed only by the reaction between the spin trap and HO˙. Ideally the spin trap should be nontoxic, devoid of physiological action, and readily distributed to the site of HO˙ production (e.g., pass through the blood-brain barrier for determination of HO˙ in the central nervous system). The reaction product should not undergo metabolism. All too often the analytical procedure is insufficient to measure the low levels of reaction products formed in vivo. Therefore it is common practice to use high levels of the spin-trap to promote reaction product formation, thereby raising the biological level of reaction products above the limit of detection of the analytical system. However, by so doing, it is easy to exceed the safe levels of the spin trap, resulting in altered physiology (see later discussion). Furthermore, many researchers do not "normalize" their data and interpret the changes in reaction product as a reflection of altered HO˙ production. This may lead to erroneous conclusions since changes may simply be the result of interanimal variability in spin-trap distribution and metabolism and/or accuracy and location of spin-trap administration (for exogenously administered spin-traps).

a). Phenylalanine. L-Phenylalanine is an essential amino acid that is normally hydroxylated in the liver by phenylalanine hydroxylase with nearly exclusive production of *p*-hydroxy-L-phenylalanine [L-tyrosine] (only 0.2% of the *m*-tyrosine is formed, which can be further hydroxylated by this enzyme, forming 3,4-dihydroxyphenylalanine [L-DOPA]; 33). The presence of the decarboxylation products *o*-tyramine and *m*-tyramine in human urine suggests that both *o*-tyrosine and *m*-tyrosine are indeed produced in vivo (34, 35). Apart from the reaction catalyzed by phenylalanine hydroxylase, *m*-tyrosine and also *o*-tyrosine can be produced from L-phenylalanine by HO˙ attack. Ishimitsu et al. (1986) used an HPLC fluorescence method to directly measure endogenous plasma levels of both *o*- and *m*-tyrosine in rat and human serum samples. They found very small amounts of the *o*- and *m*-tyrosine isomers (∼3.0 ng/ml), which were approximately 3000-fold less than *p*-tyrosine levels. Unfortunately, the levels of *o*- and *m*-tyrosine were very close to limits of detection of their assay, thereby restricting its use in conditions where HO˙ levels are elevated. Although the measurement of endogenous tyrosine isomers could theoretically reflect HO˙ production, it is difficult to understand what the in vivo levels of the tyrosine isomers really represent as their levels reflect both synthesis and catabolism. This method could be improved by the measurement of L-phenylalanine, thereby permitting normalization of the data.

Kaur and Halliwell (1994) examined the formation of tyrosine isomers in vitro following exposure of premature-infant blood samples to HO˙ (derived from ozone). They also reported that HO˙ is indiscriminate and will react with both L- and D-phenylalanine isomers. They concluded that administration of D-phenylalanine, which is not a substrate for phenylalanine hydroxylase, would therefore lead to the formation of the D-isomers of tyrosine, which can only be formed by HO˙ attack and consequently would be better indicators of HO˙ exposure in vivo (37). Although this approach is good in principle, there are two issues worth considering. First, D-phenylalanine can undergo oxidative deamination by D-amino acid oxidase to form phenylpyruvic acid, which can then be reaminated to form L-phenylalanine, thereby leading to depletion of substrate (38). Second, D-tyrosine can also undergo transamination, producing *p*-hydroxyphenylpyruvate, which can be reaminated to form L-tyrosine; this will affect the levels of the D-tyrosine isomers (39, 40). The use of stereoisomers to examine in vivo levels of HO˙ is extremely complex. Unfortunately, Kaur and Halliwell's HPLC-UV method was incapable of distinguishing between the chiral forms of the amino acids and lacks sufficient sensitivity for in vivo work. HPLC-ECD offers superior sensitivity over HPLC-UV for the measurement of tyrosine isomers (phenylalanine is not electrochemically reactive and requires derivatization) (19). The use of chiral HPLC-ECD may be better suited to address the relevance of D- and L-tyrosine isomers for the in vivo measurement of HO˙.

Others have attempted to trap HO˙ in vivo following the perfusion of L-phenylalanine through a microdialysis probe. Microdialysis is a routine sampling technique, initially used by neuroscientists to study analytes in the extracellular space of a variety of tissues (41, 42). The microdialysis probe consists of inlet and outlet tubing connected together by a length of semipermeable membrane (typically 1–3 mm × 225 μm). The probe is continuously perfused (at 1–2 μL/min) with a solution containing salts of equivalent concentration and iso-osmotic to the extracellular space [e.g., artificial cerebrospinal fluid (aCSF) in the case of central dialysis]. When placed in a tissue, dialyzable solutes can diffuse through the membrane down their concentration gradient. Molecules entering the probe are swept along in the perfusion medium and are collected outside the probe for analysis using a variety of analytical procedures. Furthermore, molecules can be added into the perfusion medium for "retrodialysis" into the tissue surrounding the probe (this has the advantage for central microdialysis of circumventing the blood-brain barrier). Liu used a double-fiber microdialysis technique in the spinal cord to demonstrate that tyrosine isomers could be formed following perfusion of phenylalanine (10 m*M*) and unmixed "Fenton reagents" through the probes (27). Once the Fenton reagents are mixed in the tissue the resulting HO˙ can then react with the phenylalanine spin-trap, and the resulting tyrosine isomers collected from the probes' outlets and measured using HPLC-fluorescence following naphthalene-2,3-dicarboxyaldehyde/cyanide derivatization. Assuming that there was a 10% passage of phenylalanine from the probe, only 10% of this appeared to produce tyrosine isomers (∼60 μ*M*) in the tissue, and only after the coadministration of the Fenton reagents and phenylalanine. Although this study is interesting, it is unclear how this approach can be used to examine basal levels of HO˙ as neither *m*- nor *o*-tyrosine could be measured before administration of the Fenton reagents and phenylalanine. Furthermore, this approach would be limited to non-catecholamine-containing tissues as the perfusion of 10 m*M* L-phenylalanine through the probe would be expected to markedly affect tyrosine hydroxylase activity, catecholamine synthesis (43, 44) and dopamine release (45).

Many food substances are intentionally irradiated to prevent spoilage, to prolong shelf-life, and to destroy pathogenic organisms. Such irradiation can produce HO˙ free radicals which, as discussed earlier, can produce tyrosine isomers from phenylalanine. It is important to monitor irradiation exposure so as not to overexpose foods (with the potential production of toxic compounds) and to differentiate irradiated from nonirradiated foods (as demanded by some consumers). However, most analytical methods lack sufficient accuracy to discriminate between the exposed and nonexposed groups and are generally inconsistent. Sontag et al. (1996) recently developed a gradient HPLC coulometric electrode array approach to examine the abundance of tyrosine isomers in foods following irradiation. Coulometric electrodes are discussed in greater detail later. Using an oxidation-reduction array they showed that the free levels of o-tyrosine and m-tyrosine in shrimp were elevated in a dose-dependent fashion following irradiation. However, such an approach is not universal as they also reported that the levels of these analytes were measurable in unirradiated, extracted chicken meat. It seems, at least for chicken, that tyrosine isomer levels may be artificially elevated during storage or when the samples are prepared using acid hydrolysis.

b). Deoxyguanosine. (2 DG) Endogenous deoxyguanosine also traps HO˙, and this has been used as an indirect indicator of HO˙ production. Schneider et al. (1989) used an HPLC method with both UV and ECD for the determination of deoxyguanosine and the DNA adduct 8-hydroxy-2′-deoxyguanosine 8-OH-2′-DG, respectively 24. They compared this approach to that using SAL as the spin-trap and concluded that the trapping of HO˙ by SAL was 22 times more effective than deoxyguanosine (Table 1). The formation of DNA adducts are discussed in greater detail later.

c). Salicylic acid. The reaction between SAL and HO˙ is more rapid than with both guanosine (reacting 22-fold more slowly [24]) and phenylalanine (reacting 5 times more slowly [19]), suggesting that if HO˙ came in simultaneous contact with an equal concentration of these three compounds it would react with SAL preferentially. However, as with other trapping agents discussed earlier, the ability of such an agent to distribute effectively to the site of HO˙ production needs also to be considered when designing a new assay or analyzing results from an established assay procedure. For instance, in studies where the trapping agent requires access to the central nervous system (CNS) compartment, the agent must be able to consistently and easily pass the blood-brain barrier; otherwise administration directly into the CNS may be required. Passage into individual cells or neurons must also not be subject to uncontrolled variation. SAL reacts with HO˙ producing 2,3- and 2,5-dihydroxybenzoic acids (DHBAs) (Figure 3). Unlike phenylalanine and 2′-deoxyguanosine, SAL is not found in tissue, and furthermore, at least one of its metabolites, 2,3-DHBA, does not occur endogenously, and once formed cannot be further metabolized (47). The question as to which of the DHBA isomers is the better indicator of HO˙ production has still not been fully resolved but has been discussed at length elsewhere (19, 48). One concern is that the 2,5-isomer can be formed in vivo by the action of cytochrome P-450. The level of the 2,3-isomer is typically much less than the 2,5-isomer, and consequently measurement of the 2,3-isomer may be beyond the limits of detection of some analytical equipment (30, 49). Several HPLC approaches have been used to determine the DHBAs using UV, ECD, ECD plus UV, and EC with fluorescence (see earlier discussion). Unfortunately, few researchers measure levels of SAL and interpret changes in DHBA levels as a reflection of altered HO˙ production (21, 50–55). This, again, may lead to erroneous conclusions since altered DHBA levels could be purely the result of interanimal variability in SAL distribution and metabolism, or of accuracy and location of SAL administration when delivered in the periphery. The latter is obviously less of a problem when SAL is perfused through a microdialysis probe into the CNS.

Due to the controversy surrounding DHBA isomers and the inability to simultaneously measure SAL, we developed a sensitive HPLC-dual coulometric electrode approach, based on the method of Floyd et al. (28), but capable of measuring DHBAs and SAL on one system and a single dual-electrode detector. The coulometric flow-through graphite working electrode has many advantages over the amperometric thin-layer approach and has been reviewed recently (56–58). Unlike the inefficient amperometric thin-layer electrode where only 5% of an analyte is expected to be analyzed, the coulometric electrode approaches 100% efficiency. This means not only

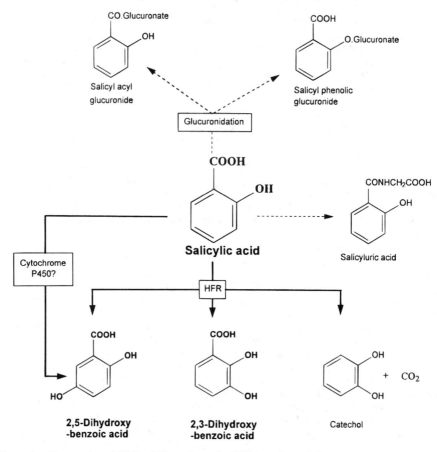

Figure 3.   The trapping of HO˙ by SAL produces the DHBAs and catechol.

improved sensitivity but also less variability and better qualitative data. For example, to measure both SAL and DHBAs on a single amperometric electrode, the oxidation potential applied to the working electrode has to be high enough to oxidize SAL. This results in a greater chance for noise (poorer signal/noise ratio) and the increased possibility of undetected coelutions occurring. However, with two coulometric electrodes placed in series, the first (upstream) can be set for the oxidation of DHBAs only while the second (downstream) can be set at a higher potential for the measurement of SAL (48). The coulometric approach therefore greatly diminishes noise, thereby improving the signal/noise ratio and, by its unique screening ability, markedly reducing the possibility of undetected coelutions taking place. Although it is possible to use two serially placed amperometric electrodes to mimic the coulometric approach, this would most likely fail due to electrode inefficiency. Unreacted analyte that escaped detection on the upstream electrode would spill over onto the downstream electrode, adversely affecting resolution. Using the coulometric approach the limits of detection for DHBAs and SAL were approximately 1 pg and 100 pg (on column), respectively. With this improved sensitivity it is now possible to greatly reduce the amount of administered SAL required for the production of DHBAs.

Although the SAL approach to measure HO˙ is probably the best and most sensitive method available it does suffer from several problems. SAL and the DHBAs have high affinities for iron (Fe^{3+}) which can affect HO˙ generation (19). Levels of SAL should be kept as low as possible so as to avoid possible toxicological effects such as acid-base, homeostatic, and neurological disturbances commonly observed with high doses of SAL routinely employed in other studies

(49, 59). Furthermore, SAL is a weak inhibitor of cyclooxygenase and may interfere with prostaglandin metabolism and can also inhibit phospholipase C (19).

While developing our application we discovered that DHBAs formed spontaneously when SAL was added to aCSF and its components (48). The levels of 2,3-DHBA were approximately 20-fold higher than the 2,5-isomer and comparable to basal DHBA levels, previously reported in the literature, measured using SAL perfusion and microdialysis sampling. Likewise, Montgomery et al. (1995) reported spontaneous formation of 3,4-DHBAs from 4-hydroxybenzoic acid (4-HBA) by various parts of their microdialysis apparatus but not when 4-HBA was placed in aCSF. Taken together, the use of SAL and microdialysis may not be practical for the in vivo measurement of HO$^\bullet$ even when metal surfaces in the microdialysis apparatus are minimized.[2]

The improved sensitivity of our approach will permit the use of lower levels of SAL and thereby overcome many of the potential problems discussed earlier when using tissue. However, apart from examining the water source and minimizing the amount of metal in the system, it is unclear how to make this approach more practical for microdialysis based experiments.

The Superoxide Free Radical Anion

The addition of one electron to oxygen yields the superoxide free radical anion (O_2^-) (Figure 1). O_2^- is produced by exposing oxygen to ionizing radiation, as part of normal metabolism (e.g., from the catabolism of xanthine to uric acid catalyzed by xanthine oxidase as well as by other flavoprotein oxidases such as aldehyde oxidase) and is deliberately produced during the respiratory burst of phagocytic cells, forming part of the body's defense system for the destruction of invading organisms. Traces of O_2^- are also formed when oxygen combines with hemoglobin and myoglobin, and, as discussed earlier, a considerable quantity of O_2^- is accidentally formed in the mitochondria when oxygen is reduced to water. Chance et al. (1979) have suggested that, assuming a 2% leakage and 10^{12} O_2 molecules consumed per cell per day, as much as 2×10^{10} hydrogen peroxide molecules and superoxide radical anions are produced per cell per day. Autooxidation of catecholamines, folates, toxins (N-methyl-4-phenyl-1,2,5,6-tetrahydropyridine [MPTP]; tetrahydroisoquinolines), drugs (haloperidol), and bipyridylium herbicides (e.g., paraquat, diquat) can all lead to production of O_2^- (62, 63).

O_2^- is a base and accepts protons, forming the more reactive hydroperoxyl radical ($HO_2^\bullet$). The pK_a of this reaction is ~4.5 such that under physiological conditions only one $HO_2^\bullet$ molecule exists for every 100 O_2^- molecules. O_2^- acts as both a mild oxidizing agent (e.g., oxidizing Fe^{2+} to Fe^{3+}, catecholamines to the semiquinone, and ascorbate to the ascorbyl radical) and a mild reducing agent (e.g., reducing Fe^{3+} to Fe^{2+} as part of the Haber-Weiss reaction discussed earlier). O_2^- is actually much less reactive than many of the free radicals found in biological systems (Table 1), yet in the past it has been blamed for much of the oxidative damage found in vivo (see ref. 10 and references therein). This destruction is now thought to be due to reaction products of O_2^- including $HO_2^\bullet$, HO$^\bullet$, and the peroxynitrite anion (ONO_2^-) (see later discussion).

$HO_2^\bullet$ synthesis from O_2^- is pH sensitive and is favored by acidic conditions. Although at physiological pH the abundance of $HO_2^\bullet$ is rather low, its levels will be elevated when the pH decreases (e.g., following ischemia, or severe acidosis, or in close proximity to the cytosolic side of the inner mitochondrial membrane). $HO_2^\bullet$ is more reactive than O_2^- and, due to its lipophilicity, would be expected to be able to cross biological membranes as effectively as H_2O_2 (64). This is important because, unlike O_2^-, it can initiate peroxidation of polyunsaturated fatty acids directly (see later discussion). However, $HO_2^\bullet$ is still a very poor initiator and there is little evidence that it can cause such damage in vivo.

The level of O_2^- is mainly controlled by both spontaneous and enzymatic dismutation:

$$2O_2^- + 2H^+ \longrightarrow H_2O_2 + O_2 \qquad (2)$$

---

[2] One possibility for DHBA production in vitro is through reaction of SAL with hydrogen peroxide. Some water stills use high energy UV light during the production of de-ionized water which can also produce hydrogen peroxide and other ROS as byproducts. In the presence of metal salts the Fenton reaction could take place, thereby promoting HO$^\bullet$ attack on SAL with the production of DHBAs. This may also explain why "basal" levels reported in the literature demonstrate such great variability.

Two forms of superoxide dismutase (SOD) exist; one, containing manganese at its active site, is found in the mitochondria, while the other, containing copper and zinc at its active site, is mainly found in the cytosol (65). Together these enzymes typically keep O_2^- below 10^{-11} M.

Recently, the level of O_2^- has been measured in real time using an O_2^--sensitive electrochemical probe (66). This sensor is based upon the original design of McNeil et al. (67). The probe consists of cytochrome c covalently bonded to a gold working electrode using 3,3′-dithiobis-(sulfosuccinimidy-propionate). This probe functions by measuring the current which flows when O_2^- is oxidized by cytochrome c (the Fe^{2+} in this protein is reoxidized to Fe^{3+} by loss of an electron to the gold electrode). This method is sensitive (with a limit of detection < 10 nM) and selective (neither $NO^{\bullet}$ nor H_2O_2 interfere) and has been used to measure the production of O_2^- in brain cell cultures (66).

Hydrogen Peroxide

Hydrogen peroxide (H_2O_2) is another ROS found in biological systems and is produced by several different enzymatic reactions, including those catalyzed by SOD, D-amino acid oxidase, amine oxidase, glycollate oxidase, and urate oxidase. H_2O_2 is also formed during redox cycling of compounds such as the catecholamines, from microsomal cytochrome P-450 (processing of fatty acids and xenobiotics), from the respiratory burst of phagocytes, and from mitochondrial respiration. H_2O_2 appears to play a role not only in phagocytic defense by the immune system but also in metabolic pathways such as the production of thyroxine in the thyroid gland (68). It is also interesting to note that H_2O_2 levels from 10^{-5} to 10^{-8} M have been reported in water (see footnote 2 earlier).

The levels of H_2O_2 are thought to be mainly controlled by enzymatic degradation brought about by catalase (Eq. 3) and several peroxidases including the selenium-dependent glutathione peroxidase (Eq. 4):

$$2H_2O_2 \longrightarrow 2H_2O + O_2 \tag{3}$$

$$H_2O_2 + 2GSH \longrightarrow GSSG + 2H_2O \tag{4}$$

$$GSSG + NADPH + H^+ \longrightarrow 2GSH + NADP^+ \tag{5}$$

The reaction catalyzed by glutathione peroxidase can proceed as reduced glutathione (GSH) is regenerated from oxidized glutathione (GSSG) by glutathione reductase (an FAD-containing enzyme) using the strong biological reducing agent, NADPH (Eq. 5). The activity of catalase and glutathione peroxidase varies from tissue to tissue. For example, both enzymes are extremely active in the liver (where considerable H_2O_2 is formed), but only the latter is active in the brain (where H_2O_2 production is much less). Within the cell, catalase appears to be in the peroxisomes whereas glutathione peroxidase is mainly found in the cytosol and matrix of the mitochondria.

H_2O_2, like O_2^-, is only a weak oxidizing agent and not very reactive in vivo (Table 1). It will react with thiols and by so doing can inactivate enzymes that contain an essential thiol group, such as the glycolytic enzyme glyceraldehyde-3-phosphate dehydrogenase. Unlike O_2^-, H_2O_2 can readily cross cell membranes, and if it encounters Fe^{2+} (or Cu^+) will form the aggressive $HO^{\bullet}$ radical capable of reacting with many chemical species found in vivo (see earlier discussion).

H_2O_2 has been measured using a variety of techniques, including peroxidase-based methods with fluorometric detection (69), chemical titration with acidified potassium permanganate, the measurement of evolved oxygen using an oxygen electrode following the addition of catalase (3), and the measurement of evolved $^{14}CO_2$ using scintillation counting when labeled 2-oxogluta-rate reacts with H_2O_2 (70). Furthermore, H_2O_2 is electrochemically active, reacting preferentially on a platinum electrode. Its measurement has been used to indirectly quantify choline and acetylcholine in both tissue and microdialysis samples (71).

Singlet Oxygen

When oxygen absorbs energy it can be excited to form singlet oxygen states ($^1\Delta GO_2$). For example, the $^1\Delta GO_2$ state that can often be encountered in biological systems has an energy 22.4 kcal above the ground state. In the laboratory $^1\Delta GO_2$ can be formed either chemically (e.g., from the reaction between H_2O_2 and hypochlorite) or photochemically (e.g., through the reaction between O_2 and dyes such as acridine orange or methylene blue). $^1\Delta GO_2$ can also be produced in biological systems by the reaction between H_2O_2 and hypochlorite resulting from phagocyte activity, the interaction between light and chlorophyll, and as a consequence of the interaction between light and rhodopsin in the eye.

The use of scavengers and the measurement of luminescence resulting from released energy as the $^1\Delta O_2$ molecule decays back to its ground state have been used to measure levels of $^1\Delta O_2$. These and other methods have been discussed in greater detail elsewhere (3). One interesting approach reacts $^1\Delta O_2$ with the spin trap 2,2,6,6-tetramethyl-4-piperidone-N-oxyl (TEMPO), with the resulting adduct being measured using HPLC-ECD (72). This technique was based on similar methods used to measure HO˙ (73) and O_2^- (74).

Reactive Nitrogen Species

Due to its electronic structure nitrogen is polyvalent (principle valences of 3 and 5) and can form a wide variety of compounds of biological relevance. Of particular interest are the oxides of nitrogen, which show complex chemical interactions and interconversions. A simplified version of this chemistry is presented in Figure 4. This section examines the biological importance and analysis of the RNS with particular emphasis on nitric oxide (NO˙), peroxynitrite anion (ONO_2^-), nitrosoperoxycarbonate anion ($ONO_2CO_2^-$), and nitrocarbonate anion ($O_2NOCO_2^-$).

Nitric Oxide

Nitric oxide (NO˙) is a small, hydrophobic, uncharged radical that readily crosses biological membranes. In the atmosphere NO˙ is a poisonous gaseous pollutant derived largely from internal

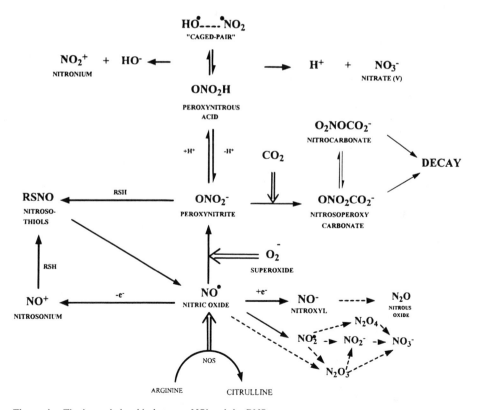

Figure 4.   The interrelationship between NO˙ and the RNS.

combustion engines, but its controlled production in vivo is extremely important and serves several functions (75–77). The primary biological roles for NO˙ are the regulation of blood pressure [its probable role as the endothelium-derived relaxing factor, EDRF (78)]; destruction of pathogens in the immune system (see earlier discussion); and a retrograde neurotransmitter with major importance in the formation of long-term memory (76). In recognition of its enormous importance, NO˙ was named "molecule of the year" in 1992 by *Science* magazine (79). Although NO˙ is now regarded as a general secondary messenger and a beneficial molecule [it inhibits propagation of lipid peroxides (80) and regulates free radical injury (81)], it also possesses a sinister side. Excessive NO˙ can directly damage iron-sulfur proteins (82) and other metalloproteins (83). NO˙ readily reacts with tyrosine and tryptophan radicals, which may play an active role in a variety of enzymatic mechanisms (e.g., ribonucleotide reductase, prostaglandin H synthase, cytochrome-*c* peroxidase) (84). NO˙ also promotes ADP-ribosylation of proteins (85) and directly damages DNA (86, 87). Furthermore, as discussed in greater detail later, NO˙ can react with O_2^- to form peroxynitrite, an aggressive molecule capable of attacking many biologically important molecules.

NO˙ is extremely labile, having an estimated half-life of <1 s in blood but as long as 6–10 s in tissue before reacting with oxygen and water forming nitrites and nitrates (75, 76). As shown in Figure 4, NO˙ is formed from L-arginine by the action of nitric oxide synthase (NOS). NO˙ is formed from the guanidino nitrogen of L-arginine in a process that consumes five electrons. Two distinct forms of NOS exist including the Ca^{2+}/calmodulin-requiring constitutive enzyme (c-NOS) (which produces relatively low levels of NO˙ and is important in neurotransmission, maintenance of vascular tone, and inhibition of platelet aggregation), and the Ca^{2+}-independent inducible enzyme (i-NOS) (primarily involved in the mediation of the cellular immune response). The important roles of these enzymes have been reviewed elsewhere (75–77). It is interesting to note that several inhibitors of NOS (mainly the guanidino substituted derivatives of arginine) occur in vivo as a result of posttranslational modification of protein-contained arginine residues by *S*-adenosylmethionine (77). These NOS inhibitors are thought to act both as regulators of NOS activity and as a reservoir of arginine for the synthesis of NO˙. However, their exact biological role is still under investigation. The principal biological action of NO˙ is on the heme prosthetic group of soluble guanylyl cyclase (SGC) (88). Activation of SGC elevates cyclic GMP, which in turn affects many biological systems and is believed to be responsible for many of the NO˙ mediated physiological effects.

Nitric oxide can be reduced to both the nitroxyl anion (NO^-) and oxidized to the nitrosonium cation (NO^+). All three show distinct chemistries; however, the exact biological significance of each of these is still being investigated (89, 90).

Several techniques have been used to monitor NO˙ both directly and indirectly. As initial observations identified the vascular endothelium as the source of endogenous NO˙ (in the form of EDRF), early attempts at demonstrating the presence of NO˙ utilized bioassay procedures (91). Taking advantage of the "humoral" nature of NO˙, Furchgott aligned in a sandwich fashion the intimal surfaces of two rabbit aortic strips, one with the endothelium intact (NO˙ donor tissue) and the other denuded of its endothelium (NO˙ biosensor tissue). Stimulation of the donor tissue with acetylcholine, which had been previously thought to increase the release of EDRF via a muscarinic receptor-mediated mechanism, led to the relaxation of the biosensor tissue. Others stimulated superfused isolated vascular tissue with intact endothelium and allowed the effluent to superfuse a "downstream" biosensor vascular tissue without endothelium (92, 93). The responses were dose-dependent, inhibited by hemoglobin, methylene blue, dithiothreitol, and hydroquinone, and dependent on the increase in the activity of SGC with a resultant increase in cyclic GMP (94–96). Obviously, a whole host of other compounds (e.g., prostacyclin, atrial natriuretic factor) can produce an endothelium-independent relaxation that would give rise to a false positive result for NO˙ in the above assay procedures. While these initial, but extremely innovative, approaches appear primitive by today's analytical standards, they were of vital importance in establishing a role for NO˙ in the vasculature and launching the tremendous interest in the biology and chemistry of this free radical.

With the knowledge that NO˙ is capable of stimulating guanylyl cyclase, some investigators have used changes in cyclic GMP as an index of nitridergic activity in tissue homogenates, cell culture systems, and microdialysates. Using commercially available (e.g., Amersham) radioimmunoassay kits to measure cyclic GMP, investigators have been able to demonstrate that such systems require L-arginine for optimal activity and are inhibited by various NOS inhibitors. For example, Palacios et al. demonstrated an approximate 40- and 80-fold increase from basal levels of cyclic GMP following incubation of cytosol from rat adrenal glands with 10 μM of the NO* generators S-nitroso-N-acetylpenicillamine (SNAP) and sodium nitroprusside (SNP), respectively. While L-arginine co-incubation potentiated the increase in cyclic GMP, co-incubation with D-arginine, NOS inhibitors (e.g., L-nitroarginine, N-iminoethyl-L-ornithine, L-nitro-monomethylarginine), or hemoglobin attenuated the response (97). While the use of cyclic GMP measurements as an indication of NO* status, especially when pharmacological tools are employed to verify NO* involvement in a given system response, has been important in expanding our understanding of the role of NO*, a lack of selectivity limits this procedure's usefulness to qualitative exploration. However, in some cases, this system may even fail qualitatively. Ignarro et al. observed an increase in cyclic GMP levels in endothelium-damaged vasculature rings following acetylcholine administration, in which the vasculature actually constricted rather than relaxed (98).

The finding that many analogs of L-arginine are capable of inhibiting NOS allows for their use in supporting a role of NO* in a particular biological response or system. While pharmacological confirmation alone fails to directly determine NO* within a given system, when it is implemented with other experimental approaches a role for NO* can be established with greater certainty. In the earlier described assay systems, and those that follow later, the use of such NOS inhibitors in vivo and in vitro is essential in assembling the required evidence that an assay procedure accurately reflects involvement of NO*.

Knowledge of the synthesis of NO* from L-arginine has led to attempts at determining the role of NO* in a particular biological system by quantitating the levels of the NOS reaction by-product L-citrulline (99). Using microdialysis techniques in rats we have demonstrated the ability of kainic acid, an excitatory amino acid neurotransmitter agonist, to increase hippocampal extracellular fluid (ECF) concentrations of L-citrulline (determined using o-phthaladehyde (OPA)/ β-mercaptoethanol (βME) precolumn derivatized microdialysates and coulometric detection) (100). Indicative of NO* involvement, the response was attenuated by pretreatment with the NOS inhibitor N-nitro-L-arginine. Others have used microdialysates containing [^{14}C]-L-arginine and determined the production of radiolabeled L-citrulline following treatments thought to increase nitridergic neurotransmission (101). As an indirect measure of NO* production this method relies on dialysis, diffusion, and uptake of radiolabeled L-arginine, followed by cellular transport, diffusion, and dialysis of radiolabeled L-citrulline. Further metabolism of citrulline is neglected in this assay. As the separation of precursor L-arginine from product L-citrulline utilized a simple column extraction, loss of radiolabeled L-citrulline metabolites, or addition of radiolabeled L-arginine metabolites distinct from genuine L-citrulline, may produce spurious results. However, this technique can be used in a supportive fashion to establish the involvement of NO* in biological systems.

Chemiluminescence-based techniques for the quantitation of NO* were initially developed for determining NO* as a pollutant in the atmosphere. NO* reacts with ozone, both in the upper atmosphere's ozone layer and in an analytical assay system, to produce a measurable chemiluminescent intermediate. The use of a photomultiplier tube and an oxygen generator is required. The NO* contained within a sample must be stripped into the gas phase by bubbling with an inert gas such as helium. Failure to adequately strip NO* into the gas phase (e.g., flow rates of inert gas lower than 8 ml/min) can lead to underestimates of NO*. A unique problem for many biological samples, especially those containing proteinaceous materials, involves foaming during the stripping procedure. The foam bubbles are capable of coating the sensing photomultiplier tube and greatly decreasing its sensitivity and accuracy. Additionally, problems associated with the sensitivity of the assay when humidity varies have been reported (102). If the assay is carried out in acidic samples in the presence of reducing agents, besides authentic NO* contributing to the signal, nitrite and labile adducts including nitrosothiols and nitrosamines will also contribute. High levels of ammonia have been reported to influence the analysis. An additional chemiluminescent-based assay system measures the reaction of NO*-generated peroxynitrite with luminol.

The ability of NO* to interact with the metal portion of various proteins accounts for its ability, on the one hand, to stimulate SGC (see earlier discussion), while, on the other hand, to inhibit the activity of others. This ability to inhibit the activity of certain enzymes in microbial systems is responsible for the cytotoxic activity of $NO^{\bullet}$.

The earliest studies in isolated tissues demonstrated the ability of hemoglobin (Hb) to dramatically attenuate nitridergic neurotransmission due to the ability of $NO^{\bullet}$ to form coordinated bonds with the iron center of this metalloprotein. When oxyhemoglobin (oxyHb) interacts with $NO^{\bullet}$, the resultant adduct, methemoglobin (metHb), is formed. This species of Hb has greatly diminished oxygen-carrying capabilities as compared to oxyHb, and when elevated, can lead to a condition known as methemoglobinemia. MetHb (normally present in the blood at less than 1% of total Hb) at levels of 15% or greater leads to cyanosis, headache, vomiting, lethargy, and confusion. Methemoglobinemia is usually caused by administration of NO_2^- and NO_3^- (as a drug, food, or their salts) in the presence of an inherited abnormal Hb molecule (which is more susceptible to oxidation and resistant to reduction) or in very young infants with diarrhea and acidosis.

In situations where oxygen tension is low (e.g., in venous blood vs. arterial blood) $NO^{\bullet}$ can interact with deoxyhemoglobin (deoxy-Hb) to form the paramagnetic adduct, nitrosylhemoglobin (NOHb). This adduct can be analyzed spectrally (103). Attempts at monitoring NOHb in the circulation as an index of $NO^{\bullet}$ exposure have generally been disappointing as the basal levels are typically below the limits of detection. Additionally, following the administration of the $NO^{\bullet}$ generator nitroglycerin to humans, only 50% of the patients exhibited measurable increases in NOHb (104). The use of Hb as a spin-trap agent to access $NO^{\bullet}$ levels is complicated by the influence of the degree of oxygenation on the adduct formed and the limitations associated with the use of an endogenous compound (see earlier discussion).

We have utilized oxyHb as an $NO^{\bullet}$ trapping agent with success in a microdialysis system. As Hb is normally not present in the ECF of the brain, we have utilized aCSF containing the nondiffusable metalloprotein oxyHb as a microdialysate solution (105). As oxyHb travels along the microdialysis probe, diffusable $NO^{\bullet}$ formed in cells adjacent to the dialysis membrane component of the probe can enter the probe compartment and react with oxyHb to form metHb. MetHb is trapped within the probe and delivered to the outlet, where it is collected for metHb analysis. As the oxidation of oxyHb to metHb is characterized by a change in its spectrophotometric absorption, the difference spectrum for metHb versus oxyHb was recorded and gave a maximum absorption at 401 nm with an isobestic point of 411 nm, which is specific for the metHb/oxyHb two-component system. This technique was able to determine the role of kainic acid and N-methyl-D-aspartic acid (NMDA) agonists on hippocampal and cerebellar $NO^{\bullet}$ production. The response was dose dependent, attenuated by NOS inhibitors and tetrodotoxin (TTX) sensitive. No interference from another possibly formed Hb derivative, carboxyHb, would be expected as its absorbance maximum is far removed from that of the metHb/oxyHb system. Care must be taken to ensure uniformity and stability of the OxyHb-containing aCSF if this "endogenous" metalloprotein is to be employed to trap $NO^{\bullet}$ in vivo.

Exogenous spin-trap agents have been utilized to determine $NO^{\bullet}$ production with some success. Lai and Komarov reported the use of a metal-chelator complex consisting of N-methyl-D-glucamine dithiocarbamate (MGD) and reduced iron (Fe^{2+}) that was capable of trapping $NO^{\bullet}$ to form the water soluble [MGD-(2)-Fe^{2+}-NO] complex (106). This complex could then be quantitated using ESR spectroscopy. When mice were pretreated with the metal-chelator complex followed by lipopolysaccharide (LPS), a treatment known to increase the release of $NO^{\bullet}$ from immune-associated cells, an increase in the $NO^{\bullet}$ signal was detected within 6 h (a time course expected for immune responses to LPS). The coadministration of a NOS inhibitor prevented the increase in the $NO^{\bullet}$ signal. Henry et al. (1993) have reviewed in detail other useful spin-trap agents.

Investigators have frequently taken advantage of the paramagnetic characteristics of NOHb (deoxyHb reaction with $NO^{\bullet}$) to investigate the role of $NO^{\bullet}$ in a number of biological systems. However, one limitation of this technique is the requirement of freeze-quenching of samples to ensure stability. A number of nitrones and nitroso spin traps have been reliably detected at room temperature in aqueous samples (108).

The oxidation of $NO^{\bullet}$ to the aqueous stable anion nitrite (NO_2^-) occurs readily under physiological conditions. In vivo formed NO_2^- can then be converted readily to nitrate (NO_3^-) such that concentrations of NO_3^- in biological fluids are typically two orders of magnitude greater than that of NO_2^-. Investigators have measured changes in the sum of NO_2^- and NO_3^- (i.e., NO_x^-) as an index of nitridergic involvement. Using the Greiss reagent (sulfanilamide and N-(1-

naphthyl)ethylenediamine dihydrochloride in phosphoric acid) and spectrophotometric detection at 540 nm, NO_x^- content can be determined in biological fluids after NO_3^- is reduced via cadmium or NO_3^- reductase to NO_2^-. Numerous studies have proven the utility of this technique in demonstrating increases in the NO_x^- signal following treatments or pathological conditions thought to involve altered $NO^{\bullet}$ status [e.g., estrogen therapy (109), immunostimulation (110), and septic shock (111)]. A recent multichannel flow injection technique utilizing the Greiss reagent and cadmium reduction reported a linearity over 2.5 orders of magnitude (25 nM to 20 μM) and may be suitable for in vivo monitoring of NO_x^- (112). Unfortunately, NO_x^- can derive from the diet, bacterial metabolism, and reactions unrelated to the L-arginine-$NO^{\bullet}$ pathway, and thus the use of NO_x^- as the sole determinant of $NO^{\bullet}$ status, would be expected to lead to spurious results.

In vivo $NO^{\bullet}$ levels have also been directly measured using voltammetry. Several different modified voltammetric sensors have been developed for tissue work, but generally suffer from several drawbacks: a lack of specificity and sensitivity, slow reaction kinetics, poor reproducibility, and difficulty in manufacture. Malinski and Taha (1992) have developed a voltammetric sensor that shows great promise. Their sensor consisted of a carbon fiber covered in a film of polymeric nickel porphyrin with selectivity resulting from a Nafion coating (to prevent passage of the negatively charged NO_2^- and NO_3^- into the probe). When used with amperometry or differential pulse voltammetry, a detection limit of 10 nM was achieved and as little as 10^{-20} mol of $NO^{\bullet}$ could be measured from a single cell. In other experiments, an appropriate reduction in $NO^{\bullet}$ signal detected was observed when inhibitors of NOS were administered, thus lending pharmacological support. It is unclear why this sensor is not more popular but this may be due to a lack of availability or difficulty in manufacture. Recently, O'Hare and O'Shea reported the development of a $NO^{\bullet}$ sensor consisting of a "robust graphite epoxy fiber" that incorporates an $NO^{\bullet}$-sensitive iron compound. This sensor is reported to be easy to produce, unaffected by peroxynitrite, and capable of measuring trace levels of $NO^{\bullet}$ in vivo (114).

Recently Friedemann et al. (115) reported on the utility of an o-phenylenediamine-modified carbon fiber coated with Nafion for the in vivo detection of $NO^{\bullet}$. The o-phenylenediamine acted to exclude larger molecules (e.g., dopamine, serotonin, etc.), while the Nafion prevented anionic species from coming into contact with the sensor. In vitro selectivities for $NO^{\bullet}$ relative to dopamine, ascorbate, and NO_2^- were >300:1, >600:1 and >900:1, respectively. Using amperometric techniques (+900 mV, 5 Hz), $NO^{\bullet}$ could be detected in acetylcholine-stimulated isolated rat renal arterioles and in vivo rat striata following <25 nl of 2 mM $NO^{\bullet}$ administered locally from a single-barrel pipette. Excellent linearity ($r^2 > .997$; 0–6 μM $NO^{\bullet}$), sensitivity (35 ± 7 nM), and response time (248 ± 41 ms) were also reported. Demonstration of the selectivity of this electrode to $NO^{\bullet}$ relative to other common extracellular constituents (e.g., H_2O_2) is required.

Strehlitz et al. (1996) recently developed an amperometric nitrite sensor that may indirectly monitor changes in $NO^{\bullet}$ levels.

Peroxynitrite

Peroxynitrite (ONO_2^-) is readily formed from the reaction between $NO^{\bullet}$ and O_2^-. In fact these compounds react at near diffusion controlled limits, and at approximately three times the dismutation rate of O_2^- by SOD (Table 1). Thus, both $NO^{\bullet}$ and O_2^- can modulate the effects of the other. For example, O_2^- can block the hypotensive effects of $NO^{\bullet}$ by diverting it to form ONO_2^-. However, as Crow and Beckman (1995) correctly point out, the probability that a particular chemical species will be attacked by another depends not only on the rate of the reaction but also on the concentration of the target. The multiplication product of the reaction rate and target concentration is called the target area, and under normal conditions, the target area of SOD exceeds the target area of $NO^{\bullet}$ by about 30-fold. However, during pathological conditions such as reperfusion following ischemia, the target area of $NO^{\bullet}$ can exceed that of SOD such that under these conditions, ONO_2^- is preferentially produced.

ONO_2^- is both a nitrating agent (see nitrotyrosine formation discussed later) and powerful oxidant. Under basic conditions ONO_2^- is remarkably stable and can even be kept safely at $-20°C$ for many weeks (117). However, it has a pK_a = 6.8 such that at physiological pH peroxynitrous acid (ONO_2H) is formed, which has a half-life of ~ 1 s at pH 7.4 and produces some of the strongest oxidants known in biological systems (90). The formation and chemistry of ONO_2^- have been reviewed in depth (90, 118, 119). Both ONO_2^- and ONO_2H attack a wide variety of biological targets, and although most reaction products can be explained by attack by either $HO^{\bullet}$ or $NO_2^{\bullet}$ (nitronium radical), it appears that neither of these compounds is actually formed to any extent in vivo. Rather, both $HO^{\bullet}$- and $NO_2^{\bullet}$-like activity may exist within the molecule, which can be explained by either a caged pair (geminate pair) or activated isomer (HOONO*) of peroxynitrous acid (118). Whatever the mechanism, attack by ONO_2H yields a complex variety of reaction products, some of which appear to be derived from $HO^{\bullet}$ and others from $NO_2^{\bullet}$. It is interesting to note that ONO_2^-/ONO_2H can be considered as a more stable form of $HO^{\bullet}$, permitting it to be carried much further and causing damage more remote than when $HO^{\bullet}$ is synthesized in situ.

ONO_2^- and ONO_2H attack many biologically important compounds. Examples include the oxidation of nonprotein and protein sulfhydryls (119), sulfides (120), deoxyribose (121), lipids (122, 123), ascorbate (124), the nitration of guanine (125, 126) and the formation of carbonyls and other oxidation products in proteins (127). ONO_2^- and ONO_2H can also hydroxylate (e.g., tyrosine, phenylalanine, and guanosine) and nitrate (formation of 3-nitrotyrosine from tyrosine) a variety of endogenous compounds (118).

Peroxynitrite levels can be determined using spectrophotometric techniques by measuring ONO_2^- absorbance at 302 nm (117), although lack of specificity and sensitivity may render this approach inadequate for in vivo investigations. Other methods indirectly measure ONO_2^-/ONO_2H levels by determining 3-nitrotyrosine concentrations using either a qualitative immunological approach (90, 128) or quantitative HPLC techniques (129, 130), and these are discussed in greater detail later.

Nitrocarbonate Anion

ONO_2^- rapidly reacts with carbon dioxide to form the unstable nitrosoperoxycarbonate anion ($ONOOCO_2^-$), which quickly rearranges to form the nitrocarbonate anion ($O_2NOCO_2^-$) (131). The rate constant for the reaction between ONO_2^- and CO_2 is sufficiently large ($3 \times 10^4\ M^{-1}$ s^{-1}; Table 1) that some have suggested that this is the predominant reaction in biological tissues (where bicarbonate/carbonate levels are >25 mM), which can account for the disappearance of ONO_2^- (132). This suggests that ONO_2^- is therefore highly unlikely to directly damage cells, although it might play a role in forming destructive cellular oxidants such as $O_2NOCO_2^-$ or one of its decomposition products.

$O_2NOCO_2^-$ reacts with tyrosine to produce both 3-nitrotyrosine and dityrosine. Recent research suggests that this reaction mechanism involves a one-electron oxidation of tyrosine yielding both tyrosyl and $NO_2^{\bullet}$ radicals as intermediary species (130). $ONO_2CO_2^-$ has a half-life <3 ms and a redox potential $> +1000$ mV and oxidizes tyrosine with a bimolecular rate constant >2.0 $\times\ 10^5\ M^{-1}\ s^{-1}$ (133, 134).

Lymar and Hurst have suggested that $ONO_2CO_2^-$ may serve two important biological functions (134). First, it acts as a scavenger of ONO_2^- and, due to its instability, will limit the site of action of ONO_2^-. Second, it may be a superior microbicide to H_2O_2, as it will not be deactivated by microbial catalase. Furthermore, unlike H_2O_2, which can diffuse a long way from its generation site and cause damage in areas remote from the site of infection, the activity of $ONO_2CO_2^-$ will be limited to the area where it is produced.

ROS AND RNS REACTION PRODUCTS: INTRODUCTION AND MEASUREMENT

As discussed earlier, various ROS and RNS and other radicals are often encountered in vivo. Although some ROS and RNS play important roles in metabolism and as part of the immune defense mechanisms, the vast majority are unwanted. Living organisms have had to develop a variety of mechanisms in an attempt to closely regulate the levels of ROS and RNS. Unfortunately, none of these mechanisms are 100% effective, and excess ROS and RNS can escape to damage a variety of important biomolecules. This section examines the damage to DNA, proteins, and lipids, and reviews the measurement of reaction products of ROS and RNS attack.

DNA Damage

Damage to DNA resulting from its interaction with ionizing radiation or attack by free radicals has been suggested to play a major role in mutagenesis, carcinogenesis, aging, and cell death

(3). Although many different ROS and RNS are produced in the cell, the attack by the hydroxyl free radical (HO•) is the most common form of radical-induced DNA damage.[3] This has been attributed by Pryor to the unique HO• structure, which possesses both high electrophilicity and high thermokinetic reactivity. Additionally, HO• is readily produced in the vicinity of DNA (135). As shown previously in Table 1, HO• readily abstracts hydrogen atoms from organic compounds ($k = 10^9 \ M^{-1} \ s^{-1}$) and adds to double bonds ($k \geq 10^{10} \ M^{-1} \ s^{-1}$), but is much less reactive with phosphate groups ($k < 10^7 \ M^{-1} \ s^{-1}$). Thus, HO• is thought to attack DNA by two main modes, first by hydrogen abstraction from the deoxyribose sugar, and second by addition to the π-bonds of the DNA bases (136).

Abstraction of hydrogen atoms from carbohydrates can lead to cleavage of the sugar-phosphate backbone. Attack on one of the strands of double-stranded DNA is not usually lethal, as the damaged DNA strand is usually held in place long enough for repair enzymes to rectify the damage. When attack by HO• molecules is excessive, damage to both DNA strands can occur. If these damages are in close proximity to each other, this can lead to permanent strand scission of the DNA molecule and potentially to cell death. Early pulse-radiolytic data suggest that about 20% of the HO• molecules attack the sugars of double-stranded DNA (see ref. 136 and references therein). The initial attack of HO• produces sugar radicals, and these have been studied using ESR techniques (138, 139).

The attack of DNA bases by irradiation and free radicals has been extensively reviewed (136, 140–142). The major mode of attack is to add to the π-bonds of the bases, at C-4 and C-8 of the purines and at C-5 and C-6 of the pyrimidines, leading to a wide variety of products (136). For example, HO• attack on 2′-deoxyguanosine (2 DG) and its 5′-monophosphate (DGMP) produces GG4OH•, GG5OH•, and GG8OH•. Furthermore, G8OH• can undergo reduction (e.g., with cellular thiols) and ring opening with the formation of formamidopyrimidine, or oxidation with rearomatization of the imidazole ring giving 7,8-dihydro-8-oxoguanine (8-oxo-G) which is tautomeric with 8-hydroxyguanine (Figure 5).

One of the major products of HO• attack on DNA is 8-hydroxy-2′-deoxyguanosine (8OH2DG). This is reported to be one of the most mutagenic legions known to date (143–145). Estimates of the daily production of oxidized adducts range from 10^4 to 10^6 molecules per cell (146), with eight 8OH2DG adducts being formed for every 10^6 normal nucleotides in rat liver nuclear DNA, while 8OH2DG levels were 16-fold higher in mitochondrial DNA (147).

DNA adducts resulting from alkylating agents are probably the best studied adducts to date and are formed by the action of exogenous agents (e.g., nitrosoureas and nitrosamines) and possibly by endogenous agents (e.g., betaine, choline, and S-adenosylmethionine) on DNA. These adducts show markedly different mutagenicity, with 7-methyl-2′-deoxyguanosine being the weakest and 3-methyl-2′-deoxyadenosine and O^6-methyl-2′-deoxyguanosine being the most potent (see ref. 137 and references therein).

Production of DNA adducts has been reported to block or slow the rate of DNA replication, which can prevent cell division or produce aberrant chromosomes (148); to misdirect incorporation of bases during replication, resulting in mutations (143, 149, 150); and to undergo hydrolysis, eventually leading to strand scission. Measurement of such adducts in tissue is currently being used as an indicator of DNA damage and exposure to mutagens.

Separation and detection of DNA adducts have proven problematic due to their low concentration and their occurrence in complex biological matrices. Various approaches have been used to measure tissue DNA adduct levels, including immunochemical methods (151), gas chromatography with mass spectrometry (GC/MS) (152, 153), GC/MS/MS (154), ^{32}P postlabeling (155–158), fluorescent labeling (157), high-performance liquid chromatography (HPLC) with UV absorption (159), and HPLC with ^{32}P postlabeling (157).[4] Many DNA adducts are electrochemically active and can be measured using HPLC with electrochemical detection (ECD). This technique is selective, sensitive, has good dynamic range, requires limited sample preparation, and does not involve the use of radioisotopes or the purchase of expensive equipment, thereby overcoming many of the issues associated with other analytical methodologies.

---

[3] While the action of ROS on DNA results in a variety of novel oxidation products, exposure of DNA to electrophiles (e.g. alkylating agents and carbon centered radicals) produces many different covalently modified bases (137).

[4] See also Cummings et. al. for a review on the use of HPLC for investigation of covalent modifications to nucleic acid (160).

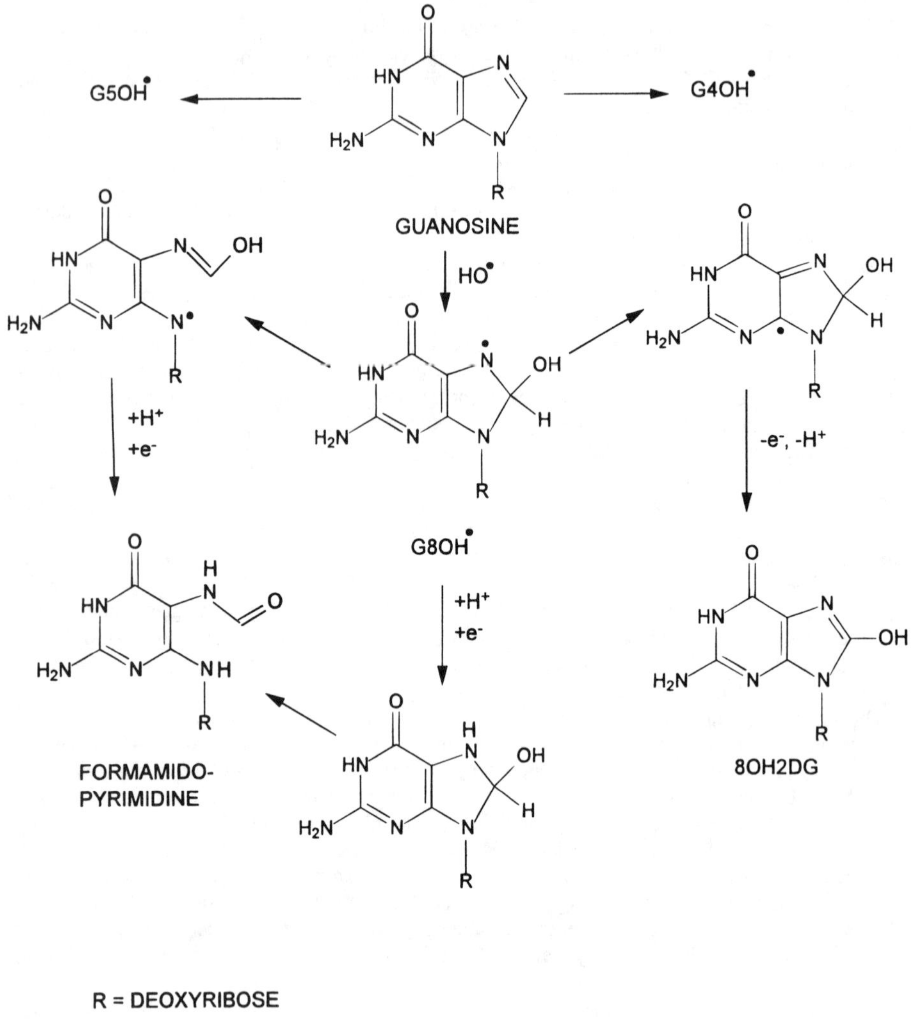

Figure 5. Proposed pathway for attack of guanosine by HO$^{\bullet}$ and the formation of 8OH2DG. Adapted from Breen and Murphy (136). Reproduced with permission of Elsevier Science Inc.

Many different HPLC-ECD methods have appeared in the literature, but most measure relatively few adducts. This is probably a result of less than ideal chromatographic and/or voltammetric resolution and limitations in the detectors used. For example, Floyd et al. were the first to report the use of HPLC-ECD to measure 8OH2DG, but measured 2DG on a UV detector (161). Van Delft et al. (1993) used a wall-jet amperometric ECD to measure the formation of *N*7-guanine in salmon sperm DNA in vitro (147). With an applied potential of 1350 mV it is unclear how long this assay would be of practical use before the working electrode would fail. Xie and Duda (1994) attempted to measure several guanine adducts and compared the use of a single-electrode thin-layer detector and a wall-jet cell. They elegantly illustrated the problem of using a single electrode to simultaneously measure a wide range of compounds with disparate voltammetric behaviors. In order to measure methyl adducts, which are more difficult to oxidize, a high potential has to be applied to the thin-layer working electrode. This compensation produces more noise, resulting in a poorer signal-to-noise (s/n) ratio. Furthermore, there is a greater probability that chromatographic interferences will occur.

Others have developed methods using dual coulometric electrodes to measure DNA adducts. As discussed earlier, these electrodes offer superior sensitivity and voltammetric selectivity and have been used to measure 8OH2DG in a variety of tissues and disease states (163–167). Ames and his group have also used a coulometric approach to measure a variety of novel DNA adducts, including 5-hydroxy-2'-deoxycytidine (168) and $1,N^2$-propano adducts of 2'-deoxyguanosine (169). We recently developed a "global" method for the measurement of guanine adducts (170). This method uses a gradient HPLC coulometric electrode array technique with low picogram sensitivity capable of simultaneously measuring 11 DNA adducts and parent bases (Figure 6). The coulometric array (CoulArray; ESA, Inc.) uses up to 16 coulometrically efficient working electrodes placed in series. This approach is similar to photodiode array except that compounds are resolved by their electrochemical (voltammetric), rather than absorbance, behavior. The coulometric array approach not only offers superior sensitivity over most other analytical techniques, but is the only electrochemical approach that produces true qualitative data (56–58). Using the coulometric array, but with a different chromatographic method, Mecocci et al. (171) measured 8OH2DG in human brain tissue and reported an age-dependent increase in its levels. Furthermore, they found that the mitochondrial ratio of 8OH2DG to 2DG was 10-fold higher than in nuclear DNA and concluded that such DNA damage may contribute to the age-dependent increase in the incidence of neurodegenerative diseases.

Others have attempted to measure 8OH2DG in urine as an indicator of oxidative stress. The measurement of 8OH2DG in urine presents several challenges to the researcher: the polar nature of 8OH2DG, the extremely low levels of its presence in urine, and the extremely complex and variable nature of the urine matrix. The complicated nature of the urine matrix makes it necessary to clean up or process the sample prior to injecting the sample into a gas or liquid chromatograph. This step can concentrate the sample and helps minimize matrix interferences. Some of the various approaches that have been used to measure urinary 8OH2DG are: HPLC-ECD following solid-phase extraction (SPE) (163, 172–176); HPLC-ECD following SPE and an immunoaffinity column (166, 177–180); HPLC-ECD following triple column switching (181–183); HPLC-ECD following SPE and triple column switching (184, 185); GC/MS following SPE and trimethylsilylation (175); GC/MS following evaporation, acetylation, and pentafluorobenzylation (186); and GC/MS following SPE, evaporation, acetylation, pentafluorobenzylation, preparative HPLC, and evaporation (187). Both analytical detection methods noted earlier, HPLC-ECD and GC/MS,

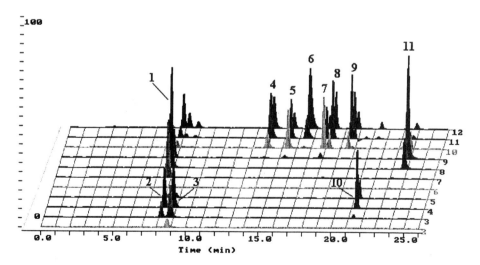

Figure 6. Gradient HPLC separation and coulometric array detection of some DNA adducts: 1–guanine; 2–8-hydroxyguanine; 3–5-hydroxy-2'-deoxycytidine; 4–3-methylguanine; 5–7-hydroxyethylguanine; 6–adenine; 7– 7-methylguanine; 8–guanosine; 9–2'-deoxyguanosine; 10–8OH2DG; 11–O^6-methylguanine. Reproduced by permission from ESA, Inc.

are very sensitive and selective. The challenge that confronts the researcher is the sample preparation method. In general, the sample preparation and purification methods for HPLC-ECD are simpler. For example, the sample preparation procedure of Teixeira et al. (186, 187) requires extensive and tedious sample evaporation/derivatization steps that take several hours to complete prior to sample introduction onto a GC/MS. Even the simpler sample preparation methods for HPLC-ECD range from fairly simple to moderately complex. Although the original methods using SPE followed by HPLC-ECD (163, 172–176) were quite simple to perform in the laboratory, the results given by a single-electrode detector may be erroneous. We have found that an 8OH2DG peak that appears to be "pure" on a single-electrode system (Figure 7A) is actually, when injected onto a coulometric array, a coelution (Figure 7B). This illustrates the danger of basing peak purity solely upon matching retention times of unknown samples to external standards. One method that takes the SPE one step further and holds perhaps the greatest promise for isolating 8OH2DG from urine is the use of SPE followed by an immunoaffinity column and HPLC-ECD (166, 177–180). Here a monoclonal antibody is isolated and selectively binds with high specificity and affinity to 8OH2DG. The method appears to be quite specific and sensitive for the determination of 8OH2DG. However, the lifetime of the columns is not known as yet, resulting in a decrease in binding efficiency over time (166). The authors state that a coelution occurs with the 8OH2DG when the column begins to lose efficiency. At this time the column is discarded. Perhaps the greatest stumbling block to researchers who wish to use these columns is the lack of commercial availability at the time of publication. Another means of sample preparation involves the use of "column switching." Here, a standard is injected onto one HPLC column and the retention time of the analyte noted. In subsequent sample injections, a valve is toggled just before the analyte elutes, resulting in a "heart-cut" that contains the peak of interest. The resulting "slug" of sample flows onto a second "trapping" column or sample loop, and finally the valve is toggled back to its original position after the analyte elutes. The "slug" containing the analyte is transferred onto a third analytical column (181–183) and then finally into a coulometric detector. This method is extremely complex, requires two detectors, two or three columns, and a switching valve, and has an extremely long run time, thereby limiting sample throughput to only a few samples per day. In addition, we have been unable to duplicate this method in our laboratory when urinary 8OH2DG concentrations were determined using a coulometric array detector and column switching. Again, as mentioned earlier, we observed that the peak eluting at the expected elution time of 8OH2DG was not pure. This demonstrates the power of coulometric array detection in measuring peak purity based on electrochemical signature as well as retention time. Others have had difficulty in reproducing these results by also using dual-channel coulometric detection (188). Another variation on the column-switching technique is to perform an SPE procedure prior to injecting the sample onto the first column (184, 185). The authors mention that this technique speeds up the analytical run time when the SPE procedure is automated. That is because the lipophilic compounds that elute for several hours remain on the disposable SPE cartridge. Again, we have been unable to duplicate the authors' work in our laboratory using coulometric array detection. For a column switching procedure to be effective, the retention mechanisms of the first and third columns must be different in order to achieve any specificity. Unfortunately, due to the polar nature of both the urine and the 8OH2DG, the retention mechanism is reverse-phase on both columns. For these reasons, we developed a simpler SPE sample preparation procedure followed by HPLC with coulometric array detection for the determination of urinary 8OH2DG concentrations (189). The SPE procedure uses a proprietary, unique stationary phase and has a unique selectivity for the 8OH2DG. Here the analyte's electrochemical signature across the coulometric array as well as its retention time verify peak purity.

Protein and Amino Acid Damage

ROS and RNS can attack proteins in a variety of ways, but most involve modification of amino acid residues. Such modifications may be harmful, especially if they occur in an essential area of the protein (e.g., the active site of an enzyme). It is fortunate that oxidative damage to proteins increases the likelihood that they will be degraded and effectively removed from the cell (see ref. 153 and references therein). However, it has been postulated that, with age, the level of protein

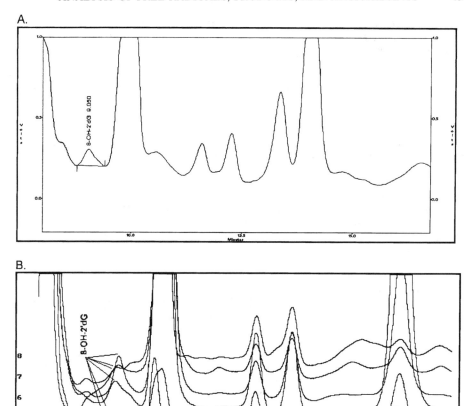

Figure 7. The measurement of 8-hydroxy-2'-deoxyguanosine (8OH2DG) in extracted human urine. Chromatogram A shows resolution of 8OH2DG (in pooled human urine) on a single-electrode detector (applied potential +400 mV). Although 8OH2DG appears to elute as a single and pure peak, it is in fact comprised of more than one analyte. Chromatogram B was produced using the same chromatographic conditions and pooled urine mix as A but with voltammetric resolution across the coulometric electrode array. The pure 8OH2DG standard responds on channels 5, 6, and 7 (+275, +335, and +380 mV) with dominant response on channel 6. As can be seen in (B) there is also response on the upstream electrodes, and, although not shown here for clarity, there is an extremely large peak at this retention time that oxidizes on the downstream (higher potential) electrodes. The ratio accuracies for 8OH2DG for the urine pool and individual urine samples were >0.9 for the preceding/dominant ratio but <0.34 for the following/dominant ratio, immediately showing that the peak is impure. Without this qualitative information from the coulometric electrode array it would be very easy to erroneously measure falsely elevated levels of 8OH2DG in urine. Reproduced with permission from ESA, Inc.

modification increases (possibly due to a disruption of antioxidant protective mechanisms), and some have even suggested that this may be one of the underlying factors involved in the development of Alzheimer's disease (190).

The reaction between H_2O_2 and a reduced transitional metal in the metal binding site of an enzyme can lead to the formation of $HO^•$, which can then immediately attack adjacent amino acid residues (191–194). Oxidation of amino acids can produce carbonyl compounds. Examples include the conversion of proline and arginine into glutamic semialdehyde, histidine into 2-oxohistidine, and lysine into lysylcarbonyl. Some amino acids are converted into others, such as proline to glutamate and histidine to asparagine (190). Aromatic amino acids can undergo hydroxylation reactions (see earlier discussion). Furthermore, cysteine is also oxidized to the disulfide cystine, and methionine is oxidized to methionine sulfoxide.

Attack on amino acids and proteins can also come from ONO_2^-, leading to: formation of protein carbonyls (195); oxidation of tryptophan, tyrosine, and cysteine residues (195,196); nitration of tyrosine and phenylalanine (197–200); formation of dityrosine (198, 200); and protein fragmentation (195). The properties of ONO_2^- have been discussed in detail earlier.

Initially carbonyl compounds were used as markers of ROS and RNS attack in vivo. For example, carbonyl compounds including the amino acid carbonyls were determined using HPLC-UV detection of their 2,4-dinitrophenylhydrazine (DNPH) derivatives (201–203). The limits of detection of carbonyl compounds have been further improved using HPLC-ECD (204). Adaptation of this method to the determination of protein/amino acid carbonyls should be readily achievable. Unfortunately, carbonyl groups may also be introduced into proteins by mechanisms other than by ROS and RNS attack (205), which raises questions as to the feasibility of this approach. Recent efforts have concentrated on the measurement of dityrosine and 3-nitrotyrosine as indicators of protein damage in vivo (Figure 8). Furthermore, 3-nitrotyrosine has also been used as an indirect measure of ONO_2^- production (206).

Dityrosine has been mainly determined in protein hydrolysates by the use of HPLC with UV absorbance (197) or fluorescence (205). It has also been measured using dabsyl chloride derivatization as well as other "more general" amino acid analytical procedures (see ref. 205 and references therein). Free dityrosine is also electrochemically active and has been quantitated along with tyrosine, 3-nitrotyrosine, and a variety of purine metabolites using gradient HPLC with coulometric array detection (207). It is also interesting to note that Beal et al. used this method to measure the formation of the DHBAs formed following the reaction between salicylic acid and $HO^•$ (see earlier discussion) (208).

Recently 3-nitrotyrosine levels were examined using a qualitative immunological approach (206, 209). The antibody, which was specific for the hydroxyl/nitro groups of 3-nitrotyrosine, could bind both free and protein-bound 3-nitrotyrosine and was not affected by other possible nitrated compounds (e.g., nitrophenylalanine, nitroguanine) or tyrosine analogs (e.g., dopamine, phosphotyrosine) (206). Although not quantitative, it does raise an important question as to the significance of free and bound 3-nitrotyrosine. This has still not been resolved.

Levels of 3-nitrotyrosine have been quantitated using HPLC with UV absorbance (199, 200) and fluorescence following precolumn derivatization with phenylisothiocyanate (195). Recently, extremely low levels of 3-nitrotyrosine and tyrosine have been accurately measured using HPLC with dual coulometric electrode detection (210) and coulometric electrode array detection (208, 210, 211).

Recently it has been found that 3-nitrotyrosine levels may be artificially elevated during sample preparation, especially when using acidic deproteinization procedures. If this is the case then other procedures will have to be found for the preparation of tissue when trying to accurately measure 3-nitrotyrosine.

Lipid Damage

Free polyunsaturated fatty acids (PUFAs) (e.g., linoleic, linolenic, and arachidonic acids) and fatty acids forming part of the structure of other lipids (e.g. triglycerides, diglycerides, phospholipids, etc.) are readily attacked by free radicals, eventually forming lipid peroxides. The consequence of such uncontrolled free radical generation is probably most familiar in rancidity of foods (212). Here lipid peroxidation leads to the formation of foul-smelling and -tasting aldehydes, organic

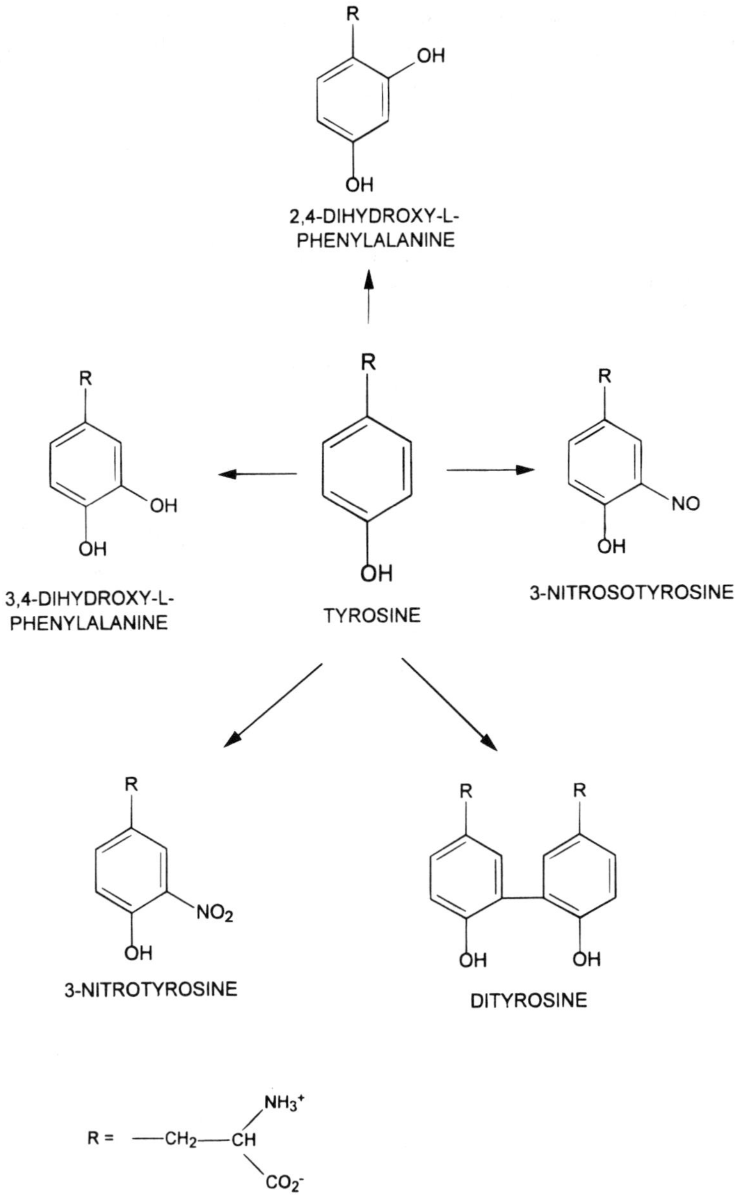

Figure 8. Some L-tyrosine derivatives potentially formed following reaction with HO· or ONO_2^-. All of these compounds are electrochemically reactive and can be determined using HPLC-ECD.

acids, and related compounds. Lipid peroxidation is also disastrous in vivo and causes impairment in membrane function (altered fluidity, changes in activity of membrane-bound enzymes and receptors, altered ionic channels, and permeability to Ca^{2+} and other ions). The products are often toxic, capable of damaging most cells in the body, and are associated with a variety of diseases such as atherosclerosis and brain damage following ischemia or trauma (3, 213).

The two major reactive species that can initiate lipid peroxidation by abstraction of a proton from the PUFA molecule are $HO^•$ and ONO_2^- (H_2O_2, O_2^- and singlet oxygen are thought to play only minor roles) (213, 214) (Figure 9). Once formed, the lipid radical can undergo several fates, but the most likely one under aerobic conditions involves reaction with oxygen to form the peroxy radical ($RO_2^•$). $RO_2^•$ can now attack other PUFAs in the membrane, thereby propagating the chain reaction of lipid peroxidation. Lipid peroxidation is finally terminated when the lipid/protein ratio decreases sufficiently such that $RO_2^•$ has to react with proteins rather than lipids, or when $RO_2^•$ encounters a chain-breaking antioxidant (CBA) capable of readily donating an hydrogen atom to form RO_2H (Eq. 6). Vitamin E (α-tocopherol) is probably the most abundant CBA found in biological membranes. Frei et al. (215) have suggested that ubiquinone (CoQ_{10}) may also play a role as a CBA, at least in mitochondrial membranes.

$$CBA-H + RO_2^• \longrightarrow RO_2H + CBA^• \tag{6}$$

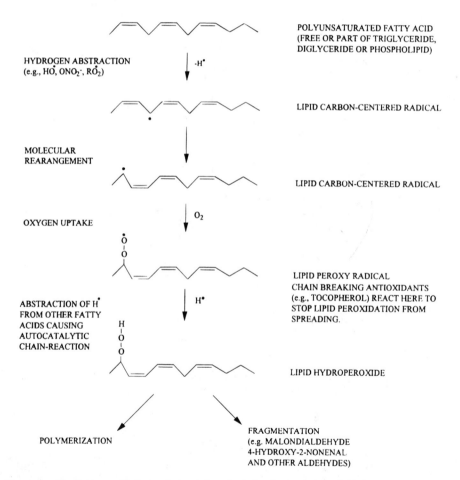

POLYUNSATURATED FATTY ACID
(FREE OR PART OF TRIGLYCERIDE,
DIGLYCERIDE OR PHOSPHOLIPID)

HYDROGEN ABSTRACTION
(e.g., $HO^•$, ONO_2^-, $RO_2^•$) $-H^•$

LIPID CARBON-CENTERED RADICAL

MOLECULAR
REARANGEMENT

LIPID CARBON-CENTERED RADICAL

OXYGEN UPTAKE O_2

LIPID PEROXY RADICAL
CHAIN BREAKING ANTIOXIDANTS
(e.g., TOCOPHEROL) REACT HERE TO
STOP LIPID PEROXIDATION FROM
SPREADING.

ABSTRACTION OF $H^•$ $H^•$
FROM OTHER FATTY
ACIDS CAUSING
AUTOCATALYTIC
CHAIN-REACTION

LIPID HYDROPEROXIDE

POLYMERIZATION

FRAGMENTATION
(e.g. MALONDIALDEHYDE
4-HYDROXY-2-NONENAL
AND OTHER ALDEHYDES)

Figure 9. The lipid peroxidation pathway. A hypothetical pathway showing initiation, propagation, and termination.

α-Tocopherol is thought to be regenerated from the tocopherol free radical by ascorbate (216, 217) reduced glutathione (218) and ubiquinol (219) (see later discussion).

Recently, Rubbo et al. (1994) reported that NO$^{\bullet}$ could not directly induce lipid peroxidation (220). However, the presence of both ROS and NO$^{\bullet}$ could either promote or inhibit lipid peroxidation, depending upon the ratio of ROS to NO$^{\bullet}$ levels. Using mass spectral analysis these authors reported a variety of lipid oxidation products including nitrito, nitro, nitrosoperoxo, and/ or nitrated lipid oxidation adducts. The formation of these compounds suggests that NO$^{\bullet}$ can act as a potent terminator of radical chain propagation reactions. Prooxidant reactions of NO$^{\bullet}$ will only occur after formation of ONO_2^- (as discussed earlier).

The detection and measurement of lipid peroxides using a wide variety of disparate methods (such as ESR, GC/MS, colorimetric and absorbance methods, chemiluminescence, and HPLC), and their advantages and disadvantages have been extensively reviewed recently (3, 213, 221, 222). HPLC-ECD has not usually been considered as a viable method for determination of lipids. However, there are a variety of HPLC-ECD methods in the literature that either measure lipids directly or as derivatives. In general HPLC-ECD is superior to the other analytical techniques, offering greater selectivity and sensitivity, and, is often easier to perform. Mulchandani and Rudolph (223) used ferrocene-linked voltammetry to measure linoleic acid and linolenic acid hydroperoxides. Lacking a separation step and having detection limits of 20 nM, this approach may not be practical for the study of ROS effects on lipid metabolism in vivo. O'Gara et al. (1989) developed a sensitive HPLC-ECD method based on the reduction of hydroperoxides by glutathione catalyzed by glutathione peroxidase (224). Stoichiometric amounts of oxidized glutathione are produced that can be measured by coulometric detection at +820 mV versus a Pd reference with a sensitivity in the low picomole range. Leukotriene B$_4$ (225), phospholipid hydroperoxides (226), and cholesterol hydroperoxides (227, 228) can be measured directly using HPLC-ECD. In each of these cases, however, the chromatographic separation of the analytes was poor. Korytowski et al. (229) measured cholesterol peroxides, fatty acid peroxides, and two synthetic phosphatidylcholine peroxides using HPLC-ECD utilizing a mercury drop electrode. Although they reported limits of detection of between 0.3 and 30 pmol, their chromatography was complicated by the use of four different isocratic conditions and the necessary, expensive, and toxic mercury drop electrode. Other approaches have used indirect HPLC-ECD methods to measure lipids. For example, prostaglandins can be measured electrochemically as their 2,4-dimethoxyanilide derivatives (230). Goldring et al. (231) measured 4-hydroxy-2-nonenal, the highly reactive breakdown product of lipid peroxidation, as its 2,4-dinitrophenylhydrazine derivative. The resulting phenylhydrazone was measured on the second electrode of a dual coulometric electrode cell (the first electrode was used to screen interferences) with limits of detection <1 pmol. Liu et al. (232) used HPLC with laser fluorescence detection to measure derivatized 4-hydroxy-2-nonenal, but with only a slight improvement in detection levels, it is hard to justify the complex ternary gradient chromatography and the expense of a laser. Adaptation of this method to a coulometric array system will allow for better selectivity and lower detection limits to be achieved. Furthermore, the use of gradient chromatography will make it possible to measure a wider variety of ROS-produced lipid fragments.

THE ANTIOXIDANTS: INTRODUCTION AND MEASUREMENT

Mention antioxidants and the first things that usually come to mind are the vitamins C, A, and E. However, these "low-molecular-mass molecules" or chain-breaking antioxidants (CBAs) are just a few in a whole multiplicity of natural defenses used by the body to combat ROS and RNS attack. Halliwell has defined an antioxidant as "any substance that, when present at low concentrations compared to those of an oxidizable substrate (e.g., lipids, proteins and DNA), significantly delays or prevents oxidation of that substrate" (see ref. 233 and references therein). This definition therefore includes not only the CBAs (e.g., ascorbic acid, tocopherol, uric acid, glutathione) but also enzymatic systems (e.g., superoxide dismutase, catalase, glutathione peroxidase) and proteins used to sequester metals capable of HO$^{\bullet}$ production (e.g., transferrin, ferritin, ceruloplasmin, hemopexin, haptoglobin, and albumin). The remainder of this chapter covers the physiological roles and measurement of the CBAs. It is very interesting to note that the very ability for the CBAs to react with free radicals (electron transfer) also renders them electrochemically active.

Ascorbic Acid

Ascorbic acid (vitamin C), essential in humans (it cannot be synthesized and has to be obtained from the diet), is a water-soluble compound with several important metabolic roles (e.g., in the production of collagen and as a cofactor for dopamine β-hydroxylase) (234). Its concentration in plasma is typically 40–140 μM (235). As an effective scavenger in vivo it can react with the O_2^-, and $HO_2^\cdot$ ($\sim 3.0 \times 10^5\ M^{-1}\ s^{-1}$; see Table 1), $HO^\cdot$ ($>10^9\ M^{-1}\ s^{-1}$), water-soluble peroxyl radical ($RO_2^\cdot$), $NO^\cdot$, nitronium radical ($NO_2^\cdot$), thiyl radicals, sulfenyl radicals, singlet oxygen, and hypochlorous acid (12, 236). Ascorbic acid is an antioxidant because the semidehydroascorbate radical (formed from ascorbic acid and a free radical) is much less reactive than most of the radicals scavenged by ascorbate. The ascorbyl radical is reduced back to ascorbic acid by enzymatic systems using either NADH or reduced glutathione (GSH) as the cofactor (237). As these are mainly intracellular, levels of ascorbyl radicals may accumulate in the extracellular fluid under periods of oxidative stress where they can damage extracellular molecules. Also, in the presence of metal ions (iron and copper), ascorbate transforms from being a strong reducing agent to being a pro-oxidant producing O_2^-, H_2O_2 and $HO^\cdot$. However, under normal conditions, iron and copper are well sequestered, as discussed earlier.

It is generally accepted, although not rigorously proven in vivo, that one of ascorbic acid's principal roles is the regeneration of membrane-confined α-tocopherol (ToH) from the α-tocopheryl radical, which is formed during destruction of lipid peroxides (see earlier and Eqs. 7 and 8). The interaction between ascorbic acid and α-tocopherol has been discussed in detail by Buettner (238) and Packer and coworkers (239–241) (Figure 10).

$$\alpha\text{-ToH} + RO_2^\cdot \longrightarrow RO_2H + \alpha\text{-To}^\cdot \qquad (7)$$

$$\alpha\text{-To}^\cdot + \text{Asc-H} \longrightarrow \alpha\text{-ToH} + \text{Asc}^\cdot \qquad (8)$$

Various approaches have been used to measure ascorbic acid, dehydroascorbic acid, and the semihydroascorbic acid radical, although there is not one assay capable of measuring all three compounds simultaneously (242). Methods include colorimetric, fluorometric, gas-liquid, enzymatic, HPLC with UV absorbance, and ECD using either amperometric or coulometric electrodes (ref. 242 and references therein). The advantages of the coulometric detection over amperometric detection have been extensively reviewed (56–58). Overall, coulometric detection offers superior measurement of ascorbic acid with a detection limit of 100 fmol on column (242). Ascorbic acid is also routinely determined as part of a global method for metabolic profiling using gradient HPLC and coulometric array technology (243, 257).

Uric Acid

Uric acid is the final product of purine metabolism in humans and primates. Catabolism of guanine yields xanthine directly, while catabolism of adenine first yields hypoxanthine and then xanthine. The conversion of hypoxanthine to xanthine and xanthine to uric acid are catalyzed by the same enzyme, xanthine oxidase, with ROS being produced as by-products (see earlier discussion). Uric acid is one of the most abundant antioxidants in the plasma and its concentration is close to its solubility limits (120–450 μM) (235). It has been hypothesized that this high circulating level may play a role in preventing aging and cancer. This may explain why humans and primates live longer than prosimians where the circulating level of uric acid is 10-fold lower (235). Although often regarded as a waste product, urate is not treated like other waste products. Production of purines is an energy-dependent process, and the loss of purines as urate is physiologically costly. Furthermore, >90% of urate is reabsorbed from the urine. It has therefore been suggested that during human evolutionary development, urate replaced ascorbate as the major water-soluble antioxidant.

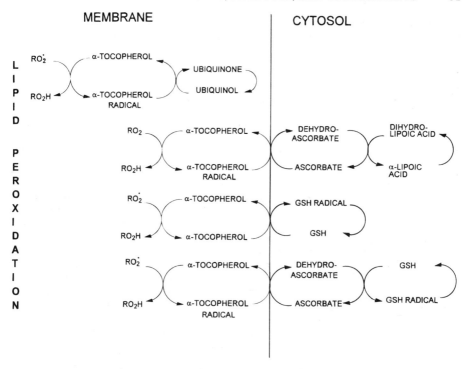

Figure 10. The cooperation between vitamin E (α-tocopherol) and various antioxidants in the prevention of lipid peroxidation. This figure shows several possible interactions between α-tocopherol and fat-soluble (membrane confined) and water-soluble (cytosolic) antioxidants. α-Tocopherol can be regenerated from its radical by reaction with ubiquinone (241), ascorbic acid (259, 260) and various thiols such as GSH (262). Ascorbic acid can be regenerated from its radical by reaction with different thiols including dihydrolipoic acid (241) and GSH (237). Terminal antioxidants are regenerated from their reaction products (e.g., GSH radical, ubiquinol, α-lipoic acid) by a variety of enzymatic and nonenzymatic mechanisms (e.g., α-lipoic acid is converted to dihydrolipoic acid by α-ketodehydrogenases using NADH).

Uric acid's principle antioxidant roles include: its ability to scavenge singlet oxygen, $RO_2^{\cdot}$, and HOCl; the binding of iron and copper in forms that do not promote free radical reactions; and protection against the formation of heme intermediates with iron in +4 and +5 valencies (235, 244). However, when uric acid reacts with $HO^{\cdot}$ or peroxyl radicals reactive uric acid radicals can be formed that can cause biological damage (245, 246). Consequently, Benzie and Strain (247) have suggested that urate acts as both a friend and a foe. Fortunately, uric acid radicals can be destroyed by reaction with ascorbic acid (246, 248). Measurement of the oxidation products of uric acid (oxonic acid, oxaluric, parabanic, cyanuric, and allantoin) may also be useful in the study of ROS damage (249).

Uric acid is usually determined using HPLC with UV (235, 250) or ECD either alone (251–253) or in conjunction with ascorbate (254–256). Uric acid is also routinely determined as part of a global method for metabolic profiling using gradient HPLC and coulometric array detection (257).

Vitamin E

The chemistry, physical properties, and role of the essential compound vitamin E in human health have been extensively reviewed (258). Vitamin E is a fat-soluble group of eight naturally occurring molecules known as the tocopherols. α-Tocopherol is a CBA and an integral part of the membrane reacting with lipid peroxides, thereby preventing lipid peroxidation (Eq. 7). As discussed earlier, α-tocopherol is regenerated from its radical either by reaction with ascorbate (259, 260) (Eq. 8), ubiquinone (261), or GSH (262) (Figure 10).

Vitamin E has been measured using HPLC-ECD either alone (263, 264) or in conjunction with ubiquinone. Simultaneous approaches use either amperometric detection (265) or the superior sensitivity and selectivity of coulometric detection used in redox mode (266, 267). Recently, Gamache et al. (1996) developed a "global" gradient HPLC coulometric array method for the simultaneous analysis of fat-soluble vitamins and related compounds (tocopherols, retinoids, carotinoids, and CoQ_{10}) and used it to study their levels in human plasma (Figure 11) (268).

Ubiquinone

Ubiquinone (2,3-dimethoxy-5-methyl-6-decaprenylbenzoquinone), also called coenzyme Q_{10} (CoQ_{10}), is a fat soluble quinone that plays an active role in shuttling electrons between NADH-Q reductase and cytochrome reductase of the respiratory chain located in the inner mitochondrial membrane. CoQ_{10} is also the entry point of high potential electrons from $FADH_2$ formed from several enzymatic reactions (e.g., succinate dehydrogenase, glycerol phosphate dehydrogenase, and fatty acyl coenzyme A dehydrogenase). CoQ_{10} is also believed to regenerate α-tocopherol from its radical formed while preventing lipid peroxidation from occurring (269) (Figure 10).

CoQ_{10} has been determined in plasma using HPLC-UV following liquid-liquid extraction (270). HPLC-ECD has been used to measure the reduced form directly (265) or the total amount following chemical reduction with sodium tetraborohydride (267, 271). Grossi et al. (1992) have developed a dual coulometric electrode technique with automated solid-phase extraction (limit of detection <5 μg/L) capable of measuring plasma levels in human subjects (272). Their approach uses one coulometric electrode to electrochemically reduce CoQ_{10} to the hydroquinone, which can then be reoxidized on the downstream coulometric electrodes (ratios of response between adjacent oxidative electrodes can then be used as a measure of peak purity). Gamache et al. (1996) routinely measure CoQ_{10} as part of their "global" fat-soluble vitamin method using gradient chromatography and coulometric array detection (Figure 11) (268).

Thiols, Disulfides, and Related Compounds

A variety of thiols (e.g., reduced glutathione), disulfides (e.g., oxidized glutathione and mixed disulfides), and related compounds (e.g., lipoamide, taurine, homocysteine, N-acetylcysteine) appear to be involved in the control of free radicals.

Glutathione

GSH is a tripeptide (γ-glutamyl-cysteinyl-glycine) with many functions in the body. Glutathione peroxidase uses GSH to convert H_2O_2 and organic peroxides (RO_2H) into water and alcohols, respectively (Eq. 9).

$$2\ GSH + RO_2H \longrightarrow GSSG + H_2O + ROH \tag{9}$$

Glutathione peroxidase is remarkable in that it has a selenium atom in its active site. GSH is regenerated from GSSG in a reaction catalyzed by glutathione reductase with reducing power provided by NADPH (Eq. 10).

$$GSSG + NADPH + H^+ \longrightarrow 2\ GSH + NADP^+ \tag{10}$$

GSH can also react with a variety of free radicals (Eq. 11) (Figure 10).

$$HO^{\bullet},\ RO^{\bullet}\ or\ RO_2^{\bullet} + GSH \longrightarrow H_2O,\ ROH,\ or\ RO_2H + GS^{\bullet} \tag{11}$$

The glutathiyl radical is less reactive than the ROS and can be readily dealt with. However, the glutathiyl radical can react with other GSH molecules, forming the very strongly reducing oxidized glutathione radical ($GSSG^{\bullet-}$) (Eq. 12), which can even reduce oxygen to form $O_2^{\bullet-}$ (Eq. 13).

$$GS^{\bullet} + GS^- \longrightarrow GSSG^{\bullet-} \tag{12}$$

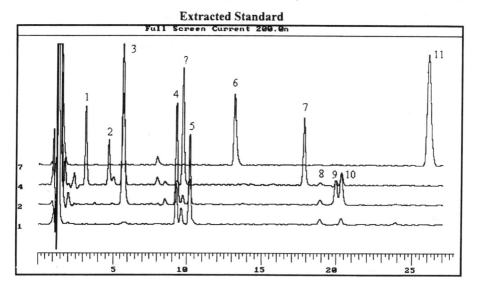

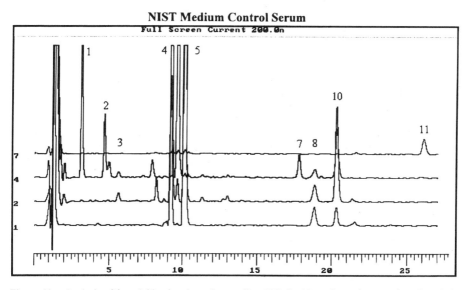

Figure 11.   Analysis of fat-soluble vitamins using gradient HPLC with coulometric array detection. Only four channels are shown for clarity. Channel 6 is set at −1000 mV to reduce compounds prior to their oxidation and measurement on channel 7 (+200 mV). This redox approach allows the measurement of vitamin K1 and coenzyme Q_{10}. The upper chromatogram presents the analysis of extracted standards (1 μg/ ml) and is shown at a gain of 200 nA. 1–Retinol; 2–retinyl acetate; 3–lutein; 4–γ-tocopherol; 5–α-tocopherol; 6–vitamin K1; 7–retinyl palmitate; 8–lycopene; 9–α-carotene; 10–β-carotene; 11–CoQ_{10}. The lower chromatogram shows the analysis of NIST medium control serum following hexane extraction, evaporation, and reconstitution as outlined earlier. This chromatogram is also presented at a gain of 200 nA. Reproduced with permission from ESA, Inc.

$$GSSG^{\cdot-} + O_2 \longrightarrow GSSG + O_2^{-} \tag{13}$$

Therefore it has been suggested that both GSH and SOD are important in cellular antioxidant defenses (ref. 238 and references therein). When SOD keeps the O_2^- level low, the glutathiyl radical dimerizes to form GSSG or other stable end products such as glutathione sulfinic acid and glutathione sulfonic acid.

GSH is also involved in the formation of α-tocopherol from its radical (see earlier discussion), acts as a sulfhydryl buffer that maintains cysteinyl residues in hemoglobin and other red blood cell proteins in the reduced state [its concentration in the red blood cell and other cells is on the order of 2mM (3)], acts as a cofactor for many enzymes (e.g., prostaglandin endoperoxide isomerase, glyoxylase), and is thought to be involved in the transport of amino acids across the plasma membrane of animal cells (273).

The most widely used technique for the measurement of GSH (and GSSG) is the enzyme recycling method (274, 275). However, plasma levels of GSH are approximately 1 μM and require more sensitive detection methods. Various HPLC techniques have been used, including fluorescence detection of the OPA derivatives (276), and direct amperometric (277–280) or direct coulometric detection (281–285). In general, the coulometric approaches offered superior sensitivity and selectivity to the dual-amperometric techniques and furthermore avoid the use of toxic and unstable mercury amalgams. With all approaches care must be exercised when collecting and storing samples to avoid artificially altering the GSH/GSSG ratio (e.g., using serine borate to inhibit γ-glutamyltransferase activity, not rupturing red blood cells, and sampling at a specific time of day to avoid possible effects of circadian rhythms).

Hill et al. (1993) used a gradient coulometric array approach to study the metabolism of a variety of S-substituted GSH conjugates formed when animals were exposed to hydroquinone (286). Their method offered excellent resolution of the conjugates, even in complex biological matrices such as urine.

Homocysteine

Homocysteine is a key metabolite in both sulfur amino acid biochemistry and in the transfer of "activated" methyl groups (in regeneration of the "methyl carrier" S-adenosylmethionine). Homocysteine is formed from methionine, can react with serine for the synthesis of cysteine, or can be converted back to methionine by transfer of a methyl group from N^5-methyltetrahydropteroyltri-L-glutamate (a folate derivative) in a reaction that involves vitamin B_{12} (cobalamin) and the enzyme tertrahydropteroylglutamate methyltransferase. It is not surprising therefore that plasma levels of homocysteine are used clinically to monitor folate and cobalamin function. Several forms of homocysteine occur in plasma, including the thiol (reduced; <0.25 μM), the free oxidized (homocystine and homocysteine-cysteine mixed disulfides; <2 μM), and the protein bound oxidized (<8 μM) (287). There is now a growing interest in the possible roles of homocysteine in control of the plasma redox thiol status in disease and oxidative stress (287).

Homocysteine has been determined using a variety of techniques, including antibody-based procedures, radioenzymatic methods, and HPLC with colorimetry, fluorescence, or ECD (see refs. 288–290 and references therein). Recently two groups have been developing HPLC-ECD assays capable of the direct measurement of several reduced and oxidized thiols (e.g., GSH, GSSG, N-acetylcysteine, methionine, homocysteine, homocystine, cysteine, and cystine) simultaneously (291, 292). As with GSH, when measuring fractionated homocyst(e)ines, care must be taken so as not to artificially alter the redox status of the plasma sample (otherwise the ratio of the levels of free-reduced and oxidized forms will be changed).

Other Sulfur-Containing Compounds

N-Acetylcysteine is used as an antioxidant both in vivo and in vitro [e.g., it is used to decrease the toxicity of diquat to hepatocytes (293) and protects against paracetamol toxicity (294)]. The action of N-acetylcysteine is thought to be through elevation of cysteine (and hence GSH) and/ or by the direct scavenging of free radical species. Arouma et al. (1989) showed that N-acetylcysteine is a powerful scavenger of hypochlorous acid and HO$^{\cdot}$, but that it reacted only slowly with H_2O_2 and totally failed to react at all with O_2^- (Table 1) (295). N-Acetylcysteine has been measured using HPLC-ECD (291, 292).

Lipoic acid (1,2-dithiolane-3-valeric acid) exists as both the reduced dithiol form (dihydrolipoic acid) and the oxidized disulfide form. It is a cofactor used by pyruvate dehydrogenase and α-ketodehydrogenase complexes in most cells, where it acts as a carrier of activated acetyl (or acyl) groups. Lipoic acid has been described as a universal antioxidant effectively reacting with $RO_2^{\cdot}$, ascorbyl radicals, $HO^{\cdot}$, $NO^{\cdot}$, tocopheryl radicals, O_2^{-}, and hypochlorous acid (239, 240, 296, 297). However, the exact role of lipoic acid in lipid peroxidation is not clear as it has been reported to be both antioxidant and pro-oxidant in nature (297). Packer has suggested that lipoic acid may function to regenerate α-tocopherol from its radical after it has reacted with lipid peroxides, thereby indirectly helping to limit lipid peroxidation (241). Lipoic acid may either act directly upon the tocopherol radical or, more likely, regenerate ascorbic acid from its radical following ascorbic acid's reaction with the tocopherol radical. This could explain why, even though there is one molecule of tocopherol per 1000–2000 (0.1 nmol/mg protein) phospholipid molecules and lipid peroxyl radicals can be generated in membranes at a rate of 1–5 nmol/mg membrane protein, there is neither total oxidative destruction of the membrane, nor depletion of α-tocopherol (241). Lipoic acid can be measured using a variety of methods including those based upon thin-layer chromatography, colorimetry, polarography, microbiology, gas chromatography (GC), GC/MS, and HPLC. HPLC-ECD offers superior sensitivity and selectivity (298–300 and references therein). Recently, HPLC with dual-coulometric detection has enabled the measurement of both reduced and oxidized lipoic acids with markedly improved sensitivities over the amperometric approaches (301).

Several acidic sulfur-containing amino acids (e.g., cysteic acid, homocysteic acid, and cysteinsulfinic acid) and γ-glutamyl di- and tripeptides occur in mammalian brain (see ref. 302 and references therein). Some are thought to be intermediary metabolites, but others may play a more sinister role as excitatory agents and neurotoxins (possibly inflicting damage by mimicking excitatory amino acid stimulation associated with oxidative stress). Orwar et al. developed HPLC-based assays (using both fluorescence and electrochemical detection) for the measurement of the o-phthaldialdehyde/β-mercaptoethanol derivatives of these compounds in microdialysate perfusates (302).

Carotenoids, Retinoids, and Related Compounds

The chemistry, physical properties, and role of carotenoids in human health have been extensively reviewed (303). Vitamin A (retinol) is formed in the intestine from its precursor β-carotene. β-Carotene and the carotenoids as a whole are abundant in green plants, carrots, and other vegetables. Following ingestion, vitamin A is stored in the liver as fatty acid esters (e.g., palmitate). Although the major role of vitamin A is in photoreception, it and the carotenoids (β-carotene, lutein, and lycopene) are all effective antioxidants and free radical scavengers (304, 305, and references therein). For example, the rate constants for the reaction of β-carotene with carbon-centered radicals and peroxyl radicals are $\sim 1.0 \times 10^4$ and 5×10^5 to $1.0 \times 10^4\ M^{-1}\ s^{-1}$, respectively (Table 1) (306).

There are several reports in the literature using HPLC-ECD for the measurement of the carotenoids. Bryan et al. (1991) used normal-phase HPLC with amperometry to measure a variety of retinoids, including vitamin A palmitate, and reported a lower detection limit of 1 ng on column (307). MacCrehan and Schonberger (308) compared HPLC-ECD and HPLC-UV techniques in the measurement of retinol and its isomers in human serum. They reported that HPLC-ECD was markedly more sensitive than the spectrophotometric approach and measured a lower detection limit of <200 pg on column. Others have measured vitamin A using amperometry and have examined its plasma levels in control patients and those with liver cirrhosis (309, 310). MacCrehan and Schonberger (311) developed a gradient method capable of simultaneously measuring retinol, α-tocopherol, and β-carotene. Although the method produced data comparable to those obtained from HPLC-UV/visible, its full potential could not be realized due to poor voltammetric resolution (the single thin-layer electrode had to be driven at 900 mV vs. Ag/AgCl reference in order to detect retinol) and an inability to use more sensitive gain ranges, probably a result of baseline noise and drift during gradient elution. Gamache et al. (1996) routinely measure a variety of carotenoids and retinoids in biological samples as part of their "global" fat-soluble vitamin method using gradient chromatography and coulometric array detection (Figure 11) (268).

The exact biological role of the carotenoids and retinoids is at present unclear. Although they play a role in prevention of lipid peroxidation, this appears mainly to be under the jurisdiction of the tocopherols. It may be that the significance of the carotenes and retinoids only becomes apparent when the α-tocopherol system is compromised or under periods of prolonged oxidative stress. However, although β-carotenes appear to be beneficial by playing an antioxidant role, they also act as pro-oxidants (312–314).

Other Phytochemicals as Antioxidants

Many different hydroxylated metabolites are found in plants and many are thought to act as antioxidants. These include phenols, polyphenols, flavonoids, and possibly lignans.

Phenols

A wide variety of free and conjugated phenols exists in plants, and some have been tested as antioxidants in vitro. For example, ferulic acid was shown to inhibit lipid peroxidation in phospholipid liposomes and to protect deoxyribose against $HO^•$ attack (315). Similarly, thymol, carvacrol, and 6-gingerol were found to decrease lipid peroxidation and to effectively scavenge peroxyl radicals, whereas zingerone was only weakly effective (316). The $HO^•$ scavenging ability of salicylic acid and tyrosine were discussed in detail earlier. At present, it is unclear whether consumption of food plants high in phenols will be of benefit in controlling ROS levels in humans as it is doubtful that the effective circulating levels of these compounds will be reached. However, plant phenols may prove useful in preventing food spoilage and may be preferred alternatives to the synthetic antioxidants presently added to foods. As discussed in the next section, plant phenols have previously been measured along with the polyphenols and flavonoids using gradient HPLC with coulometric array detection.

Polyphenols and Flavonoids

Polyphenols are an abundant and complex group of secondary metabolites found in higher plants and include three families of compounds: the proanthocyanidins, the derivatives of gallic acid, and the derivatives of hexahydroxydiphenic acid. Possibly the most important group of phenols in food is the flavonoids, which consist of the catechins, proanthocyanins, anthocyanidins and the flavones, and flavonols and their glycosides. The flavonoids are thought to possess a range of medicinal properties and act as antibacterial, antitumor and antiviral agents (317). Furthermore, flavonoids also possess antiradical and antioxidant properties (318–320). However, not all flavonoids are equally potent. Hydroxylation of the B ring in positions 2′, 3′, and 4′ promotes radical scavenging capability and inhibition of peroxidation of lipids, while hydroxylation of the 7-position increases the potency of inhibition of superoxide dismutase (321). Similarly, van Acker et al. (322) reported that the B ring must contain a catechol structure for efficient scavenging activity, while a 3-hydroxyl moiety in conjunction with a C2-C3 double bond also promotes free radical scavenging. Recently, Saija et al. (1995) compared the scavenging ability of quercetin, rutin, hesperetin, and naringenin and studied their ability to protect membranes (323). They concluded that the ability of a flavonoid to protect against lipid peroxidation is dependent not only upon its structure but also on its ability to interact with and penetrate the lipid bilayers.

The catechins are widely distributed in plants and are especially abundant in green tea leaves (making up approximately 30% of the dry leaf weight). They are reported to give tea its antioxidant and anticancer properties. Although tea contains many potential radical scavengers, it is thought that the primary antioxidant in green tea is epigallocatechin (EGC) (324). Epicatechin (EC), epicatechin gallate (ECG), and epigallocatechin gallate (EGCG) also quench free radical chain reactions that may damage membrane lipids and play a role in inhibiting mutagenicity by preventing DNA damage (315, 325, 326). Polymeric proanthocyanidins consist of chains of flavan-3-ol units and can complex strongly with proteins and carbohydrates, but their possible role as antioxidants has yet to be fully examined (327,328). Anthocyanins are responsible for plant pigmentation and, like the proanthocyanidins, their potential antioxidant role is as yet undetermined. Flavones, flavanols, and their glycosides are distributed widely in plants. It has been estimated that humans consume approximately 1 g of these compounds daily. The most common and most bioactive member is quercetin. Quercetin and other polyphenols, such as ellagic acid and chlorogenic acid, may play a major role in carcinogenesis by both reducing the availability of carcinogens and interfering with their biotransformation in the liver (329, 330).

Several different approaches have been used to measure polyphenolics and flavonoids, including those based on GC/MS and HPLC, the latter with a variety of different detection systems (visible, UV, and photo diode array [PDA]) (331, 332, and references therein). These techniques are often inadequate due to sample complexity or trace substance levels. Although HPLC-ECD offers superior sensitivity over other techniques, as previously discussed, single-electrode approaches generally suffer from poor voltammetric resolution (56–58). Gradient elution with coulometric array detection is ideally suited for the measurement of phytochemicals. The coulometric array offers the stability of HPLC-UV detectors with better resolution than HPLC-PDA. The difference between one flavonoid and another sometimes is simply the shift of an hydroxyl group from one position to another. This would not be expected to affect the compound's UV absorbance spectrum, but it can markedly change its electrochemical voltammetric behavior. Coulometric array detection, therefore, can achieve greater resolution of analogous flavonoids and polyphenolics than PDA (57).

A number of researchers have reported the use of coulometric arrays to measure polyphenolics and flavonoids. Gamache et al. (333) developed a gradient method with an array of 16 coulometric electrodes capable of analyzing 27 compounds, including phenols, polyphenols, and flavonoids, in 30 min in a 20 μL sample (Figure 12). Achilli et al. (334) developed a similar gradient method capable of resolving 34 analytes in 45 min. Uchida et al. (335) used a coulometric array to examine the antioxidant levels during fermentation, storage, and packaging, and discussed the role of phytochemicals as antioxidants in controlling beer stability.

Recently, Acworth et al. (336) developed a gradient coulometric array method capable of measuring EC, ECG, EGC, and EGCG within 30 min and used it to examine urinary levels of these compounds following green tea consumption. Lee et al. (337) adapted this method to measure levels of these compounds in plasma and urine of human subjects consuming green tea.

Phytoestrogens and Lignans

Two groups of diet-derived diphenolic compounds, the isoflavonic phytoestrogens (e.g., genistin, genistein, daidzin, daidzein, and equol) and the lignans (enterolactone and enterodiol), are often found in human urine and tissues. They possess several purported biological activities. For example, they can prevent some forms of hormone-dependent cancers (338 and references therein), probably by binding to estrogen receptors, thereby blocking the action of more potent estrogens (339). Genistein, by interfering with the transmission of signals from growth factor receptors through the inhibition of tyrosine kinase and other enzymes, can affect cell division and tumor growth (339, 340). More interestingly, genistein may also inhibit cytochrome P-450 enzymes (341), thereby reducing the production of H_2O_2 during the catabolism of xenobiotics. It appears that some of the biological effects reported for these compounds may result from their dual role in ROS control by inhibiting ROS production and by reacting with them if produced.

Several different techniques have been used to measure lignans and phytoestrogens, including GC (342), GC/MS (343, 344), HPLC-MS (345) and HPLC-UV (327, 328). Setchell et al. (345) compared isocratic HPLC-UV and isocratic HPLC-ECD methods and concluded that electrochemical measurement was 300- to 1000-fold more sensitive than the UV approach. Recently, Gamache et al. (268) developed a sensitive, gradient coulometric array method capable of measuring a variety of phytoestrogens (coumestrol, daidzein, daidzin, equol, and quercetin), mammalian lignans, estrogens (estrone, estradiol, and estriol), and the synthetic estrogen analog diethylstilbestrol simultaneously (Figure 13). It is interesting to note that estrogens can also act as antioxidants independent of their ability to bind to the estrogen receptor (348–350).

α-Ketoacids

A number of different α-ketoacids including pyruvate, oxaloacetate, α-ketoglutarate, and the branched-chain keto (2-oxo) acids (the transamination products of the branched-chain amino

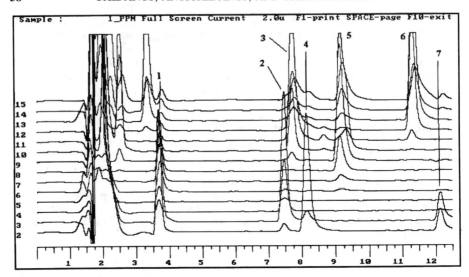

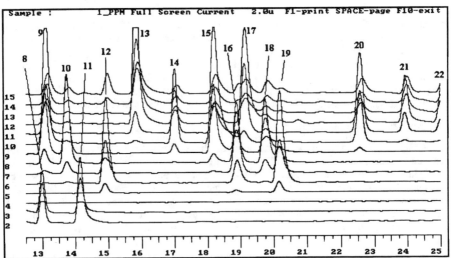

Figure 12. Analysis of a variety of phenol, polyphenol, and flavonoid standards using gradient chromatography and coulometric array detection. 1–Gallic acid; 2–protocatechuic acid; 3–4-aminobenzoic acid; 4–gentisic acid; 5–tryptophan; 6–4-hydroxybenzoic acid; 7–chlorogenic acid; 8–catechin; 9– 4-hydroxyphenylacetic acid; 10–vanillic acid; 11–caffeic acid; 12– syringic acid; 13–4-hydroxycoumarin; 14– vanillin; 15–coumaric acid; 16–syringaldehyde; 17– umbelliferone; 18–eugenol; 19–ferulic acid; 20–scopoletin; 21–narirutin; 22–naringin. The chromatogram is presented at a gain of 2.0 μA. Reproduced with permission from ESA, Inc.

acids [isoleucine, leucine, and valine]) take part in intermediary metabolism. There has recently been renewed interest in the α-ketoacids due to their antioxidant capability (Eq. 14).

$$R-COCO_2H + H_2O_2 \longrightarrow R-CO_2H + H_2O + CO_2 \tag{14}$$

This reaction appears to be biologically important as the α-ketoacids protect against H_2O_2-induced cytotoxicity (351) and H_2O_2-induced damage to DNA (352, 353). H_2O_2 can also react with other biologically important macromolecules such as proteins (see earlier discussion). It is interesting to note that many enzymes that are vulnerable to oxidative damage (e.g., pyruvate

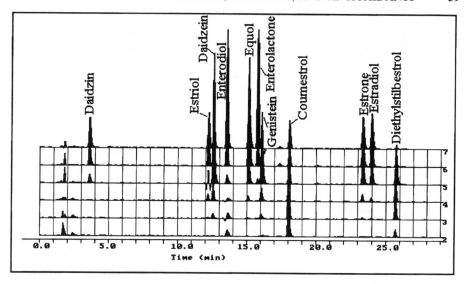

Figure 13. Analysis of phytoestrogens, mammalian lignans, estrogens, and the synthetic estrogen analog diethylstilbestrol using gradient HPLC with coulometric array detection. Analytes were resolved voltammetrically using an eight-channel coulometric electrode array (channels 1–8: +260, +320, +380, +440, +500, +560, +620, −300 mV vs. Pd). The chromatogram shows resolution of all metabolites within 27 min. This method has recently been adapted to include the measurement of xenoestrogens such as bisphenol A and bisphenol B. Reproduced with permission from ESA, Inc.

dehydrogenase, pyruvate kinase, isocitrate dehydrogenase, and lactate dehydrogenase) happen to be associated with α-ketoacids (producing them or using them as substrates). Nath et al. have suggested that the α-ketoacids may help to protect these enzymes during periods of oxidative stress (353). It is probably not a coincidence that many of these enzymes and the α-ketoacids are associated with the mitochondrion, an organelle that can produce ROS in abundance.

The tissue concentrations of the α-ketoacids are typically in the micromolar range as compared to nanomolar concentrations of H_2O_2. This suggests that if α-ketoacids come into contact with H_2O_2 the latter will be destroyed. In this respect, α-ketoacids should also be considered along with catalase and glutathione peroxidase as the major factors controlling intracellular levels of H_2O_2.

One condition where α-ketoacids may be important is in exercise. Physical activity can increase oxygen consumption by 10- to 15-fold over resting conditions, and this increase in oxygen demand can result in oxidative stress and may lead to tissue damage, especially to muscle and liver. However, the amount of damage depends on many factors including exercise intensity and duration, the tissue being examined, and the training state of the individual (354). The possible mechanisms involved in controlling and preventing such damage have been discussed in detail elsewhere (355). It is interesting to note that the levels of α-ketoacids may increase in muscle during (or following) aerobic exercise. For example, the oxidation of glucose gives rise to pyruvate, which then feeds into the Krebs cycle. Also, branched-chain amino acids can be transaminated to produce the branched-chain ketoacids. These are not further metabolized in muscle (356). Consequently, the α-ketoacid pool would be expected to be increased. Aerobic muscular activity therefore results in both increased levels of ROS and α-ketoacids. It is tempting to speculate that this represents a self-regulating protective mechanism to the presence of ROS.

Melatonin

Melatonin (5-methoxy-*N*-acetyltryptamine; MLT) is a pineal gland neurohormone involved with the timing of reproductive activity in many species. The photoperiod (e.g., fewer daylight hours in winter) influences the synthesis and release of this neurohormone so as to optimize the timing of reproductive fertility. For example, sheep pineal glands respond to the short photoperiod of winter, leading to an enhancement of their reproductive cycle such that conception is most likely to occur in the winter with offspring born in the spring. On the other hand, MLT plays an opposite role in species with a very short gestation period (e.g., hamsters) who are generally long-day breeders, since survival of the species' offspring is enhanced if born during the spring or summer, a time close to conception. Some have suggested that MLT plays a role in the onset of puberty in humans via influences on the control of gonadotropin-releasing hormones (357). MLT plasma concentrations are high in prepubescent children but eventually fall below a critical value, not due to decreased MLT production, but rather due to increased body mass. This in turn may initiate the onset of puberty. While the photoperiod is probably less important for entraining the reproductive system in humans, the influence of exposure to light/dark probably is involved with MLT production and sleep patterns. In normal people, endogenous MLT levels rise during the evening, peaking during the normal hours of sleep (358). Numerous studies have investigated the utility of MLT in enhancing the quality of sleep in blind individuals who lack the normal photoperiod input to the retina, and also in alleviating the severity of jet lag. MLT is currently sold for self-medication aimed at decreasing sleep onset time.

MLT is also known to be a potent and effective free radical scavenger of $HO^•$ (359). The free radical scavenging potency of MLT (IC50 = 21 μM) is much greater than that of mannitol (IC50 = 123 μM) and GSH (IC50 = 286 μM) (360). The oxidation of MLT to 5-methoxy-N-acetyl-N-formylkynuramine by $HO^•$ yields a compound that is not subject to reduction or regeneration. Neither MLT nor its metabolites are characterized by pro-oxidant actions. The high lipophilicity of MLT makes it an excellent candidate to scavenge $HO^•$ radicals at their site of production and provide protection against highly reactive species.

Various techniques have been utilized to measure MLT in plasma, urine, and tissue. An assay procedure capable of measuring MLT in many samples quickly involves radioimmunoassay (RIA) (361, 362). This technique, which utilizes antiserum typically raised to an MLT complex in rabbit, allows for a reasonable estimate of MLT; however, as with all RIA procedures, it suffers from an inherent lack of absolute selectivity. A competitive solid-phase enzyme immunoassay for MLT has recently been described that allows for detection of as little as 1.0 fmol (363). Despite the problems of selectivity, RIA is a very popular method employed by researchers to estimate MLT levels. The use of GC-negative ion chemical ionization mass spectrometry of the pentafluoropropionyl spirocyclic derivatives of MLT allows for greatly improved analyte selectivity detection (364, 365). However, the cost associated with the instrumentation can prove prohibitive for many researchers. Less expensive techniques utilize HPLC coupled to fluorescence, amperometric, and coulometric detection (366–370). MLT and several other metabolites have been measured in a variety of tissues using gradient HPLC and coulometric array detection (371).

MLT has been measured in microdialysates obtained from birds and rats using HPLC-ECD (372, 373). Recently, differential-pulse voltammetry with pretreated carbon fiber microelectrodes has been used in the rat pineal and suprachiasmatic nucleus to monitor MLT changes (374).

CONCLUSIONS

Anaerobic organisms are typically restricted to environments where the level of oxygen is reduced or even absent. These organisms use a variety of compounds as the terminal electron acceptor and produce a moderate amount of energy to drive metabolism. Aerobic organisms, however, use oxygen as the terminal electron acceptor. These organisms have evolved to efficiently capture the major production of energy resulting from the reduction of oxygen to water. Proof of the favorability of oxygen reduction is readily apparent in the great diversity of both single-cell and multicellular aerobic species now living on the earth. However, the success of aerobes is not without a price. Partially reduced oxygen species (often refered to as the reactive oxygen species) and several different nitrogen oxides (the reactive nitrogen species) occur in vivo and represent a major challenge to living organisms.

The ROS and RNS show markedly different reactivities. Of these, $HO^•$ appears to be the most aggressive, reacting with many biologically important molecules. H_2O_2, $NO^•$, and O_2^- show rapid reactivity with only a few molecules. $RO_2^•$, $RO^•$, $NO_2^•$, and ONO_2^- show intermediary reactivities. The body has several lines of defense in dealing with the ROS and RNS. Some reactive species such as HO˙ once formed are too reactive to control effectively, so mechanisms exist to prevent their formation (chelation of metals) or deal with their reaction products (e.g., excission and excretion of DNA adducts). Others (e.g., O_2^-) are much less energetic and are usually controlled enzymatically (e.g., by superoxide dismutase) or are rendered ineffective by their reaction with other molecules such as the antioxidants. Effective control of the ROS and RNS may be important in the prevention of premature aging and diseases.

ROS and RNS attack on macromolecules produces a wide variety of products. The attack of DNA by HO˙ results in the formation of several uniquely modified DNA bases, and their measurement by HPLC-ECD is often used to access exposure of tissue to ROS and RNS. Damage to proteins includes production of protein carbonyls, which can be measured using HPLC-ECD as their dinitrophenylhydrazone derivatives. Furthermore, proteins also produce a variety of modified amino acids (e.g., 3-nitrotyrosine and dityrosine), which can be measured directly using HPLC-ECD. Both free and protein-bound modified amino acids are used as indicators of tissue exposure to ROS and RNS. However, increasing evidence suggests that the very procedures used to prepare a sample (including protein and DNA hydrolysis) may artificially affect amino acid modification. Lipid peroxidation initiated by HO˙ and ONO_2^- produces a variety of lipid carbonyls (e.g., malondialdehyde and 4-hydroxy-2-nonenal) and nitrogen-containing lipid adducts. Lipid carbonyls can be measured using HPLC-ECD following their derivatization with dinitrophenylhydrazine. Lipid peroxidation can be prevented by CBAs.

With the recent excitement surrounding the potential benefits of antioxidants (the free radical scavengers) in humans there is currently, and will continue to be, a greater need for sensitive and selective analytical procedures for the quantitation of these compounds in their native sources as well as in biological fluids and tissues. As many of these compounds exist as part of a larger, more complex biochemical matrix, the challenges facing the analyst will be enormous. Particular attention must be directed at the establishment of assay procedures possessing adequate sensitivity, selectivity, and reproducibility, despite the presence of numerous potential coelutants. The "health food industry" has been capitalizing on the sale of antioxidants without any apparent quality assurance, since the use of such "dietary supplements" is not regulated for effectiveness or safety by the Food and Drug Administration as a result of the Dietary Supplement Health and Education Act passed by Congress in 1994. For example, manufacturers are not required to present evidence of purity or even to demonstrate that any of the claimed antioxidant is present in the product. As a result, there will be ample opportunities for antioxidant analyses in the future as consumers demand information about the purity of such products. As research continues into the mechanisms of antioxidant actions, the need for sensitive and selective analytical procedures will grow.

REFERENCES

1. Balentine, J. ed. 1982. *Pathology of Oxygen Toxicity.* London: Academic Press.
2. Nichols, C. W., and Lambersten, C. J. 1969. Effects of high oxygen pressure on the eye. *N. Engl. J. Med.* 281:25–30.
3. Halliwell, B., and Gutteridge, J. M. C. eds. 1989. *Free Radicals in Biology and Medicine.* Oxford: Clarendon Press.
4. Gerschman, K., Gilbert, D. L., Nye, S. W., Dwyer, P., and Fenn, W. O. 1954. Oxygen poisoning and x-irradiation: A mechanism in common. *Science* 119:623–626.
5. Fridovich, I. 1986. Superoxide dismutases. *Methods Enzymol.* 58:61–97.
6. Colton, C. A., and Gilbert, D. L. 1987. Production of superoxide anions by a CNS macrophage, the microglia. *FEBS Lets.* 223:284–288.
7. Babior, B. M. 1978. Oxygen-dependent microbial killing by phagocytes. Part 1. *N. Engl. J. Med.* 298:645–668.
8. Halliwell, B. 1994. Free radicals, antioxidants, and human disease: curiosity, cause, or consequence? *Lancet* 344:721–724.
9. Ames, B .N., Shigenaga, M. K., and Hagen, T. M. 1993. Oxidants, antioxidants, and the degenerative diseases of aging, *Proc. Natl. Acad. Sci. U.S.A.* 90:7915–7922.
10. Halliwell, B., and Gutteridge, J. M. C. 1990. Role of free radicals and catalytic metal ions in human disease: An overview. *Methods Enzymol.* 186:1–85.
11. Kehrer, J. P. 1993. Free radicals as mediators of tissue injury. *Crit. Rev. Toxicol.* 23:21–48.
12. Halliwell, B. 1994. Free radicals and antioxidants: A personal view. *Nutr. Rev.* 52:253–265.

13. Halliwell, B., and Gutteridge, M. C. 1992. Biologically relevant metal ion-dependent hydroxyl radical generation. *FEBS Lett.* 307:108–112.

14. Fahn, S., and Cohen, G. 1992. The oxidant stress hypothesis in Parkinson's disease: Evidence supporting it. *Ann. Neurol.* 32:804–812.

15. Ceballos, I., Lafon, M., Javoy-Agid, F., Hirsch, E., Nicole, A., Sinet, P. M., and Agid, Y. 1990. Superoxide dismutase and Parkinson's disease. *Lancet* 335:1035–1036.

16. Pryor, W. A., and Squadrito, G. L. 1995. The chemistry of peroxynitrite: A product from the reaction of nitric oxide with superoxide. *Am. J. Physiol.* 268:L699–L722.

17. Candeias, L. P., Patel, K. B., Stratford, M. R. L., and Wardman, P. 1993. Free hydroxyl radicals are formed on the reaction between the neutrophil-derived species, superoxide anion and hypochlorous acid. *FEBS Lett.* 333:151–153.

18. Pryor, W. A. 1986. Oxy-radicals and related species: Their formation lifetimes, and reactions. *Annu. Rev. Physiol.* 48:657–663.

19. Kaur, H., and Halliwell, B. 1994. Detection of hydroxyl radicals by aromatic hydroxylation. *Methods Enzymol.* 233:67–82.

20. Floyd, R. A. 1983. Hydroxyl free-radical spin-adduct in rat brain synaptosomes. Observation on the reduction of the nitroxide. *Biochim. Biophys. Acta* 756:204–206.

21. Takemura, G., Onodera, T., and Ashraf, M. 1992. Quantification of hydroxyl radical and its lack of relevance to myocardial injury during early reperfusion and graded ischemia in rat hearts. *Circ. Res.* 71:96–105.

22. Richmond, R., Halliwell, B., Chauhan, J., and Darbue, A. 1981. Superoxide-dependent formation of hydroxyl radicals: Detection of hydroxyl radicals by the hydroxylation of aromatic compounds. *Anal. Biochem.* 118:328–335.

23. Halliwell, B., Grootveld, M., and Gutteridge, J. M. C. 1988. Methods for the measurement of hydroxyl radicals in biomedical systems: Deoxyribose degradation and aromatic hydroxylation. *Methods Biochem. Anal.* 33:59–90.

24. Schneider, J. E., Browning, M. N., Zhu, X., Eneff, K. L., and Floyd, R. A. 1989. Characterization of hydroxyl free radical mediated damage to plasmid pBR322 DNA. *Mutat. Res.* 214:23–31.

25. Maskos, Z., Rush, J. D., and Koppenol, W. H. 1990. The hydroxylation of the salicylate anion by a Fenton reaction and T-radiolysis: A consideration of the respective mechanisms. *Free Radic. Biol. Med.* 8:153–162.

26. Takemura, G., Onodera, T., Millard, R. W., and Ashraf, M. 1993. Demonstration of hydroxyl radical and its role in hydrogen peroxide induced myocardial injury: Hydroxyl radical dependent and independent mechanisms. *Free Radic. Biol. Med.* 15:13–25.

27. Liu, D. 1993. Generation and detection of hydroxyl radical In vivo in rat spinal cord by microdialysis administration of Fenton's reagents and microdialysis sampling. *J. Biochem. Biophys. Methods* 27:281–291.

28. Floyd, R. A., Watson, J. J., and Wong, P. K. 1984. Sensitive assay of hydroxyl free radical formation utilizing high pressure liquid chromatography with electrochemical detection of phenol and salicylate hydroxylation products *J. Biochem. Biophys. Methods* 10:221–235.

29. Giovanni, A., Liang, L. P., Hastings, T. G., and Zigmond, M. J. 1995. Estimating hydroxyl radical content in rat brain using systemic and Iintraventricular salicylate: Impact of methamphetamine. *J. Neurochem.* 64:1819–1825.

30. Hall, E. D., Andrus, P. K., Althaus, J. S., and VonVoigtlander, P. F. 1993. Hydroxyl radical production and lipid peroxidation parallels selective post-ischemic vulnerability in gerbil brain. *J. Neurosci. Res.* 34:107–112.

31. Smith, S. L., Andrus, P. K., Zhang, J.-R., and Hall, E. D. 1994. Direct measurement of hydroxyl radicals, lipid peroxidation, and blood–brain barrier disruption following unilateral cortical impact head injury in the rat. *J. Neurotrauma* 11:393–404.

32. Grammas, P., Liu, G.-J, Wood, K., and Floyd, R. A. 1993. Anoxia/ reoxygenation induces hydroxyl free radical formation in brain microvessels. *Free Radic. Biol. Med.* 14:553–557.

33. Goodwin, B.L. 1979. Phenylalanine hydroxylase. In *Aromatic Amino Acid Hydroxylase and Mental Disease,* ed. M. Youdim, pp. 5–24. Oxford: John Wiley and Sons.

34. Perry, T. L., Hestrin, M., MacDougall, L., and Hansen, S. 1966. Urinary amines of intestinal bacterial origin. *Clin. Chim. Acta* 14:116–123.

35. Boulton, A. A., Dyck, L. E., and Durden, D. A. 1974. Hydroxylation of beta-phenylethylamine in the rat. *Life Sci.* 15:1673–1683.

36. Ishimitsu, S., Fujimoto, S., and Ohara, A. 1986. Determination of *m*-tyrosine and *o*-tyrosine in human serum by high-performance liquid chromatography with fluorometric detection. *J. Chromatogr.* 378:222–225.

37. Kaur, H., and Halliwell, B. 1994. Aromatic hydroxylation of phenylalanine as an assay for hydroxyl

radicals: Measurement of hydroxyl radical formation from ozone and in blood from premature babies using improved HPLC methodology. *Anal. Biochem.* 220:11–15.

38. Konno, R., and Yasumura, Y. 1984. Involvement of D-amino acid oxidase in D-amino acid utilization in the mouse. J. Nutr. 114:1617–1621.

39. Friedman, M., and Gumbmann, M. R. 1984. The nutritive value and safety of D-phenylalanine and D-tyrosine in mice. *J. Nutr.* 114:2089–2096.

40. Anonymous 1996. Dietary D-tyrosine as an antimetabolite in mice. *Nutr. Rev.* 43:156–158.

41. Ungerstedt, U. 1984. Measurement of neurotransmitter release by intracranial dialysis. In *Measurement of Neurotransmitter Release In Vivo*, ed. C. A. Marsden, pp. 81–105. New York: John Wiley and Sons.

42. Robinson, T. E., and Justice, J. B., Jr., eds. 1991. *Microdialysis in the Neurosciences. Techniques in The Behavioral and Neural Sciences*, Vol. 7. Amsterdam: Elsevier.

43. Ikeda, M., Levitt, M., and Undenfriend, S. 1967. Phenylalanine as a substrate and inhibitor of tyrosine hydroxylase. *Arch. Biochem. Biophys.* 120:420–427.

44. Katz, I., Lloyd, T., and Kaufman, S. 1976. Studies on phenylalanine and tyrosine hydroxylation by rat brain tyrosine hydroxylase. *Biochim. Biophys. Acta* 445:567–578.

45. During, M. J., Acworth, I. N., and Wurtman, R. J. 1988. Phenylalanine administration influences dopamine release in the rat's corpus striatum. *Neurosci. Lett.* 93:91–95.

46. Sontag, G., Bernweiser, I., and Krach, C. 1996. *HPLC with Electrode Array Detection and Its Applications in Analytical Food Chemistry.* Progress in HPLC, Vol. 7, eds. I. Acworth, M. Naoi, S. Parvez, and S. H. Parvez. Amsterdam: VS Press.

47. Ghiselli, A., Laurenti, O., De Mattia, G., Maiani, G., and Ferro-Luzzi, A. 1992. Salicylate hydroxylation as an early marker of in vivo oxidative stress in diabetic patients. *Free Radic. Biol. Med.* 13:621–626.

48. McCabe, D. R., Maher, T. J., and Acworth, I. N. 1996. Improved method for the estimation of hydroxyl free radical levels in vivo based on liquid chromatography with electrochemical detection. *J. Chromatogr.* in press.

49. Powell, S. R., Hall, D., Aiuto, L., Wapnir, R. A., Teichberg, S., and Tortolani, A. J. 1994. Zinc improves postischemic recovery of isolated rat hearts through inhibition of oxidative stress. *Am. J. Physiol.* 266:H2497–H2507.

50. Udassin, R., Ariel, I., Haskel, Y., Kitrossky, N., and Chevion, M. 1991 Salicylate as an in vivo free radical trap: Studies on ischemic insult to the rat intestine. *Free Radic. Biol. Med.* 10:1–6.

51. Globus, M. Y.-T., Busto, R., Lin, B., Schnippering, H., and Ginsberg, M. D. 1995. Detection of free radical activity during transient global ischemia and recirculation: Effects of intraischemic brain temperature modulation, *J. Neurochem.* 65:1250–1256.

52. Globus, M. Y.-T., Alonso, O., Dietrich, W. D., Busto, R., and Ginsberg, M. D. 1995. Glutamate release and free radical production following brain injury: Effects of posttraumatic hypothermia. *J. Neurochem.* 65:1704–1711.

53. Olano, M., Song, D., Murphy, S., Wilson, D. F., and Pastuszko, A. 1995. Relationships of dopamine, cortical oxygen tension, and hydroxyl radicals in brain of newborn piglets during posthypoxic recovery. *J. Neurochem.* 65:1205–1212.

54. Shen, Y., and Sangiah, S. 1995. N+, K+-ATPase, glutathione, and hydroxyl free radicals in cadmium chloride testicular toxicity in mice. *Arch. Environ. Contam. Toxicol.* 29:174–179.

55. Patthy, M., Kiraly, I., and Sziraki, I. 1995. Separation of dihydroxybenzoates, indicators of in vivo hydroxyl radical formation, in the presence of transmitter amines and some metabolites in rodent brain, using high-performance liquid chromatography with electrochemical detection. *J. Chromatogr. B.* 664:247–252.

56. Acworth, I. N., and Gamache, P. G. 1996. The coulometric electrode array for use in HPLC analysis. *Am. Lab.* 5:33–38.

57. Acworth, I. N., and Bowers, M. 1996. An introduction to HPLC-based electrochemical detection: From single electrode to multi-electrode Aarrays. In *Progress in HPLC*, Vol. 6, eds. I. N. Acworth, M. Naoi, H. Parvez, and S. Parvez, pp. 1–55. Amsterdam: VS Press.

58. Svendsen, C. N. 1993. Multi-electrode array detectors in high-performance liquid chromatography: A new dimension in electrochemical analysis. *Analyst* 118:123–129.

59. Brooks, P. M., and Day, R. O. 1991. Nonsteroidal anti-inflammatory drugs–Differences and similarities. *N. Engl. J. Med.* 324:1716–1725.

60. Montgomery, J., Ste-Marie, L., Boismenu, D., and Vachon, L. 1995. Hydroxylation of aromatic compounds as indices of hydroxyl radical production: A cautionary note revisited. *Free Radic. Biol. Med.* 19:927–933.

61. Chance, B., Sies, H., and Boveris, A. 1979. Hydroperoxide metabolism in mammalian organs. *Physiol. Rev.* 59:527–605.

62. Kehrer, J. P. 1993. Free radicals as mediators of tissue injury and disease. *Crit. Rev. Toxicol.* 23:21–48.

63. Acworth, I. N., and Bailey, B., eds. 1995. *The Handbook of Oxidative Metabolism.* Chelmsford, MA: ESA, Inc.

64. Halliwell, B., and Gutteridge, J. M. C. 1986. Oxygen free radicals and iron in relation to biology and medicine: Some problems and concepts. *Arch. Biochem. Biophys.* 246:501–514.
65. Fridovich, I. 1989. Superoxide dismutases. An adaptation to a paramagnetic gas. *J. Biol. Chem.* 264:7761–7767.
66. Manning, P., McNeil, C. J., Hillhouse, E. W., and Cooper, J. M. 1996. Direct, real time measurement of free radical production by brain cells in culture *Anal. Chem.* Vol 68 1858–1864.
67. McNeil, C. J., Greenough, K. R., Weeks, P. A., and Self, C. H. 1992. Electrochemical sensors for direct reagentless measurement of superoxide production by human neutrophils. *Free Radic. Biol. Med.* 17:399–406.
68. Dupuy, C., Virion, A., Ohayon, R., Kamiewski, J. M., Deme, D., and Pommier, J. 1991. Mechanism of hydrogen peroxide formation catalyzed by NADPH oxidase in thyroid plasma membrane. *J. Biol. Chem.* 266:3739–3743.
69. Corbett, J. T. 1989. The scopoletin assay for hydrogen peroxide: A review and a better method. *J. Biochem. Biophys. Methods* 256:297–308.
70. Varma, S. D. 1989. Radio-isotopic determination of sub-nanomolar amounts of peroxides. *Free Radic. Res. Commun.* 5:359–368.
71. Greaney, M. D., Marshall, D. L., Bailey, B. A., and Acworth, I. N. 1993. Improved method for the routine analysis of acetylcholine release in vivo: Quantitation in the presence and absence of esterase inhibitor. *J. Chromatogr.* 622:125–135.
72. Motohashi, N., and Mori, I. 1988. High-performance liquid chromatography-electrochemical detection of singlet oxygen by reaction with 2,2,6,6-tetramethyl-4-piperidone. *J. Chromatogr.* 465:417–421.
73. Pritsos, C. A., Constantinides, P. P., Tritton, T. R., Heimbrrok, D. C., and Sartorelli, A. C. 1985. Use of high-performance liquid chromatography to detect hydroxyl and superoxide radicals generated from mitomycin c. *Anal. Biochem.* 150:294–299.
74. Floyd, R. A., Lewis, C. A., and Wong, P. K. 1984. High-pressure liquid chromatography-electrochemical detection of oxygen free radicals. *Methods Enzymol.* 105:231–237.
75. Feldman, P. L., Griffith, O. W., and Stuehr, D. J. 1993. The surprising life of nitric oxide. *Chem. Eng. News* December:26–38.
76. Snyder, S. H., and Bredt, D. S. 1992. Biological roles of nitric oxide. *Sci. Am.* May:68–77.
77. Kostka, P. 1992. Free radicals (nitric oxide). *Anal. Chem.* 67:411R–416R.
78. Moncada, S., Palmer, R. M. J., and Higgs, E. A. 1991. Nitric oxide: Physiology, pathophysiology, and pharmacology. *Pharmacol. Rev.* 43:109–141.
79. Koshland, D. E. 1992. The molecule of the year. *Science* 258:1861.
80. Rubbo, H., Radi, R., Trujillo, M., Telleri, R., Kalynararaman, B., Barnes, S., Kirk, M., and Freeman, B. A. 1995. Nitric oxide regulation of superoxide and peroxynitrite-dependent lipid peroxidation: Formation of novel nitrogen-containing oxidized lipid derivatives. *J. Biol. Chem.* 269:26066–26075.
81. Rubbo, H., Darley-Usmar, V., and Freeman, B. 1996. Nitric oxide regulation of tissue free radical injury. *Chem. Res. Toxicol.* 9:809–820.
82. Moncada, S., and Higgs, A. 1993. The L-arginine-nitric oxide pathway. *N. Engl. J. Med.* 329:2002–2011.
83. Radi, R. 1996. Reactions of nitric oxide with metalloproteins. *Chem. Res. Toxicol.* 9:828–835.
84. Eiserich, J. P., Butler, J., Van Der Vleit, A., Cross, C. E., and Halliwell, B. 1995. Nitric oxide rapidly scavenges tyrosine and tryptophan radicals. *Biochem. J.* 310:745–749.
85. Brune, D., Dimmeler, S., Molina, V., and Lapetina, E. G. 1994. Nitric oxide: A signal for ADP-ribosylation of proteins. *Life Sci.* 54:61–70.
86. Wink, D. A., Kasprzak, K. S., Maragos, C. M., Elespura, R. K., Misra, M., Dunams, T. M., Cebula, T. A., Koch, W. H., Andrews, A. W., and Allen, J. S. 1991. DNA deaminating ability and genotoxicity of nitric oxide and its progenitors. *Science* 254:1001–1003.
87. Nguyen, T., Brunson, D., Crespi, C. L., Penman, B. W., Wishnok, J. S. and Tannenbaum, S. R. 1992. DNA damage and mutation in human cells exposed to nitric oxide in vitro. *Proc. Natl. Acad. Sci. U.S.A.* 89:3030–3034.
88. Craven, P. A., and DeRubertis, F. R. 1978. Restoration of responsiveness of purified guanylate cyclase to nitrosoguanidine, nitric oxide and related activators by heme and hemeproteins. Evidence for the involvement of paramagnetic nitrosyl-heme complex in enzyme activation. *J. Biol. Chem.* 253:8433–8443.
89. Stamler, J. S., Singel, D. J., and Loscalzo, J. 1992. Biochemistry of nitric oxide and its redox-activated forms. *Science* 258:1898–1902.
90. Crow, J. P., and Beckman, J. S. 1995. Reactions between nitric oxide, superoxide, and peroxynitrite: Footprints of peroxynitrite in vivo. *Adv. Pharmacol.* 34:17–43.
91. Furchgott, R. F. 1984. The role of endothelium in responses of vascular smooth muscle to drugs. *Annu. Rev. Pharmacol. Toxicol.* 24:175–197.
92. Rubanyi, G. M., Lorenz, R. R., Vanhoutte, P. M. 1985. Bioassay of endothelium-derived relaxing factor(s): Inactivation by catecholamines. *Am. J. Physiol.* 249:H95–H101.

93. Griffith, T. M., Edwards, D. H., Lewis, M. J., Newby, A. C., and Henderson, A. H. 1984. The nature of endothelium-derived vascular relaxant factor. *Nature* 308:645–647.

94. Rapaport, R. M., and Murad, F. 1983. Agonist induced endothelium-dependent relaxation in rat thoracic aorta may be mediated through cyclic GMP. *Circ. Res.* 52:352–357.

95. Martin, W., Villani, G. M., Jothianandan, D., and Furchgott, R. F. 1985. Selective blockade of endothelium-dependent and glyceryl trinitrate-induced relaxation by hemoglobin and by methylene blue in the rabbit aorta. *J. Pharmacol. Exp. Ther.* 232:708–716.

96. Cocks, T. M., Angus, J. A., Campbell, J. H., and Campbell, G. R. 1985. Release and properties of endothelium-derived relaxation factor (EDRF) from endothelial cells in culture. *J. Cell Physiol.* 123:310–320.

97. Palacios, M., Knowles, R. G., Palmer, R. M. J., and Moncada, S. 1989. Nitric oxide from L-arginine stimulates the soluble guanylate cyclase in adrenal glands. *Biochem. Biophys. Res. Commun.* 165:802–809.

98. Ignarro, L. J., Burke, T. M., Wood, K. S., Wolin, M. S., and Kadowitz, P. J. 1984. Association between cGMP accumulation and acetylcholine-elicited relaxation of bovine intrapulmonary artery. *J. Pharmacol. Exp. Ther.* 228:682–690.

99. Moncada, S., and Palmer, R. M. J. 1990. The L-arginine: Nitric oxide pathway in the vessel wall. In *Nitric Oxide from L-Arginine: A Bioregulatory System*, eds. S. Moncada and E. A. Higgs, pp. 19–33. Amsterdam: Elsevier.

100. Wang, S. S., and Maher, T. J. 1992. Changes in hippocampal ECF amino acid levels in anesthetized rats during kainic acid-induced seizures: A microdialysis study. *Pharmacologist* 34:190.

101. Bhardwaj, A., Northington, F. J., Koehler, R. C., Stiefel, T., Hanley, D. F., and Traystman, R. J. 1995. Adenosine modulates *N*-methyl-D-aspartate-stimulated hippocampal nitric oxide production in vivo. *Stroke* 26:1627–1633.

102. Aoki, T. 1990. Continuous flow determination of nitrite with membrane separation/chemiluminescence detection. *Biomed. Chromatogr.* 4:128–130.

103. Wennmalm, A., Benthin, G., and Petersson, A.-S. 1992. Dependence of the metabolism of nitric oxide (NO) in healthy human whole blood on the oxygenation of its red cell haemoglobin. *Br. J. Pharmacol.* 106:507–508.

104. Cantilena, L. R., Smith, R. P., Frasur, S., Kruszyna, H., Kruszyna, R., and Wilcox, D. E. 1992. Nitric oxide hemoglobin in patients receiving nitroglycerin as detected by electron paramagnetic resonance spectroscopy. *J. Lab. Clin. Med.* 120:902–907.

105. Balcioglu, A., and Maher, T. J. 1993. Determination of kainic acid-induced release of nitric oxide using a novel hemoglobin trapping technique with microdialysis. *J. Neurochem.* 61:2311–2313.

106. Lai, C.-S., and Komarov, A. M. 1994. Spin trapping of nitric oxide produced in vivo in septic shock mice. *FEBS Lett.* 345:120–124.

107. Henry, Y., Lepoivre, M., Drapier, J.-C., Ducrocq, C., Boucher, J.-C., and Guissani, A. 1993. EPR characterization of molecular targets for NO in mammalian cells and organelles. *FASEB J.* 7:1124–1134.

108. Arroya, C. M., Forray, C., El-Fakahany, E. E., and Rosen, G. M. 1990. Receptor-mediated generation of an EDRF-like intermediate in a neuronal cell line detected by spin trapping techniques. *Biochem. Biophys. Res. Commun.* 170:1177–1183.

109. Rosseli, M., Imthurm, B., Macas, E., Keller, P. J., and Dubey, R. K. 1994. Circulating nitrite/nitrate levels increase with follicular development: Indirect evidence for estradiol mediated NO release. *Biochem. Biophys. Res. Commun.* 202:1543–1552.

110. Hibbs, J. B., Westenfelder, C., Taintor, R., Vavrin, Z., Kabiltz, C., Baranowski, R. L., Ward, J. H., Menlove, R. L., McMurry, M. P., Kushner, J. P., and Samlowski, W. E. 1992. Evidence for cytokine-inducible nitric oxide synthesis from L-arginine in patients receiving interleukin-2 therapy. *J. Clin. Invest.* 89:867–877.

111. Shi, Y., Li, H.-Q., Shen, C.-K., Wang, J.-H., Qin, S.-W., Liu, R., and Pan, J. 1993. Plasma nitric oxide levels in newborn infants with sepsis. *J. Pediatr.* 123:435–438.

112. Pratt, P. F., Nithipatikom, K., and Campbell, W. B. 1995. Simultaneous determination of nitrate and nitrite in biological samples by multichannel flow injection analysis. *Anal. Biochem.* 231:383–386.

113. Malinski, T., and Taha, Z. 1992. Nitric oxide release from a single cell measured *in situ* by a porphyrinic-based microsensor. *Nature* 358, 676–678.

114. O'Hare, T. and O'Shea, M. 1996. Internet communications.

115. Friedemann, M. N., Robinson, S. W., and Gerhardt, G. A. 1996. *o*-Phenylenediamine-modified carbon fiber electrodes for the detection of nitric oxide. *Anal. Chem.* 68:2621–2628.

116. Strehlitz, B., Grundig, B., Schumacher, W., Kroneck, P. M. H., Vorlop, K.-D., and Kotte, H. 1996. A nitrite sensor based on a highly sensitive nitrite reductase mediator-coupled amperometric detection. *Anal. Chem.* 68:807–816.

117. Beckman, J. S., Chen, J., Ischiropoulos, H., and Crow, J. P. 1994. Oxidative chemistry of peroxynitrite. *Methods Enzymol.* 233:229–240.

118. Pryor, W. A., and Squadrito, G. L. 1995. The chemistry of peroxynitrite: A product from the reaction of nitric oxide with superoxide. *Am. J. Physiol.* 268:L699–L722.
119. Radi, R., Beckman, J. S., Bush, K. M., and Freeman, B. A. 1991. Peroxynitrite oxidation of sulfhydryls. *J. Biol. Chem.* 266:4244–4250.
120. Pryor, W. A., Jin, X., and Squadrito, G. L. 1994. One- and two-electron oxidations of methionine by peroxynitrite. *Proc. Natl. Acad. Sci. U.S.A.* 91:11173–11177.
121. Beckman, J. S., Beckman, T. W., Chen, J., Marshall, P. A., and Freeman, B. A. 1990. Apparent hydroxyl radical production by peroxynitrite: Implications for endothelial injury from nitric oxide and superoxide *Proc. Natl. Acad. Sci. U.S.A.* 87:1620–1624.
122. Radi, R., Beckman, J. S., Bush, K. M., and Freeman, B. A. 1991. Peroxynitrite-induced membrane lipid peroxidation: The cytotoxic potential of superoxide and nitric oxide. *Arch. Biochem. Biophys.* 288:481–487.
123. Rubbo, H., Radi, R., Trujillo, M., Telleri, R., Kalyanaraman, B., Barnes, S., Kirk, M., and Freeman, B. A. 1994. Nitric oxide regulation of superoxide and peroxynitrite-dependent lipid peroxidation. *J. Biol. Chem.* 269:26066–26075.
124. Bartlett, D., Church, D. F., Bounds, P. L., and Koppenol, W. H. 1995. The kinetics of oxidation of L-ascorbic acid by peroxynitrite. *Free Radic. Biol. Med.* 18:85–92.
125. Yermilov, V., Rubio, J., Becchi, M., Friesen, M. D., Pignatelli, B., and Ohshima, H. 1995. Formation of 8-nitroguanine with peroxynitrite in vitro. *Carcinogenesis* 16:2045–2050.
126. Yermilov, V., Rubio, J., and Ohshima, H. 1995. Formation of 8-nitroguanine in DNA treated with peroxynitrite in vitro and its rapid removal from DNA by depurination. *FEBS Lett.* 376:207–210.
127. Ischiropoulos, H., and Al-Mehdi, A. B. 1995. Peroxynitrite-mediated oxidative protein modifications. *FEBS Lett.* 364:279–282.
128. Beckman, J. S., Ye, Y. Z., Anderson, P. G., Chen, J., Accavitti, M. A., Tarpey, M. M., and White, C. R. 1994. Extensive nitration of protein tyrosines in human atherosclerosis detected by immunohistochemistry. *Biol. Chem. Hoppe-Seyler* 375:81–88.
129. Crow, J. P., and Ischiropoulos, H. 1996. Detection and quantitation of nitrotyrosine residues in proteins: An in vivo marker of peroxynitrite, *Methods Enzymol.*
130. Maruyama, W., Hashizume, Y., Matsubara, K., and Naoi, M. 1996. Indentification of 3-nitro-L-tyrosine, a product of nitric oxide and superoxide, as an indicator of oxidative stress in the human brain. *J. Chromatogr. B* 676:153–158.
131. Uppa, R. M., Squadrito, G. L., and Pryor, W. A. 1996. Acceleration of peroxynitrite oxidations by carbon dioxide. *Arch. Biochem. Biophys.* 327:335–343.
132. Lymar, S. V., and Hurst, J. K. 1995. Rapid reaction between peroxynitrite ion and carbon dioxide: Implications for biological activity. *J. Am. Chem. Soc.* 117:8867–8868.
133. Lymar, S. V., Jiang, Q., and Hurst, J. K. 1996. Mechanism of carbon dioxide-catalyzed oxidation of tyrosine by peroxynitrite. *Biochemistry.*
134. Lymar, S. V. and Hurst, J. K. Carbon dioxide: Physiological catalyst for peroxynitrite-mediated cellular damage or cellular protectant. *Chem. Res. Toxicol.* 9:845–850.
135. Pryor, W. A. 1988. Why is the hydroxyl radical the only radical that commonly adds to DNA? Hypothesis: It has a rare combination of high electrophilicity, high thermochemical reactivity, and a mode of production that can occur near DNA. *Free Radic. Biol. Med.* 4:219–223.
136. Breen, A. P., and Murphy, J. A. 1995. Reactions of the oxyl radicals with DNA. *Free Radic. Biol. Med.* 18:1033–1077.
137. Marnett, L. J., and Burcham, P. C. 1993. Endogenous DNA adducts: Potential and paradox. *Chem. Res. Toxicol.* 6:771–785.
138. Hiraoka, W., Kuwabara, M., Sato, F., Matsuda, A., and Ueda, T. 1990. Free-radical reactions induced by OH-radical attack on cytosine-related compounds: A study by a method combining ESR, spin trapping and HPLC. *Nucleic Acid Res.* 18:1217–1223.
139. Kuwabara, M., Ohshima, H., Sato, F., Ono, A., and Matsuda, A. 1993. Spin trapping detection of Precursors of hydroxyl-radical-induced DNA damage: Identification of precursor radicals of DNA strand breaks in oligo(dC)$_{10}$ and oligo (dT)$_{10}$. *Biochemistry* 32:10599–10606.
140. Steenken, S. 1989. Purine bases, nucleosides and nucleotides: Aqueous solution redox chemistry and transformation reactions of their radical cations and e$^-$ and OH adducts. *Chem. Rev.* 89:503–520.
141. von Sonntag, C., ed. 1987. *The Chemical Basis of Radiation Biology*, pp. 116–166. London: Taylor and Francis.
142. Wiseman, H., and Halliwell, B. 1996. Damage to DNA by reactive oxygen and nitrogen species: Role in inflammatory disease and progression of cancer. *Biochem. J.* 313:17–29.
143. Shibutani, S., Takeshita, M., and Grollman, A. P. 1991. Insertion of specific bases during DNA synthesis past the oxidation-damaged 8-oxo-dG. *Nature* 349:431–434.
144. Ohtsuka, E., and Nishimura, S. 1987. Misreading of DNA templates containing 8-hydroxydeoxyguanosine as the modified base and at adjacent residues. *Nature* 327:77–99.

145. Ngheim, Y., Cabrera, M., Cupples, C. G., and Miller, J. H. 1989. The *mut*Y gene: A mutator locus in *E. coli* that generates G.C. T.A. transversion. *Proc. Natl. Acad. Sci. U.S.A.* 85:2709–2713.

146. Ames, B. N., and Gold, L. S. 1991. Endogenous mutagens and the causes of aging and cancer. *Mutat. Res. Fundam. Mol. Mech. Mutagen.* 250:3–16.

147. Richter, C., Park, J.-W., and Ames, B. N. 1988. Normal oxidative damage to mitochondrial and nuclear DNA is extensive. *Proc. Natl. Acad. Sci. U.S.A.* 85:6465–6467.

148. Moore, P., and Strauss, B. S. 1979. Sites of inhibition of in vitro DNA synthesis in carcinogen and UV-treated ΦX174DNA. *Nature* 278:664–666.

149. Echols, H., and Goodman, M. F. 1991. Fidelity mechanisms in DNA replication. *Ann. Rev. Biochem.* 60:477–511.

150. Strauss, B. S. 1991. The "A Rule" of mutagenic specificity: A consequence of DNA polymerase bypass of non-instructional lesions? *BioEssays* 13:79–84.

151. van Delft, J. H. M., van Winden, M. J. M., van den Ende, A. M. C., and Baan, R. A. 1993. Determining N7-alkylguanine adducts by immunochemical methods and HPLC with electrochemical detection: Applications in animal studies and in monitoring human exposure to alkylating agents. *Environ. Health Perspect.* 99:25–32.

152. Dizdaroglu, M., and Gajewski, E. 1990. Selected-ion mass spectrometry: Assays of oxidative DNA damage. *Methods Enzymol.* 186, 530–544.

153. Dizdaroglu, M. 1994. Chemical determination of oxidative damage by gas chromatography-mass spectrometry. *Methods Enzymol.* 234:3–16.

154. Cushnir, J. R., Naylor, S., Lamb, J. H., and Farmer, P. B. 1993. Tandem mass spectrophotometric approaches for the analysis of alkylguanines in human urine. *Org. Mass Spectrom.* 28:552–558.

155. Randeroth, K., Reddy, M. V., and Gupta, R. C. 1981. ^{32}P-Labeling for DNA damage. *Proc. Natl. Acad. Sci. U.S.A.* 78:6126–6129.

156. Gupta, R. C., Reddy, M. V., and Randerath, K. 1982. ^{32}P-Postlabeling analysis of non-radioactive aromatic carcinogenic DNA adducts. *Carcinogenesis* 3:1081–1092.

157. Cadet, J., Odin, F., Mouret, J.-F., Polverelli, M., Audic, A., Giacomoni, P., Favier, A., and Richard, M.-J. 1992. Chemical and biochemical postlabeling methods for singling out specific oxidative DNA lesions. *Mutat. Res.* 275:343–354.

158. Reddy, M. V., and Randerath, K. 1986. Nuclease P1-mediated enhancement of sensitivity of ^{32}P-postlabeling test for structurally diverse DNA adducts. *Carcinogenesis* 7:1543–1551.

159. Cathcart, R., Schwiers, E., Saul, R. L., and Ames, B. N. 1984. Thymine glycol and thymidine glycol in human and rat urine: A possible assay for oxidative DNA damage. *Proc. Natl. Acad. Sci. U.S.A.* 81:5633–5637.

160. Cummings, J., French, R. C., and Smyth, J. F. 1993. Application of high-performance liquid chromatography for the recognition of covalent nucleic acid modification with anticancer drugs. *J. Chromatogr.* 618:251–276.

161. Floyd, R. A., Watson, J. J., Wong, P. K., Altmiller, D. H., and Rickard, R. C. 1986. Hydroxyl free radical adduct of deoxyguanosine: Sensitive detection and mechanisms of formation. *Free Radic. Res. Commun.* 1:163–172.

162. Xie, F., and Duda, C. 1994. Determination of an oxidation product of guanine in DNA by microbore liquid chromatography. *Curr. Sep.* 13:18–21.

163. Shigenaga, M. K., Park, J.-W., Cundy, K.-C., Gimeno, C. J., and Ames, B. N. 1990. In vivo oxidative DNA damage: Measurement of 8-hydroxy-2'deoxyguanosine in human urine by HPLC with electrochemical detection. *Methods Enzymol.* 186:521–530.

164. Takeuchi, T., and Morimoto, K. 1993. Increased formation of 8-hydroxy-deoxyguanosine, an oxidative DNA damage, in lymphoblasts from Fanconi's anemia patients due to possible catalase deficiency. *Carcinogenesis* 14:1115–1120.

165. Fischer-Nielson, A., Jeding, I. B., and Loft, S. 1994. Irradiation-induced formation of 8-hydroxy-2'-deoxyguanosine and its prevention by scavengers. *Carcinogenesis* 15:1609–1612.

166. Shigenaga, M. K., Aboujaoude, E. N., Chen, Q., and Ames, B. N. 1994. Assays of oxidative DNA damage biomarkers 8-oxo-2'-deoxyguanosine and 8-oxoguanine in nuclear DNA and biological fluids by high performance liquid chromatography and electrochemical detection. *Methods Enzymol.* 234:16–33.

167. Adachi, S., Zeisig, M., and Moller, L. 1995. Improvements in the analytical method for 8-hydroxydeoxyguanosine in nuclear DNA. *Carcinogenesis* 16:253–258.

168. Wagner, J. R., Hu, C.-C., and Ames, B. N. 1992. Endogenous oxidative damage of deoxycytidine in DNA. *Proc. Natl. Acad. Sci. U.S.A.* 89:3380–3384.

169. Douki, T., and Ames, B. N. 1994. An HPLC-EC assay for 1,N^2-propano adducts of 2'-deoxyguanosine with 4-hydroxynonenal and other α,β-unsaturated aldehydes. *Chem. Res. Toxicol.* 7:511–518.

170. Gamache, P. H., McCabe, D. R., Parvez, H., Parvez, S., and Acworth, I. N. 1997. The measurement of markers of oxidative damage, antioxidants and related compounds using HPLC and coulometric

array analysis. In *Progress in HPLC*, Vol. 6, eds. I. N. Acworth, M. Naoi, H. Parvez, and S. Parvez. pp. 99–126. Amsterdam: VS Press.

171. Mecocci, P., MacGarvey, U., Kaufman, A., Koontz, D., Shoffner, J., Wallace, D., and Beal, M. F. 1993. Oxidative damage to mitochondrial DNA shows marked age-dependent increases in human brain. *Ann. Neurol.* 34:609–616.

172. Shigenaga, M. K., Gimeno, C. J., and Ames, B. N. 1989. Urinary 8-hydroxy-2'-deoxyguanosine as a biological marker of in vivo oxidative damage, *Proc. Natl. Acad. Sci. U.S.A.* 86:9697–9701.

173. Vigue, C. A., Frei, B., Shigenaga, M. K., Ames, B. N., Packer, L., and Brooks, G. A. 1993. Antioxidant stress and indexes of oxidative stress during consecutive days of exercise. *J. Appl. Physiol.* 75:566–573.

174. Faux, S. P., Gao, M., Aw, T. C. and Braithwaite, R. A. 1994. Molecular epidemiological studies in workers exposed to chromium-containing compounds. *Clin. Chem.* 40:1454–1456.

175. Lunec, J., Herbert, K., Blount, S., Griffiths, H. R., and Emery, P. 1994. 8-Hydroxydeoxyguanosine, a marker of oxidative DNA damage in systemic lupus erythematosus. *FEBS Lett.* 348:131–138.

176. Brown, R. K., McBurney, A., Lunec, J., and Kelly, F. J. 1995. Oxidative damage to DNA in patients with cystic fibrosis. *Free Radic. Biol. Med.* 18:801–806.

177. Fraga, C. G., Shigenaga, M. K., Park, J.-W., Degan, P. and Ames, B. N. 1990. Oxidative damage to DNA during aging: 8-Hydroxy-2'-deoxyguanosine in rat organ DNA and urine. *Proc. Natl. Acad. Sci. U.S.A.* 87:4533–4537.

178. Shigenaga, M. K. and Ames, B. N. 1991. Assays for 8-hydroxy-2'-deoxyguanosine: A biomarker of in vivo oxidative DNA damage. *Free Radic. Biol. Med.* 10:211–216.

179. Degan, P., Shigenaga, M. K., Park, E.-M., Alperin, P. E., and Ames, B. N. 1991. Immunoaffinity isolation of urinary 8-hydroxy-2'-deoxyguanosine and 8-hydroxyguanine and quantification of 8-hydroxy-2'-deoxyguanosine in DNA by polyclonal antibodies. *Carcinogenesis* 12:865–871.

180. Park, E.-M., Shigenaga, M. K., Degan, P., Korn, T. S., Kitzler, J. W., Wehr, C. M., Kolachana, P., and Ames, B. N. 1992. Assay of excised oxidative DNA lesions: Isolation of 8-oxoguanine and its nucleoside Derivatives from biological fluids with a monoclonal antibody column. *Proc. Natl. Acad. Sci. U.S.A.* 89:3375–3379.

181. Loft, S., Vistisen, K., Ewertz, M., Tjonneland, K., Overvad, K., and Poulsen, E. 1992. Oxidative DNA damage estimated by 8-hydroxyguanosine excretion in humans: Influence of smoking, gender and body mass index. *Carcinogenesis* 13:2241–2247.

182. Loft, S., Larsen, P. N., Rasmussen, A., Fischer-Nielsen, A., Bondesen, S., Kirkegaard, P., Rasmussen, L. S., Ejlersen, E., Tornøe, K., Bergholdt, R., and Poulsen, H. E. 1995. Oxidative DNA damage after transplantation of the liver and small intestine in pigs. *Transplantation* 59:16–20.

183. Verhagen, H., Poulsen, H. E., Loft, S., van Poppel, G., Willems, M. I., and van Bladeren, P. J. 1995. Reduction of oxidative DNA-damage in humans by Brussels sprouts. *Carcinogenesis* 16:969–970.

184. Lagorio, S., Tagesson, C., Forastiere, F., Iavarone, I., Axelson, O., and Carere, A. 1994. Exposure to benzene and urinary concentrations of 8-hydroxyguanosine, a biological marker of oxidative damage to DNA. *Occup. Environ. Med.* 51:739–743.

185. Tagesson, C., Kallberg, M., Klintenberg, C., and Starkhammar, H. 1995. Determination of urinary 8-hydroxydeoxyguanosine by automated coupled-column high performance liquid chromatography: A powerful technique for assaying in vivo oxidative DNA damage in cancer patients. *Eur. J. Cancer* 31A:934–940.

186. Teixeira, A. J. R., Gommers-Ampt, J. H., van de Werken, G., Westra, J. G., Stavenuiter, J. F. C., and de Jong, A. P. J. M. 1993. Method for the analysis of oxidized nucleosides by gas chromatography/mass spectrometry. *Anal. Biochem.* 214:474–483.

187. Teixeira, A. J .R., Ferreira, M. R., van Dijk, W. J., van de Werken, G. and de Jong, A. P. J. M. 1995. Analysis of 8-hydroxy-2'-deoxyguanosine in rat urine and liver DNA by stable isotope gas chromatography/mass spectrometry. *Anal. Biochem.* 226:307–319.

188. McCabe, D. R. A simple electrode array method for the analysis of 8OH2DG in human urine. ESA Application Note. In Preparation.

189. Smith, C. D., Carney, J. M., Tatsumo, T., Stadtman, E. R., Floyd, R. A., and Markesbery, W. R. 1992. Protein oxidation in aging brain. *Ann. N.Y. Acad. Sci.* 663:110–119.

190. Gutteridge, J. M. C., and Wilkins, S. 1983. Copper salt-dependent hydroxyl radical formation. Damage to proteins acting as antioxidants. *Biochem. Biophys. Acta* 759:38–41.

191. Levine, R. L., Oliver, C. N., Fulks, R. M., and Stadtman, E. R. 1981. Turnover of bacterial glutamine synthetase. Oxidative inactivation precedes proteolysis. *Proc. Natl. Acad. Sci. U.S.A.* 78:2120–2124.

192. Levine, R. L. 1983. Oxidative modification of glutamine synthetase. I. Inactivation is due to loss of one histidine residue. *J. Biol. Chem.* 258:11823–11827.

193. Amici, A., Levine, R. L., and Stadtman, E. R. 1989. Conversion of amino acid residues in proteins and amino acid homopolymers to carbonyl derivatives by metal catalyzed oxidation reactions. *J. Biol. Chem.* 264:3341–3346.

194. Ischiropoulos, H., and Al-Mehdi, A. B. 1995. Peroxynitrite-mediated oxidative protein modifications. *FEBS Lett.* 364:279–282.

195. Radi, R., Beckman, J. S., Bush, K. M. and Freeman, B. A. 1991. Peroxynitrite oxidation of sulfhydryls. *J. Biol. Chem.* 266:4244–4250.

196. Ischiropoulos, H., Zhu, L., Chen, J., Tsai, M., Martin, J. C., Smith, C. D., and Beckman, J. S. 1992. Peroxynitrite-mediated tyrosine nitration catalyzed by superoxide dismutase. *Arch. Biochem. Biophys.* 298:431–437.

197. van der Vleit, A., O'Neill, C. A., Halliwell, B., Cross, C. E., and Kaur, H. 1994. Aromatic hydroxylation and nitration of phenylalanine and tyrosine by peroxynitrite: Evidence for hydroxyl radical production from peroxynitrite. *FEBS Lett.* 339:89–92.

198. Kaur, H., and Halliwell, B. 1994. Evidence for nitric oxide-mediated oxidative damage in chronic inflammation: Nitrotyrosine in serum and synovial fluid from rheumatoid patients. *FEBS Lett.* 350:9–12.

199. van der Vleit, A., Eiserich, J. P., O'Neill, C. A., Halliwell, B., and Cross, C. E. 1995. Tyrosine modification by reactive nitrogen species. *Arch. Biochem. Biophys.* 319:341–349.

200. Fung, K., and Grosjean, D. 1981. Determination of nanogram amounts of carbonyls as 2,4-dinitrophenyl-hydrazones by high-performance liquid chromatograph *Anal. Chem.* 53:168–171.

201. Ayene, S. I., Al-Mehdi, A. B., and Fisher, A. B. 1993. Inhibition of lung tissue oxidation during ischemia/reperfusion by 2-mercaptopropionylglycine. *Arch. Biochem. Biophys.* 303:307–312.

202. Smith, C. D., Carney, J. M., Starke-Reed, P. E., Oliver, C. N., Stadtman, E. R., Floyd, R. A., and Markesbery, W. R. 1991. Excess brain protein oxidation and enzyme dysfunction in normal aging and in Alzheimer's disease. *Proc. Natl. Acad. Sci. U.S.A.* 88:10540–10543.

203. Chiavari, G., and Bergamini, C. 1985. High-performance liquid chromatography of carbonyl compounds as 2,4-dinitrophenylhydrazones with electrochemical detection. *J. Chromatogr.* 318:427–432.

204. Guilivi, C., and Davies, K. J. A. 1994. Dityrosine: A marker for oxidatively modified proteins and selective proteolysis. *Methods Enzymol.* 233:363–371.

205. Crow, J. P., and Beckman, J. S. 1995. Reactions between nitric oxide, superoxide, and peroxynitrite: Footprints of peroxynitrite in vivo. *Adv. Pharmacol.* 34:17–43.

206. Milbury, P. E., and Matson, W. R. 1995. Tyrosine and purine derived indicators of oxidative stress and free radical action. Abstr. 2–104. Annual Meeting of the Oxygen Society, Pasadena, CA.

207. Beal, M. F., Ferrante, R. J., Henshaw, R., Matthews, R. T., Chan, P. H., Kowall, N. W., Epstein, C. J., and Schulz, J. B. 1995. 3-Nitropropionic acid neurotoxicity is attenuated in copper/zinc superoxide dismutase transgenic mice. *J. Neurochem.* 65:919–922.

208. Crow, J. P., and Ischiropoulos, H. 1996. Detection and quantitation of nitrotyrosine residues in proteins: An in vivo marker of peroxynitrite. *Methods Enzymol.*, submitted.

209. Maruyama, W., Hashizume, Y., Matsubara, K., and Naoi, M. 1996. Identification of 3-nitro-L-tyrosine, a product of nitric oxide and superoxide, as an indicator of oxidative stress in the human brain. *J. Chromatogr. B* 676:153–158.

210. Schulz, J. B., Matthews, R. T., Muqit, M. M. K., Browne, S. E., and Beal, M. F. 1995. Inhibition of neuronal nitric oxide synthase by 7-nitroindazole protects against MPTP-induced neurotoxicity in mice. *J. Neurochem.* 64:936–939.

211. Halliwell, B., Murcia, M. A., Chirico, S., and Aruoma, O. I. 1995. Free radicals and antioxidants in food and in vivo: What they do and how they work. *Crit. Rev. Food Sci. Nutr.* 35:7–20.

212. Halliwell, B., and Chirico, S. 1993. Lipid peroxidation: Its mechanism, measurement, and significance. *Am. J. Clin. Nutr.* 57:715S–725S.

213. Radi, R., Beckman, J. S., Bush, K. M., and Freeman, B. A. 1991. Peroxynitrite-induced membrane lipid peroxidation: The cytotoxic potential of superoxide and nitric oxide. *Arch. Biochem. Biophys.* 288:481–484.

214. Frei, B., Kim, M. C., and Ames, B. N. 1990. Ubiquinol-10 is an effective lipid-soluble antioxidant at physiological concentrations. *Proc. Natl. Acad. Sci. U.S.A.* 87:4879–4883.

215. Esterbauer, H., Striegl, G., Puhl, H., and Rotheneder, M. 1989. Continuous monitoring of in vitro oxidation of human low density lipoprotein. *Free Radic. Res. Commun.* 6:67–75.

216. Tappel, A. L. 1968. Will antioxidant nutrients slow the aging process? *Geriatrics* 23:97–105.

217. Sies, H., and Murphy, M. E. 1991. Role of tocopherols in the protection of biological systems against oxidative damage. *J. Photochem. Photobiol. B* 8:211–224.

218. Kagan, V. E., Serbinova, E. A., and Packer, L. 1990. Antioxidant effects of ubiquinones in microsomes and mitochondria are mediated by tocopherol recycling. *Biochem. Biophys. Res. Commun.* 169:851–857.

219. Rubbo, H., Radi, R., Trujillo, M., Telleri, R., Kalyanaraman, B., Barnes, S., Kirk, M., and Freeman, B. A. 1994. Nitric oxide regulation of superoxide and peroxynitrite-dependent lipid peroxidation. *J. Biol. Chem.* 269:26066–26075.

220. Pryor, W. A. 1989. On the detection of lipid hydroperoxides in biological samples. *Free Radic. Biol. Med.* 7:177–178.

221. Jialal, I., and Devaraj, S. 1996. Low-density lipoprotein oxidation, antioxidants, and atherosclerosis: A clinical biochemistry perspective. *Clin. Chem.* 42:498–506.
222. Mulchandani, A., and Rudolph, D. C. 1995. Amperometric determination of lipid hydroperoxides. *Anal. Biochem.* 225:277–282.
223. O'Gara, C. Y., Maddipati, K. R., and Marnett, L. J. 1989. A sensitive electrochemical method for quantitative hydroperoxide determination. *Chem. Res. Toxicol.* 2:295–300.
224. Hermann, T., Steinhilber, D., and Roth, H. 1987. Determination of leukotriene B_4 by high-performance liquid chromatography with electrochemical detection. *J. Chromatogr.* 416:170–175.
225. Yamada, K., Terao, J., and Matsushita, S. 1987. Electrochemical detection of phospholipid hydroperoxides in reverse-phase high-performance liquid chromatography. *Lipids* 22:125–128.
226. Korytowski, W., Bachowski, G. J., and Girotti, A. W. 1993. Analysis of cholesterol and phospholipid hydroperoxides by high-performance liquid chromatography with mercury drop electrochemical detection. *Anal. Biochem.* 213:111–119.
227. Arai, H., Terao, J., Abdalla, D. S. P., Suzuki, T., and Takama, K. 1996. Coulometric detection in high-performance liquid chromatographic analysis of cholesteryl ester hydroperoxides. *Free Radic. Biol. Med.* 20:365–371.
228. Korytowski, W., Geiger, P. G., and Girotti, A. W. 1995. High-performance liquid chromatography with mercury cathode electrochemical detection: Application to lipid hydroperoxide analysis. *J. Chromatogr.* 670:189–197.
229. Knopse, J., Steinhilber, D., Herrmann, T., and Roth, H. 1988. Picomole determination of 2,4-dimethoxy-analides of prostaglandins by high-performance liquid chromatography with electrochemical detection. *J. Chromatogr.* 442:444–450.
230. Goldring, C., Casini, A., Maellero, E., Bello, B. D., and Comporti, M. 1993. Determination of 4-hydroxynonenal by high performance-liquid chromatography with electrochemical detection. *Lipids* 28:141–145.
231. Liu, Y.-M., Miao, J.-R., and Toyo'oka, T. 1996. Determination of 4-hydroxy-2-nonenal by precolumn derivatization and liquid chromatography with laser fluorescence detection. *J. Chromatogr. A* 719:450–456.
232. Halliwell, B., Aeschbach, R., Loliger, J., and Aruoma, O. I. 1995. The characterization of antioxidants. *Food Chem. Toxicol.* 33:601–617.
233. Levine, M. 1986. New concepts in the biology and biochemistry of ascorbic acid. *N. Engl. J. Med.* 314:892–902.
234. Ames, B. N., Cathcart, R., Schwiers, E., and Hochstein, P. 1981. Uric acid provides an antioxidant defense in humans against oxidant- and radical-caused aging and cancer: A hypothesis. *Proc. Natl. Acad. Sci. U.S.A.* 78:6858–6862.
235. Rose, R. C., and Bode, A. M. 1993. Biology of free radical scavenger: An evaluation of ascorbate. *FASEB. J.* 7:1135–1142.
236. Halliwell, B., Murcia, M. A., Chiroco, S., and Aruoma, O. I. 1995. Free radicals and antioxidants in food and in vivo: What they do and how they work. *Crit. Rev. Food Sci. Nutr.* 35:7–20.
237. Buettner, G. R. 1993. The pecking order of free radicals and antioxidants: Lipid peroxidation, α-tocopherol, and ascorbate. *Arch. Biochem. Biophys.* 300:535–543.
238. Kagan, V. E., Shvedova, A., Serbinova, E., Khan, S., Swanson, C., Powell, R., and Packer, L. 1992. Dihydrolipoic acid–universal antioxidant both in the membrane and in the aqueous phase: Reduction of peroxyl, ascorbyl and chromanoxyl radicals. *Biochem. Pharmacol.* 44:1637–1649.
239. Kagan, V. E., Serbinova, E. A., Forte, T., Scita, G., and Packer, L. 1992. Recycling of vitamin E in human low density lipoproteins. *J. Lipid Res.* 33:385–397.
240. Packer, L. 1994. Antioxidant properties of lipoic acid and its therapeutic effects in prevention of diabetes complications and cataracts. *Ann. N.Y. Acad. Sci.* 738:257–264.
241. Washko, P. W., Welch, R. W., Dhariwal, K. R., Wang, Y., and Levine, M. 1992. Ascorbic acid and dehydroascorbic acid analysis in biological samples. *Anal. Biochem.* 204:1–14.
242. Gamache, P., Ryan, E., and Acworth, I. N. 1993. Analysis of phenolic and flavonoid compounds in juice beverages using high-performance liquid chromatography with coulometric array detection. *J. Chromatogr.* 635:143–150.
243. Halliwell, B., and Gutteridge, J. M. C. 1990. The antioxidants of human extracellular fluids. *Arch. Biochem. Biophys.* 280:1–8.
244. Kittridge, K. J., and Willson, R. L. 1984. Uric acid substantially enhances the free radical-induced inactivation of alcohol dehydrogenase. *FEBS Lett.* 170:162–164.
245. Aruoma, O. I., and Halliwell, B. 1989. Inactivation of alpha 1-antiproteinase by hydroxyl radicals. The effect of uric acid. *FEBS Lett.* 244:776–80.
246. Benzie, I. F. F., and Strain, J. J. Uric acid: Friend or foe? *Redox Rep.* 2:231–234.
247. Maples, K. R., and Mason, R. P. 1988. Free radical metabolites of uric acid. *J. Biol. Chem.* 263:1709–1712.

248. Kaur, H., and Halliwell, B. 1990. Action of biologically-relevant oxidizing species upon uric acid. Identification of uric acid oxidation products. *Chem. Biol. Interact.* 73:235–247.

249. Tang-Liu, D. D.-S., and Riegelman, S. 1982. An automated HPLC assay for simultaneous quantitation of methylated xanthines and uric acids in urine. *J. Chrom. Sci.* 20:155–159.

250. Aoki, T., Yoshiura, M., Iwamoto, T., and Irayama, K. 1984. Postmortem changes of uric acid in various rat tissues: Determination of uric acid by reversed-phase high-performance liquid chromatography with electrochemical detection. *Anal. Biochem.* 143:113–118.

251. Iwamoto, T., Yoshiura, M., and Irayama, K. 1983. A simple, rapid and sensitive method for the determination of rat serum uric acid by reversed-phase high-performance liquid chromatography with electrochemical detection, *J. Chromatogr.* 278:156–159.

252. Roch-Ramel, F., Diezi-Chomety, F., Roth, L., and Weiner, I. M. 1980. A micropuncture study of urate excretion by cebus monkeys employing high-performance liquid chromatography with amperometric detection of urate. *Pflugers Arch.* 383:203–207.

253. Honegger, C. G., Langeman, H., Krenger, W., and Kempf, A. 1989. Liquid chromatographic determination of common water-soluble antioxidants in human tissue. *J. Chromatogr.* 487:463–468.

254. Irayama, K., Yoshiura, M., Iwamoto, T., and Ozaki, Y. 1984. Simultaneous determination of uric and ascorbic acids in human serum by reversed-phase high-performance liquid chromatography with electrochemical detection. *Anal. Biochem.* 141:238–243.

256. Shirachi, D. Y., and Omaye, S. T. 1992. The simultaneous measurement of uric acid and ascorbic acid in the lateral ventricles of freely moving rats by brain microdialysis and electrochemical detection. *Proc. West Pharmacol. Soc.* 35:161–163.

256. Rizzo, V., Melzi D'Eril, G., Achilli, G., and Cellerino, G. P. 1991. Determination of neurochemicals in biological fluids by using an automated high-performance liquid chromatographic system with a coulometric array detector. *J. Chromatogr.* 536:229–236.

257. Diplock, A. T., Machlin, L. J., Packer, L., and Pryor, W. A., eds. 1989. *Vitamin E Biochemistry and Health Implications.* New York: Ann. N.Y. Acad. Sci., vol. 570.

258. Esterbauer, H., Striegl, G., Puhl, H., and Rotheneder, M. 1989. Continuous monitoring of in vitro oxidation of human low density lipoprotein. *Free Radic. Res. Commun.* 6:67–75.

259. Tappel, A. L. 1968. Will antioxidant nutrients slow the aging process? *Geriatrics* 23:97–105.

260. Kagan, V. E., Serbinova, E. A., and Packer, L. 1990. Antioxidant Effects of ubiquinones in microsomes and mitochondria are mediated by tocopherol recycling. *Biochem. Biophys. Res. Commun.* 169:851–857.

261. Sies, H., and Murphy, M. E. 1991. Role of tocopherols in the protection of biological systems against oxidative damage. *J. Photochem. Photobiol.* B 8:211–224.

262. Castle, M. C., and Cook, W. J. 1985. Measurement of vitamin E in serum and plasma by high performance liquid chromatography and electrochemical detection. *Ther. Drug Monit.* 7:364–368.

263. Vatassery, G. T., Smith, W. E., and Quach, H. T. 1993. A liquid chromatographic method for the simultaneous determination of α-tocopherol and tocopherolquinone in human red blood cells and other biological samples where tocopherol is easily oxidized during sample treatments. *Anal. Biochem.* 214:426–430.

264. Lang, J. K., and Packer, L. 1987. Quantitative determination of vitamin E and oxidized and reduced coenzyme Q by high-performance liquid chromatography. *J. Chromatogr.* 385:109–117.

265. Murphy, M. E., and Kehrer, J. P. 1987. Simultaneous measurement of tocopherols and tocopheryl quinones in tissue fractions using high-performance liquid chromatography with redox cycling electrochemical detection. *J. Chromatogr.* 421:71–82.

266. Edlund, P. O. 1988. Determination of coenzyme Q_{10}, α-tocopherol and cholesterol in biological samples by coupled-column liquid chromatography with coulometric and ultraviolet detection. *J. Chromatogr.* 425:87–97.

267. Gamache, P. H., McCabe, D. R., Parvez, H., Parvez, S., and Acworth, I. N. 1997. The measurement of markers of oxidative damage, anti-oxidants and related compounds using HPLC and coulometric array analysis. In *Progress in HPLC*, Vol. 6, eds. I. N. Acworth, M. Naoi, H. Parvez, and S. Parvez, pp. 99–126. Amsterdam: VS Press.

268. Frei, B., Kim, M. C., and Ames, B. N. 1990. Ubiquinol-10 is an effective lipid-soluble antioxidant at physiological concentrations. *Proc. Natl. Acad. Sci. U.S.A.* 87:4879–4883.

269. Takada, M., Ikenoya, S., Yuzuriha, T., and Katayama, K. 1982. Studies on reduced and oxidized coenzyme Q (ubiquinones). II. The determination of oxidation-reduction levels of coenzyme Q in mitochondria, microsomes and plasma by high-performance liquid chromatography. *Biochem. Biophys. Acta* 679:308–314.

270. Okamoto, T., Fukanaga, Y., Ida, Y., and Kishi, T. 1988. Determination of reduced and total ubiquinones in biological materials by liquid chromatography with electrochemical detection. *J. Chromatogr.* 430:11–19.

271. Grossi, G., Bargossi, A. M., Fiorella, P. L., Piazzi, S., Battino, M., and Bianchi, G. P. 1992. Improved

high-performance liquid chromatographic method for the determination of coenzyme Q_{10} in plasma. *J. Chromatogr.* 593:217–226.

272. Meister, A., and Anderson, M. E. 1983. Glutathione. *Annu. Rev. Biochem.* 52:711–760.
273. Tietze, F. 1969. Enzymic method for quantitative determination of nanogram amounts of total and oxidized glutathione: Applications to mammalian blood and other tissues. *Anal. Biochem.* 27:502–522.
274. Griffith, O. W. 1980. Determination of glutathione and glutathione disulfide using glutathione reductase and 2-vinylpyridine. *Anal. Biochem.* 106:207–212.
275. Michelet, F., Guerguen, R., Leroy, P., Wellman, M., Nicolas, A., and Siest, G. 1995. Blood and plasma glutathione measured in healthy subjects by HPLC: Relation to sex, aging, biological variables, and life habits. *Clin. Chem.* 41:1509–1517.
276. Ozcimder, M., Louter, A. J. H., Lingeman, H., Voogt, W. H., Frei, R. W., and Bloemendal., M. 1991. Determination of oxidized, reduced and protein-bound glutathione in eye lenses by high-performance liquid chromatography and electrochemical detection. *J. Chromatogr.* 570:19–28.
277. Stein, A. F., Dills, R. L., and Klaassen, C. D. 1986. High-performance liquid chromatographic analysis of glutathione and its thiol and disulfide degradation. *J. Chromatogr.* 381:259–270.
278. Demaster, E. G., Shirota, F. N., Redfern, B., Goon, D. J., and Nagasawa, H. T. 1984. Analysis of hepatic reduced glutathione, cysteine, and homocysteine by cation-exchange high-performance liquid chromatography with electrochemical detection. *J. Chromatogr.* 308:83–91.
279. Richie, J. P., and Lang, C. A. 1987. The determination of glutathione, cyste(e)ine, and other thiols and disulfides using high-performance liquid chromatography with dual electrochemical detection. *Anal. Biochem.* 163, 9–15.
280. Harvey, P. R., Ilson, R. G., and Strasberg, S. M. 1989. The simultaneous determination of oxidized and reduced glutathiones in liver tissue by ion-pairing reversed-phase high-performance liquid chromatography with a coulometric detector. *Clin. Chem. Acta* 180:203–212.
281. Shimada, K., Oe, T., and Nambara, T. 1987. Sensitive ferrocene reagents for derivatization of thiol compounds in high-performance liquid chromatography with dual-electrode coulometric detection. *J. Chromatogr.* 419:17–25.
282. Smith, N. C., Dunnett, M., and Mills, P. C. 1995. Simultaneous quantitation of oxidized and reduced glutathione in equine biological fluids by reversed-phase high-performance liquid chromatography using electrochemical detection. *J. Chromatogr. B* 673:35–41.
283. Carro-Ciampi, G., Hunt, P. G., Turner, C. J., and Wells, P. G. 1988. A high-performance liquid chromatographic assay for reduced and oxidized glutathione in embryonic, neonatal, and adult tissue using a porous graphite electrochemical detector. *J. Pharmacol. Methods* 19:75–83.
284. Krien, P. M., Margou, V., and Kermici, M. 1992. Electrochemical determination of femtomole amounts of free and oxidized glutathione: Application to human hair follicles. *J. Chromatogr.* 567:255–261.
285. Hill, B. A., Kleiner, H. E., Ryan, E. A., Dulik, D. M., Monks, T. J., and Lau, S. S. 1993. Identification of multi-S-substituted conjugates of hydroquinone by HPLC-coulometric electrode array analysis and mass spectrometry. *Chem. Res. Toxicol.* 6:459–469.
286. Ueland, P. M. 1995. Homocysteine species as components of plasma redox thiol status. *Clin. Chem.* 41:340–342.
287. Andersson, A., Lindgren, A., and Hultberg, B. 1995. Effects of thiol oxidation and thiol export from erythrocytes on determination of redox Sstatus of homocysteine and other thiols in plasma from healthy subjects and patients with cerebral infarction. *Clin. Chem.* 41:361–366.
288. Shipchandler, M. T., and Moore, E. G. 1995. Rapid, fully automated measurement of plasma homocyst-(e)ine with the Abbott Imx® analyzer. *Clin. Chem.* 41:991–994.
289. Jacobsen, D. W., Gatautis, V. J., Green, R., Robinson, K., Savon, S. R., Secic, M., Ji, J., Otto, J. M., and Taylor, L. M. 1994. Rapid HPLC determination of total homocysteine and other thiols in serum and plasma: Sex differences and correlation with cobalamin and folate concentrations in healthy subjects. *Clin. Chem.* 40:873–881.
290. Achilli, G., and Cellerino, G. P. 1995. In *Handbook of Oxidative Metabolism*, eds. I. N. Acworth and B. Bailey. ESA, Inc.
291. Cotgreave, I. A., Sandy, M. S., Berggren, M., Moldeus, P. W., and Smith, M. T. 1987. *N*-Acetylcysteine and glutathione-dependent protective effects of PZ51 (Ebselen) against diquat-induced cytotoxicity in isolated hepatocytes. *Biochem. Pharmacol.* 36:2899–2904.
292. Miners, J, O., Drew, R., and Birkett, D. J. 1984. Mechanism of action of paracetamol protective agents in mice in vivo. *Biochem. Pharmacol.* 33:2995–3000.
293. Arouma, O. I., Halliwell, B., Hoey, B. M., and Butler, J. 1989. The antioxidant action of *N*-acetylcysteine: Its reaction with hydrogen peroxide, hydroxyl radical, superoxide and hypochlorous acid. *Free Radic. Biol. Med.* 6:593–597.
294. Packer, L., Witt, E. H., and Tritschler, H. J. 1995. Alpha-lipoic acid as a biological antioxidant. *Free Radic. Biol. Med.* 19:227–250.
295. Scott, B. C., Aruoma, O., Evans, J. P., O'Neill, C., van der Vliet, A., Cross, C. E., Tritschler, H., and

Halliwell, B. 1994. Lipoic and dihydrolipoic acids as antioxidants: A critical evaluation. *Free Radic. Res.* 20:119–133.

296. Teichert, J., and Preiss, R. 1992. HPLC-methods for the determination of lipoic acid and its reduced form in human plasma. *Int. J. Clin. Pharm. Ther. Toxicol.* 30:511–512.

297. Kamata, K., and Akiyama, K. 1990. High-performance liquid chromatography with electrochemical detection for the determination of thioctic and thioctic acid amide. *J. Pharm. Biomed Anal.* 8:453–456.

298. Teichert, J., and Preiss, R. 1995. Determination of lipoic acid in human plasma by high-performance liquid chromatography with electrochemical detection. *J. Chromatogr. B* 672:277–281.

299. The Measurement of Oxidized and Reduced Lipoic Acid. Application Note. ESA, Inc. In preparation.

300. Orwar, O., Folestad, S., Einarsson, S., Andine, P., and Sandberg, M. 1991. Automated determination of neuroactive acidic sulfur-containing amino acids and γ-glutamyl peptides using liquid chromatography with fluorescence and electrochemical detection. *J. Chromatogr.* 566:39–55.

301. Canfield, L. M., Krinsky, N. I., and Olson, J. A., eds. 1993. *Carotenoids in Human Health*. New York: Ann. N.Y. Acad. Sci., vol. 691.

302. Frei, B., and Ames, B. N. 1991. Small molecule antioxidant defenses in human extracellular fluids. In *The Molecular Biology of Free Radic. Systems*, ed. J. Scandalios. Cold Spring Harbor, NY: Cold Spring Harbor Laboratory Press.

303. Rice-Evans, C. A., and Diplock, A. T. 1993. Current status of antioxidant therapy. *Free Radic. Biol. Med.* 15:77–96.

304. Ozhogina, O. A., and Kasaikina, O. T. 1995. β-Carotene as an interceptor of free radicals. *Free Radic. Biol. Med.* 19:575–581.

305. Bryan, P. D., Honigberg, I. L., and Meltzer, N. M. 1991. Electrochemical detection of retinoids using normal phase HPLC. *J. Liq. Chromatogr.* 14:2287–2295.

306. MacCrehan, W. A., and Schonberger, E. 1987. Reversed-phase high-performance liquid chromatographic separation and electrochemical detection of retinol and its isomers. *J. Chromatogr.* 417:65–78.

307. Wring, S. A., Hart, J. P., and Knight, D. W. 1988. Voltammetric behavior of all-*trans*-retinol (vitamin A1) at a glassy carbon electrode and its determination in human serum using high-performance liquid chromatography with electrochemical detection. *Analyst* 113:1785–1789.

308. Wring, S. A., and Hart, J. P. 1989. Electrochemical determination of all-*trans*-retinol, and correlation with retinol binding protein in liver cirrhosis. *Med. Lab. Sci.* 46:367–369.

309. MacCrehan, W. A., and Schonberger, E. 1987. Determination of retinol, α-tocopherol, and β-carotene in serum by liquid chromatography with absorbance and electrochemical detection. *Clin. Chem.* 33:1585–1592.

310. Pryor, W. A., Strickland, T., and Church, D. F. 1988. Comparison of the efficiencies of several natural and synthetic antioxidants in aqueous sodium dodecyl sulfate micelle solutions. *J. Am. Chem. Soc.* 110:2224–2229.

311. Krinsky, N. I. 1989. Antioxidant functions of carotenoids. *Free Radic. Biol. Med.* 7:617–635.

312. Burton, G. W., and Ingold, K. U. 1983, β-Carotene: An unusual type of lipid antioxidant. *Science* 224:569–573.

313. Scott, B. C., Butler, J., Halliwell, B., and Aruoma, O. I. 1993. Evaluation of the antioxidant actions of ferulic acid and catechins. *Free Radic. Res. Commun.* 19:241–253.

314. Aeschbach, R., Loliger, J., Scott, B. C., Murcia, A., Butler, J., Halliwell, B., and Aruoma, O. I. 1994. Antioxidant actions of thymol, cavacrol, 6-gingerol, zingerone and hydroxytyrosol. *Food Chem. Toxicol.* 32:31–36.

315. Harborne, J. B. ed. 1988. *The Flavonoids: Advances in Research Since 1980*. London: Chapman and Hall.

316. Robak, J., and Glyglewski, R. J. 1988. Flavonoids are scavengers of the superoxide anion. *Biochem. Pharmacol.* 37:837–841.

317. Bors, W., and Saran, M. 1987. Radic. scavenging by flavonoids antioxidants. *Free Radic. Res. Commun.* 2:289–294.

318. Bors, W., Heller, W., Michel, C., and Saran, M. 1990. Flavonoids as antioxidants: Determination of antiradical efficiencies. *Methods Enzymol.* 186:343–355.

319. Cotelle, N., Bernnier, J.-L., Catteau, J.-P., Pommery, J., Wallet, J.-C., and Gaydou, E. M. 1995. Antioxidant properties of hydroxy-flavones. *Free Radic. Biol. Med.* 20:35–43.

320. Van Acker, S. A. B. E., van den Berg, D.-J., Tromp, M. N. J. L., Griffioen, D. H., van Bennekom, W. P., van der Vijhg, W. J. F., and Bast, A. 1996. Structural aspects of antioxidant activity of flavonoids. *Free Radic. Biol. Med.* 20:331–342.

321. Saija, A., Scalese, M., Lanza, M., Marzullo, D., Bonina, F., and Castelli, F. 1995. Flavonoids as antioxidant agents: Importance of their interaction with biomembranes. *Free Radic. Biol. Med.* 19:481–486.

322. Lunder, T. L. 1992. Catechins of green tea. In *Phenolic Compounds in Food and Their Effects on Health*, eds. M.-T. Huang, C.-T. Ho, and C. Y. Lee, pp. 114–120. ACS Symposium Series, developed

from a symposium sponsored by the Division of Agricultural and Food Chemistry of the American Chemical Society at the Fourth Chemical Congress of North America. Washington, DC: ACS Press.

323. Osawa, T. 1992. Phenolic antioxidants in dietary plants as antimutagens. In *Phenolic Compounds in Food and Their Effects on Health*, eds. M.-T. Huang, C.-T. Ho, and C. Y. Lee, pp. 135–149. ACS Symposium Series, developed from a Symposium Sponsored by the Division of Agricultural and Food Chemistry of the American Chemical Society at the Fourth Chemical Congress of North America. Washington, DC: ACS Press.

324. Osawa, T., Ramarathnam, N., Kawakishi, S., and Namiki, M. 1992. Antioxidant defense systems generated by phenolic plant constituents. In *Phenolic Compounds in Food and Their Effects on Health*, eds. M.-T. Huang, C.-T. Ho, and C. Y. Lee, pp. 122–143. ACS Symposium Series, developed from a Symposium Sponsored by the Division of Agricultural and Food Chemistry of the American Chemical Society at the Fourth Chemical Congress of North America. Washington, DC: ACS Press.

325. Butler, L. G., and Rogler, J. C. 1992. Biochemical mechanisms of the antinutritional effects of tannins. In *Phenolic Compounds in Food and Their Effects on Health*, eds. M.-T. Huang, C.-T. Ho, and C. Y. Lee, pp. 298–304. ACS Symposium Series, developed from a Symposium Sponsored by the Division of Agricultural and Food Chemistry of the American Chemical Society at the Fourth Chemical Congress of North America. Washington, DC: ACS Press.

326. Haslam, E., Lilley, T. H., Warminski, E., Liao, H., Cai, Y., Martin, R., Gaffney, S. H., Goulding, P. N., and Luck, G. 1992. Polyphenol complexation. In *Phenolic Compounds in Food and Their Effects on Health*, eds. M.-T. Huang, C.-T. Ho, and C. Y. Lee, pp. 8–50. ACS Symposium Series, developed from a Symposium Sponsored by the Division of Agricultural and Food Chemistry of the American Chemical Society at the Fourth Chemical Congress of North America. Washington, DC: ACS Press.

327. Huan, M.-T., and Ferraro, T. 1992. Phenolic compounds in food and cancer prevention. In *Phenolic Compounds in Food and Their Effects on Health*, eds. M.-T. Huang, C.-T. Ho, and C. Y. Lee, pp. 8–34. ACS Symposium Series, developed from a Symposium Sponsored by the Division of Agricultural and Food Chemistry of the American Chemical Society at the Fourth Chemical Congress of North America. Washington, DC: ACS Press.

328. Stavric, B., Matula, T. I., Klassen, R., Downie, R. H., and Wood, R. J. 1992. Effect of flavonoids on mutagenicity and bioavailability of xenobiotics in foods. In *Phenolic Compounds in Food and Their Effects on Health*, eds. M.-T. Huang, C.-T. Ho, and C. Y. Lee, pp. 239–249. ACS Symposium Series, developed from a Symposium Sponsored by the Division of Agricultural and Food Chemistry of the American Chemical Society at the Fourth Chemical Congress of North America. Washington, DC: ACS Press.

329. Ho, C.-T., Lee, C. Y, and Huang, M.-T., eds. 1992. *Phenolic Compounds in Food and Their Effects on Health*. ACS symposium series, Vols. 506 and 507, developed from a symposium sponsored by the Division of Agricultural and Food Chemistry of the American Chemical Society at the Fourth Chemical Congress of North America. Washington, DC: ACS Press.

330. Nagy, S., and Wade, R. L., eds. 1995. *Methods to Detect Adulteration of Fruit Juice Beverages*. Auburndale, FL: Agscience, Inc.

331. Gamache, P., Ryan, E., and Acworth, I. N. 1993. Analysis of phenolic and flavonoid compounds in juice beverages using high-performance liquid chromatography with coulometric array detection. *J. Chromatogr.* 635:143–150.

332. Achilli, G., Cellerino, G. P., Gamache, P. H., and Melzi d'Eril., G. V. 1993. Identification and determination of phenolic constituents in natural beverages and plant extracts by means of a coulometric electrode array system. *J. Chromatogr.* 632:111–117.

333. Uchida, M., Kataoka, Y., and Ono, M. 1992. Determination of reducing substances in beer and wort by HPLC with 16 electrode electrochemical detection and its application to the practical brewing. *2nd BCOA Meeting*, St Louis, MO.

334. Acworth, I. N., Kuo, M., Ryan, E., and Gamache, P. H. 1994. Analysis of catechins, phenolic and polyphenolic compounds in tea using HPLC and coulometric array detection. *Pittsburgh Conference on Analytical Chemistry and Applied Spectroscopy*, Chicago. Paper no. 391aP.

335. Lee, M.-J., Wang, Z.-Y, Li, H., Chen, L., Sun, Y., Gobbo, S., Balentine, B. A., and Yang, C. S. 1995. Analysis of plasma and urinary tea polyphenols in human subjects. *Cancer Epidemiol. Biomark. Prevent.* 4:393–399.

336. Steinmetz, K. A., and Potter, J. D. 1991. Vegetables, fruit, and cancer. *Cancer Causes Control* 2:427–440.

337. Messina, M., and Messina, V. 1991. Increasing use of soyfoods and their potential role in cancer prevention. *J. Am. Diet. Assoc.* 91:836–840.

338. Adlercreutz, H. 1990. Western diet and western diseases: Some hormonal and biochemical mechanisms and associations. *Scand. J. Clin. Lab. Invest.* 201:3–23.

339. Messina, M., and Barnes, S. 1991. The role of soy products in reducing risk of cancer. *J. Natl. Cancer Inst.* 83:541–546.

340. Fotsis, T., Heikkinen, R., Adlercreutz, H., Axelson, M., and Setchell, K. D. R. 1982. Capillary gas chromatographic method for the analysis of lignans in human urine. *Clin. Chim. Acta* 121:361–371.
341. Adlercreutz, H., Fotsis, T., Lampe, J., Wahala, K., Makela, T., Brunow, G., and Hase, T. 1993. Quantitative determination of lignans and isoflavonoids in plasma of omnivorous and vegetarian women by isotope dilution gas chromatography-mass spectrometry. *Scand. J. Clin. Invest.* 53:5–18.
342. Setchell, K. D. R., Lawson, A. M., McLaughlin, L. M., Patel, S., Kirk, D. N., and Axelson, M. 1983. Measurement of enterolactone and enterodiol, the first mammalian lignans, using stable isotope dilution and gas chromatography mass spectrometry. *Biomed. Mass. Spectrom.* 10:227–235.
343. Setchell, K. D. R., Welsh, M. B., and Lim, C. K. 1987. High-performance liquid chromatographic analysis of phytoestrogens in soy protein preparations with ultraviolet, electrochemical and thermospray mass spectrometric detection. *J. Chromatogr.* 386:315–323.
344. Wang, G., Kuan, S., Francis, O. J., Ware, G. M., and Carmen, A. S. 1990. A simplified HPLC method for the determination of phytoestrogens in soybean and its processed products. *J. Agric. Food Chem.* 38:185–190.
345. Supko, J. G., and Philips, L. R. 1995. High-performance liquid chromatography assay for genistein in biological fluids. *J. Chromatogr.* 666:157–167.
346. Keaney, J. F., Jr., Shwaery, G. T., Xu, A., Nicolosi, R. J., Loscalzo, J., Foxall, T. L., and Vita, J. A. 1994. 17 Beta-estradiol preserves endothelial vasodilator function and limits low-density lipoprotein oxidation in hypercholstrolemic swine. *Circulation* 89:2251–2259.
347. Goodman, Y., Bruce, A. J., Cheng, B., and Mattson, M. P. 1996. Estrogens attenuate and corticosterone exacerbates excitotoxicity, oxidative injury, and amyloid β-peptide toxicity in hippocampal neurons. *J. Neurochem.* 66:1836–1844.
348. Behl, C., Widmann, M., Trapp, T., and Holsboer, F. 1995. 17 β-Estradiol protects neurons from oxidative stress-induced cell death in vitro. *Biochem. Biophys. Res. Commun.* 216:473–482.
349. Salahudeen, A. K., Clark, E. C., and Nath, K. A. 1991. Hydrogen peroxide-induced renal injury. A protective role for pyruvate in vitro and in vivo. *J. Clin. Invest.* 88:1886–1893.
350. Nath, K. A., Enright, H., Nutter, L. M., Fischereder, M. F., Zhou, J. N., and Hebbel, R. P. 1994. The effect of pyruvate on oxidant injury to isolated and cellular DNA. *Kidney Int.* 45:166–176.
351. Nath, K. A., Ngo, E. O., Hebbel, R. P., Croatt, A. J., Zhou, B., and Nutter, L. M. 1995. α-Ketoacids scavenge H_2O_2 in vitro and in vivo and reduce menadione-induced DNA injury and cytotoxicity. *Am. J. Physiol.* 268:C227–C236.
352. Witt, E. H., Reznick, A. Z., Viguie, C. A., Starke-Reed, P., and Packer, L. 1992. Exercise, oxidative damage and effects of antioxidant manipulation. *J. Nutr.* 122:766–773.
353. Ji, L. L. 1995. Oxidative stress during exercise: Implication of antioxidant nutrients. *Free Radic. Biol. Med.* 18:1079–1086.
354. Ji, L. L., Miller, R. H., Nagle, F. J., Lardy, H. A., and Stratman, F. W. 1987. Amino acid metabolism during exercise in trained rats: The potential role of carnitine in the metabolic fate of branched-chain amino acids. *Metabolism* 36:748–752.
355. Silman, R. E. 1993. Melatonin: A contraceptive for the nineties. *Eur. J. Obstet. Gynecol. Reprod. Biol.* 49:3–9.
356. Dawson, D., and Encel, N. 1993. Melatonin and sleep in humans. *J. Pineal Res.* 15:1–12.
357. Poeggeler, B., Reiter, R. J., Tan, D.-X., Chen, L.-D., and Manchester, L. C. 1993. Melatonin, hydroxyl radical-mediated oxidative damage, and aging: A hypothesis. *J. Pineal Res.* 14:151–168.
358. Reiter, R. J., Melchiorri, D., Sewerynek, E., Poeggeler, B., Barlow-Walden, L., Chuang, J.-I., Ortiz, G. G., and Acuna-Castroviejo, D. 1995. A review of the evidence supporting melatonin's role as an antioxidant. *J. Pineal Res.* 18:1–11.
359. Wetterberg, L., Eriksson, O., Friberg, Y., and Vangbo, B. 1978. A simplified radioimmunoassay for melatonin and its application to biological fluids. Preliminary observations on the half-life of plasma melatonin in man. *Clin. Chim. Acta* 86:169–177.
360. Brzezinski, A., Seibel, M. M., Lynch, H. J., Deng, M.-H., and Wurtman, R. J. 1987. Melatonin in human preovulatory follicular fluid. *J. Clin. Endocrinol. Metab.* 64:865–867.
361. Yie, S. M., Johansson, E., and Brown, G. M. 1993. Competitive solid-phase enzyme immunoassay for melatonin in human and rat serum and rat pineal gland. *Clin. Chem.* 39:2322–2325.
362. Best, S. A., Midgley, J. M., Huang, W., and Watson, D. G. 1993. The Determination of 5-hydroxytryptamine, related indolealkylamines and 5-hydroxyindoleacetic acid in the bovine eye by gas chromatography-negative ion chemical ionization mass spectrometry. *J. Pharm. Biomed. Anal.* 11:323–333.
363. Dubbels, R., Reiter, R. J., Klenke, E., Goebel, A., Schnakenberg, E., Ehlers, C., Schiwara, H. W., and Schloot, W. 1995. Melatonin in edible plants identified by radioimmunoassay and high performance liquid chromatography-mass spectrometry. *J. Pineal Res.* 18:28–31.
364. Hattori, A., Migitaka, H., Iigo, M., Itoh, M., Yamamoto, K., Ohtani-Kaneko, R., Hara, M., Suzuki, T., and Reiter, R. J. 1995. Identification of melatonin in plants and its effects on plasma melatonin levels and binding to melatonin receptors in vertebrates. *Biochem. Mol. Biol. Int.* 35:627–634.

365. Brammer, G. L. 1994. Duodenum is not a consistent source of melatonin in rats. *Life Sci.* 55:775–787.
366. Vieira, R., Miguez, J., Lema, M., and Aldegunde, M. 1992. Pineal and plasma melatonin as determined by high-performance liquid chromatography with electrochemical detection. *Anal. Biochem.* 205:300–305.
367. Vitale, A. A., Ferrari, C. C., Aldanan, H., and Affanni, J. M. 1996. Highly sensitive method for the determination of melatonin by normal-phase high-performance liquid chromatography with fluorometric detection. *J. Chromatogr. B* 681:381–384.
368. Harumei, T., Akutsu, H., and Matsushima, S. 1996. Simultaneous determination of serotonin, *N*-acetylserotonin and melatonin in the pineal gland of juvenile golden hamster by high-performance liquid chromatography with electrochemical detection. *J. Chromatogr. B* 675:152–156.
369. Siuciak, J. A., Gamache, P. H., and Dubochovich, M. L. 1992. Monoamines and their precursors and metabolites in the chicken brain, pineal and retina: Regional distribution and day/night variations. *J. Neurochem.* 58:722–729.
370. Hasegawa, M., Goto, M., Oshima, I., and Ebihara, S. 1994. Application of in vivo microdialysis to pineal research in birds: Measurement of circadian rhythms of melatonin. *Neurosci. Biobehav. Rev.* 18:537–540.
371. Azekawa, T., Sano, A., Aoi, K., Sei, H., and Morita, Y. 1990. Concurrent on-line sampling of melatonin in pineal microdialysates from conscious rat and its analysis by high-performance liquid chromatography with electrochemical detection. *J. Chromatogr.* 530:47–55.
372. Crespi, F., Ratti, E., and Trist, D.G. 1994. Melatonin, a hormone monitorable in vivo by voltammetry? *Analyst* 119:2193–2197.
373. Halliwell, B., Aeschbach, R., Loliger, J., and Aruoma, O. I. 1995. The characterization of antioxidants. *Food Chem. Toxicol.* 7:601–617.
374. Aruoma, O. I., Halliwell, B., Hoey, B. M., and Butler, J. 1989. The antioxidant action of *N*-acetylcysteine: Its reaction with hydrogen peroxide, hydroxyl radical, superoxide, and hydrochlorous acid. *Free Radic. Biol. Med.* 6:593–597.
375. Aruoma, O. I., Laughton, M. J., and Halliwell, B. 1989. Carnosine, homocarnosine and anserine: Could they act as antioxidants in vivo? *Biochem. J.* 264:863–869.
376. Bartlett, D. B., Church, D. F., Bounds, P. L., and Koppenol, W. H. 1994. The kinetics of the oxidation of L-ascorbic acid by peroxynitrite. *Free Radic. Biol. Med.* 18:85–92.
377. Aruoma, O. I., Halliwell, B., Hoey, B. M., and Butler, J. 1988. The antioxidant action of taurine, hypotaurine and their metabolic precursors. *Biochem. J.* 256:251–255.
378. Denicola, A., Souza, J. M., Gatti, R. M., Augusto, O., and Radi, R. 1995. Desferrioxamine inhibition of the hydroxyl radical-like reactivity of peroxynitrite: Role of the hydroxaminc groups. *Free Radic. Biol. Med.* 19:11–19.
379. Breen, A. P., and Murphy, J. A. 1995. Reactions of the oxyl radicals with DNA. *Free Radic. Biol. Med.* 18:1033–1077.
330. Hiller, K. D., Hodd, P. L., and Wilson, R. L. 1983. Antiimflammatory drugs: Protection of a bacterial virus as an in vitro biological measure of free radical activity. *Chem. Biol. Interact.* 47:293–305.
381. Vieira, A. J. S. C., Candeias, L. P., and Steenken, S. 1993. Hydroxyl radical induced damage to the purine bases of DNA: In vivo studies. *J. Chem. Phys.* 90:881–897.
382. Beckman, J. S. 1994. Peroxynitrite versus hydroxyl radical: The role of nitric oxide in superoxide-dependent cerebral injury. *Ann. N.Y. Acad. Sci.* 738:69–75.
383. Bielski, B. H. J., Cabelli, D. E., Arudi, R. L., and Ross, A. B. 1985. Reactivity of perhydroxyl/superoxide radicals in aqueous solution. *J. Phys. Chem. Ref. Data* 14:1041–1100.
384. Buettner, G. R. 1993. The pecking order of free radicals and antioxidants: Lipid peroxidation, α-tocopherol, and ascorbate. *Arch. Biochem. Biophys.* 300:535–543.
385. Alvarez, B., Denicola, A., and Radi, R. 1995. Reaction between peroxynitrite and hydrogen peroxide: Formation of oxygen and slowing of peroxynitrite decomposition. *Chem. Res. Toxicol.* 8:859–864.
386. van der Vleit, A., Eiserich, J. P., O'Neill, C. A., Halliwell, B., and Cross, C. E. 1995. Tyrosine modification by reactive nitrogen species. *Arch. Biochem. Biophys.* 319:341–349.
387. Ischiropoulos, H., Zhu, L., Chen, J., Tsai, M., Martin, J. C., Smith, C. D., and Beckman, J. S. 1992. Peroxynitrite-mediated tyrosine nitration catalyzed by superoxide dismutase. *Arch. Biochem. Biophys.* 298:431–437.
388. Scarpa, M., Rigo, A., Maiorino, M., Ursini, F., and Gregolin, C. 1984. Formation of α-tocopherol radical and recycling of α-tocopherol by ascorbate during peroxidation of phosphatidylcholine. *Biochim. Biophys. Acta* 801:215–219.
389. Mukai, K., Kikuchi, S., and Urano, S. 1990. Stopped-flow kinetic study of the regeneration reaction of tocopheroxyl radical by reduced ubiquinone-10 in solution. *Biochim. Biophys. Acta* 1035:77–82.
390. Neta, P., Huie, R. E., and Ross, A. B. 1990. Rate constants for reactions of peroxyl radicals in fluid solutions. *J. Phys. Chem. Ref. Data* 19:413–513.
391. Arouma, O. I., Halliwell, B., Hoey, B. M., and Butler, J. 1989. The antioxidant action of *N*-acetylcysteine:

Its reaction with hydrogen peroxide, hydroxyl radical, superoxide and hypochlorous acid. *Free Radic. Biol. Med.* 6:593–597.

392. Koppenol, W. H., and Butler, J. 1985. Energetics of interconversion reactions of oxy radicals. Adv. *Free Radic. Biol. Med.* 1:91–131.

393. Koppenol, W. H. 1990. Oxyradical reactions: From bond-dissociation energies to reduction potentials. *FEBS Lett.* 264:165–167.

394. Surdhar, P. S., and Armstrong, D. A. 1986. Redox potentials of some sulfur containing radicals. *J. Phys. Chem.* 90:5915–5917.

395. Surdhar, P. S., and Armstrong, D. A. 1987. Reduction potentials and exchange reactions of thiyl radicals and disulfide anion radicals *J. Phys. Chem.* 91:6532–6537.

396. Koppenol, W. H. 1976. Reactions involving singlet oxygen and the superoxide anion. *Nature* 262:420–421.

397. Simic, M. G., and Jovanovic, S. V. 1989. Antioxidation mechanisms of uric acid. *J. Am. Chem. Soc.* 111:5778–5782.

398. Scurlock, R., Rougee, M., and Bensasson, R. V. 1990. Redox properties of phenols, their relationship to singlet oxygen quenching and to their inhibitory effects on benzo(a)pyrene-induced neoplasia. *Free Radic. Res. Commun.* 8:251–258.

399. Simic, M. G. 1990. Pulse radiolysis in study of oxygen radicals. *Methods Enzymol.* 186:89–100.

400. Njus, D., and Kelly, P. M. 1991. Vitamins C and E donate single hydrogen atoms in vivo. *FEBS Lett.* 284:147–151.

401. Craw, M. T., and Depew, M. C. 1985. Contributions of electron spin resonance spectroscopy to the study of vitamins C, E and K *Rev. Chem. Intermed.* 6:1–31.

402. Golumbic, C., and Mattill, H. A. 1940. The oxidation of vitamin E. *J. Biol. Chem.* 134:535–541.

403. Koppenol, W. H. 1987. Thermodynamics of reactions involving oxyradicals and hydrogen peroxide. *Bioelectrochem. Bioenerg.* 18:3–11.

404. Koppenol, W. H. 1987. *CRC Critical Reviews, Membrane Oxidation*, ed. Vigo-Pelfrey, Vol. 1, pp. 1–13. Boca Raton, FL: CRC Press.

405. Farrington, J. A., Land, E. J., and Swallow, A. J. 1980. The one electron reduction potentials of NAD. *Biochim. Biophys. Acta* 590:273–276.

406. Williams, N. H., and Yandell, J. K. 1982. Outer-sphere electron-transfer reactions of ascorbic anions. *Aust. J. Chem.* 35:1133–1144.

407. Stellwagen, E. 1978. Haem exposure as the determinate of oxidation-reduction potential of haem proteins. *Nature* 275:73–74.

408. DeVries, S., Berden, J. A., and Slater, E. C. 1980. Properties of a semiquinone anion located in the QH$_2$:cytochrome c oxidoreductase segment of the mitochondrial respiratory chain. *FEBS Lett.* 122:143–148.

409. Reed, C. A. 1982. *The Biological Chemistry of Iron*, eds. H. D. Dunford, D. Dolphin, K. N. Raymond, and L. Sieker, pp. 25–42. London: D. Reidel.

410. Segel, I. H. 1976. *Biochemical Calculations*. New York: John Wiley and Sons.

411. Watt, G. D., Frankel, R. B., and Papaefthymios, G. C. 1985. Reduction of mammalian ferritin. *Proc. Natl. Acad. Sci. U.S.A.* 82:3640–3643.

412. Anderson, R. F. 1976. Electron transfer and equilibrium between pyridiryl radicals and FAD. *Ber. Bunsenges. Phys. Chem.* 80:969–972.

413. Anderson, R. F. 1983. Energetics of the one-electron reduction steps of riboflavin, FMN and FAD to their fully reduced forms. *Biochim. Biophys. Acta* 722:158–162.

414. Wardman, P. 1989. Reduction potentials of one-electron couples involving free radicals in aqueous solution. *J. Phys. Chem. Ref. Data* 18:1637–1755.

415. Harris, D. C., Rinehart, A. L., Hereld, D., Schwartz, R. W., Burke, F. P., and Salvador, A. P. 1985. Reduction potential of iron in transferrin. *Biochim. Biophys. Acta* 838:295–301.

416. Ahmad, R., and Armstrong, D. A. 1984. The effect of pH and complexation on redox reactions between RS˙, radicals and flavins. *Can. J. Chem.* 62:171–177.

417. Mitchell, J. B., Samuni, A., Krishna, M. C., DeGraff, W. G. Ahn, M. S., Samuni, R., and Russo, A. 1990. Biologically active metal-independent superoxide dismutase mimics. *Biochemistry* 29:2802–2807.

418. Schwartz, H. A. 1981. Free radicals generated by radiolysis of aqueous solutions. *J. Chem. Educ.* 58:101–105.

419. Lymar, S. V., and Hurst, J. K. 1995. Role of compartmentation in promoting toxicity of leukocyte-generated strong oxidants. *Chem. Res. Toxicol.* 8:833–840.

Chapter Three

VITAMIN E

Sharon Landvik

Vitamin E refers to a group of eight naturally occurring compounds: alpha-, beta-, gamma-, and delta-tocopherols and tocotrienols. Alpha-tocopherol, especially the naturally occurring *d*-alpha-tocopherol, has the highest biological activity (1). Vitamin E is the major lipid-soluble, chain-breaking antioxidant in body tissues, protecting cell membranes at an early stage of free radical attack (2, 3). Highly unstable free radicals attack the polyunsaturated fatty acids of membrane phospholipids, and can damage both the structure and function of cell membranes in a chain reaction unless they are quenched by an antioxidant (4, 5). Free radicals have been implicated in development of a number of degenerative diseases and conditions (6). The potential protective role of vitamin E in preventing or minimizing free radical damage associated with cardiovascular disease, cancer, premature aging, cataract, and air pollution has been under investigation. This review of vitamin E discusses absorption and transport, functions, sources, requirements and safety, and current research findings on efficacy of vitamin E.

ABSORPTION AND TRANSPORT

Absorption of vitamin E is dependent on an individual's ability to absorb fat. Tocopherol is absorbed into the lymphatic system from the intestinal tract and enters the blood as a component of the chylomicrons. The majority of vitamin E in plasma is in low-density lipoproteins (LDL) (1). Alpha-tocopherol accounts for almost 87% of the total tocopherol concentration in adult human plasma (7). There is a high correlation between serum tocopherol and total lipid concentrations. There are considerable variations in vitamin E concentrations in body tissues; the highest levels are in lipid-rich cell fractions (Table 1) (8). Vitamin E levels are most frequently measured in plasma and range from 0.5 to 1.6 mg/dl in normal populations (9).

Table 1. *d*-Alpha-tocopherol concentration of human tissues

Tissue	μg/g fresh weight
Adipose	150
Adrenal	132
Pituitary	40
Testis	40
Platelets	30
Heart	20
Muscle	19
Liver	13
Ovary	11
Plasma	9.5
Uterus	9
Kidney	7
Erythrocytes	2.3

CLINICAL DEFICIENCY

Clinical vitamin E deficiency has been documented in individuals with a chronic malabsorption syndrome, patients on total parenteral nutrition, and premature infants (2). Conditions interfering with normal digestion, absorption, or transport of fat have been associated with low serum vitamin E levels (9). Serum vitamin E levels can be below 20% of normal in patients with malabsorption syndromes such as celiac disease, biliary atresia, and cystic fibrosis. Serum vitamin E levels are often unmeasurable in patients with abetalipoproteinemia (5).

Severe and chronic vitamin E deficiency can lead to a progressive neurological syndrome, indicating the importance of vitamin E in optimal development and maintenance of the function and integrity of the nervous system and skeletal muscle (10). Characteristics of this syndrome include progressive neuropathy with absent or altered reflexes, ataxia, limb weakness, and sensory loss in the legs and arms. Improvement of neurological function has been documented with appropriate vitamin E therapy; progressive neurological damage may be prevented in children with prolonged cholestatic disease by early initiation of vitamin E therapy (11, 12).

Plasma vitamin E concentrations are frequently low in patients on total parenteral nutrition. Parenteral lipid emulsions contain primarily gamma- and delta-tocopherol homologs, which are much less biologically active forms than alpha-tocopherol. Plasma vitamin E levels cannot be maintained with high intakes of gamma and delta isomers; thus, alpha-tocopherol supplementation is required for patients on total parenteral nutrition (13, 14). Low-birth-weight premature infants are also vulnerable to vitamin E deficiency due to inadequate body stores, impaired absorption, and reduced transport capacity in the blood because of low levels of low-density lipoproteins at birth (15).

FUNCTIONS

Vitamin E is nature's most effective lipid-soluble antioxidant. The body's requirement for vitamin E is not met by other exogenous or endogenous antioxidants, even in cases of severe vitamin E deficiency (16). Vitamin E protects polyunsaturated fatty acids in cell membranes that are important for membrane function and structure (17, 18). Increased vitamin E intakes have been shown to enhance immune response. Vitamin E regulates platelet aggregation by inhibiting platelet cyclooxygenase activity and thus decreases prostaglandin (thromboxane) production. Vitamin E also has a role in mitochondrial function, nucleic acid and protein metabolism, and hormonal production, protects vitamin A from destruction in the body, and spares selenium (8, 19).

SOURCES

The richest dietary sources of vitamin E are vegetable oils (primarily soybean, sunflower, and corn oils), nuts, and sunflower seeds. Whole grains and wheat germ are also major sources. Foods of animal origin are generally low in vitamin E. The amount of vitamin E (as *d*-alpha-tocopherol) supplied by the "normal" U.S. diet is estimated to be 7.4–9 mg/day (11.0–13.4 IU) (8).

Vitamin E is the exception to the paradigm that synthetic and natural vitamins are equivalent because their molecular structures are identical. Natural-source vitamin E (*R,R,R*-alpha-tocopherol), derived from vegetable oils, is a single stereoisomer. Synthetic vitamin E (all-*rac*-alpha-tocopherol, also known as *dl*-alpha-tocopherol) is a mixture of eight stereoisomers, only one of which is *d*-alpha-tocopherol. The other seven isomers have different molecular configurations, all with lower biological activity than *d*-alpha-tocopherol (20).

Results of animal bioassays and human studies have demonstrated that the biological potency of natural forms of vitamin E is higher than that from synthetic sources. Research data also suggest that the lungs, red blood cells, blood plasma, and brain demonstrate preferential uptake of natural-source vitamin E compared to one of the isomers in synthetic vitamin E. Physiological differences between natural and synthetic vitamin E relate to the preferential retention of *d*-alpha-tocopherol in blood and tissues compared to other tocopherols (21, 22).

REQUIREMENTS

Adult vitamin E requirements may vary at least fivefold, depending on dietary and lifestyle habits and/or tissue composition from previous intake patterns. Even greater variability in vitamin

E requirements has been demonstrated in animals on a high polyunsaturated fatty acid intake. Serum or plasma vitamin E concentrations are generally considered to be the most convenient and useful measurement of vitamin E status. Individuals with plasma vitamin E levels of less than 0.5 mg/dl are usually considered to be vitamin E deficient (8). Daily vitamin E intakes of 10–30 mg in healthy adults will maintain serum vitamin E levels in the normal range (9). In a group of well-nourished adult subjects, mean plasma vitamin E levels were 1.06 mg/dl at baseline and doubled to 2.03 mg/dl after daily supplementation with 800 IU vitamin E for 8 wk (23).

An evaluation of vitamin E requirements should consider whether to focus on vitamin E intakes that are adequate to prevent deficiency symptoms and allow normal physiological function or on higher vitamin E intakes to prevent oxidative damage (2, 24). In a study of healthy adults, daily supplementation of 1000 IU vitamin E for 10 days significantly reduced breath pentane excretion. Based on these results, it may be inferred that there are undesirably high levels of lipid peroxidation in the body that can be decreased by vitamin E supplementation, which may be significant based on research showing involvement of free radical damage in body processes and certain diseases and the role of vitamin E in controlling or preventing lipid peroxidation (3, 25).

CARDIOVASCULAR DISEASE

It has been well established that cholesterol deposited in arteries originates primarily from LDL, and that elevated LDL levels are associated with an increased risk for atherosclerosis. In an early stage in development of atherosclerosis, there is accumulation in the arteries of foam cells, which are macrophages that have taken up oxidized LDL. These foam cells are a key component of the fatty streak lesion (26). LDL is an important target of free radicals, and LDL oxidation is considered to be an important event in development of atherosclerosis (27). Cell and animal data support the hypothesis that oxidative modification of LDL leads to their enhanced uptake by macrophages, leading to conversion of macrophages into foam cells, and that antioxidants may be protective against LDL oxidation (28–30).

In isolated-cell studies, cell-mediated oxidation of LDL was largely prevented over 24 h when high quantities of vitamin E were added to the culture medium. Supplementation of plasma with increasing levels of vitamin E before LDL isolation resulted in a proportional increase in the duration of the lag phase during which there was no detectable oxidative modification of LDL (26, 31). In a study of WHHL rabbits, vitamin supplementation with low dose vitamin E for 6 months significantly decreased the maximal oxidation rate of LDL (32).

In studies of healthy, nonsmoking subjects supplemented with vitamin E, there was a significant increase in resistance of LDL isolated from plasma to induced oxidation (33–37). Resistance of LDL to oxidation also increased significantly in a group of smokers supplemented with vitamin E (38), and in smokers supplemented with vitamins C and E, beta-carotene, and selenium (39).

The minimal dose of vitamin E required to protect LDL from oxidation was evaluated in a study of healthy young adults. Resistance of LDL to oxidation increased in a dose-dependent manner. The maximum rate of LDL oxidation was significantly decreased only after daily supplementation with 400 or 800 IU vitamin E (40). In another study evaluating the effectiveness of varying doses of vitamin E on induced LDL oxidation in healthy men, there was no significant effect of daily supplementation with 60 or 200 IU vitamin E for 8 wk compared to baseline values. Groups that received at least 400 IU vitamin E per day showed a decreased susceptibility of LDL to oxidation and significant inverse correlations between plasma and LDL alpha-tocopherol levels and the oxidation rate (41).

Studies in animals have also investigated possible protective effects of vitamin E on the development and progression of atherosclerosis. In specific types of hens and rabbits that are susceptible to development of atherosclerosis, early aortic lesion development was significantly inhibited by vitamin E supplementation (42–45). Prevention and regression of induced atherosclerosis by vitamin E were studied in male monkeys on an atherosclerosis-promoting diet. Stenosis progressed more rapidly and to a greater extent in unsupplemented monkeys compared to vitamin E-treated monkeys. Stenosis in the group of animals with established atherosclerosis significantly decreased from 33% to 8% after 8 mo of vitamin E therapy (46). In a study comparing the effectiveness of low-dose vitamin E or probucol for 6 mo on progression of atherosclerosis in WHHL rabbits, the average area of aorta covered with plaques was 58.7% in controls, 62.7%

in probucol-treated animals, and 48.9% in vitamin E-supplemented rabbits at the end of the study; the differences were not statistically significant (32).

A number of recent studies have investigated the relationship between plasma antioxidant levels and incidence of coronary heart disease in adult populations. In the WHO/MONICA study, which compared plasma antioxidant concentrations of middle-aged men (40–59 yr of age) from different European populations, there was a high inverse correlation between age-specific mortality from ischemic heart disease and lipid-standardized plasma vitamin E levels (47). In contrast, there was no consistent association between serum selenium, vitamin A, or vitamin E levels and risk of death from coronary heart disease in prospective epidemiological studies in Finland and the Netherlands (48, 49). However, data from the two studies in Finland and the Netherlands should be considered with reservation due to methodological problems including lack of standardization for cholesterol and triglycerides (50).

In a study in Poland, plasma vitamin E concentrations were significantly lower in patients with stable and unstable angina compared to healthy controls (51). In a study in the United Kingdom, there was a significant inverse relationship between plasma vitamin E levels and angina risk. According to the researchers, the results suggest that some populations with a high coronary heart disease incidence may benefit from a diet rich in antioxidants, particularly vitamin E (52). In a study of Finnish men and women initially free from heart disease, the relative risk of heart disease mortality was 32% lower for men and 65% lower for women in the highest tertile of dietary vitamin E intake compared to the lowest tertile (53).

In a study of 39,910 male health professionals in the United States, there was a 36% decreased relative risk of coronary heart disease in men who took more than 60 IU vitamin E per day compared to men who took less than 7.5 IU daily. Men consuming at least 100 IU vitamin E per day for at least 2 yr had a 37% lower relative risk of coronary heart disease than men who did not use vitamin E supplements (54).

Results of an 8-year study of 87,245 healthy nurses in the United States showed that women in the top fifth of vitamin E intake had a 34% lower relative risk of major coronary disease than women in the lowest fifth, after adjustment for age and smoking. Women taking vitamin E supplements of more than 100 mg per day for at least 2 years had a 48% lower relative risk of coronary heart disease. No protective effect was seen in women whose only source of vitamin E was from the diet. The researchers in both Harvard-based studies (health professionals and nurses) noted that these data and other evidence suggest that vitamin E supplements may decrease heart disease risk (55).

In a double-blind trial of male smokers in Finland, there was little difference in total incidence or mortality from coronary heart disease between the vitamin E-supplemented group and the placebo group. Rates of coronary heart disease were slightly lower in the vitamin E-supplemented group than in the placebo group (602 vs. 637 cases). Rates of ischemic stroke were slightly decreased among vitamin E-supplemented subjects (56 vs. 67 cases), and rates of hemorrhagic stroke were slightly increased compared to the placebo group (66 vs. 44 cases). However, implications of this study are limited because a criterion for entry into the study was a history of having smoked at least 20 cigarettes daily for 36 yr before the study was begun. (Smoking is an extremely strong risk factor for coronary heart disease, stroke and many cancers.) The daily dose of vitamin E (50 mg) was lower than that observed to be protective in prospective studies, and the length of follow-up was relatively short (5–8 yr) (56).

The association between vitamin E and vitamin C intake and coronary artery disease progression was recently evaluated in middle-aged men with previous coronary bypass graft surgery. Study results showed that men who took at least 100 IU vitamin E per day from supplements showed significantly less coronary artery lesion progression for all lesions and mild to moderate lesions than men who took less than 100 IU per day. Vitamin E did not have a significant effect on severe lesions (57).

Results of recent studies suggest that vitamin E and other antioxidants may be important factors or even preventers of coronary heart disease (50). Documentation of the specific protective role of the antioxidants against coronary heart disease is awaiting results of controlled intervention trials currently in progress and other planned studies (50, 52).

CANCER

Cancer is believed to be the result of external factors combined with a hereditary disposition for cancer. Approximately 75–80% of all human cancers are estimated to be environmentally induced, and dietary factors are considered to play a major role. Research evidence has demonstrated that free radicals frequently have a role in the process of cancer initiation and promotion. Based on the results of cell culture and animal research, it appears that vitamin E and other antioxidants may alter cancer incidence and growth through their action as anticarcinogens, quenching free radicals or reacting with their products. Although these studies do not provide conclusive documentation and a significant effect of vitamin E is not seen in all experimental models, the majority of these studies show a protective effect for vitamin E in regard to cancer risk in some sites (58–64).

Controlled human studies on the antioxidants and cancer are limited in number. However, the majority of available epidemiological data suggest that vitamin E and other antioxidants decrease the incidence of certain cancers (Table 2). Since the range of dietary vitamin E intake within a population may be narrow, a protective effect of vitamin E may not be fully demonstrated in all epidemiological studies evaluating serum vitamin E concentrations and cancer risk (65).

Several epidemiological studies in Finland have demonstrated an inverse correlation between serum antioxidant concentrations and subsequent cancer risk (65–68). Low blood levels of beta-carotene, vitamin C, or vitamin E were associated with increased risk for certain types of cancer in a study in Switzerland (69). However, in follow-up of a blood pressure risk study in the United States, it was concluded that there was no relationship between blood antioxidant levels and subsequent cancer risk, although blood vitamin E levels were somewhat lower in subjects who later developed cancer, but seemed to relate to cholesterol concentrations (70).

In a study of men of Japanese ancestry in Hawaii, there was a significant correlation between serum beta-carotene levels and subsequent lung cancer risk but not between serum vitamin A or vitamin E levels and subsequent risk of lung, bladder, or gastrointestinal cancers (71). In another study in the United States, serum beta-carotene and vitamin E concentrations showed a protective association with lung cancer (72). In two studies in England, men who were diagnosed as having cancer within 1 yr after blood collection had significantly lower mean serum vitamin E levels than controls. Women with low plasma vitamin E levels had a significantly increased risk of breast cancer (73, 74).

Results of a U.S. study demonstrated that newly diagnosed lung cancer patients had significantly lower average serum vitamin E and carotenoid concentrations than controls (75). Blood vitamin E levels were significantly lower in subjects who subsequently developed lung cancer than in controls in another study in the United States (76). Mean blood vitamin E and selenium levels were also significantly lower in a group of patients in Japan than in controls (77). In a study in Finland, the age-adjusted relative risk of lung cancer in the lowest tertile of intake compared to the highest tertile was 2.5 for carotenoids and 3.1 for vitamin C and vitamin E, using dietary intake data based on dietary history interviews (78). Dietary beta-carotene intake was associated with a significant decrease in lung cancer risk, and use of vitamin E supplements was also protective against lung cancer in a study in the United States (79).

Mean plasma vitamin E and beta-carotene concentrations were significantly decreased in women with cervical dysplasia or cancer compared with controls in a U.S. study (80). In a study of the relation of diet to invasive cervical cancer risk, a high dietary intake of vitamins C and E was associated with a significantly decreased risk of cervical cancer. Vitamin A and vitamin E supplementation was associated with a slightly reduced risk of cervical cancer (81).

Plasma vitamin C and vitamin E concentrations were significantly lower in a group of newly diagnosed oral cancer patients than in matched controls in a study in India (82). In a U.S. study, individuals who took vitamin E supplements had a significantly decreased risk of oral and pharyngeal cancer (83). The relationship between diet and incidence of oral and pharyngeal cancer among black Americans was evaluated in a multicenter study. A lower oral and pharyngeal cancer risk was associated with an increased intake of fiber and carotene and vitamins C and E in men and vitamin C and fiber in women (84).

In a study in Italy, gastric cancer risk increased with increasing intake of nitrites and protein and decreased with increasing intake of vitamins C and E, beta-carotene, and vegetable fat (85).

Table 2. Epidemiological studies of antioxidants and subsequent cancer risk

Cancer site	Number of subjects with cancer	Country	Findings	Reference
All sites	51	Finland	11.4-Fold adjusted risk of fatal cancer with low blood vitamin E and selenium levels	65
All sites	453 males	Finland	0.7 Adjusted relative risk of cancer in 2 highest quintiles of blood vitamin E levels	66
Reproductive organs	313 females	Finland	1.6-Fold cancer risk with serum vitamin E levels in the lowest quintile, 10-fold risk of breast cancer with low serum selenium and vitamin E levels	67
Upper gastrointestinal tract	150	Finland	Relative cancer risk of 2.2 in the 3 lowest quintiles of serum vitamin E and 3.3 in the lowest quintile of serum selenium	68
Colorectal			No inverse correlation between serum vitamin E and selenium levels and colorectal cancer risk	
Stomach	129 males	Switzerland	Beta-carotene, vitamin C, and Vitamin E blood levels were lower in cancer cases than in controls	69
Colon			Vitamin E blood levels were low in cancer cases compared to controls	
All sites	111	United States	No relationship between blood antioxidant levels and subsequent cancer risk	70
Lung	284 males	United States	Relative cancer risk of 3.4 with serum beta-carotene levels in the lowest quintile	71
Bladder, gastrointestinal tract			No association between serum vitamin A or E levels and cancer risk	
Nine primary sites	436	United States	Serum beta-carotene and vitamin E had a protective association with lung cancer	72
Breast	39 females	United Kingdom	Fivefold greater cancer risk for women with vitamin E blood levels in the lowest fifth than in the highest fifth	73
All sites	271 males	United Kingdom	Serum vitamin E levels were significantly lower in patients diagnosed with cancer within 1 yr after blood collection but not in other cancer subjects	74
Lung	59	United States	Serum vitamin E and carotenoid levels were significantly lower in cancer patients than in controls	75
Lung	99	United States	2.5-Fold higher cancer risk with serum vitamin E levels in the lowest quintile than in the highest quintile	76

Table 2. Epidemiological studies of antioxidants and subsequent cancer risk—Continued

Cancer site	Number of subjects with cancer	Country	Findings	Reference
Lung	55	Japan	Blood vitamin E selenium levels were significantly lower in cancer patients than in controls	77
Lung	117 males	Finland	Adjusted risk of cancer in nonsmokers in the lowest tertile of intake compared to the highest tertile was 2.5 for carotenoids and 3.1 for vitamins C & E	78
Lung	413 nonsmokers	United States	Dietary beta-carotene intake and vitamin E supplements were associated with decreased cancer risk	79
Cervix	116 females	United States	Blood beta-carotene and vitamin E levels were significantly lower in cancer cases than in controls	80
Cervix	189 females	United States	High vitamin C and vitamin E intake were associated with a significantly lower cancer risk	81
Mouth	24	India	Plasma vitamin C and vitamin E levels were significantly lower in cancer patients than in controls	82
Mouth and pharynx	1,103	United States	Vitamin E supplementation was associated with a significantly reduced cancer risk	83
Mouth and pharynx	190	United States	Lower cancer risk was associated with increased intake of carotene, fiber and vitamins C and E in males and of vitamin C and fiber in females	84
Stomach	1,016	Italy	Fivefold difference in cancer risk between high vitamins C and E intake & low protein and nitrite intake and low vitamins C and E intake and high protein and nitrite intake	85
Colon	212	United States	68% Lower relative cancer risk for subjects in the highest quintile of total vitamin E intake compared to the lowest quintile	86

The relative risk of subsequent development of colon cancer was 68% lower for subjects in the highest quintile of total vitamin E intake compared to the lowest quintile in a U.S. study. The decreased risk was primarily related to the use of vitamin E supplements (86).

Based on accumulating evidence from animal and human studies, it has been suggested that vitamin E and other antioxidants merit continued active research evaluation utilizing epidemiological studies and randomized placebo-controlled clinical trials to evaluate the role of antioxidant supplements in prevention of cancer (87). Three such large clinical trials have been recently completed, with varied results.

In a 5-yr study of 29,584 adults in Linxian, China, there was a significant 9% reduction in overall mortality in the group supplemented with vitamin E, beta-carotene, and selenium. The decreased mortality was primarily due to a 13% reduction in cancer (primarily esophageal and gastric cancer) (88).

A clinical trial of 864 patients who had previously had a colorectal adenoma removed tested the efficacy of beta-carotene and vitamins C and E in preventing the occurrence of new adenomas over a 4-yr period. Study results did not show any evidence that either beta-carotene or vitamins C and E decreased the incidence of colorectal adenomas. The researchers noted that the results are in apparent conflict with the reduced risk of invasive colorectal cancers suggested by epidemiological studies and that these studies may have detected an effect that occurs only after adenomas develop. Most of the detected adenomas in the present study were small, and only a small fraction of these adenomas would ever progress to cancer if untreated (89).

In a trial of 29,133 male smokers 50–69 yr of age from Finland who had smoked an average of 20.4 cigarettes per day for an average of 35.9 yr, there was no decrease in lung cancer incidence in men supplemented daily with 50 mg synthetic vitamin E. The incidence of prostate cancer was lower in the vitamin E-supplemented group than in the group on placebo (56). However, results of this study should be viewed with caution as it is unlikely that antioxidants or other nutrients could prevent damage associated with the high-risk behavior of these heavy smokers over a period of many years. Results of additional intervention trials currently in progress should provide more conclusive documentation of the specific role of vitamin E and other antioxidants in cancer prevention.

Free-radical-mediated cell damage has been implicated in development of pathological changes associated with aging (90). It has been suggested that free radical production may be a contributory factor in the depressed immune response documented in aged animals and that improved antioxidant status may have an immunostimulatory effect (91).

The effect of vitamin E supplementation (800 mg per day for 30 days) or placebo on cell-mediated immune response was evaluated in 32 healthy adults 60 yr of age or older. There was a significant improvement in delayed-type hypersensitivity skin test response in the vitamin E-supplemented group. Immune response was enhanced in most but not all of the vitamin E-supplemented subjects (92).

A study of 30 elderly patients who had been hospitalized for over 3 mo evaluated cell-mediated immune function prior to and following daily antioxidant supplementation with 8000 IU vitamin A, 100 mg vitamin C, and 50 mg vitamin E for 28 days. Cell-mediated immune function improved in the antioxidant-supplemented group but was unchanged in the group on placebo. As noted by the researchers, the results suggest that antioxidant supplementation with slightly higher than RDA levels can improve cell-mediated immunity. Additional research is required to determine whether long-term supplementation is associated with decreased morbidity in long-stay patients (93).

The effects of antioxidants on free radical concentrations were evaluated in a study in Poland of 100 subjects, 60–100 yr of age. Average blood malondialdehyde levels decreased 26% in subjects receiving 200 IU vitamin E, 13% in the group supplemented with 400 mg vitamin C, and 25% for combined supplementation with vitamins E and C (94).

In nursing home patients in Finland, there was a marked improvement in general condition after 2 mo in vitamin E- and selenium-supplemented subjects, which continued throughout the 1-yr study (95). In another study in Finland, levels of serum thiobarbituric acid reactants were initially higher in a group of elderly nursing home patients but declined to levels of younger controls after supplementation with vitamins E, C, and B_6, selenium, beta-carotene, and zinc for 3 mo (96).

Research is continuing on the protective role of vitamin E and other antioxidants in the aging process. Current research data suggest that free radicals significantly influence the aging process, that free-radical-mediated damage can be controlled with adequate antioxidant defenses, and that optimal antioxidant intake may lead to a healthier life.

CATARACT

Age is considered to be a major risk factor in cataract development. However, it is not clear whether cataracts develop as a result of cumulative insults over a lifetime or whether decreased resistance or repair capacity of the lens—or the aging process itself—increases an individual's susceptibility to cataract formation (97, 98). The lens of the eye is very susceptible to light-induced lipid peroxidation, and oxidation is believed to be an early and significant event in development of most cases of senile cataract (97, 99).

Animal studies have demonstrated that vitamin E can arrest and reverse cataract development to some extent. In isolated animal lenses and animal models, vitamin E delayed or minimized cataract development induced by experimental oxidative stress (100–109). Recent epidemiological data have also suggested a relationship between cataract risk and antioxidant status.

In a comparison of self-reported vitamin supplementation by 175 subjects with cataracts and 175 individually matched cataract-free subjects, significantly more supplementary vitamins C and E were taken by the cataract-free group. In the group of subjects who took only vitamin E supplements, cataract risk was 56% lower than in the group who did not take vitamin E. In subjects using vitamin C supplements alone, there was a 70% decrease in cataract risk compared to subjects who did not take vitamin C (99).

The correlation between antioxidant status and senile cataract was studied in 112 subjects 40–70 yr of age. Study results suggest that high plasma concentrations of at least two of the three antioxidant vitamins (vitamins E and C and carotenoids) were associated with a significantly decreased risk of cataract development compared to low plasma concentrations of one or more of these vitamins. The odds ratio for senile cataract (controlled for age, race, sex, and diabetes) was 0.2 for subjects with a high serum antioxidant status (110).

The association between serum concentrations of vitamin E, beta-carotene, and selenium and subsequent risk of senile cataract was evaluated over a 15-yr period in a study in Finland. Low serum vitamin E and beta-carotene concentrations predicted an increased cataract risk. The odds ratio for cataract risk was 2.6 for patients with serum vitamin E and beta-carotene concentrations in the lowest third. The researchers concluded that low serum vitamin E and beta-carotene concentrations are risk factors for senile cataract and that controlled studies on the role of antioxidant vitamins in cataract prevention are warranted (111).

AIR POLLUTION

Cigarette smoke contains numerous substances that are known to be oxidants or free radicals (112). Nitrogen dioxide and ozone are present in very high concentrations in polluted environments and can initiate free radical reactions that lead to lung damage (113). A protective role for vitamin E against the damaging effects of smoke and smog has been demonstrated in animal studies (114–118).

Protective effects of vitamin E against pollution damage have also been evaluated in humans. In a study of the effects of daily vitamin E supplementation (600 mg) on red blood cell susceptibility of red blood cells to ozone-related free radical damage, vitamin E significantly protected red blood cells at the highest levels of hydrogen peroxide-induced stress but not at lower levels (119). Vitamin E supplementation was not protective against biological responses to short-term ozone in other studies of Los Angeles residents exposed to photochemical smog. However, according to the researchers, it is possible that vitamin E supplementation may have a protective effect on human lung tissue, where free radical concentrations may be higher than in the blood (120, 121).

Since smokers inhale high levels of free radicals in the gaseous and tar phases of tobacco, an increased antioxidant intake may be beneficial in the lung's defense against free radical damage (122). In a study of young adult smokers, the lower respiratory tract fluid of smokers was deficient in vitamin E compared to nonsmokers. Vitamin E supplementation (2400 IU per day for 3 wk) led to increased vitamin E levels in lower respiratory tract fluid. However, vitamin E levels remained much lower than baseline levels of nonsmokers, demonstrating that vitamin E utilization may be increased in lung cells of smokers (112).

In another study, red blood cells of smokers showed increased peroxidation when incubated with hydrogen peroxide compared to nonsmokers; this effect was inhibited when smokers were supplemented with vitamin E (1000 IU per day for 2 wk). Plasma levels of free radical products were much higher in unsupplemented smokers than in vitamin E-supplemented smokers or nonsmokers. As noted by the researchers, study results suggest that cigarette smoking affects antioxidant status (122).

Initial breath pentane output was significantly higher in a group of smokers than in nonsmokers, although both groups had similar plasma vitamin E levels. When smokers were supplemented with vitamin E (800 mg per day for 2 wk), breath pentane excretion decreased significantly, but remained significantly higher than in nonsmokers. The researchers concluded that a normal plasma vitamin E concentration does not prevent the increased peroxidation observed in smokers but that lipid peroxidation can be significantly decreased by vitamin E supplementation. It was further noted that the recommended daily allowance for vitamin E may be insufficient for individuals exposed to cigarette smoke (123). Quitting smoking is obviously the best approach to eliminate smoke-related damage and disease. However, for individuals who continue to smoke and for passive smokers, research results suggest that vitamin E can help protect the lungs from damage.

SAFETY

Research in rats has demonstrated that the acute and chronic toxicities of oral vitamin E are very low and that vitamin E is not mutagenic, teratogenic, or carcinogenic. Studies in a number of animal species have shown that oral vitamin E intakes of up to 200 mg/kg body weight were well tolerated, without evidence of side effects (124).

In double-blind, placebo-controlled human studies, there were very few observed side effects, and no specific side effect was consistently seen in all studies. Side effects of oral vitamin E intake were also uncommon in other clinical studies that were not necessarily double-blind. The majority of side effects attributed to vitamin E were reported in letters to the editor as uncontrolled studies or individual case studies. Most of these reported side effects were not observed in the larger, well-controlled clinical trials (124, 125). However, it should be noted that high oral intakes of vitamin E can exacerbate the blood coagulation defect of vitamin K deficiency due to malabsorption or anticoagulant therapy. Thus, high levels of vitamin E intake are contraindicated in such conditions (125). Blood evaluations in studies of normal adults on vitamin E supplementation did not demonstrate any blood coagulation abnormalities (126, 127).

Based on a review of animal and human data on safety and tolerance of oral vitamin E, it may be concluded that vitamin E is safe at levels commonly consumed. Except for interaction with vitamin K in patients on anticoagulant therapy, there are no specific side effects associated with oral vitamin E intake.

SUMMARY

Free-radical-mediated cell damage has been implicated in the development of a number of degenerative conditions and diseases. Research evidence has shown that increased intakes of vitamin E and other antioxidants may help protect the body from cumulative oxidative damage. As research continues on the specific role of vitamin E and other antioxidants in preventing or delaying development of certain degenerative diseases, increased intake of these antioxidants can provide protection from the increasingly high free radical levels present in the environment and associated with lifestyle habits.

REFERENCES

1. Bjorneboe, A., Bjorneboe, G., and Drevon, C. 1990. Absorption, transport and distribution of vitamin E. *J. Nutr.* 120:233–242.
2. Horwitt, M. K. 1986. Interpretations of requirements for thiamin, riboflavin, niacin-tryptophan, and vitamin E plus comments on balance studies and vitamin B_6. *Am. J. Clin. Nutr.* 44:973–985.
3. Van Gossum, A., Kurian, R., Whitwell, J., and Jeejeebhoy, K. N. 1988. Decrease in lipid peroxidation measured by breath pentane output in normals after oral supplementation with vitamin E. *Clin. Nutr.* 7:53–57.
4. Oski, F. A. 1980. Vitamin E—A radical defense. *N. Engl. J. Med.* 303:454–455.
5. Fritsma, G. A. 1983. Vitamin E and autoxidation. *Am. J. Med. Technol.* 49:453–456.
6. Cross, C. E. 1987. Oxygen radicals and human disease. *Ann. Intern. Med.* 107:526–545.

7. Gonzalez, M. J. 1990. Serum concentrations and cellular uptake of vitamin E. *Med. Hypotheses* 32:107–110.

8. Machlin, L. J. 1991. Vitamin E. In *Handbook of Vitamins*, pp. 99–144. New York: Marcel Dekker.

9. Carpenter, D. 1985. Vitamin E deficiency. *Semin. Neurol.* 5:283–287.

10. Sokol, R. J. 1988. Vitamin E deficiency and neurologic disease. *Annu. Rev. Nutr.* 8:351–373.

11. Satya-Murti, S., Howard, L., Krohel, G., and Wolf, B. 1986. The spectrum of neurologic disorder from vitamin E deficiency. *Neurology* 36:917–921.

12. Vitamin E deficiency and neurologic dysfunction. 1986. *Nutr. Rev.* 44:268–269.

13. Vandewoude, M. G., Vandewoude, M. F. J., and DeLeeuw, I. H. 1986. Vitamin E status in patients on parenteral nutrition receiving intralipid. *J. Parenteral Enteral. Nutr.* 10:303–305.

14. Kelly, F. J., and Sutton, G. L. J. 1989. Plasma and red blood cell vitamin E status of patients on total parenteral nutrition. *J. Parenteral Enteral. Nutr.* 13:510–515.

15. Lloyd, J. K. 1990. The Importance of vitamin E in human nutrition. *Acta Pediatr. Scand.* 79:6–11.

16. Ingold, K. U., Webb, A. C., Witter, D., Burton, G. W., Metcalfe, T. A., and Muller, D. P. R. 1987. Vitamin E remains the major lipid-soluble, chain-breaking antioxidant in human plasma even in individuals suffering severe vitamin E deficiency. *Arch. Biochem. Biophys.* 259:224–225.

17. Burton, G. W., Joyce, A. and Ingold, K. U. 1982. First proof that vitamin E is major lipid-soluble, chain-breaking antioxidant in human blood plasma. *Lancet* 2:327.

18. Inhibition of free radical chain oxidation by alpha-tocopherol and other plasma antioxidants. 1988. *Nutr. Rev.* 46:206–207.

19. Watson, R. R., and Leonard, T. K. 1986. Selenium and vitamins A, E and C: Nutrients with cancer prevention properties. *J. Am. Diet. Assoc.* 86:505–510.

20. Howitt, M. K. 1986. The promotion of vitamin E. *J. Nutr.* 116:1371–1377.

21. Cheng, S. C., Burton, G. W., Ingold, K. U., and Foster, D. O. 1987. Chiral discrimination in the exchange of alpha-tocopherol stereoisomers between plasma and red blood cells. *Lipids* 22:469–473.

22. Ingold, K. U., Burton, G. W., and Foster, D. O. 1987. Biokinetics of and discrimination between dietary *RRR*- and *SRR*-alpha-tocopherols in the male rat. *Lipids* 22:163–172.

23. Willet, W. C., Stampfer, M. J., Underwood, B. A., Taylor, J. O., and Hennekens, C. H. 1983. Vitamins A, E, and carotene: Effects of supplementation on their plasma levels. *Am. J. Clin. Nutr.* 38:559–566.

24. Jacobson, H. N. 1987. Dietary standards and future developments. *Free Radic. Biol. Med.* 3:209–213.

25. Lemoyne, M., Van Gossum, A., Kurian, R., Ostro, M., Axler, J., and Jeejeebhoy, K. N. 1987. Breath pentane analysis as an index of lipid peroxidation: A functional test of vitamin E status. *Am. J. Clin. Nutr.* 46:267–272.

26. Esterbauer, H., Dieber-Rotheneder, M., Waeg, G., Striegl, G., and Jurgens, G. 1990. Biochemical, structural and functional properties of oxidized low-density lipoprotein. *Chem. Res. Toxicol.* 3:77–92.

27. Sato, S., Niki, E., and Shimasaki, H. 1990. Free radical-mediated chain oxidation of low-density lipoprotein and its synergistic inhibition by vitamin E and vitamin C. *Arch. Biochem. Biophys.* 279:402–405.

28. Esterbauer, H., Jurgens, G., Quehenberger, O., and Koller, E. 1987. Autoxidation of human low-density lipoprotein: Loss of polyunsaturated fatty acids and vitamin E and generation of aldehydes. *J. Lipid Res.* 28:495–509.

29. Esterbauer, H., Rotheneder, M., Striegl, G., Waeg, G., Ashy, A., Sattler, W., and Jurgens, G. 1989. Vitamin E and other lipophilic antioxidants protect LDL against oxidation. *Food Sci. Technol.* 91:316–324.

30. Steinberg, D., Parthasarathy, S., Carew, T. E., Khoo, J. C., and Witztum, J. L. 1989. Beyond cholesterol-modifications of low-density lipoprotein that increase its atherogenicity. *N. Engl. J. Med.* 320:915–924.

31. Jessup, W., Rankin, S. M., DeWhalley, C. V., Hoult, J. R. S., Scott, J., and Leake, D. S. 1990. Alpha-tocopherol consumption during low-density lipoprotein oxidation. *Biochem. J.* 265:399–405.

32. Kleinveld, H. A., Demacker, P. N. M., and Stalenhoef, A. F. H. 1994. Comparative study on the effect of low-dose vitamin E and probucol on the susceptibility of LDL to oxidation and the progression of atherosclerosis in Watanabe heritable hyperlipidemic rabbits. *Arterioscler. Thromb.* 14:1386–1391.

33. Abbey, M., Nestel, P. J. and Baghurst, P. A. 1993. Antioxidant vitamins and low-density lipoprotein oxidation. *Am. J. Clin. Nutr.* 58:525–532.

34. Rifici, V. A., and Khachadurian, A. K. 1993. Dietary supplementation with vitamins C and E inhibits in vitro oxidation of lipoproteins. *J. Am. Coll. Nutr.* 12:631–637.

35. Jialal, I., and Grundy, S. M. 1993. Effect of combined supplementation with alpha-tocopherol, ascorbate and beta-carotene on low-density lipoprotein oxidation. *Circulation* 88:2780–2786.

36. Reaven, P. D., Khouw, A., Beltz, W. F., Parthasarathy, S., and Witztum, J. L. 1993. Effect of dietary antioxidant combinations in humans. Protection of LDL by vitamin E but not by beta-carotene. *Arterioscler. Thromb.* 13:590–600.

37. Suzukawa, M., Ishikawa, T., Yoshida, H., and Nakamura, H. 1995. Effect of in vivo supplementation

with low-dose vitamin E on susceptibility of low-density lipoprotein and high-density lipoprotein to oxidative modification. *J. Am. Coll. Nutr.* 14:46–52.

38. Princen, H. M. G., Van Poppel, G., Vogelezang, C., Buytenhek, R., and Kok, F. J. 1992. Supplementation with vitamin E but not beta-carotene in vivo protects low-density lipoprotein from lipid peroxidation in vitro. Effect of cigarette smoking. *Arterioscler. Thromb.* 212:554–562.

39. Nyyssonen, K., Porkkala, E., Salonen, R., Korpela, H., and Salonen, J. T. 1994. Increase in oxidation resistance of atherogenic serum lipoproteins following antioxidant supplementation: A randomized double-blind placebo-controlled clinical trial. *Eur. J. Clin. Nutr.* 48:633–642.

40. Princen, H. M. G., Van Duyvenvoorde, W., Buytenhek, R., van der Laarse, A., van Poppel, G., Leuven, J. A. G., and van Hinsbergh, V. W. M. 1995. Supplementation with low doses of vitamin E protects LDL from lipid peroxidation in men and women. *Arterioscler. Thromb. Vasc. Biol.* 15:325–333.

41. Jialal, I., Fuller, C. J., and Huet, B. A. 1995. The effect of alpha-tocopherol supplementation on LDL oxidation. *Arterioscler. Thromb. Vasc. Biol.* 15:190–198.

42. Smith, T. L., and Kummerow, F. A. 1989. Effect of dietary vitamin E on plasma lipids and atherogenesis in restricted ovulator chickens. *Atherosclerosis* 75:105–109.

43. Wojeicki, J., Rozewicka, L., Barcew-Wiszniewska, B., Samockowiec, L., Juzwiak, S., Kadlubowska, D., Tustanowski, S., and Juzyszyn, Z. 1991. Effect of selenium and vitamin E on the development of experimental atherosclerosis in rabbits. *Atherosclerosis* 87:9–16.

44. Williams, R. J., Motteram, J. M., Sharp, C. H., and Gallagher, P. J. 1992. Dietary vitamin E and the attenuation of early lesion development in modified Watanabe rabbits. *Atherosclerosis* 94:153–159.

45. Willingham, A. K., Bolanos, C., Bohannan, E., and Canedella, R. J. 1993. The effects of high levels of vitamin E on the progression of atherosclerosis in the Watanabe heritable hyperlipidemic rabbit. *J. Nutr. Biochem.* 4:651–654.

46. Verlangieri, A. J., and Bush, M. J. 1992. Effects of d-alpha-tocopherol supplementation on experimentally induced primate atherosclerosis. *J. Am. Coll. Nutr.* 11:130–137.

47. Gey, K. F., and Puska, P. 1989. Plasma vitamins E and A inversely correlated to mortality from ischemic heart disease in cross-cultural epidemiology. *Ann. N.Y. Acad. Sci.* 570:268–282.

48. Salonen, J. T., Salonen, R., and Penttilae, I. 1985. Serum fatty acids, apolipoproteins, selenium and vitamin antioxidants and the risk of death from coronary artery disease. *Am. J. Cardiol.* 56:226–231.

49. Kok, F., De Bruijn, A. M., and Vermeeren, R. 1987. Serum selenium, vitamin antioxidants and cardiovascular mortality: A 9 year follow-up study in the Netherlands. *Am. J. Clin. Nutr.* 45:462–468.

50. Gey, K. F., Puska, P., Jordan, P., and Moser, U. K. 1991. Inverse correlation between plasma vitamin E and mortality from ischemic heart disease in cross-cultural epidemiology. *Am. J. Clin. Nutr.* 53:326S–334S.

51. Sklodowska, M., Wasowicz, W., Gromadzinska, J., Miroslaw, W., Strzelczyk, M., Malczyk, J., and Goch, J. H. 1991. Selenium and vitamin E concentrations in plasma and erythrocytes of angina pectoris patients. *Trace Elem. Med.* 8:113–117.

52. Riemersma, R. A., Wood, D. A., Macintyre, C. C. A., Elton, R. A., Gey, K. F., and Oliver, M. F. 1991. Risk of angina pectoris and plasma concentrations of vitamins A, C and E and carotene. *Lancet* 337:1–5.

53. Knekt, P., Reunanen, A., Jarvinen, R., Seppanen, R., Heliovaara, M., and Aromaa, A. 1994. Antioxidant vitamin intake and coronary mortality in a longitudinal population study. *Am. J. Epidemiol.* 139:1180–1189.

54. Rimm, E. B., Stampfer, M. J., Ascherio, A., Giovannucci, E., Colditz, G. A., and Willett, W. C. 1993. Vitamin E consumption and the risk of coronary heart disease in men. *N. Engl. J. Med.* 328:1450–1456.

55. Stampfer, M. J., Hennekens, C. H., Manson, J. E., Colditz, G. A., Rosner, B., and Willett, W. C. 1993. Vitamin E consumption and risk of coronary disease in women. *N. Engl. J. Med.* 328:1444–1449.

56. Alpha-Tocopherol, Beta-Carotene Cancer Prevention Study Group. 1994. The effect of vitamin E and beta-carotene on the incidence of lung cancer and other cancers in male smokers. *N. Engl. J. Med.* 330:1029–1035.

57. Hodis, H. N., Mack, W. J., LaBree, L., Cashin-Hemphill, L., Sevanian, A., Johnson, R., and Azen, S. P. 1995. Serial coronary angiographic evidence that antioxidant vitamin intake reduces progression of coronary artery atherosclerosis. *J. A. M. A.* 273:1849–1854.

58. Borek, C., Ong, A., Mason, H., Donahue, L., and Biaglow, J. E. 1986. Selenium and vitamin E inhibit radiogenic and chemically induced transformation in vitro via different mechanisms. *Proc. Natl. Acad. Sci. U.S.A.* 83:1490–1494.

59. Horvath, P. M., and Ip, C. 1983. Synergistic effect of vitamin E and selenium in the chemoprevention of mammary carcinogenesis in rats. *Cancer Res.* 43:5335–5341.

60. Odeleye, O. E., Eskelson, C. D., Mufti, S. I., and Watson, R. R. 1992. Vitamin E inhibition of lipid peroxidation and ethanol mediated promotion of esophageal tumorigenesis. *Nutr. Cancer* 17:223–234.

61. Trickler, D., and Shklar, G. 1987. Prevention by vitamin E of experimental oral carcinogenesis. *J. Natl. Cancer Inst.* 78:165–167.

62. Shklar, G., Schwartz, J., Trickler, D., and Niukian, K. 1987. Regression by vitamin E of experimental oral cancer. *J. Natl. Cancer Inst.* 78:987–992.

63. Shklar, G., Schwartz, J., Trickler, D., and Reid, S. 1989. Regression of experimental cancer by oral administration of combined alpha-tocopherol and beta-carotene. *Nutr. Cancer* 12:321–325.

64. Shklar, G., Schwartz, J., Trickler, D., and Reid, S. 1990. Prevention of experimental cancer and immunostimulation by vitamin E (immunosurveillance). *J. Oral Pathol. Med.* 19:60–64.

65. Knekt, P., Aromaa, A., Maatela, J., Aaran, R., Nikkari, T., Hakama, M., Hakulinen, T., Peto, R., Saxen, E., and Teppo, L. 1988. Serum vitamin E and risk of cancer among Finnish men during a ten-year follow-up. *Am. J. Epidemiol.* 127:28–41.

66. Salonen, J. T., Salonen, R., Lappetelainen, R., Maenpaa, P., Alfthan, G., and Puska, P. 1985. Risk of cancer in relation to serum concentrations of selenium and vitamins A and E: Matched case-control analysis of prospective data. *Br. Med. J.* 290:417–420.

67. Knekt, P. 1988. Serum vitamin E level and risk of female cancers. *Int. J. Epidemiol.* 17:281–286.

68. Knekt, P., Aromaa, A., Maatela, J., Alfthan, G., Aaran, R., Teppo, L., and Hakama, M. 1988. Serum vitamin E, serum selenium and the risk of gastrointestinal cancer. *Int. J. Cancer* 42:846–850.

69. Stahelin, H. B., Rosel, F., Buess, E., and Brubacher, G. 1984. Cancer, vitamins, and plasma lipids: Prospective basel study. *J. Natl. Cancer Inst.* 73:1463–1468.

70. Willett, W. C., Polk, B. F., Underwood, B. A., Stampfer, M. J., Pressel, S., Rosner, B., Taylor, J. O., Schneider, K., and Hames, C. G. 1984. Relation of serum vitamins A and E and carotenoids to the risk of cancer. *N. Engl. J. Med.* 310:430–434.

71. Nomura, A. M. Y., Stemmermann, G. N., Heilbrun, L. K., Salkeld, R. M., and Vuilleumier, J. P. 1985. Serum vitamin levels and the risk of cancer of specific sites in men of Japanese ancestry in Hawaii. *Cancer Res.* 45:2369–2372.

72. Comstock, G. W., Helzlsouer, K. J., and Bush, T. L. 1991. Prediagnostic serum levels of carotenoids and vitamin E as related to subsequent cancer in Washington County, Maryland. *Am. J. Clin. Nutr.* 53:260S–264S.

73. Wald, N. J., Boreham, J., Hayward, J. L., and Bulbrook, R. D. 1984. Plasma retinol, beta-carotene and vitamin E levels in relation to the future risk of breast cancer. *Br. J. Cancer* 49:321–324.

74. Wald, N. J., Thompson, S. G., Densem, J. W., Boreham, J., and Bailey, A. 1987. Serum vitamin E and subsequent risk of cancer. *Br. J. Cancer* 56:69–72.

75. Le Gardeur, B. Y., Lopez, S. A., and Johnson, W. D. 1990. A case-control study of serum vitamins A, E, and C in lung cancer patients. *Nutr. Cancer* 14:133–140.

76. Menkes, M. S., Comstock, G. W., Vuilleumier, J. P., Helsing, K. J., Rider, A. A., and Brookmeyer, R. 1986. Serum beta-carotene, vitamins A and E, selenium, and the risk of lung cancer. *N. Engl. J. Med.* 315:1250–1254.

77. Miyamoto, H., Araya, Y., Ito, M., Isobe, H., Dosaka, H., Shimizu, T., Kishi, F., Yamamoto, I., Honma, H., and Kawakami, Y. 1987. Serum selenium and vitamin E concentrations in families of lung cancer patients. *Cancer* 60:1159–1162.

78. Knekt, P., Jarvinen, R., Seppanen, R., Rissanen, A., Aromaa, A., Heinonen, O. P., Albanes, D., Heinonen, M., Pukkala, E., and Teppo, L. 1991. Dietary antioxidants and the risk of lung cancer. *Am. J. Epidemiol.* 134:471–479.

79. Mayne, S. T., Janerich, D. T., Greenwald, P., Chorost, S., Tucci, S., Zaman, M. B., Melamed, M. R., Kiely, M., and McKneally, M. F. 1994. Dietary beta-carotene and lung cancer risk in U.S. nonsmokers. *J. Natl. Cancer Inst.* 86:33–38.

80. Palan, P. R., Mikhail, M. S., Basu, J., and Romney, S. L. 1991. Plasma levels of antioxidant beta-carotene and alpha-tocopherol in uterine cervix dysplasias and cancer. *Nutr. Cancer* 15:13–20.

81. Verreault, R., Chu, J., Mandelson, M., and Shy, K. 1989. A case-control study of diet and invasive cervical cancer. *Int. J. Cancer* 43:1050–1054.

82. Manoharan, S., and Nagini, S. 1994. Lipid peroxidation and antioxidant status in oral cancer patients. *Med. Sci. Res.* 22:291–292.

83. Gridley, G., McLaughlin, J. K., Block, G., Blot, W. J., Gluch, M., and Fraumeni, J. F. 1992. Vitamin supplement use and reduced risk of oral and pharyngeal cancer. *Am. J. Epidemiol.* 135:1083–1092.

84. Gridley, G., McLaughlin, J. K., Block, G., Blot, W. J., Winn, D. M., Greenberg, R. S., Schoenburg, J. B., Preston-Martin, S., Austin, D. F., and Fraumeni, J. F. 1990. Diet and oral and pharyngeal cancer among blacks. *Nutr. Cancer* 14:219–225.

85. Buiatti, E., Palli, D., Decarli, A., Amadori, D., Avellini, C., Bianchi, S., Bonaguri, C., Cipriani, F., Cocco, P., Giacosa, A., Marubini, E., Miracci, C., Puntoni, R., Russo, A., Vindigni, C., Fraumeni, J. F., and Blot, W. J. 1990. A case-control study of gastric cancer and diet in Italy. II. Association with nutrients. *Int. J. Cancer* 45:896–901.

86. Bostick, R. M., Potter, J. D., McKenzie, D. R., Sellers, T. A., Kushi, L. H., Steinmetz, K. A., and Folsom, A. R. 1993. Reduced risk of colon cancer with high intake of vitamin E: The Iowa Women's Health Study. *Cancer Res.* 53:4230–4237.

87. Hennekens, C. H., Stampfer, M. J., and Willett, W. 1984. Micronutrients and cancer chemoprevention. *Cancer Detect. Prev.* 7:147–158.

88. Blot, W. J., Li, J. Y., Taylor, P. R., Guo, W., Dawsey, S., Wang, G. Q., Yang, C. S., Zheng, S. F., Gail, M., Li, G. Y., Yu, Y., Liu, B. Q., Tangrea, J., Sun, Y. H., Liu, F., Fraumeni, J. F., Zhang, Y. H., and Li, B. 1993. Nutrition intervention trials in Linxian, China: Supplementation with specific vitamin, mineral combinations, cancer incidence and disease-specific mortality in the general population. *J. Natl. Cancer Inst.* 85:1483–1492.

89. Greenberg, G. R., Baron, J. A., Tosteson, T. D., Freeman, D. H., Beck, G. J., Bond, J. H., Colacchio, T. A., Coller, J. A., Frankl, H. D., Haile, R. W., Mandel, J. S., Nierenberg, D. W., Rothstein, R., Snover, D. C., Stevens, M. M., Summers, R. W., and van Stolk, R. U. 1994. A clinical trial of antioxidant vitamins to prevent colorectal adenoma. *N. Engl. J. Med.* 331:141–147.

90. Harman, D. 1984. Free radical theory of aging: The free radical diseases. *Age* 7:111–131.

91. Meydani, S. N., Meydani, M., Barkland, P. M., Liu, S., Miller, R. A., Cannon, J. G., Rocklin, R., and Blumberg, J. B. 1989. Effect of vitamin E supplementation on immune responsiveness of the aged. *Ann. N.Y. Acad. Sci.* 570:283–290.

92. Meydani, S. N., Barklund, M. P., Liu, S., Meydani, M., Miller, R. A., Cannon, J. G., Morrow, F. D., Rocklin, R., and Blumberg, J. B. 1990. Vitamin E supplementation enhances cell mediated immunity in healthy elderly subjects. *Am. J. Clin. Nutr.* 52:557–563.

93. Penn, N. D., Purkins, L., Kelleher, J., Heatley, R. V., Mascie-Taylor, B. H., and Belfield, P. W. 1991. The effect of dietary supplementation with vitamins A, C and E on cell mediated immune function in elderly long stay patients: A randomized controlled trial. *Age Ageing* 20:169–174.

94. Wartanowicz, M., Panczenko-Kresowska, B., Ziemlanski, S., Kawalska, M., and Okolska, G. 1984. The effect of alpha-tocopherol and ascorbic acid on the serum lipid peroxide level in elderly people. *Ann. Nutr. Metab.* 28:186–191.

95. Tolonen, M., Halme, M., and Sarna, S. 1985. Vitamin E and selenium supplementation in geriatric patients. *Biol. Trace Element Res.* 7:161–168.

96. Tolonen, M., Sarna, S., Halme, M., Tuominen, S. E. J., Westermarck, T., Nordberg, U. R., Keinonen, M., and Schrijver, J. 1988. Antioxidant supplementation decreases TBA reactants in serum of elderly. *Biol. Trace Element Res.* 17:221–228.

97. Bunce, G. E., and Hess, J. L. 1988. Cataract—What is the role of nutrition in lens health? *Nutr. Today* 23:6–12.

98. Taylor, A. 1989. Associations between nutrition and cataract. *Nutr. Rev.* 47:225–234.

99. Robertson, J. M., Donner, A. P., and Trevithick, J. R. 1989. Vitamin E intake and risk of cataracts in humans. *Ann. N.Y. Acad. Sci.* 570:372–382.

100. Trevithick, J. R., Creighton, M. O., Ross, W. M., Stewart-DeHaan, P. J., and Sanwal, M. 1981. Modelling cortical cataractogenesis. 2. In vitro effects on the lens of agents preventing glucose- and sorbitol-induced cataracts. *Can. J. Ophthalmol.* 16:32–38.

101. Ross, W. M., Creighton, M. O., Inch, W. R., and Trevithick, J. R. 1983. Radiation cataract formation diminished by vitamin E in rat lenses in vitro. *Exp. Eye Res.* 36:645–653.

102. Stewart-DeHaan, P. J., Creighton, M. O., Sanwal, M., Ross, W. M., and Trevithick, J. R. 1981. Effects of vitamin E on cortical cataractogenesis induced by elevated temperature in intact rat lenses in medium 199. *Exp. Eye Res.* 32:54–60.

103. Creighton, M. O., Ross, W. M., Stewart-DeHaan, P. J., Sanwal, M., and Trevithick, J. R. 1985. Modelling cortical cataractogenesis. VII. Effects of vitamin E treatment on galactose-induced cataracts. *Exp. Eye Res.* 40:213–222.

104. Ross, W. M., Creighton, M. O., Stewart-DeHaan, P. J., Sanwal, M., Hirst, M., and Trevithick, J. R. 1982. Modelling cortical cataractogenesis: 3. In vivo effects of vitamin E on cataractogenesis in diabetic rats. *Can. J. Opthalmol.* 17:61–66.

105. Creighton, M. O., Sanwal, M., Stewart-DeHaan, P. J., and Trevithick, J. R. 1983. Modelling cortical cataractogenesis. V. Steroid cataracts induced by solumedrol partially prevented by vitamin E in vitro. *Exp. Eye Res.* 37:65–76.

106. Gupta, P. P., Pandey, D. J., Sharma, A. L., Srivastava, R. K., and Mishra, S. S. 1984. Prevention of experimental cataract by alpha-tocopherol. *Indian J. Exp. Biol.* 22:620–622.

107. Varma, S. D., Chand, D., Sharma, Y. R., Kuck, J. F., and Richards, R. D. 1984. Oxidative stress on lens and cataract formation: Role of light and oxygen. *Current Eye Res.* 3:35–57.

108. Bhuyan, K. C., Bhuyan, D. K., and Podos, S. M. 1982. The role of vitamin E in therapy of cataract in animals. *Ann. N.Y. Acad. Sci.* 393:169–171.

109. Ross, W. M., Creighton, M. O., and Trevithick, J. R. 1990. Radiation cataractogenesis induced by neutron or gamma irradiation in the rat lens is reduced by vitamin E. *Scanning Microsc.* 4:641–650.

110. Jacques, P. F., Chylack, L. T., McGandy, R. B., and Hartz, S. C. 1988. Antioxidant status in persons with and without senile cataract. *Arch. Ophthalmol.* 106:337–340.

111. Knekt, P., Heliovaara, M., Rissanen, A., Aromaa, A., and Aaran, R. K. 1992. Serum antioxidant vitamins and risk of cataract. *Br. Med. J.* 305:1392–1394.
112. Pacht, E. R., Kaseki, H., Mohammed, J. R., Cornwell, D. G., and Davis, W. B. 1986. Vitamin E in the alveolar fluid of cigarette smokers. *J. Clin. Invest.* 77:789–796.
113. Pryor, W. A. 1991. Can vitamin E protect humans against the pathological effects of ozone in smog? *Am. J. Clin. Nutr.* 53:702–722.
114. Sevanian, A., Hacker, A. D., and Elsayed, N. 1982. Influence of vitamin E and nitrogen dioxide on lipid peroxidation in rat lung and liver microsomes. *Lipids* 17:269–277.
115. Chow, C. K., Plopper, C. G., and Dungworth, D. L. 1979. Influence of dietary vitamin E on the lungs of ozone-exposed rats. *Environ. Res.* 20:309–317.
116. Chow, C. K., Chen, L. H., Thacker, R. R., and Griffith, R. B. 1984. Dietary vitamin E and pulmonary biochemical responses of rats to cigarette smoking. *Environ. Res.* 34:8–17.
117. Elsayed, N. M., Kass, R., Mustafa, M. G., Hacker, A. D., Ospital, J. J., Chow, C. K., and Cross, C. E. 1988. Effect of dietary vitamin E level on the biochemical response of rat lung to ozone inhalation. *Drug-Nutr. Interact.* 5:373–386.
118. Morgan, D. L., Dorsey, A. F., and Menzel, D. B. 1985. Erythrocytes from ozone-exposed mice exhibit decreased deformability. *Fundam. Appl. Toxicol.* 5:137–143.
119. Calabrese, E. J., Victor, J., and Stoddard, M. A. 1985. Influence of dietary vitamin E on susceptibility to ozone exposure. *Bull. Environ. Contam. Toxicol.* 34:417–422.
120. Posin, C. I., Clark, K. W., Jones, M. P., Buckley, R. D., and Hackney, J. D. 1979. Human biochemical response to ozone and vitamin E. *J. Toxicol. Environ. Health* 5:1049–1058.
121. Hackney, J. D., Linn, W. S., Buckley, R. D., Jones, M. P., Wightman, L. H., and Karuza, S. K. 1981. Vitamin E supplementation and respiratory effects of ozone in humans. *J. Toxicol. Environ. Health* 7:383–390.
122. Duthie, G. G., Arthur, J. R., James, W. P. T., and Vint, H. M. 1989. Antioxidant status of smokers and nonsmokers—effects of vitamin E supplementation. *Ann. N.Y. Acad. Sci.* 570:435–438.
123. Hoshino, E., Shariff, R., Van Gossum, A., Allard, J. P., Pichard, C., Kurian, R., and Jeejeebhoy, K. N. 1990. Vitamin E suppresses increased lipid peroxidation in cigarette smokers. *J. Parenteral Enteral Nutr.* 14:300–305.
124. Bendich, A., and Machlin, L. J. 1988. Safety of oral intake of vitamin E. *Am. J. Clin. Nutr.* 48:612–619.
125. Kappus, H., and Diplock, A. T. 1992. Tolerance and safety of vitamin E: A toxicology position report. *Free Radic. Biol. Med.* 13:55–74.
126. Farrell, P. M., and Bieri, J. G. 1975. Megavitamin E supplementation in man. *Am. J. Clin. Nutr.* 28:1381–1386.
127. Tsai, A. C., Kelley, J. J., Peng, B., and Cook, N. 1978. Study on the effect of megavitamin E supplementation in man. *Am. J. Clin. Nutr.* 31:831–837.

Chapter Four

VITAMIN E AND POLYUNSATURATED FATS: ANTIOXIDANT AND PRO-OXIDANT RELATIONSHIP

Melanie A. Banks

INTRODUCTION

Knowledge of the antagonistic relationship between oxygen free radicals and antioxidants has become the foundation for our understanding of the pathogenesis of many of the chronic diseases that are associated with aging, including cancer, atherosclerotic heart disease, respiratory disease, diabetes, and others. Endogenously generated toxic oxygen metabolites, as well as environmental pollutants (e.g., ozone) and ionizing radiation, contribute to the oxidant burden on living organisms, while antioxygenic nutrients and metabolites counteract it (1). Vitamin E and polyunsaturated fats (PUFA) are one such anti- and pro-oxidant pair, and the balance between these components appears to be of major importance in determining cellular susceptibility to free radical injury. The role of vitamin E in health and disease has recently been the subject of a major review (2). Therefore, this article focuses on certain key aspects of the relationship between vitamin E and PUFA, rather than attempting to serve as a comprehensive treatise. In addition to its well-known role as an antioxidant, other possible functions of vitamin E are discussed. These other functions may also be operative in the influence of vitamin E on chronic disease development. Furthermore, what is presently known about the extracellular and intracellular transport of vitamin E to its loci of action is addressed. Finally, aspects of research into the mechanisms of action of vitamin E that deserve further attention are proposed.

VITAMIN E

Vitamin E is the major lipid-soluble terminator of the propagation phase of free-radical-induced lipid peroxidation in biological systems (3). Vitamin E activity is exhibited by the naturally occurring compounds comprising the tocopherol and tocotrienol classes. The chemical structure of these compounds consists of a chromanol ring to which a phytyl side chain is attached. The tocopherols differ from the tocotrienols in that the former contain a saturated phytyl group, while the latter have an unsaturated one (4). While the tocotrienols have been demonstrated to possess more potent antioxidant activity in vitro, the tocopherols exhibit more potent antioxidant activity in vivo (5). In fact, within the tocopherol series, whose α, β, γ, and δ members differ in the number and position of the methyl groups on the chroman ring, α-tocopherol has the

The author is grateful to Beth Tomczak (staff member) and Kevin Thomas (medical student), both of LECOM, for their assistance with the technical aspects in the preparation of this chapter.

greatest biological activity and is the most abundant form of vitamin E in tissues. With respect to stereoisomeric forms (which result from the chiral nature of the phytyl tail), RRR-α-tocopherol has the highest biological potency (4).

Interestingly, 6-hydroxychroman derivatives of vitamin E with shorter (C11, C6, and C1) isoprenoid side chains have been experimentally determined to possess both higher antioxidant activity and higher intramembrane mobility in biomembrane preparations. The increased mobility of these molecules enhances their ability to scavenge oxygen free radicals. However, the 16-carbon side chain in the α-tocopherol molecule is apparently necessary for its retention in specific membrane areas, and shortening the side chain length by only one isoprenoid unit causes a significant loss in the molecule's vitamin activity (6). Physicochemical studies have shown that the hydrophobic phytyl tail of the molecule is positioned in parallel with the unsaturated fatty acid side chains of membrane phospholipids, while the hydrophilic chromanol head group is positioned next to the polar head groups of the lipids that are near the membrane surface (3).

Products of the oxidation of α-tocopherol as a result of its interaction with peroxidized lipids include α-tocopheryl p-quinone and a dihydroxytocopheryl dimer. The quinone has been the most frequently detected species in biological samples, but simple chemical oxidation of the vitamin and the auto-oxidation of unsaturated oils have been shown to result in the formation of a high proportion of the dimer. Vitamin E is thought to reduce fatty acyl hydroperoxy radicals (ROO$\cdot$) to the more stable hydroperoxides (ROOH), which are then presumably reduced to their corresponding hydroxy acids by a non-Se-dependent glutathione peroxidase. The tocopheroxy radical (E-TO$\cdot$) that is generated as a consequence of this process can undergo two further chemical alterations: (1) an oxidation that involves opening of the chroman ring, forming tocopheryl quinone, and (2) an intramolecular electron shift, forming a carbanion that reacts with an identical molecule to yield the tocopheryl dimer. Neither of these stable oxidation products have significant vitamin E activity under physiological conditions. Whether the α-tocopheroxy radical is recycled in vivo by reduction is not certain. It is, however, capable of being reduced in vitro by various biological agents, including vitamin C (7).

Vitamin E is generally affirmed to function in concert with other antioxygenic nutrients to protect biological tissues from oxidant-induced damage. While exhibiting additive or even synergistic interactions with many other nutrients and metabolites including selenium, the sulfur amino acids, β-carotene, vitamin C, coenzyme Q, and synthetic antioxidants and/or food additives such as butylated hydroxyanisole (BHA), butylated hydroxytolune (BHT), and phenylenediamine (DPPD), the antioxidant action of this vitamin is also antagonized by PUFA, iron and other transition elements, and vitamin A. An additional confounding factor when attempting to study the interrelationships between vitamin E and other anti- and pro-oxidant compounds lies in the fact that several "antioxidants" can exhibit pro-oxidant behavior under different physiological and nonphysiological conditions. For example, β-carotene, vitamin C, and even vitamin E itself can act as pro-oxidants, depending upon their individual concentrations and the microenvironment (e.g., pO_2, presence of metal ions and chelators, concentrations of other reactants) in a given system (7–12).

In molecular terms, vitamin E reduces not only peroxy (ROO$\cdot$), but also hydroxyl (OH$\cdot$) and superoxide ($O_2^{\cdot-}$) radicals and singlet oxygen ($^1\Delta O_2$) within biological membranes. Selenium is a component of the enzyme glutathione peroxidase, which reduces hydrogen peroxide. Oxidized glutathione (GSSG), which is generated in the reduction of hydrogen peroxide to water, can be reconverted to reduced glutathione (GSH) by the enzyme glutathione reductase, which requires NADPH as a cofactor. As a component of glutathione (γ-glutamylcysteinylglycine), the sulfur amino acid cysteine contributes its sulfhydryl group to the glutathione redox cycle. The lipid-soluble provitamin β-carotene is a more effective scavenger of singlet oxygen than vitamin E, and exhibits its antioxidant activity only at low oxygen tensions, while vitamin E functions best at high concentrations of oxygen. Vitamin C (ascorbic acid), a water-soluble antioxidant, can not only scavenge superoxide and hydroxyl radicals and singlet oxygen, but also can reduce metal ions, sulfhydryl compounds and quinones. Vitamin C is hypothesized to act at the aqueous/membrane interface to regenerate vitamin E by reducing the tocopheroxy radical (E-0$\cdot$). Coenzyme Q is thought to act in a manner similar to vitamin E; subcellularly, it is localized in the mitochondria (inner membrane), while vitamin E concentrations are highest in the mitochondria (both inner and outer membrane) and the endoplasmic reticulum (although it is present in other subcellular membrane fractions, including plasma membrane preparations). It is of relevance that both mitochondrial and microsomal fractions are rich in highly polyunsaturated fatty acids, especially arachidonic acid (C20:4) and docosahexaenoic acid (C22:6) (3). Iron and other transition metals such as copper catalyze the formation of hydroxyl radical from hydrogen peroxide, as well as the decomposition of lipid hydroperoxides (ROOH) to yield alkoxy (RO$\cdot$) radicals. The mechanism for the antagonism between vitamin A and vitamin E is not known, but an excess intake of vitamin A is known to decrease the uptake and tissue deposition of vitamin E. Paradoxically, β-carotene may exhibit a synergistic interaction with vitamin E through these molecules' differential effects in scavenging singlet oxygen, while as a precursor of vitamin A, β-carotene may exhibit an antagonistic interaction with vitamin E. Despite its varied interactions with other anti- and pro-oxidant nutrients and metabolites, the primary determinant of the vitamin E status of experimental animals and of humans is the intake of PUFA. The dietary vitamin E requirement can vary by fivefold depending upon the fatty acid composition and content of the diet (7–9).

There have been many attempts over the years to define the vitamin E requirement in the diet as a function of PUFA intake (7), and a few recent attempts to predict it in model systems based upon the concept of a "total antioxidant pool" (9). Data from already completed and ongoing epidemiological studies should prove to be extremely useful in providing information on the dietary (or indirectly, serum) levels of vitamin E, along with nutrients that interact with vitamin E, which are optimal for overall health and prevention of chronic disease. One overriding problem in reaching a scientific consensus on the recommended dietary allowance (RDA) of vitamin E for humans has been, and probably will continue to be, the consideration of whether the recommendation should reflect use of the vitamin only to prevent nutritional deficiency symptoms (e.g., erythrocyte hemolysis) or to also provide for its pharmacological action (e.g., prevention of ozone-induced pulmonary injury). The levels of vitamin E required for the latter function may well be two orders of magnitude higher than for the former. Fortunately, though, in the case of vitamin E (as opposed to other micronutrients), supplementation at extremely high doses has been shown not to result in any significant clinical side effects (13). It should be noted that in a recent Food and Drug Administration conference on antioxidant nutrients and cancer and cardiovascular disease, increased vitamin E supplementation in the human diet was still not recommended (14).

Several factors contribute to the inability to quantitate the vitamin E requirement of animals and of humans based on dietary PUFA intake. Although the amount of vitamin E required to stabilize PUFA in the membranes of tissues is greatly influenced by the degree of unsaturation of the lipids, PUFA are not deposited in tissues in exact proportion to their occurrence in the diet. Additionally, the extent of incorporation of dietary fatty acids into tissues varies from tissue to tissue within the same organism, and even between subcellular organelle fractions within an homogenate of a given tissue (7, 15). Furthermore, once deposited in tissues, PUFA can be modified by elongation and/or desaturation and can be catabolized to varying degrees depending on the metabolic needs of the organism. Studies in animals have confirmed a nonlinear relationship between PUFA intake and the vitamin E requirement. For example, data have shown that when the cellular membranes are already loaded to capacity with PUFA, the intake of linoleic acid, which is the main PUFA in the diet, has little effect on the vitamin E requirement. The National Research Council recommended allowances for vitamin E for individual animal species have been formulated solely on the basis of experimental evidence accumulated from studies utilizing diets of differing PUFA content wherein deficiency symptoms were induced and/or abolished by feeding different amounts of the vitamin. The 1974 RDA for humans, 10 mg (IU) of α-tocopherol equivalents per day, was formulated to meet the requirements of those individuals in the U.S. population with the highest PUFA intakes, based on the finding that 0.5 mg (IU) of α-tocopherol equivalents per gram of PUFA was characteristic of U.S. dietary intake and on the knowledge that there is no significant evidence of vitamin E deficiency in our population (7). The RDA for vitamin E has recently been raised to 15 mg (IU) of α-tocopherol equivalents per day (16).

Dietary sources of vitamin E include animal fats and vegetable oils. Because they contain predominantly saturated fatty acids, animal fats as a class have a lower vitamin E content than vegetable oils, which are much higher in polyunsaturated fatty acid content. In addition, the vitamin E in animal fats is comprised almost exclusively of α-tocopherol, whereas that in vegetable oils is mainly γ-tocopherol, which has only about 10% as much biological activity. In mixed diets, in fact, the γ form predominates. In general, the vitamin E content of fats and oils correlates well with the amount needed to stabilize their component fatty acids against oxidative rancidity. There may be an interesting exception, however, in the case of marine oils. It has been suggested that while in an aquatic environment, with its lower temperatures and oxygen tensions, the higher chain length polyunsaturated fats in the tissues of these organisms are stable to oxidation, whereas when these fats are placed in a nonaquatic environment, rapid deterioration occurs (7).

The classical signs and symptoms of vitamin E deficiency in animals include erythrocyte hemolysis, encephalomalacia, muscle degeneration (nutritional muscular dystrophy), liver necrosis, testicular atrophy in males, gestation resorption in females, and pancreatic atrophy. Many of these disorders can be partially or fully ameliorated by selenium supplementation (7,17). Vitamin E deficiency in humans is rare, presumably because of the wide variety of foods that humans typically consume and because they possess relatively large fat stores (which also contain vitamin E) that are difficult to deplete. Nonetheless, in certain individuals who are genetically predisposed to defects in vitamin E transport (familial isolated vitamin E deficiency) (4) or in individuals with fat malabsorption syndromes (18), vitamin E status may be compromised. The effects of vitamin E deficiency in humans consist of neuromuscular problems and increased erythrocyte hemolysis (18–20). It is now apparent that many of the consequences of vitamin E deficiency in animals and in human can be attributed to lipid peroxidative damage to the macromolecular structures within the cells of the target tissue(s). However, other pathophysiological mechanisms of vitamin E deficiency-induced disease cannot be ruled out, such as a direct effect of vitamin E on cellular differentiation and proliferation. How and why specific tissues become the targets of vitamin E deficiency-related damage, whereas others appear to be protected in this regard, remains an issue for future investigation.

Clues to the possible functions of vitamin E other than in an antioxidant capacity can be garnered from both the current and prior scientific literature. Vitamin E has been shown to bind to the nuclei of hepatic cells—specifically, to the nucleoprotein complex and probably to nonhistone phosphoproteins, which have a high affinity for DNA and may play an important role in the modulation of gene expression. Comparison of RNA synthesized in the livers of vitamin E-deficient and -sufficient rats led to the conclusion that the synthesis of specific classes of RNA were stimulated by the presence of vitamin E. A high-affinity nuclear-associated receptor for vitamin E has also been identified in hepatic cells. Thus, vitamin E appears to exert a hormone-like effect by binding to nuclear receptors, interacting with specific sites in the genome, inducing the transcription of RNA, and presumably initiating a further series of biochemical events (21, 22).

Another possible function of vitamin E has been elucidated in experiments with protein kinase C (PKC), an enzyme that is involved in the control of cellular signal transduction and that influences cellular growth and differentiation. The activity of isolated brain PKC is significantly decreased by concentrations of α-tocopherol that are less than physiological. Vitamin E seems to exert its effect on PKC through an enzyme-ligand interaction. α-tocopherol also inhibits PKC activity and cell proliferation in cultured vascular smooth muscle cells; it apparently acts by preventing translocation of the enzyme from the cytosol (inactive form) to the membrane (active form). The latter results could have relevance to the vascular smooth muscle cell proliferation seen in atherosclerotic heart disease (23).

The anticancer action of vitamin E is also being explored at the molecular level. Initial investigations in this area focused on the ability of vitamin E to inhibit lipid peroxidation, with the view that oxidant injury to macromolecules in the cell (particularly to nucleic acids) could lead to mutation, carcinogenesis, and/or tumor promotion (24). Current investigations in this area have also focused on the possible roles of vitamin E in modulating the immune response and in direct cytotoxicity, which may be involved in cancer prevention and cancer regression, respectively. In an animal model of human oral cancer, oral administration of vitamin E has been shown to completely prevent the development of carcinomas, and injection of vitamin E close to the site of the tumors has been shown to result in the regression of established carcinomas. Both of these effects were accompanied by an accumulation of lymphocytes and macrophages that expressed tumor necrosis factors. In studies using cell lines derived from human cancers, both a lung-derived squamous carcinoma (SCC-25) and a tongue-derived oral carcinoma (SK-MES) responded with decreased cellular viability and proliferation after incubation with α-tocopherol, though to differing degrees. In addition, vitamin E exhibited antitumor effects in a malignant melanoma cell line (A375) and in a breast cancer cell line (MCF-7), though the effects were much stronger in the former than in the latter. In cell culture studies, an immunomodulatory mechanism of action on the part of vitamin E can be excluded; therefore, vitamin E must act by directly altering the cellular metabolism of the tumor cells in some manner (25). In the author's own experience, vitamin E deficiency in an animal model of human pancreatic cancer accelerated tumor growth, whereas vitamin E deficiency in noncarcinogen-treated animals resulted in pancreatic acinar cell atrophy and pseudo-ductular proliferation (or acinar cell dedifferentiation). Since in this particular model of pancreatic ductular carcinoma the early stages involve a loss of acinar cells and a proliferation of ductular ones, it is tempting to speculate that the lack of vitamin E enhanced an effect of the nitrosamine carcinogen on cellular differentiation and proliferation (26, 27).

Vitamin E may also exert its effects in a manner other than, or in addition to, as an antioxidant in the phenomena of cell senescence, death, and removal. Senescent cell antigen (SCA) is generated by modification of protein band 3, a ubiquitous membrane protein that serves as an anion transporter and that attaches the plasma membrane to the cell cytoskeleton. It is expressed on aged cells and signals their removal by phagocytosis. In erythrocytes, oxygen free radical damage results in alterations of the lipid and protein components of the membrane, leading to a breach in membrane integrity and hemolysis of the cell. Indeed, erythrocyte hemolysis is a characteristic sign of vitamin E deficiency. However, the structural changes that take place in the membranes of cells exposed to oxidants such as hydrogen peroxide are more pronounced and less specific than those that occur in aged cells in situ and in vitamin E-deficient erythrocytes, despite the fact that the latter cannot adequately scavenge free radical compounds; both aged and vitamin E-deficient erythrocytes specifically express SCA (19). This evidence suggests that oxidant injury in itself may not be the only mechanism by which cellular aging occurs and that, again, vitamin E may play other roles beside being an antioxidant.

Lastly, vitamin E may function above and beyond its well-accepted antioxidant role by maintaining the physical structure of the cellular membranes (a membrane stabilizer) and by modifying membrane fluidity and the associated physicochemical activity of membrane-associated proteins. Of course, in its antioxidant role alone, vitamin E may also stabilize the membrane structurally by preventing chemical modification of the component lipids and proteins. Physical stabilization of membranes by α-tocopherol is believed to take place through its direct interaction with the phospholipids in the lipid bilayer, including van der Waals-type bonding between the methyl groups on the phytyl side chain and the chromanol ring and the unsaturated fatty acids residues and also hydrogen bonding between the chromanol ring and the phosphate groups of the lipids. Reflecting this function are the effects of vitamin E on membrane fluidity. Short-chain homologs of α-tocopherol have increased mobility in the membrane when compared to the parent compound with its 16C side chain, and in fact, the 1C α-tocopherol derivative actually increases the permeability of the membrane bilayer such that cellular damage results. Interestingly, the incorporation of a higher proportion of polyunsaturated fatty acids into cellular membranes generally also increases membrane fluidity at physiological temperature. Vitamin E at low concentrations has been demonstrated to increase membrane fluidity, but at high concentrations its effect is unclear; reported decreases in membrane fluidity may be due to the use of very fluidized liposomes and nonphysiologic concentrations of the vitamin. Effects of membrane fluidity on membrane protein function, as with, for example, the insulin receptor and the Na^+/K^+-ATPase enzyme/pump, are well documented (3, 28, 29).

In summary, the experimental evidence just cited indicates that vitamin E may both directly and indirectly play a multiplicity of roles in the repertoire of cellular metabolism. Most prominent among these are its roles as an antioxidant, as a membrane stabilizer, and as a putative regulator of cellular growth, differentiation, and senescence.

POLYUNSATURATED FATS AND LIPID PEROXIDATION

Fatty acids are major components of the structure of phospholipids, which comprise the most abundant class of lipids within the cellular membranes. Sphingolipids are also a prominent member of the lipid bilayer of biological membranes, but only in highly specialized tissues such as the heart and brain; they too contain fatty acids as a major component of their structure. Of the three major classes of fatty acids in nature, saturated, monounsaturated, and polyunsaturated, only the PUFA, because of their unique structural characteristic of having the double bonds separated by methylene groups, are subject to oxidative decomposition, or what is now commonly known as lipid peroxidation.

In the most widely accepted scheme, lipid peroxidation is initiated by free radical (R$\cdot$) attack on a methylene carbon of a polyunsaturated fatty acid; it is propagated by reaction of the resultant carbon-centered lipid radical (L$\cdot$) with molecular oxygen to form a lipid peroxy radical (LOO$\cdot$), which then reacts with another PUFA to form a lipid hydroperoxide (ROOH) and another carbon-centered lipid radical; this is terminated by the reaction of two like or different free radical species. Vitamin E functions in this last regard. The products of the chain-reactive process are conjugated lipid dienes, aldehydes, polymers, and hydrocarbons. Each of these products can be assayed in an attempt to quantitate the overall process. Lipid peroxidation can take place in the absence or presence of enzymes, and the latter route plays an important role in metabolic pathways such as the arachidonic acid cascade, whereby prostaglandins and leukotrienes are generated, and the cytochrome P-450 system, whereby detoxification of xenobiotic compounds occurs. In vivo, lipid peroxidation is often initiated by redox cycling of iron and the enzymatic or nonenzymatic generation of reactive oxygen species (30–33).

Lipid peroxidation can be both initiated and amplified by iron and other transition metals. Iron can initiate lipid peroxidation through a Fenton-like reaction of the metal itself (Fe^{2+}) or of an iron complex with hydrogen peroxide; this results in generation of the hydroxyl radical (OH$\cdot$). On the other hand, iron or iron-containing compounds such as heme can increase the rate of lipid peroxidation by converting lipid hydroperoxides (LOOH) to reactive alkoxy (LO$\cdot$) or peroxy (LOO$\cdot$) radicals, which can initiate further oxidative lipid decomposition. Redox cycling of iron from the Fe^{2+} to the Fe^{3+} state and back promotes lipid peroxidation, and it has been suggested that both species of the element are required for this process to take place. Extracellular and intracellular sources of iron that could participate in the reactions of lipid peroxidation include hemoglobin, ferritin, transferrin, and low-molecular-weight iron-sequestering organic compounds. It has been suggested that the intracellular iron-storage protein ferritin is the most likely biological candidate for a source of iron in lipid peroxidative reactions. Ferritin can initiate lipid peroxidation through release of reduced iron via attack by the superoxide anion radical (31, 32, 34).

Biochemical reactions that are known to generate oxygen free radicals through enzymatic processes and the toxic metabolites that they produce include: urate oxidase/H_2O_2, D-amino acid oxidase/H_2O_2, glycolate oxidase/H_2O_2, and L-hydroxyacid oxidase/H_2O_2, all of which are located in peroxisomes; xanthine oxidase/$O_2^{\cdot-}$ and H_2O_2, which is widely distributed in the cell; aldehyde oxidase/$O_2^{\cdot-}$ and H_2O_2, which is present in the (liver) cytosol; and monoamine oxidase/H_2O_2, which is located in the outer mitochondrial membrane. In addition, polymorphonuclear leukocytes and macrophages enzymatically produce $O_2^{\cdot-}$ by the action of NADPH oxidase, which is located in the plasma membrane. Nonenzymatic reactions that produce toxic superoxide anion include the auto-oxidation of thiols, hydropterins, flavins, hemoglobin, and the catechol epinephrine, and also flavoproteins, ubiquinone, and the cytochromes, which are associated with the mitochondrial and microsomal electron transport systems. Both $O_2^{\cdot-}$ and H_2O_2 are less reactive than other oxygen free radical species and can diffuse away from the site at which they were generated; H_2O_2 can even cross biological membranes. In contrast, both OH$\cdot$ and $^1\Delta O_2$ are highly reactive and are utilized at their sites of production. Enzymatic transformation of $O_2^{\cdot-}$ to H_2O_2 by superoxide dismutase takes place in the mitochondria and the cytosol, while H_2O_2 is removed and converted to H_2O by the actions of catalase in the peroxisomes and glutathione peroxidase in the cytosol (33). It is significant to note that in general the enzymatic reactions that utilize molecular oxygen show little increase in oxygen free radical production with increases in pO_2, most likely because they are saturated with oxygen under normal conditions. Due to this fact, and due to their limited distribution within the cell, these enzyme activities can be predicted to contribute in only a minor fashion to the toxicity of hyperbaric oxygen. However, the nonenzymatic (auto-oxidative) reactions for which molecular oxygen serves as a substrate probably contribute significantly to the toxic effects of hyperoxia, since the rate of intracellular generation of oxygen free radicals is in this case directly proportional to the pO_2. "Redox cycling," in which continued production of toxic oxygen metabolites and their continued reduction by mitochondrial or microsomal respiratory chain carriers proceed simultaneously, may result from exposure to hyperbaric oxygen and also is a suggested mechanism for the cytotoxicity of several pharmacological agents, including adriamycin, bleomycin, and daunomycin, as well as the herbicide paraquat (33).

The macromolecular consequences of lipid peroxidation within the cellular membranes are many. Reaction of hydroxyl radicals with carbohydrates leads to chain breaks in structural molecules like hyaluronic acid. A similar reaction involving the polyribose backbone can result in DNA strand breaks. Modification of nucleic acids may result in genetic defects, and several base modifications from oxidative injury have been described; these include 8-hydroxyguanosine, thymine glycol, and 5-hydroxymethyluracil. Fragmentation of proteins may occur through reaction of hydroxyl radicals with proline and histidine residues, and the sulfhydryl groups on cysteine and methionine residues can be oxidatively modified by oxygen metabolities such as hydrogen peroxide. Also, the unsaturated aldehydes produced from lipids themselves as a result of lipid peroxidation can react with the amine groups (e.g., lysine residues) of proteins to form Schiff bases which effectively cross-link lipids and proteins. In addition to producing deleterious functional changes in all classes of cellular macromolecules, lipid peroxidation can ultimately lead to such extensive membrane damage that cellular viability is threatened and cell lysis occurs. It has been suggested that an influx of Ca^{2+} as a result of membrane leakage is the ultimate event that precipitates cell death from lipid peroxidative damage (30).

While polyunsaturated fatty acids, specifically linoleic acid (C18:2), have long been recognized as being essential in the human diet, it has not been until recently that individual requirements for both omega-6 (ω-6) and omega-3 (ω-3) PUFA have been demonstrated. The amount of linoleic acid (ω-6) and linolenic acid (ω-3) combined that is required in the human diet for prevention of fatty acid deficiency symptoms (scaly dermatitis and growth impairment) is surprising low: 3% of total calories. Most humans under normal circumstances consume much more than the required amount of these fatty acids, so that a fatty acid deficiency has rarely been seen except, for example, in severe malnutrition and in early attempts at total parenteral nutrition. Linoleic acid is required in the human diet because we do not possess the necessary enzymes to elongate and desaturate the lower chain-length fatty acid precursor(s). This particular fatty acid, in addition to having a structural role in cellular membranes, is a substrate for the synthesis of arachidonic acid (C20:4), a metabolic precursor of the prostaglandins, leukotrienes, and related compounds that exhibit hormone-like activities (35, 36).

Perhaps because it is required in such low amounts and because it is widely available in plant-derived food sources, the essentiality of α-linolenic acid (C18:3), an ω-3 fatty acid [(to be distinguished from γ-linolenic acid (C18:3), an ω-6 fatty acid and the product of elongation and desaturation of linoleic acid], has only recently been identified. Higher chain length products of α-linolenic acid, eicosapentaenoic acid (C20:5; EPA), and docosohexaenoic acid (C22:6; DHA) are now known to be vital components of neural tissue, and though very difficult to induce, an ω-3 fatty acid deficiency has been shown to result in learning and behavioral deficits in experimental animals and in neurological problems in humans. While it has been assumed that dietary supplementation of α-linolenic acid can effectively substitute for ingestion of its metabolic end products, EPA and DHA, in humans, this has not been definitively established. In the rat animal model, the conversion of α-linolenic acid to DHA and EPA is significant, but complementary studies in humans have demonstrated that this conversion is much more limited (37–42).

Current dietary fat recommendations for the U.S. populace reflect the concern that increased calorie consumption and the concomitant increase in calories from fat (since fat is of much higher caloric density than the other macronutrients, carbohydrate and protein) are related to increased mortality from chronic diseases of all causes, as associated with obesity. Epidemiologic studies and experimental human and animal studies both have indicated that increased dietary fat consumption contributes to increased risks of cardiovascular disease (CVD) and cancer (CA). In general, dietary fat is known to promote both CVD and CA development. But specifically, PUFA consumption is known to mitigate cardiovascular risk while enhancing cancer risk. With respect to CVD, the most recent evidence suggests that not all saturated fatty acids are "bad" and that monounsaturated fatty acids are not "neutral" but rather beneficial (43–51). Especially relevant are recent studies in which the dictum that PUFA consumption decreases CVD has been challenged; PUFA are now known to contribute to the oxidation of low-density lipoprotein (LDL), which has been suggested to enhance the risk of CVD. Furthermore, the role of antioxidant nutrients such as vitamin E, β-carotene and vitamin C in counteracting LDL oxidation has become the focus of much research activity, in both animal and human studies (52–54). With respect to CA, the role of individual fatty acids has not been as widely investigated; however, the results of experimental studies in animals suggest caution in recommending increased PUFA intake to decrease CVD risk at the expense of CA risk (55–57). The widely accepted current recommendations for dietary fat consumption have been formulated keeping in mind that heart disease and cancer are the first and second leading causes of death in the United States, respectively. We are advised to consume no more than 30% of total calories as fat, and of that, no more than one-third each of saturated, monounsaturated, and polyunsaturated fats.

ROLES OF DIETARY FAT AND VITAMIN E IN CHRONIC DISEASES

Coronary Heart Disease

Arteriosclerosis is the underlying cause of coronary heart disease (CHD), which is a leading cause of death in the United States and other industrialized countries. The final result of CHD is myocardial infarction, a sudden and life-threatening event, but the development of the disease occurs slowly over a period of many years. Risk factors for the development of CHD include smoking, hypertension, genetic predisposition, and one or more alterations in lipid metabolism, including elevated total plasma cholesterol and triglyceride levels, elevated low-density lipoprotein (LDL) cholesterol, and depressed high-density lipoprotein (HDL) cholesterol levels (45, 58).

The consumption of a diet low in animal protein and fat (high in vegetable content) or of Mediterranean-style foods (high in olive oil, legumes, and seafood) is associated with lower risk of CHD. While increased consumption of ω-6 PUFA versus saturated fatty acids has long been thought to decrease total plasma cholesterol and to elevate the HDL/LDL ratio, it has only been recently that increased consumption of monounsaturated fatty acids and of the higher chain length ω-3 PUFA from marine oils has also been shown to result in favorable lipid profiles with respect to CHD. One relevant fact that has largely been ignored in past analyses but has emerged of late is that PUFA-rich vegetable oils and olive oil (which is a rich source of monounsaturated as well as polyunsaturated fatty acids) and fresh seafood are all good sources of vitamin E when compared to animal fats. It may well be that at least some of the protective value of consuming vegetable and marine fats for mitigation of CHD risk that was previously attributed to the fatty acid composition of these food sources alone is attributable to increased consumption of vitamin E (45, 59).

Chronic marginal deficiency of vitamin E in humans has been reported to cause arteriosclerotic lesions and cardiomyopathy. In a recent epidemiological study a lower risk of both CHD and stroke was seen in U.S. male and female professionals of both sexes who were among persons in the highest quintile of vitamin E intake/supplementation. In the multicultural, European WHO/MONICA study the plasma antioxidant levels of middle-aged (40–49 years old) males were correlated with their risk of CHD according to population region. The data indicated that vitamin E was the strongest predictor, in an inverse manner, of the risk of CHD mortality, but also that vitamin C, vitamin A, and its precursor β-carotene interact synergistically with vitamin E. In fact, vitamin E status was a far greater predictor of CHD risk than total plasma cholesterol and blood pressure levels, which are classical risk factors for CHD; vitamin E alone accounted for 60% of the differences in mortality. In studies with experimental animals, vitamin E deficiency has been shown to result in a breakdown of vascular endothelial cells and the formation of fibrous and calcified lesions of the aorta (rats and primates), as well as myocardial necrosis and thrombosis (piglets). Vitamin E supplementation in cholesterol-fed primates and in triglyceride-fed rabbits has been shown to reduce atheromatosis of the aorta, and in the Watanabe heritable hyperlipidemic rabbit to reduce the plaque surface of the aorta, as well as in middle-aged hens to counteract spontaneous arteriosclerosis. Conversely, feeding peroxidized diets has been shown to result in sudden death accompanied by degenerative changes in heart muscle in the pig. Thus there appears to be a direct association between vitamin E, PUFA/lipid peroxidation, and risk of developing atherosclerosis and CHD (59, 60).

The most recent advance in our understanding of the mechanism(s) by which vitamin E may mitigate the development of CHD comes from studies with oxidized LDL. It is hypothesized that oxidatively modified LDL plays an important role in atherogenesis by triggering the deposition of cholesterol in arteries, which is a requisite for the formation of atherosclerotic plaque, the substance responsible for narrowing of the arterial lumen and subsequent myocardial ischemia and/or infarction. Oxidized LDL is taken up and degraded by the scavenger receptors in macrophages that have been recruited to the intima of arteries lying beneath the endothelium; the deposition of cholesterol in these cells as cholesterol esters in the cytoplasm leads to their conversion to foam cells, which form fatty streak lesions. Fatty streaks develop into the more advanced atherosclerotic plaque through recruitment of lymphocytes, proliferation of smooth muscle cells, and calcification and fibrosis. The putative steps in this process have been reviewed elsewhere (45, 58). Free radical injury to plasma lipoproteins including LDL and the arterial cells may be due to endogenously generated oxygen radicals and/or propagated radical formation from peroxidation of PUFA. Transient hypoxia followed by reoxygenation, such as may be induced by vasoconstriction, also induces the intracellular production of free radicals as a result of the reintroduction of molecular oxygen. Cell types that have been reported to oxidatively modify LDL in vitro include endothelial cells from rabbits, human arteries and veins, bovine and human smooth muscle cells, and human macrophages; incubation of LDL in serum-free medium in the presence of iron or copper results in similar, if not identical, changes in the structure and biological properties of LDL as those seen in cell incubations (45, 52, 53, 58, 59).

Several of the oxygen free radical species generated by cells could be responsible for LDL oxidation in vivo. However, catalase, which detoxifies H_2O_2, and mannitol, a hydroxyl radical scavenger, both fail to inhibit the oxidative modification of LDL, suggesting that these two species are not involved in the process. Superoxide anion is a likely candidate, since it is produced by both arterial smooth muscle cells and macrophages; however, inhibition of LDL oxidation by superoxide dismutase has not been consistent in experimental studies. It is possible that endothelial cells induce LDL oxidation via the action of lipoxygenase in the arachidonic acid cascade, since inhibitors of this enzyme have been shown to suppress the process by 70–80%. That oxidative modification of LDL actually takes place in vivo is supported by studies using monoclonal antibodies and immunocytochemical techniques in experimental animals (52).

In vitro studies have shown that during LDL oxidation, the chemical composition of the lipoprotein changes. The earliest change is the consumption of vitamin E, followed by the disappearance of other antioxidants including the carotenoids (β-carotene and lycopene); this is defined as the "lag phase." Second, PUFAs are rapidly oxidized in the order of docosahexaenoic acid, arachidonic acid, and linoleic acid (more highly to less highly polyunsaturated); this is defined as the "propagation phase," due to the appearance of conjugated dienes (lipid peroxidation products). Finally, the lipid hydroperoxides generated during the propagation phase begin to be cleaved into aldehydes and other products; this is defined as the "decomposition phase." It is of significance that the form of LDL taken up by the macrophage scavenger receptor has been identified as that which appears during the decomposition phase of LDL oxidation (52, 53).

The major antioxidant endogenously present in LDL is α-tocopherol; an average of six molecules is estimated to be present per LDL particle. The low LDL content of other antioxidants (e.g., 0.3 β-carotene molecules per LDL molecule) suggests that α-tocopherol is the only significant antioxidant therein. Interestingly, of the PUFA bound to LDL (average of 48% of the total fatty acids), linoleic acid accounts for about 86%, while arachidonic acid (approximately 12%) and docosahexaenoic acid (approximately 2%) account for much more minor amounts. Of further interest is the fact that oleic acid (a monounsaturated fat) enriched LDL is resistant to oxidation (52, 53, 61).

It has been observed in many studies that addition of high concentrations of vitamin E and other antioxidants (including BHT and probucol, a hyperlipidemic drug) prevents the oxidative modification of LDL by cells in culture and its subsequent uptake by macrophages. In vitro incubation of LDL with vitamin E-loaded plasma resulted in an approximately fivefold enrichment of LDL with vitamin E, which increased the lag phase of LDL oxidation; the correlation coefficient for the lengthening of the lag phase with LDL vitamin E content was 0.83. Vitamin

E supplementation also resulted in resistance to LDL oxidation in an ex vivo study. However, in both types of studies, the effect was not consistent among subjects, in that some displayed a relative lack of response to the effects of vitamin E supplementation (i.e., no statistical correlation of length of the lag phase with LDL vitamin E content). It has been suggested that factors that determine the effectiveness of protection against LDL oxidation, independent of vitamin E, might include the PUFA content and the amount of other antioxidants, among others (53).

In conclusion, it appears that the PUFA/vitamin E pro- and antioxidant relationship may be intimately involved in the genesis and progression of CHD. It should be noted that vitamin E offers therapeutic promise, especially in the form of its synthetic water-soluble analogue, Trolox (a carboxylic acid derivative of the chroman ring of vitamin E), in mitigating or even ameliorating the oxidant-induced damage in reperfusion/reoxygenation injury due to spontaneous recovery from myocardial infarction and in interventional procedures such as balloon angioplasty or cardiopulmonary bypass surgery (62, 63).

Cancer

Cancer is the second leading cause of death in the United States, outled only by cardiovascular disease. Although no direct cause-and-effect relationship has been established, dietary factors have been estimated to account for a large proportion of the possible environmental causes of human cancer. Increased total caloric consumption enhances cancer development, but of the macronutrients, fat has been identified as the most likely culprit. Evidence from both epidemiological and experimental studies suggests that increased consumption of dietary fat in general is associated with increased cancer occurrence at all sites, but especially of the breast, prostate, and colon. In animal experiments, increased consumption of polyunsaturated fats (ω-6) in particular has been shown to increase the incidence of spontaneous and chemically induced tumors, particularly at lower levels of total fat intake; this may be related to the requirement of small amounts of essential fatty acids for growth. In a few epidemiological studies that were designed primarily to study the effects of high fat diets and elevated serum cholesterol on atherosclerosis, the incidence of cancer was found to be greater in subjects who consumed high levels of ω-6 polyunsaturated fats (to lower serum cholesterol levels). Likewise, comparison of the increased per capita dietary fat intake and cancer mortality in the United States over a period of approximately sixty years (1909–1972) revealed significant positive correlations for both total fat and vegetable fat (which is predominately unsaturated) but not for animal fat (which is predominately saturated). Omega-3 fatty acids seem to exert an anticancer effect by antagonizing ω-6 fatty acid metabolism. Dietary fat has most often been demonstrated to exert its effect on tumor development in animal models during the promotional stage of carcinogenesis, but its involvement in tumor initiation cannot be ruled out. Certainly, the pro-oxidant action of (polyunsaturated) fat can be reconciled with a tumor-promoting function (43, 44, 55–57, 64–68).

Experimental attempts to modulate chemical carcinogenesis with vitamin E date back to as early as 1934. Early studies yielded conflicting results, most likely because ill-defined diets and poorly characterized animal models were employed. Later studies gave more consistent findings wherein dietary vitamin E, often in high doses, has been shown to protect against chemically induced cancers of the skin, forestomach, colon, mammary gland, and oral mucosa. Conversely, vitamin E deficiency has been shown to enhance a chemically induced cancer of the pancreas. The variable anticancer effects of vitamin E in various animal models may well depend upon the type and dosage of the administered carcinogen as well as other dietary modifying factors, such as the level of selenium and the level and types of dietary fat (27, 69–71). Vitamin E can affect the activation (through the mixed-function oxygenase system) and binding of a carcinogen to cellular macromolecules (i.e., tumor initiation, as well as tumor promotion). Protection against lipid peroxidation is one mechanism by which an antioxidant such as vitamin E might exert effects on tumor promotion, but other mechanisms, such as through the modulation of protein kinase C activity and through immunoenhancement, have also been demonstrated (25, 67,68, 72). In addition, a direct effect of vitamin E on gene expression (21, 22) can be considered.

Combined evidence from different types of epidemiological studies conducted in the past decade points to an anticancer role for vitamin E in humans. In general, observational (geographic, case-control, and cohort) studies, wherein vitamin E exposure has been assessed on the basis of dietary recall information·or blood vitamin E levels, have demonstrated weak to moderate inverse associations between vitamin E and lung cancer (especially among nonsmokers), esophageal, gastric and colon cancers, and breast cancer. These studies may be confounded by the inherent difficulties in accurately interpreting vitamin E consumption from dietary recalls and in accurately assessing vitamin E status from stored blood samples. However, these results have encouraged the initiation of many large-scale intervention trials in which the effects of vitamin E alone or in conjunction with other putative inhibitory antioxidant nutrients, including β-carotene, vitamin C, and selenium, are being studied in relation to cancer risk at specific sites (skin, lung, esophagus, colon) or at all sites. Some of these studies are being conducted on high-risk groups such as those with a history of adenomatous colon polyps, smokers, and tin miners, and those with prior basal-cell skin carcinomas. Preliminary results of these clinical trials should be forthcoming in a few years (72).

Respiratory Disease

Occupational exposure to various lung toxicants and cigarette smoking are the most frequent causes of chronic bronchitis and emphysema. Of course, smoking is associated with the development of other chronic diseases, primarily lung cancer and coronary heart disease. However, cigarette smoke is also a concentrated source of numerous environmental free-radical-generating pollutants, such as nitrogen dioxide (NO_2), that are directly delivered to the lung and that are capable of initiating lung oxidant injury. Ozone (O_3), an oxidant gas with similar lung-damaging properties to NO_2, is a component of the nearly ubiquitous environmental hazard known as smog; it is produced as a result of the photochemical decomposition of molecular oxygen and is the most potent oxidant substance to which human beings are routinely exposed. For this reason, ozone may be presumed to play a role in the development of chronic lung disease (73–75).

Exposure of the lung to oxidant gases such as NO_2, O_3, and hyperbaric oxygen results in pulmonary injury that is characterized by damage to the epithelium of conducting airways, the terminal bronchioles, and the proximal and/or distal alveoli. The pulmonary responses to exposure to oxidant gases can be broadly classified into two phases: (1) an initial injury phase, in which bronchiolar ciliated cells and alveolar type I cells (primarily epithelial in character) are morphologically damaged, and in which biochemical alterations, such as oxidation of unsaturated fatty acids and of sulfhydryl-containing amino acids and proteins, occur; and (2) a secondary repair phase, in which proliferation of Clara cells and apparent dedifferentiation of alveolar type II (surfactant-containing) cells to replace the damaged ciliated cells and type I cells (respectively) occur, and in which an induction of antioxidant enzyme activities, including those of the glutathione redox cycle, occurs (76–79).

As a consequence of ozone exposure, unsaturated fatty acids within the lung are oxidized; this has been demonstrated to result from the insertion of ozone into the double bond(s) to form a trioxide derivative, which subsequently decomposes to form a diradical, which can then ultimately form a hydroxyhydroperoxide. In this way, ozone promotes lipid peroxidation within the lung; however, ozone-induced recruitment of inflammatory and phagocytic cells to the lung also occurs, and superoxide anion, hydrogen peroxide, and hypochlorous anion (neutrophils) that are released by these cells may well contribute to the oxidant burden (75).

Animal experimentation, despite different levels and duration of exposure, has shown that vitamin E protects against lung injury from oxidant gases such as ozone; vitamin E-deficient rats are more susceptible to morbidity (as estimated by lung pathology) and mortality from ozone exposure than those fed adequate vitamin E. Vitamin E has been demonstrated to react with the peroxidized lipids in a free radical scavenging mode, rather than trapping ozone directly, in its role as a protector against ozone toxicity to the lung. Lung vitamin E levels rise in response to ozone exposure in vitamin E-supplemented but not vitamin E-deficient rats, and mobilization of labeled vitamin E from other body sites to the lung has been demonstrated in the former but not the latter. Vitamin E supplementation in animals also conferred protection against ozone-induced increases in the enzymes glutathione peroxidase, glutathione reductase, and superoxide dismutase, among others; however, the magnitude of the protective effect plateaued at a certain level of supplementation (74, 75).

Though less extensively studied, vitamin C (75) and taurine, a membrane stabilizer and putative antioxidant (77–79), have also been shown to protect against ozone exposure in animal studies. The interactions of these and other antioxidant nutrients with vitamin E in potential protection against oxidative lung injury from various sources needs to be investigated.

Interestingly, no significant differences have been detected in the plasma vitamin E levels between cigarette smokers and nonsmokers, and there has been no significant correlation between plasma vitamin E levels and pulmonary function tests or lung cytogenetic changes in smokers. However, the levels of α-tocopherol were lower and the levels of α-tocopheryl quinone were higher in bronchoalveolar lavage fluid obtained from smokers, and higher levels of vitamin E were detected in alveolar macrophages harvested from smokers versus nonsmokers (73, 76). The results from these studies mesh well with those from animal studies of the role of vitamin E in lung oxidant injury; both indicated increased mobilization of vitamin E to the lung under conditions of oxidant stress.

Trials involving controlled, short-term exposure to low levels of ozone have not yielded much information with respect to vitamin E. This might be expected, however, since ethical considerations preclude the use of high-ozone and/or repeated exposures that would result in experimental lung injury in humans (75).

Diabetes

A role for vitamin E in the prevention and treatment of long-term complications from diabetes has received intense attention of late. While the degree of hyperglycemia and the extent of diabetic complications have been confirmed to be directly correlated, the causal mechanism(s) have only recently begun to be elucidated. Advanced glycosylation end products (AGEs) are believed to be responsible for the development of the pathologies associated with long-standing diabetes (80).

The oxidation of glucose and other monosaccharides by transition metals such as Cu^{2+} (which increases in plasma concentration with age and in diabetes) can generate H_2O_2 and OH·, as well as ketoaldehyde products. The ketoaldehydes and reactive oxygen species can cause protein fragmentation and the nonenzymatic formation of glycoprotein adducts; this process may be amplified by O_2^- generation and by lipid peroxidation and its metabolic consequences (81–83). The formation and tissue accumulation of these fluorescent AGEs bear a striking resemblance to the formation and deposition of lipid-protein adducts and fluorescent pigments (e.g., lipofuscin) that have been associated with vitamin E deficiency and aging (84).

The AGEs have been proposed to enhance diabetes-related pathology, including retinopathy, nephropathy, neuropathy, and microvascular complications (which may lead to cardiovascular disease) by way of three general mechanisms: (1) alteration of protein function in target tissues through glycosylation by glucose, fructose, and glycolytic intermediates; (2) alteration of the interaction between soluble mediators such as cytokines and hormones with intracellular receptors; and (3) alteration of signal transduction pathways involving ligands on the extracellular matrix (80, 85).

Vitamin E has been shown to reduce protein glycosylation both in vitro and in vivo (86, 87). Interestingly, vitamin E supplementation has also been shown to improve insulin action in nondiabetics and in Type II (adult-onset) diabetics (88). It should be noted that another antioxidant vitamin, vitamin C, and agents that prevent the formation of Schiff bases (from the reaction of aldehydes with amines), such as aminoguanidine and vitamin B_6, also have been shown to be potentially useful in this regard (86, 89, 90).

Other Chronic Diseases

Therapeutic use of vitamin E has been suggested to modify other chronic diseases in which lipid peroxidative injury seems to play a role, such as arthritis (91), certain dermatological disorders such as inflammatory and fibrotic skin lesions (92), certain neurological diseases such as multiple sclerosis and Parkinson's disease (93), and chronic alcoholism (94), in addition to the known diseases that are directly caused by vitamin E deficiency.

ABSORPTION, TRANSPORT, AND DISTRIBUTION OF VITAMIN E

The absorption of dietary vitamin E parallels that of dietary lipids; thus bile salts and pancreatic enzymes are important components of the process. However, the efficiency of absorption decreases as large amounts of vitamin E are administered; reportedly, 50–70% is absorbed at the usual dietary levels (<1 mg), but only about 10% is absorbed with pharmacological doses as low as 200 mg. The absorption of vitamin E takes place in the proximate small intestine by passive diffusion; there is no discrimination between the α and γ forms of tocopherol in the process. Once absorbed, the vitamin is transported with the predominant portion of ingested fat by the lymphatic vessels to the circulation, where it is carried by the lipoproteins, primarily HDL and LDL in humans. Rapid exchange of tocopherol with erythrocytes takes place in the plasma. There is no evidence for the existence of a specific vitamin E binding/carrier protein in plasma, in contrast to other fat-soluble vitamins (4, 95).

α-tocopherol is taken up by most tissues, including liver, lung, heart, skeletal muscle, and adipose tissue. In general, tissues with greater lipid content tend to concentrate more vitamin E. The liver is a major repository of vitamin E, but adipose tissue stores are much higher yet less available for use by other tissues. Like the pool in adipose tissue, that in skeletal muscle is difficult to deplete. Intracellularly, α-tocopherol is especially abundant in the mitochondrial, microsomal, and plasma membranes. The major route of excretion of tocopherol metabolites is via the feces, probably in conjunction with the bile acid metabolites (95).

After degradation of chylomicron remnants that have been formed in and secreted by the intestine, α-tocopherol is packaged into nascent VLDL, which is secreted into the bloodstream by the liver. Spontaneous exchange of α-tocopherol between different lipoproteins has been documented. Thus, α-tocopherol is redistributed during lipoprotein metabolism. Discrimination between α and γ isomers, and between stereoisomers of α-tocopherol has been shown to occur at the level of VLDL secretion, and is thought to be facilitated by a unique (30 kD) hepatic α-tocopherol binding protein that has been isolated, purified, and partially characterized. This mechanism is most likely responsible for the fact that although γ-tocopherol generally predominates in the diet, α-tocopherol is preferentially incorporated into tissues. Demonstrated mechanisms for uptake of α-tocopherol into tissues include: (1) by the action of lipoprotein lipase in adipose tissue and muscle and (2) by LDL receptor-dependent and -independent pathways in other tissues (4, 96).

The intracellular transport of α-tocopherol between the plasma membrane and subcellular organelles to its sites of action is even less well understood than the interorgan transport of the vitamin. However, intracellular vitamin E binding proteins (EBPs) were identified as early as 1972 (97) and have been the focus of renewed research interest lately. A low-molecular-weight (14.2 kD) cytosolic EBP has recently been described in heart and liver; it specifically binds α-tocopherol in preference to the γ isomer and enhances the transfer of α-tocopherol from liposomes to mitochondria. Furthermore, this EBP has been shown to differ in identity from an intracellular fatty acid-binding protein by immunochemical means (96). This intracellular EBP may not be an exclusive one, however. Also recently, the author has identified a much higher molecular weight (>100 kD) EBP in heart and erythrocytes (98, 99), and this EBP may be similar or identical to one previously isolated from liver. It should be noted in this regard that more than one EBP has been isolated from the cytosol of cultured smooth muscle cells (100). Further characterization of the properties of the EBPs isolated thus far is necessary in order to understand their role(s) in vitamin E transport and availability within the cell.

SUGGESTED AREAS FOR FURTHER RESEARCH

Many aspects of the biochemistry of vitamin E warrant further investigation, in view of the vitamin's antioxidant and other functions that appear to be involved in the prevention of human disease. More specific assessments of vitamin E requirements are needed, with respect to fatty acid composition (especially with respect to ω-3 PUFAs) of the diet and the relationships between and with other antioxidant nutrients including β-carotene, vitamin C, and the sulfur amino acids. Superimposed on this is the need to define pharmacological doses of vitamin E that have therapeutic benefit in the treatment of specific chronic diseases. Special consideration in this regard should be given to the question of whether vitamin E under any circumstances acts as a pro-oxidant in vivo, and to the question of whether supplementation of the vitamin produces significant toxic side effects and (if so), at what dosage. Furthermore, attention should be given to understanding the mechanisms by which vitamin E is transported into cells and between subcellular organelles, since this may elucidate clues to the utility of vitamin E therapy. Finally, putative mechanisms of action of vitamin E other than as an antioxidant should be explored; the use of molecular biology techniques to identify the products of vitamin E-induced gene expression would be extremely helpful in that search.

REFERENCES

1. Ames, B. N., Shigenaga, M. K., and Hagen, T. M. 1993. Oxidants, antioxidants, and the degenerative diseases of aging. *Proc. Natl. Acad. Sci. U.S.A.* 90:7915–7922.
2. Packer, L., and Fuchs., J., eds. 1993. *Vitamin E in Health and Disease.* New York: Marcel Dekker.
3. Zimmer, G., Thürich, T., and Scheer, B. 1993. Membrane fluidity and vitamin E. In *Vitamin E in Health and Disease*, eds. L. Packer and J. Fuchs, pp. 207–222. New York: Marcel Dekker.
4. Traber, M. G., Cohn, W., and Muller, D. P. R. 1993. Absorption, transport, and delivery to tissues. In *Vitamin E in Health and Disease*, eds. L. Packer and J. Fuchs, pp. 35–51. New York: Marcel Dekker.
5. Serbinova, E. A., Tsuchiya, M., Goth, S., Kagan, V. E., and Packer, L. 1993. Antioxidant action of α-tocopherol and α-tocotrienol in membranes. In *Vitamin E in Health and Disease*, eds. L. Packer and J. Fuchs, pp. 235–243. New York: Marcel Dekker.
6. Kagan, V. E., Serbinova E. A., Packer, L., Zhelev, Z. Z., Bakolova, R. A., and Robarov, S. R. 1993. Intermembrane transfer of α-tocopherol and its homologs. In *Vitamin E in Health and Disease*, eds. L. Packer and J. Fuchs, pp. 117–179. New York: Marcel Dekker.
7. Draper H. H. 1993. Interrelationships of vitamin E with other nutrients, In *Vitamin E in Health and Disease*, eds. L. Packer and J. Fuchs, pp. 53–61. New York: Marcel Dekker.
8. Blakely, S. R., Mitchell, G. V., Jenkins, M. L. Y., Grundel E. 1993. Effects of β-carotene on vitamin E. In *Vitamin E in Health and Disease*, eds. L. Packer and J. Fuchs, pp. 63–68. New York: Marcel Dekker.
9. Tappel, A. 1993. Combinations of vitamin E and other antioxygenic nutrients in protection of tissues. In *Vitamin E in Health and Disease*, eds. L. Packer and J. Fuchs, pp. 313–325. New York: Marcel Dekker.
10. Reed, D. J. 1993. Interaction of vitamin E, ascorbic acid, and glutathione in protection against oxidative damage. In *Vitamin E in Health and Disease*, eds. L. Packer and J. Fuchs, pp. 269–281. New York: Marcel Dekker.
11. Bowry, V. W., Ingold, K. U., and Stocker, R. 1992. Vitamin E in human low-density lipoprotein—When and how this antioxidant becomes a pro-oxidant. *Biochem J.* 288:341–344.
12. Mukai, K. 1993. Synthesis and kinetic study of antioxidant and prooxidant actions of vitamin E derivatives. In *Vitamin E in Health and Disease*, eds. L. Packer and J. Fuchs, pp. 97–119. New York: Marcel Dekker.
13. Bendich, A., and Machlin, L. J. 1993. The safety of oral intake of vitamin E: Data from clinical studies from 1986 to 1991. In *Vitamin E in Health and Disease*, eds. L. Packer and J. Fuchs, pp. 411–416. New York: Marcel Dekker.
14. Food Drug Administration. 1993. FDA Public Conference on Antioxidant Nutrients and Cancer and Cardiovascular Disease. Washington, DC, pp. 1–19.
15. Berlin E., McClure, D., Banks, M. A., and Peters, R. C. 1994. Heart and liver fatty acid composition and vitamin E content in miniature swine fed diets containing corn and menhaden oils. *Comp. Biochem. Physiol.* 109A:53–61.
16. Packer, L. 1993. Vitamin E: Biological activity and health benefits: Overview. In *Vitamin E in Health and Disease*, eds. L. Packer and J. Fuchs, pp. 977–982. New York: Marcel Dekker.
17. National Academy of Sciences. 1972. *Nutrient Requirements of Laboratory Animals*, 2nd rev. ed. Washington, DC: National Academy of Sciences.
18. Sokol, R. J. 1993. Vitamin E deficiency and neurological disorders. In *Vitamin E in Health and Disease*, eds. L. Packer and J. Fuchs, pp. 815–849. New York: Marcel Dekker.
19. Kay, M. M. B. 1993. Vitamin E deficiency causes appearance of an aging antigen and accelerated cellular aging and removal. In *Vitamin E in Health and Disease*, eds. L. Packer and J. Fuchs, pp. 287–296. New York: Marcel Dekker.
20. Meydani S. N., and Tengerdy, R. P. 1993. Vitamin E and the immune response. In *Vitamin E in Health and Disease*, eds. L. Packer and J. Fuchs, pp. 549–561. New York: Marcel Dekker.
21. Nair, P. P., Patnaik, R. N., and Hauswirth, J. W. 1978. Cellular transport and binding of d-α-tocopherol. In *Tocopherol, Oxygen and Biomembranes*, eds. C. deDuve and O. Hayaishi, pp. 121–130. Amsterdam: Elsevier/North Holland Biomedical Press.

22. Patnaik, R., Kessie, G., Nair, P. P., and Biswal, N. 1984. Vitamin E binding proteins in mammalian cells. In *Vitamins, Nutrition and Cancer*, ed. K. Prasad, pp. 105–117. Basel: Karger.
23. Azzi, A. M., Bartoli, G., Boscoboinik, D., Henely, C., and Szewczyk, A. 1993. α-Tocopherol and protein kinase C regulation of intracellular signaling. In *Vitamin E in Health and Disease*, eds. L. Packer and J. Fuchs, pp. 371–383. New York: Marcel Dekker.
24. Borek, C. 1990. Vitamin E as an anticarciongen. *Ann. N. Y. Acad. Sci.* 570:417–420.
25. Shklar, G., and Schwartz, J. L. 1993. Effects of vitamin E on oral carcinogenesis and oral cancer. In *Vitamin E in Health and Disease*, eds. L. Packer and J. Fuchs, pp. 497–511. New York: Marcel Dekker.
26. Banks, M. A., Martin, W. G., and Hinton, D. E. 1987. Long-term histological observations in the liver and pancreas of vitamin E and selenium deficient Syrian golden hamsters. *J. Nutr. Growth Cancer* 4:109–128.
27. Banks, M. A., Martin, W. G., Shinozuka, H., and Hinton, D. E. 1987. Effect of vitamin E deficiency on pancreatic, hepatic and biliary tumors induced by *N*-nitrosobis(2-oxopropyl)amine in the Syrian golden hamster. *J. Nutr. Growth Cancer* 4:221–237.
28. Berlin, E., Bhathena, S. J., Judd, J. T., Nair, P. P., Peters, R. C., Bhagavan, H. N., Ballard-Barbash, R., and Taylor, P. R. 1992. Effects of omega-3 fatty acid and vitamin E supplementation on erythrocyte membrane fluidity, tocopherols, insulin binding, and lipid composition in adult men. *J. Nutr. Biochem.* 3:392–399.
29. Awad, A. B. 1986. Effect of dietary fats on membrane lipids and functions. *J. Environ. Pathol. Toxicol.* 7:1–14.
30. Thomas, J. A. 1994. Oxidative stress, oxidant defense, and dietary constituents. In *Modern Nutrition in Health and Disease*, 8th ed., Vol 1, eds. M. E. Shils, J. A. Olson, and M. Shike, pp. 501–512. Philadelphia: Lea & Febiger.
31. Gutteridge, J. M. C. 1988. Lipid peroxidation: Some problems and concepts. In *Oxygen Radicals and Tissue Injury: Proceedings of an Upjohn Symposium*, ed. B. Halliwell, pp. 9–19. Bethesda, MD: Federation of American Societies for Experimental Biology.
32. Borg, D. C., and Schiaich, K. M. 1988. Iron and iron-derived radicals. In *Oxygen Radicals and Tissue Injury: Proceedings of an Upjohn Symposium*, ed. B. Halliwell, pp. 20–26. Bethesda, MD: Federation of American Societies for Experimental Biology.
33. Fisher, A. B. 1988. Intracellular production of oxygen-derived free radicals. In *Oxygen Radicals and Tissue Injury: Proceedings of an Upjohn Symposium*, ed. B. Halliwell, pp. 34–42. Bethesda, MD: Federation of American Societies for Experimental Biology.
34. Aust, S. D. 1988. Sources of iron for lipid peroxidation in biological systems. In *Oxygen Radicals and Tissue Injury: Proceedings of an Upjohn Symposium*, ed. B. Halliwell, pp. 27–33. Bethesda, MD: Federation of American Societies for Experimental Biology.
35. Zlotkin, S. H. 1991. Neonatal nutrition. In *Nutritional Biochemistry and Metabolism with Clinical Applications*, 2nd ed., ed. M. C. Linder, pp. 349–371. Norwalk, CT: Appleton & Lange.
36. Linscheer, W. G., and Vergroesen, A. J. 1993. Lipids. In *Modern Nutrition in Health and Disease*, 8th ed., Vol 1, eds. M. E. Shils, J. A. Olson, and M. Shike, pp. 47–88. Philadelphia: Lea & Febiger.
37. Neuringer, M., and Connor, W. E. 1986. ω-3 Fatty acids in the brain and retina: Evidence for their essentiality. Nutr. Rev. 44:285–294.
38. Nettleton, J. A. 1991. ω-3 Fatty acids: Comparison of plant and seafood sources in human nutrition. *J. Am. Diet. Assoc.* 91:331–337.
39. Johnston, P. V., and Fritsche, K. L. 1986. Linolenate metabolism. *Nutr. Rev.* 44:315–316.
40. Simopoulos, A. P. 1986. Historical perspective, conclusions and recommendations, and actions. In *Health Effects of Polyunsaturated Fatty Acids in Seafoods*, eds. A. P. Simopoulos, R. R. Kifer, and R. E. Martin, pp. 3–29. Orlando, FL: Academic Press.
41. Salem, N. Jr., Kim, H. Y., and Yergey, J. A. 1986. Docosahexaenoic acid. In *Health Effects of Polyunsaturated Fatty Acids in Seafoods*, eds. A. P. Simopoulos, R. R. Kifer, and R. E. Martin, pp.263–317. Orlando, FL: Academic Press.
42. Dratz, E. A., and Deese, A. J. 1986. Role of docosahexaenoic acid in biological membranes. In *Health Effects of Polyunsaturated Fatty Acids in Seafoods*, eds. A. P. Simopoulos, R. R. Kifer, and R. E. Martin, pp. 319–351. Orlando, FL: Academic Press.
43. Committee on Diet, Nutrition and Cancer. 1983. Diet, nutrition and cancer—Executive summary of the Report of the Committee on Diet, Nutrition, and Cancer. *Cancer Res.* 43:3018–3023.
44. National Dairy Council. 1983. Diet, nutrition, and cancer. *Dairy Council Digest* 54:31–36.
45. Gotto, A. M., Jr. 1991. The Role of Lipids in Coronary Heart Disease. Kalamazoo, MI: Upjohn Company.
46. Denke, M. A., and Grundy, S. M. 1991. Effects of fats high in stearic acid on lipid and lipoprotein concentrations in men. *Am. J. Clin. Nutr.* 54:1036–1040.
47. Vessby, B. 1994. Implications of long-chain fatty acid studies. *INFORM* 5:182–185.
48. Yu, S., Derr, T., Etherton, T. D., and Kris-Etherton, P. M. 1995. Plasma cholesterol-predictive equations

demonstrate that stearic acid is neutral and monounsaturated fatty acids are hypocholesterolemic. *Am. J. Clin. Nutr.* 61:1129–1139.

49. Sundram, K., Hayes, K. C., and Siru, O. H. 1995. Both dietary 18:2 and 16:0 may be required to improve the serum LDL/HDL cholesterol ratio in normocholesterolemic men. *J. Nutr. Biochem.* 6:179–187.
50. Hayes, K. C., Pronczuk, A., and Khosla, P. 1995. A rationale for plasma cholesterol modulation by dietary fatty acids: Modeling the human response in animals. *J. Nutr. Biochem.* 6:188–194.
51. Grundy, S. M. 1986. Effects of fatty acids on lipoprotein metabolism in man. In *Health Effects of Polyunsaturated Fatty Acids in Seafoods*, eds. A. P. Simopoulos, R. R. Kifer, and R. E. Martin, pp. 153–171. Orlando, FL: Academic Press.
52. Luc, G., and Fruchart, J. C. 1993. Lipoprotein oxidation and atherosclerosis. In *Vitamin E in Health and Disease*, eds. L. Packer and J. Fuchs, pp. 635–647. New York: Marcel Dekker.
53. Esterbauer, H., Puhl, H., Waeg, G., Krebs, A., and Dieber-Rotheneder, M. 1993. The role of vitamin E in lipoprotein oxidation. In *Vitamin E in Health and Disease*, eds. L. Packer and J. Fuchs, pp. 649–671. New York: Marcel Dekker.
54. Jackson, R. L., Ku, G., and Thomas, C. E. 1993. Antioxidants: A biological defense mechanism for the prevention of atherosclerosis. *Med. Res. Rev.* 13:161–182.
55. National Research Council. 1982. Lipids. In *Diet Nutrition and Cancer*, pp. 73–105. Washington, DC: National Academy Press.
56. Enig, M. G., Munn, R. J., and Keeny, M. 1978. Dietary fat and cancer trends-A critique. *Fed. Proc.* 37:2215–2220.
57. Hopkins, G. J., and West, C. E. 1976. Minireview-Possible roles of dietary fats in carcinogenesis. *Life Sciences* 19:1103–1116.
58. Wissler, R. W. 1992. Important points in the pathogenesis of atherosclerosis. In *Atherosclerosis*, ed. A. Gotto. Kalamazoo, MI: Upjohn Company.
59. Gey, K. F. 1993. Vitamin E and other essential antioxidants regarding coronary heart disease: Risk assessment studies. In *Vitamin E in Health and Disease*, eds. L. Packer and J. Fuchs, pp. 589–633. New York: Marcel Dekker.
60. Willingham, A. K., Bolanos, C., Bohannan, E., Cendella, R. J. 1993. The effects of high levels of vitamin E on the progression of atherosclerosis in the Watanabe heritable hyperlipidemic rabbit. *J. Nutr. Biochem.* 4:651–654.
61. Reaven, P., Parthasarathy, S., Grasse, B. J., Miller, E., Almazan, F., Mattson, F. H., Khoo, J. C., Steinberg, D., and Witzum, J. L. 1991. Feasibility of using an oleate-rich diet to reduce the susceptibility of low-density lipoprotein to oxidative modification in humans. *Am. J. Clin. Nutr.* 54:701–706.
62. Mickle, D. A. G. 1993. Antioxidant therapy in cardiac surgery, In *Vitamin E in Health and Disease*, eds. L. Packer and J. Fuchs, pp. 673–680. New York: Marcel Dekker.
63. Lucchesi, B. R. 1990. Myocardial Reoxygenation Injury. Kalamazoo, MI: Upjohn Company.
64. Klurfeld, D. M. 1995. Fat effects in experimental tumorigenesis. *J. Nutr. Biochem.* 6:201–205.
65. O'Connor, T. P., Roebuck, B. D., Peterson, F., and Campbell, T. C. 1985. Effect of dietary intake of fish oil and fish protein on the development of L-azaserine-induced preneoplastic lesions in rat pancreas. *J. Natl. Cancer Inst.* 75:959–962.
66. Karmali, R. A., Marsh, J., and Fuchs, C. 1984. Effect of omega-3 fatty acids on growth of rat mammary tumor. *J. Nat. Cancer Inst.* 73:457–461.
67. Slaga, T. J. 1984. Multistage skin carcinogenesis: A useful model for the study of the chemoprevention of cancer. *Acta Pharmacol. Toxicol.* 55:107–124.
68. Shamberger, R. J. 1972. Increase of peroxidation in carcinogenesis. *J. Natl. Cancer Inst.* 48:1491–1497.
69. National Research Council. 1982. Vitamins. In *Diet, Nutrition and Cancer*, pp. 139–161. Washington, DC: National Academy Press.
70. Birt, D. F. 1986. Update on the effects of vitamins A, C, and E and selenium on carcinogenesis. *Proc. Soc. Exp. Biol. Med.* 183:311–320.
71. Knekt, P. 1993. Epidemiology of vitamin E: Evidence for anticancer effects in humans, In *Vitamin E in Health and Disease*, eds. L. Packer and J. Fuchs, pp. 517–527. New York: Marcel Dekker.
72. Chen J., Goetchius M. P., Campbell, T. C., and Combs, G. F., Jr. 1982. Effects of dietary selenium and vitamin E on hepatic mixed-function oxidase activities and in vivo covalent binding of aflatoxin B$_1$ in rats. *J. Nutr.* 112:324–331.
73. Chow, C. K. 1993. Vitamin E and cigarette smoking-induced oxidative damage. In *Vitamin E in Health and Disease*, eds. L. Packer and J. Fuchs, pp. 683–697. New York: Marcel Dekker.
74. Elsayed, N. M. 1993. Modulation of pulmonary vitamin E by environmental oxidants. In *Vitamin E in Health and Disease*, eds. L. Packer and J. Fuchs, pp. 699–709. New York: Marcel Dekker.
75. Pryor, W. A. 1993. The role of vitamin E in the protection of in vitro systems and animals against the effects of ozone. In *Vitamin E in Health and Disease*, eds. L. Packer and J. Fuchs, pp. 713–736. New York: Marcel Dekker.

76. Duthie, G. G. 1993. Antioxidant status in smokers. In *Vitamin E in Health and Disease*, eds. L. Packer and J. Fuchs, pp. 711–713. New York: Marcel Dekker.

77. Banks, M. A., Porter, D. W., Martin, W. G., and Castranova, V. 1990. Effects of in vitro ozone exposure on peroxidative damage, membrane leakage, and taurine content of rat alveolar macrophages. *Toxicol. Appl. Pharmacol.* 105:55–65.

78. Banks, M. A., Porter, D. W., Martin, W. G., and Castranova, V. 1991. Ozone-induced lipid peroxidation and membrane leakage in isolated rat alveolar macrophages: Protective effects of taurine. *J. Nutr. Biochem.* 2:308–313.

79. Banks, M. A., Porter, D. W., Martin, W. G., and Castranova, V. 1992. Taurine protects against oxidant injury to rat alveolar pneumocytes. In *Taurine Nutritional Value and Mechanisms of Action*, eds. J. B. Lombardini, S. W. Schaffer, and J. Azuma, pp. 341–354. New York: Plenum Press.

80. Brownlee, M. 1994. Glycation and diabetic complications. I 43:836–841.

81. Hunt, J. V., Dean, R. T., and Wolff, S. P. 1988. Hydroxyl radical production and autoxidative glycosylation-Glucose autoxidation as the cause of protein damage in the experimental glycation model of diabetes mellitus and ageing. *Biochem. J.* 256:205–212.

82. Baynes, J. W. 1991. Role of oxidative stress in development of complications in diabetes. *Diabetes* 40:405–412.

83. Ceriello, A., Giugliano, D., Quatraro, A., Dello Ruso, P., and Lefebvre, P. J. 1991. "Metabolic control may influence the increased superoxide generation in diabetic serum." *Diabet. Med.* 8:540–542.

84. Katz, M. L., Robison, W. G. 1985. Nutritional Influences on Autooxidation, Lipofuscin Accumulation, and Aging. In: *Free Radicals, Aging, and Degenerative Diseases*, ed. Johnson, J. E., Alan R. Liss, Inc., New York, NY.

85. Curcio, F., Ceriello, A. 1992. Decreased Cultured Endothelial Cell Proliferation in High Glucose Medium is Reversed by Antioxidants: New Insights on the Pathophysiological Mechanisms of Diabetic Vascular Complications. In *Vitro Cellular Development Biology* 28A:787–790.

86. Ceriello, A., Quatraro, A., Giug.iano, D. 1992. New Insights on Non-enzymatic Glycosylation May Lead to Therapeutic Approches for the Prevention of Diabetic Complications. *Diabet. Med.* 9:297–299.

87. Ceriello, A., Giugliano, D., Quatraro, A., Donzella, C., Dipalo, G., Lefbvre, P. J. 1991. Vitamin E Reduction of Protein Glycosylation in Diabetes — New Prospect for Prevention of Diabetic Complications? *Diabetes Care* 14:68–72.

88. Paolisso, G., D'Amore, A., Giugliano, D., Ceriello, A., Varricchio, M, D'Onofrio, F. 1993. Pharmacologic Doses of Vitamin E Improve Insulin Action in Healthy Subjects and Noninsulin-dependent Diabietic Patients. *Am. J. Clin. Nutr.* 57:650–656.

89. Khatami, M., Suldan, Z., David, I., Li, W., Rockey, J. H. 1988. Inhibitory Effects of Pyridoxal Phosphate, Ascorbate and Aminoguanidine on Nonenzymatic Glycosylation. *Life Science* 43, 1725–1731.

90. Khatami, M. 1990. Role of Pyridoxal Phosphate/Pyridoxine in Diabetes-Inhibition of Nonenzymatic Glycosylation. In Vitamins B6, Ed. Dakshinamurti, K., *Ann. N. Y. Acad. Sci.* 585:502–504.

91. Blankenhorn, G., Clewing, S. 1993. Human Studies of Vitamin E and Rheumatic Inflammatory Disease. In *Vitamin E in Health and Disease*, eds. Packer, L. Fuchs, J. Marcel Dekker, Inc., New York, NY, pp. 563–575.

92. Fuchs, J., Packer, L. 1993. Vitamin E in Dermatological Therapy. In *Vitamin E in Health and Disease*, eds. Packer, L., Fuchs, J., Marcel Dekker, Inc., New York, NY, pp. 739–763.

93. Westermarck, R., Antila, E., Atroshi, F. 1993. Vitamin E Therapy in Neurological Diseases. In *Vitamin E in Health and Disease*, eds. Packer, L., Fuchs, J., Marcel Dekker, Inc., New York, NY, pp. 799–806.

94. Noromann, R., Rovach, H. 1993. Vitamin E Disturbances During Alcohol Intoxication. In *Vitamin E in Health and Disease*, eds. Packer, L., Fuchs, J. Marier Dekker, Inc., New York, NY, pp. 935–945.

95. Farrell, P. M., Robers, R. J. 1994. Vitamin E. In *Modern Nutrition in Health and Disease*, Eighth ed., vol 1. Eds. Shils, M. E., Olson, J. A., Shike M. 1994, Lea & Febiger, Philadelphia, PA, pp. 326–341.

96. Dutta-Roy A. K., Gordon, M. J., Campbell, F. M., Duthie, G. G., James, W. P. T. 1994. Vitamin E Requirements, Transport, and Metabolism: Role of in Vitamin E in Health and Disease. In *Vitamin E in Health and Disease*, α-Tocopherol-Binding Proteins," *J. Nutr. Biochem.* 5: 562–570.

97. Rajaram, O. V., Fatterpaker, P., Sreeninvasan, A. 1973. Occurrence of α-Tocopherol Binding Protein in Rat Liver Cell Sap. *Biochem. Biophys. Res. Commun.* 52:459–465.

98. Banks, M. A., Peters, R. C., Berlin, E. 1994 (Abst.). Purification and Partial Characterization of Vitamin E Binding Proteins from Rat and Miniature Swine Heart Cytosol. *FASEB J.* 8:A1353.

99. Banks, M. A., Berlin, E., Peters, R. C. 1994 (Abst.). Isolatin of a High Molecular Weight Vitamin E Binding Protein From Rat and Human Erythrocyte Lysates. *American Chemical Society Book of Abstracts*, Part I, Biological chemistry, #79.

100. Nalecz, K. A., Nalecz, M. J., Azzi, A. 1992. Isolation of TYocoherol-Binding Proteins from the Cytosol of Smooth Muscle A7r5 cells. *Eur. J. Biochem.* 209:37–42.

Chapter Five

α-TOCOPHEROL, β-CAROTENE, AND OXIDATIVE MODIFICATION OF HUMAN LOW-DENSITY LIPOPROTEIN

Hazel T. Bowen and Stanley T. Omaye

Atherosclerotic cardiovascular disease continues to be a leading cause of morbidity and mortality in adults in industrialized countries, despite a decline in mortality from coronary artery disease and stroke in the last two and a half decades (1). Epidemiological data gathered in the last 15 years have revealed that development of coronary artery disease can be attributed only in part to the well-established, "classical" risk factors: that is, to elevated total cholesterol, LDL cholesterol and LDL/HDL ratio, hypertension, and a history of smoking (for review see ref. 2). Interaction of a number of variables is currently considered to influence the disease process (3), among which dietary factors appear to have notable impact. The possible effects of diet on the risk of coronary heart disease have been of interest for the last four decades. It is well known that fruits and vegetables produce a great number of biologically active organic compounds, including antioxidants, that may significantly affect human health. The purpose of this review is threefold: to briefly discuss epidemiological evidence that links antioxidants, such as beta-carotene and vitamin E, to cardiovascular effects; to summarize the "oxidation hypothesis" of atherosclerosis and its implication that natural antioxidants may be able to prevent or slow the progression of atherosclerosis; and to review recent studies that test the ability of vitamin E and beta-carotene to inhibit the oxidation of low density lipoprotein in vitro.

CLINICAL DATA

A growing body of evidence from a variety of epidemiologic studies and clinical trials (4–11) supports the role of antioxidants in the prevention and treatment of atherosclerotic cardiovascular disease. The most compelling epidemiologic findings to date come from two recent prospective studies. The Nurses' Health Study (12–14) and the Health Professionals' Follow-up Study (15) examined the relationship between antioxidant intake and the risk of coronary artery disease in a population of greater than 87,000 women over an 8-year follow-up, and close to 40,000 men over a 4-year follow-up, respectively. The relative risk for coronary heart disease in participants with the highest total intake of vitamin E (including vitamin E supplements) compared to those with the lowest total intake was 0.66 for women and 0.64 for men (12, 15). Women in the highest quintile of beta-carotene consumption had a 22% risk reduction for coronary events (p, trend = .02) compared with those having the lowest consumption (13). There was a 39% risk reduction (p, trend = .01) for stroke in the highest intake category (14).

An inverse association existed between men in the highest quintile of beta-carotene intake and risk for coronary artery disease, and a relative risk of 0.75 (p, trend = .04) for those in the highest intake category. No such relationship was found for vitamin C. Former smokers with a dietary intake of beta-carotene in the highest quintile had a significant decrease in risk; among current smokers, the reduction in risk was significant even in the third quintile group (15).

Another prospective cohort study of 1299 elderly Massachusetts residents found that dietary beta-carotene was associated with decreased risk for cardiovascular mortality (16). During a 4.75-year follow-up, after controlling for age, sex, smoking, alcohol consumption, cholesterol intake, and functional status, those in the highest quartile of dietary beta-carotene intake had a relative risk for death from cardiovascular causes of 0.57 (p, trend = .02). For fatal myocardial infarction, the relative risk was 0.32 (p, trend = .02).

An analysis of a subgroup of the Physicians' Health Study provided additional evidence for a protective effect of beta-carotene (17). Of 333 men with angina pectoris and/or coronary revascularization randomized to receive 50 mg of beta-carotene every other day or a placebo, those assigned to beta-carotene experienced a 51% reduction in all major coronary events (defined in this study as myocardial infarction, revascularization, or cardiac death), and a 54% (p, trend = .014) decrease in all major vascular events (defined as stroke, myocardial infarction, revascularization, or cardiac death). The protective effects of beta-carotene in this study and vitamin E in the Nurses' Health study and the Health Professionals' Follow-up Study weren't evident until after the second year of follow-up. This delay in the protective effect is consistent with the theory that antioxidant intake slows the progression of atherosclerosis. The results from these studies are summarized in Table 1.

A recently published study that examined the association of supplementary vitamin C and E intake and coronary heart disease progression (18) found during a 2-year follow-up that the subjects, 188 men who had undergone previous coronary artery bypass grafting who comprised a subgroup of the Cholesterol Lowering Atherosclerosis Study (CLAS), taking vitamin E supplements of 100 IU or more daily, demonstrated less coronary artery lesion progression as determined by quantitative coronary angiography than subjects with a daily supplementary vitamin E intake of less than 100 IU. No benefit was found for use of vitamin C, used either exclusively or in conjunction with supplementary vitamin E or multivitamins.

Not all epidemiologic data support the effectiveness of antioxidant supplementation in the

Table 1. Summary of trends in selected clinical trials

Study	Endpoint	Antioxidant	Relative risk[a]
Nurses' Health			
	CHD	E (total intake)	0.66[d]
		E (suppl. >2 yr)	0.59[d]
	coronary events	β-Carotene	0.88
	stroke	E	0.71[a]
		β-Carotene	0.71[d]
Health Professionals'			
Follow-up	CHD	E (total intake)	0.64[d]
		E (100–149 IU/d)	0.54[d]
		E (suppl. >2 yr)	0.59[d]
		β-Carotene	0.71[d]
		β-Carotene (smokers)	0.30
Massachusetts Elderly			
	Cardiovascular mortality	β-Carotene (dietary intake)	0.57[f]
	Fatal myocardial infarction	β-Carotene (dietary intake)	0.032[f]
Physicians' Health			
	Coronary events[b]	β-Carotene (50 mg on alternate days)	0.49
	Vascular events[c]	β-Carotene (50 mg on alternate days)	0.46

[a] Highest vs. lowest intake.
[b] Defined as myocardial infarction, revascularization, or cardiac death.
[c] Defined as stroke myocardial infarction, revascularization, or cardiac death.
[d] Adjusted for age and smoking.
[e] Adjusted for age and coronary risk factors.
[f] Adjusted for age, sex, smoking, alcohol consumption, cholesterol consumption, and functional status.

prevention and treatment of coronary heart disease. A randomized trial from Finland in which the incidence of lung cancer was the primary endpoint showed that neither beta-carotene nor alpha-tocopherol supplements had an effect on mortality from cardiovascular disease (19). Some methodologic concerns have been raised concerning this study, however. The alpha-tocopherol supplement used was reported to have low bioavailability (20), and the level of supplementation was below that which observational studies have suggested is necessary to decrease cardiovascular risk (12, 15).

Several large randomized trials now underway are evaluating the effectiveness of higher doses of antioxidant supplements in the prevention and treatment of coronary heart disease. In the remaining segments of the Physicians' Health Study, scheduled to be completed this year, beta-carotene is being tested in 22,000 men. The Women's Health Study is testing aspirin, vitamin E, and beta-carotene in 40,000 women. A companion study to the Women's Health Study, the Women's Antioxidant Cardiovascular Disease Study, is designed to investigate the effectiveness of antioxidant supplements beta-carotene, vitamin E, and vitamin C for secondary prevention in 8000 women with existing cardiovascular disease. The Carotene and Retinol Efficacy Trial Study is testing the efficacy of beta-carotene and retinoic acid supplementation for prevention of cancer and cardiovascular disease in 18,000 U.S. male asbestos workers. Finally, a combination of antioxidant vitamins, including beta-carotene, vitamin C, and vitamin E, in addition to selenium and zinc, is being tested in 15,000 French men and women (21). Over the next several years, a substantial amount of data on the efficacy of antioxidant supplements in the prevention of primary, as well as secondary, coronary heart disease should be forthcoming as results from these studies becomes available. And for the first time, information regarding the safety of larger doses of antioxidant vitamins given over long periods of time, a long-time concern of researchers, will be available.

OXIDIZED LDL AND ATHEROSCLEROSIS

Numerous studies done in the last two and a half decades have provided the background for development of the current theory of atherogenesis in which oxidatively modified low-density lipoprotein (LDL) is implicated in atheroma formation and progression. According to the "oxidation hypothesis of atherosclerosis," one of the initial events in the formation of the atherosclerotic lesion is the subendothelial accumulation of blood monocytes in the vascular intima (Figure 1). Monocytes apparently adhere to and, subsequently migrate through, areas of increased permeability in the endothelium; blood lipoproteins are also able to pass through these lesion-prone sites, resulting in increased concentrations of lipoproteins and monocyte accumulation in the subintimal space. These monocytes differentiate into macrophages, which avidly take up and degrade LDL, resulting in the buildup of cholesterol esters in the cytoplasm of the macrophages. These lipid-filled cells, or foam cells, form the fatty streak of early atherosclerotic plaques (22–24). Fatty streaks then may progress to mature plaques containing foam cells derived from smooth muscle cells and macrophages, smooth muscle cells, T-lymphocytes, and macrophages surrounding a necrotic lipid core (25, 26). Although an elevated plasma concentration of LDL cholesterol is a primary risk factor for the development of atherosclerosis, the mechanism by which cells accumulate cholesterol isn't immediately evident, since cultured macrophages exposed to high concentrations of LDL do not accumulate cholesterol (27, 28). Removal of the majority of LDL from the plasma is mediated by the LDL or B/E receptor (29, 30); there are a limited number of these receptors expressed by macrophages, and they are down-regulated in the presence of high concentrations of cholesterol (31). Because it was demonstrated that macrophage-derived foam cells do develop in vivo in the presence of elevated levels of LDL, Brown and Goldstein (31) proposed that LDL must undergo some type of modification in order for it to be taken up rapidly enough to generate foam cells. Subsequent experiments confirmed that chemically modified LDL, which is no longer recognized by the B/E receptor, is rapidly taken up by macrophages by a specific receptor termed the scavenger or acetyl-LDL receptor (33, 34). The scavenger receptor isn't feedback-inhibited and, it thereby provides a pathway for the uncontrolled uptake of modified LDL, resulting in foam cell formation.

Evidence suggests that some form of oxidative step is likely to be involved in the modification of LDL (31, 32) resulting in its recognition by the scavenger receptor–a critical step in the

Lipoproteins and monocytes accumulate at lesion-prone sites in subintima of artery

$\downarrow$

Reactive oxygen species are generated by endothelial cells, smooth muscle cells, and macrophages within subendothelium

$\downarrow$

Intimal lipoproteins are oxidatively modified by reactive oxygen species; modified LDL exerts cytotoxic effects resulting in foam cell necrosis

$\downarrow$

Macrophages take up oxidized lipoproteins via unregulated scavenger receptor resulting in foam cell formation (early fatty streak)

$\downarrow$

Smooth muscle cells migrate to and proliferate in arterial intima (mature atherosclerotic lesion)

$\downarrow$

Arterial lumen narrows as a result of intimal thickening (partial blockage of the artery)

Figure 1.   Proposed sequence of events in the initiation of the atherosclerotic lesion.

development of atherosclerosis. It is likely that LDL oxidation occurs primarily in the microenvironment of the subintimal space of the artery, where it is excluded from the protection of plasma antioxidants, and is in close proximity to oxidizing cells. The four major cell types within the artery wall, endothelial cells, smooth muscle cells, and monocytes and macrophages (35–38), have been shown to oxidize LDL in vitro, under serum-free conditions, to a form recognized by the scavenger receptor. Since cell-conditioned media alone can't sustain oxidation (39), it may be necessary for physical contact to be maintained between the cells and LDL.

Direct evidence that LDL oxidation occurs in vivo is demonstrated by the presence of autoantibodies against oxidized LDL (oLDL) in human and rabbit atherosclerotic lesions (40). In tissue specimens of patients with established carotid atherosclerosis, antibodies specific to malondialdehyde-modified apo-B100 were found to react with antigens associated with smooth muscle cells, macrophage-derived foam cells, and within the necrotic lipid core of atherosclerotic lesions (41). (Malondialdehyde is a highly reactive by-product of lipid peroxidation capable of forming Schiff bases.) Other findings supporting the existence of oLDL in vivo are the presence of increased thiobarbituric acid-reactive substances (TBARS) in the serum of atherosclerotic patients (42), and the presence of lipid peroxides and fluorescent compounds characteristic of lipid peroxidation in atherosclerotic plaques (42–44). Perhaps the strongest support for the presence of oLDL in vivo is found in a number studies using animal models in which antioxidants were shown to retard the progression of atherosclerotic lesions (45–48).

EFFECT OF ANTIOXIDANT VITAMINS ON OXIDATIVE MODIFICATION OF LDL

LDL can be oxidized in the presence of transition metals (copper and iron), in cell-free systems, and by incubation with endothelial cells, smooth muscle cells, and macrophages (35–38, 49). The use of in vitro copper-catalyzed oxidation systems may be appropriate models for demonstrating events occurring in atherosclerotic lesions because the presence of catalytic copper ions in such lesions has been demonstrated (50). Dietary antioxidants, including vitamins C and E, and beta-carotene, have been shown to inhibit oxidation of LDL by cells and in cell-free systems

(36, 37, 51–53). Auto-oxidation of LDL as measured by increase in aldehydes and decrease of polyunsaturated fatty acids (PUFAs) has been shown to occur when LDL is depleted of its endogenous antioxidants (54). However, some investigators have suggested that lipid peroxidation begins immediately after the consumption of ubiquinol-10, and prior to depletion of alpha-tocopherol and the carotenoids (55). The disappearance of LDL's antioxidants begins with ubiquinol-10, alpha- and gamma-tocopherol followed by the carotenoids, with beta-carotene disappearing last (55, 56) (Figure 2). The degradation of endogenous antioxidants is followed by a propagating lipid peroxidation chain reaction indicated by increase in diene absorption (56, 57). The sequence of these events suggests that the primary element in predicting the resistance of LDL to oxidation (as determined by the lag phase) may be its antioxidant content, although many studies have failed to demonstrate a clear correlation between the lag period and consumption of antioxidants (53, 58–60). Recent investigations have found that a number of other factors may affect the oxidative susceptibility of LDL, including fatty acid composition (61, 62), core lipid structure (63), size and density of the LDL (64), and the presence of preformed hydroperoxides (65).

A great deal of work in the last 5 yr has been devoted to determining the protective effect of endogenous antioxidants contained in LDL. Early studies, in which vitamin E was added in excess (from 5- to 80-fold higher than endogenous vitamin E within the LDL particle) to the incubation (oxidation) medium, demonstrated that cell-mediated oxidative modification of LDL could be inhibited up to 24 h, depending on the concentration of vitamin E added (36, 37, 58, 59). It is unclear, however, whether the primary effect was on the LDL or on the cells. Esterbauer et al. (reviewed in ref. 65), however, analyzed a large number of unsupplemented individual LDL samples (59), and found that there were no significant correlations between lag phase and vitamin E, or lag phase and total LDL antioxidant content. Similar results were obtained by Jessup et al. (53) in cell-mediated and cell-free oxidation systems. Dieber-Rotheneder et al. (66) conducted a supplementation trial with similar results. Prior to supplementation there was little correlation between vitamin E content and lag phase ($r^2 = 0.145$). During supplementation the duration of the lag phase was significantly longer, but there continued to be large interindividual variation. A correlation coefficient between the vitamin E content of LDL and lag phase of 0.51 was obtained during supplementation. It was concluded from these results that variables other than vitamin E influence the oxidative susceptibility of LDL. An equation was derived by Esterbauer et al. (56) describing the relationship between alpha-tocopherol and lag phase for the LDL of a given donor:

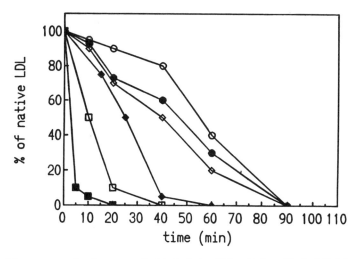

Figure 2. Disappearance of endogenous antioxidants during LDL oxidation ubiquinol-10 (closed squares); alpha-tocopherol (open squares); gamma-tocopherol (closed squares); lycopene (open triangles); lutein/zeaxanthin (closed circles); beta-carotene (open circles). Adapted from ref. 56 with permission.

$$y = kx + a \tag{1}$$

The lag phase in copper ion-mediated oxidation (y) depends on content of vitamin E (x), a subject-specific efficiency constant of vitamin E (k), and a vitamin E-independent variable (a). When results from Dieber-Rotheneder's ex vivo supplementation (66) and Esterbauer et al.'s in vitro loading studies (59) were analyzed according to this equation, a linear correlation between vitamin E content and lag phase ($r^2 = 0.95$–0.99; $p < .001$) was obtained (56). An approach used by our lab that has been successful in diminishing the effect of between-subject variation and determining the correlation between LDL vitamin E content and lag phase is to pool the plasma of a number of healthy individuals (>15), and enrich it with alpha-tocopherol in vitro prior to separation of the LDL (Table 2).

Jialal and Grundy (67) conducted an ex vivo study of oral supplementation with 800 IU alpha-tocopherol, compared with a placebo, in 24 male volunteers over a 12-wk period. They found a significant reduction in TBARS activity and conjugated dienes in time-course curves of LDL modification at 6 and 12 wk for the supplemented group, as well as a significant decrease in the rate of oxidation. These observations suggest that alpha-tocopherol supplementation significantly decreases the propagation phase of LDL oxidation, and support previous studies reporting that enrichment of LDL with vitamin E decreases its susceptibility to oxidation. In this study, the lag phase wasn't measured, so the large interindividual variation found in previous studies wasn't observed. Recently, a placebo-controlled randomized trial conducted by Jialal et al. examined kinetics of LDL oxidation, including lag phase, oxidation rate, and time-course curves. These indices were used to determine the oxidative resistance of LDL (68). In this trial, 48 male volunteers were assigned to receive either a placebo or alpha-tocopherol at dosages of 60, 200, 400, 800, or 1200 IU per day for 8 wk. Significant differences in TBARS activity and conjugated dienes were observed among subjects receiving supplements 400 IU and greater compared with those receiving the placebo. This effect was not seen in those receiving 60 or 200 IU supplements, however. The lag phases and oxidation rates for those receiving 60 and 200 IU were not significantly different from those receiving placebo. For subjects receiving at least 400 IU, the lag phase increased with increasing alpha-tocopherol dosage, and the oxidation rates for those receiving 800 and 1200 IU showed a significant inverse correlation with dosage. Large interindividual variation in lag phase, computed from conjugated dienes and TBARS, was observed, in agreement with previous studies (53, 59, 66).

It has been suggested that anywhere from one-third (60) to 80% (69) of the antioxidant capacity of LDL can be attributed to its alpha-tocopherol content. Because beta-carotene is the last endogenous antioxidant to be consumed before the propagating lipid peroxidation begins (56), it might be predicted that it plays a supporting role in the antioxidant defense of LDL. In order to distinguish the antioxidant effect of beta-carotene from that of alpha-tocopherol, it is necessary to load the LDL with it (59). Initial in vitro trials found no effect of beta-carotene on LDL oxidation (49,70), contradicted by results obtained by Jialal et al. (52). This discrepancy

Table 2. Correlation between lag phase and in vitro antioxidant enrichments

Beta-carotene (µg/mg LDL protein)	Lag time (min)	r^2
0.145	36.5	
0.243	37.5	
0.323	40.0	
0.615	38.8	
		0.478
Alpha-tocopherol (µg/mg LDL protein)		
3.1	30.0	
3.7	35.0	
5.3	41.3	
6.7	46.3	
		0.951

may have been due to the varying solubility of beta-carotene in organic solvents. When ethanol is used as the solvent, the procedure followed in one of the studies (49), minimal amounts of beta-carotene dissolve, and therefore very little can become incorporated within the LDL's lipid core. Conclusions about the protective effect of beta-carotene on LDL oxidation using this procedure are not valid. A number of papers describe the use ethanol to solubilize beta-carotene prior to extraction into hexane. Standard curves based on many of these protocols are inaccurate in their estimation of beta-carotene concentrations. Jialal et al. (52) first dissolved beta-carotene in hexane, then diluted it with ethanol prior to adding it to the oxidation medium. Although the purity of the stock solution was assayed by HPLC, the level of incorporation into the LDL was not determined. Whether or not beta-carotene is incorporated in the LDL particle is clearly pertinent in assessing its performance as an antioxidant. Using a cell-free system, Jialal found beta-carotene to be a more potent inhibitor of LDL oxidation on a molar basis than alpha-tocopherol as determined by total TBARS activity and conjugated dienes at the end of 24 h. Beta-carotene was also found to inhibit LDL oxidation in the cell-mediated system as measured by macrophage degradation and electrophoretic mobility. Because the antioxidant activity of beta-carotene is most pronounced at low oxygen tension (71), the investigators conclude that it could complement the antioxidant activity of alpha-tocopherol, which is more effective at high oxygen tension.

The ability of beta-carotene to prevent Cu^{2+}-catalyzed oxidation of Lp(a), a lipoprotein with chemical and physical properties similar to those of LDL, as well as LDL, was also examined (72). Beta-carotene was added to the oxidation medium in a solvent of 1 part dimethylsulfoxide (DMSO) and 10 parts ethanol in a concentration (21.5 μg/ml), approximately 100 times that in native LDL. Because beta-carotene is quite sparingly soluble in ethanol and DMSO, as discussed earlier, the percentage increase in Lp(a) and LDL beta-carotene, found to be 32% and 19%, respectively, was somewhat surprising. However, beta-carotene was found to inhibit oxidation of both Lp(a) (by 44%) and LDL (by 29%) compared with controls as determined by TBARS activity.

Evidence for the protective effect of beta-carotene in supplementation studies appears to contradict the findings of in vitro studies. Although Abbey et al. (73) found a significant positive correlation between LDL beta-carotene and lag time, the relationship was not shown to be independent of alpha-tocopherol. In a prospective double-blind trial testing the ability of supplementation with 60 mg beta-carotene, 1600 mg alpha-tocopherol, and 2 g ascorbate to inhibit oxidative modification of LDL, no protective effect of beta-carotene was demonstrated despite an approximately 20-fold enrichment above baseline levels of beta-carotene in LDL (74). On the contrary, a significant increase in the rate of macrophage degradation of LDL was associated with beta-carotene supplementation. These results are consistent with the observation of Gaziano et al. (75) that LDL isolated after beta-carotene supplementation exhibited a significantly decreased lag phase compared with LDL isolated prior to supplementation. These results suggest a possible pro-oxidant effect of beta-carotene at higher concentrations at ambient oxygen pressure. Interestingly, in vitro loading of LDL, either by addition of beta-carotene to the LDL preparation or by incubation in plasma prior to LDL isolation, did not lead to increased oxidative susceptibility, although no protective effect was observed.

Results from our in vitro loading experiments using a Cu^{2+}-mediated oxidation system contradicted this observation, however. A linear relationship between LDL beta-carotene concentration and rate of oxidation was observed, and there was an exponential relationship between TBARS concentration and LDL beta-carotene concentration. In addition, a linear correlation ($r^2 = 0.995$; $p < .0001$) was observed between initial beta-carotene concentration and amount of beta-carotene oxidized in 60 min. These results suggest that, under these conditions, beta-carotene is participating in the propagating reaction.

Jialal and Grundy (76) tested supplements containing a combination of 1 g ascorbate, 800 mg alpha-tocopherol, and 30 mg beta-carotene for their efficacy in increasing LDL oxidation resistance in vitro. Although the combined supplement resulted in a significant inhibition of LDL oxidation as determined by a twofold increase in lag time and a 40% decrease in oxidation rate, there was no added benefit when compared with alpha-tocopherol supplementation alone.

Results from these experiments should be interpreted with caution, however. When in vitro oxidation studies are conducted, it is assumed that, in vivo, antioxidants act to directly protect LDL. Although direct protection may be one of the mechanisms involved, it cannot be assumed that antioxidant activity elsewhere does not affect LDL resistance to oxidation. It has been shown that beta-carotene is a relatively inefficient antioxidant at ambient oxygen pressures but, when present in low concentrations at physiologic oxygen tension, it is a very effective antioxidant (71). Since these experiments were conducted under conditions of high oxygen tension, generalizations from in vitro studies to in vivo conditions regarding the effectiveness of beta-carotene as a protective antioxidant for LDL are precluded.

CONCLUSIONS

A large body of epidemiologic evidence now supports the role of antioxidant vitamins including ascorbate, alpha-tocopherol, and beta-carotene as protective agents against atherosclerotic cardiovascular disease. These vitamins have been shown to protect low-density lipoprotein against oxidative modification in ex vivo supplementation studies and in vitro loading studies, suggesting that the protective effect against atherosclerosis associated with their intake may be mediated by their antioxidant activity. Data emerging from in vitro and ex vivo studies, however, has not shown a clear correlation between LDL content of vitamin E or beta-carotene and decrease in the oxidative susceptibility of LDL. The consistently observed large interindividual variation in LDL susceptibility cannot been explained in terms of antioxidant content alone; compositional and structural factors including density and size of the LDL particle, fatty acid and cholesterol composition, and core lipid structure must also be considered as potential influences on LDL resistance to oxidation. The functions of these compounds other than their antioxidant activity may also mediate their protective effect against atherosclerosis. Alpha-tocopherol has been shown to inhibit protein kinase C activity, resulting in inhibition of vascular smooth muscle cell proliferation, an effect unrelated to its radical scavenging properties (77, 78), a clearly antiatherogenic effect. Two studies found that beta-carotene supplementation increases levels of HDL (79, 80), which, itself, is known to be a potent inhibitor of LDL oxidation (81). The oxidation hypothesis of atherogenesis and its implication that antioxidant vitamins can potentially prevent and/or treat atherosclerotic cardiovascular disease should be viewed as a preliminary step in understanding the biochemical mechanisms responsible for the development of this disease.

REFERENCES

1. Committee on Diet and Health, Food and Nutrition Board, Commission on Life Sciences, National Research Council. *Recommended Dietary Allowances.* 1989. Washington, DC: National Academy Press.
2. Gey, K. F. 1993. Vitamin E and other essential antioxidants regarding coronary heart disease: Risk assessment studies. In *Vitamin E in Health and Disease*, eds. L. Packer and J.Fuchs, pp. 589–633. New York: Marcel Dekker.
3. Steinberg, D., and Witztum, J. L. 1990. Lipoproteins and atherogenesis. Current concepts *J. A. M. A.* 264:3074–3121.
4. Gey, K. F., Stahelin, H. B., Puska, P., and Evans, A. 1987. Relationship of plasma vitamin C to mortality from ischemic heart disease. *Ann. N. Y. Acad. Sci.* 498:110–123.
5. Gey, K. F., Brubacher, G. B., and Stahelin, H. B. 1987. Plasma levels of antioxidant vitamins in relation to ischemic heart disease and cancer. *Am. J. Clin. Nutr.* 5:1368–1377.
6. Gey, K. F., and Puska, P. 1989. Plasma vitamins E and A inversely related to mortality from ischemic heart disease in cross-cultural epidemiology. *Ann. N. Y. Acad. Sci.* 570:268–282.
7. Gey, K. F., Puska, P., Jordan, P., and Ulrich, K. M. 1991. Inverse correlation between vitamin E and mortality from ischemic heart disease in cross-cultural epidemiology. *Am. J. Clin. Nutr.* 53:326S–34S.
8. Riemersma, R. A., Oliver, M., Elton, A., Alfthan, G., Vartianen, M. Salo, M., Rubba, P., Mancivi, H., Georgi, H. Vuilleumier, J., and Gey, K. F. 1990. Plasma antioxidants and coronary heart disease: Vitamins C and E and selenium. *Eur. J. Clin. Nutr.* 44:143–150.
9. Street, D. A., Comstock, G. W., Salkeld, R. M., Schuep, W., and Klag, M. 1991. A population based case-control study of serum antioxidants and myocardial infarction. *Am. J. Epidemiol.* 134:719–720.
10. Riemersma, R. A., Wood, D. A., Macintyre, C. C. H., Elton, R., Gey, K. F., and Oliver, M. F. 1991. Risk of angina pectoris and plasma concentrations of vitamins A, C, E, and carotene. *Lancet* 337:1–5.
11. Riemersma, R. A., Wood, D. A., Macintyre, C. C. H., Elton, R., Gey, K. F., and Oliver, M. F. 1989. Low plasma vitamin E and C: Increased risk of angina in Scottish men. *Ann. N. Y. Acad. Sci.* 570: 291–295.
12. Stamfer, M. J., Hennekens, C. H., Manson, J. E., Colditz, G. A., Rossner, B., and Willett, W. C. 1993. Vitamin E consumption and the risk of coronary disease in women. *N. Engl. J. Med.* 328:1444–1449.

13. Manson, J. E., Stamfer, M. J., Willett, W. C., Colditz, G. A., Rosner, B., Speizer, F. E., and Hennekens, F. E. 1991. A prospective study of antioxidant vitamins and incidence of coronary heart disease in women. *Circulation* 84(4, Suppl. II):2186(abstr.).

14. Manson, J. E., Stamfer, M. J., Willett, W. C., Colditz, G. A., Speizer, F. E., and Hennekens, C. H. 1993. Antioxidant vitamins and incidence of stroke in women. *Circulation* 87:2(abstr.).

15. Rimm, E. B., Stamfer, M. J., Ascherio, A., Giovannucci, E., Colditz, G. A., and Willett, W. C. 1993. Vitamin E consumption and the risk of coronary heart disease in men. *N. Engl .J. Med.* 328:1450–1456.

16. Gaziano, J. M., Manson, J. E., Branch, L. G., LaMott, F., Colditz, G. A., Buring, J. E., and Hennekens, C. H. 1992. Dietary beta-carotene and decreased cardiovascular mortality in an elderly cohort. *J. Am. Coll. Cardiol.* 19(3, suppl. A): 377A.(abstr.).

17. Gaziano, J. M., Manson, J. E., Ridker, P. M., Buring, J. E., and Hennekens, C. H. 1990. Beta-carotene therapy for chronic stable angina. *Circulation* 82(suppl 3):201.

18. Hodis, H. N., Mack, W. J., LaBree, L., Cashin-Hemphill, L., Sevanian, A., Johnson, R., and Azen, S. P. 1995. Serial coronary angiographic evidence that antioxidant vitamin intake reduces the progression of coronary atherosclerosis. *J. A. M. A.* 273:1849–1854.

19. Alpha-Tocopherol, Beta-Carotene Cancer Prevention Study Group. 1994. The effect of vitamin E and beta-carotene on the incidence of lung cancer and other cancers in male smokers. *N. Engl .J. Med.* 330:1029–1035.

20. Hoffman, R. M., and Harinder, S. G. 1995. Antioxidants and the prevention of coronary heart disease. *Arch. Intern. Med.* 155:241–246.

21. Gaziano, J. M. 1994. Antioxidant vitamins and coronary artery disease risk. *J. A. M. A.* 97(suppl 3A):18–21.

22. Schaffner, T., Taylor, K., Bartucci, E. J., Fischer-Dzoga, K., Beenson, J. H., Glagov, S., and Wissler, R. 1980. Arterial foam cells with distinctive immunomorphic and histochemical features of macrophages. *Am. J. Pathol.* 100:57–80.

23. Gerrity, R. G. 1981. The role of the monocyte in atherogenesis. I. Transition of blood-borne monocytes into foam cells in fatty lesions. *Am. J. Pathol.* 103:181–190.

24. Gerrity, R. G. 1981. The role of the monocyte in atherogenesis. II. Migration of foam cells from atherosclerotic lesions. *Am. J. Pathol.* 103:191–200.

25. Ross, R. 1986. The pathenogenesis of atherosclerosis: An update. *N. Engl. J. Med.* 314:488–500.

26. Ross, R. 1993. The pathenogenesis of atherosclerosis: An update. A perspective for the 1990's. *Nature* 362:801–809.

27. Brown, M. S., and Goldstein, J. L. 1983. Lipoprotein metabolism in the macrophage: Implications for cholesterol deposition in atherosclerosis. *Annu. Rev. Biochem.* 52:223–261.

28. Goldstein, J. L., Ho, Y. K., Basu, S. K., and Brown, M. S. 1979. Binding site on macrophages that mediates uptake and degradation of acetylated low density lipoprotein, producing massive cholesterol deposition. *Proc. Natl. Acad. Sci. U.S.A.* 76:333–337.

29. Kesaniemi, Y. A., Witztum, J. L., and Steinbrecher, U. P. 1983. Receptor-mediated catabolism of low density lipoprotein in man: Quantitation using glucosylated low density lipoprotein. *J. Clin. Invest.* 71:950–959.

30. Pittman, R. C., Carew, T. E., Attie, A., D., Witztum, J. L., Watanabe, Y., and Steinberg, D. 1982. Receptor-dependent and receptor-independent degradation of low density lipoprotein in normal rabbits and receptor-deficient mutant rabbits. *J. Biol. Chem.* 257:7994–8000.

31. Steinberg D., Parthasarathy, S., Carew, T. E., Khoo, J. C., and Witztum, J. L. 1989. Beyond cholesterol: Modifications of low-density lipoprotein that increase its atherogenicity. *N. Engl. J. Med.* 320:915–924.

32. Henriksen, T., Mahoney, E. M., and Steinberg, D. 1983. Enhanced macrophage degradation of biologically modified low density lipoprotein. *Arteriosclerosis* 3:149–159.

33. Mahley, R. W., Innerarity, T. L., Weisgraber, K. H., and Oh, S. Y. 1979. Altered metabolism (in vivo and in vitro) of plasma lipoproteins after selective chemical modification of lysine residues of the apoproteins. *J. Clin. Invest.* 64:743–750.

34. Parthasarathy, S., and Rankin, S. M. 1992. Atherogenic effects of oxidatively modified LDL. *Prog. Lipid. Res.* 31:127–143.

35. Heinecke, J. W., Rosen, H., and Chait, A. 1984. Iron and copper promote modification of low density lipoprotein by human arterial smooth muscle cells in culture. *J. Clin. Invest.* 74:1890–1894.

36. Morel, D. W., DiCorleto, P. E., and Chisholm, G. M. 1984. Endothelial and smooth muscle cells alter low density lipoprotein in vitro by free radical oxidation. *Arteriosclerosis* 4:357–364.

37. Cathcart, M. K., Morel, D. W., and Chisolm, G. M. 1985. Monocytes and neutrophils oxidize low density lipoprotein making it cytotoxic. *J. Leukocyte Biol.* 38:341–350.

38. Parasarathy, S., Printz, D. J., Boyd, D., Joy, L., and Steinberg, D. 1986. Macrophage oxidation of low density lipoprotein generates a modified form recognized by the scavenger receptor, *Arteriosclerosis* 6:505–510.

39. Parthasarathy, S., and Steinberg, D. 1992. Cell-induced oxidation of LDL. *Curr. Opin. Lipidol.* 3:313–317.

40. Yla-Herttuala, S., Palinski, W., Butler, S. W., Picard, S., Steinberg, D., and Witztum, J. L. 1994. Rabbit and human athrosclerotic lesions contain IgG that recognizes epitopes of oxidized LDL. *Arterioscler. Thromb.* 14:32–40.

41. Holvoet, P., Perez, G., Bernar, H., Brouwers, E., Vanloo, B., Rosseneu, M., and Collen, D. 1994. Stimulation with a monoclonal antibody (mAb4E4) of scavenger receptor-mediated uptake of chemically modified low density lipoproteins by THP-1-derived macrophages enhances foam cell generation, *J. Clin. Invest.* 93:89–98.

42. Ledwozyw, A., Michalak, J., Stepian, A., and Kadziolka, A. 1986. The relationship between plasma triglycerides, cholesterol, total lipid and lipid peroxidation products during human atherosclerosis. *Clin. Chim. Acta* 155:275–284.

43. Haberland, M. E., Fong, D., and Cheng, L. 1988. Malondialdehyde-altered protein occurs in atheroma of Watanabe heritable hyperlipidemic rabbits. *Science* 241:215–218.

44. Palinski, W., Rosenfeld, M., Yla-Herttuala, S., Gurtner, G., Socher, S., Butler, S., Parasarathy, S., Carew, T. E., Steinberg, D., and Witztum, J. 1989. Low density lipoprotein undergoes oxidative modification in vivo. *Proc. Natl. Acad. Sci. U.S.A.* 86:1372–1376.

45. Carew, T. E., Schwenke, D. C., and Steinberg, D. 1987. Antiatherogenic effect of probucol unrelated to its hypocholesterolemic effect: Evidence that antioxidants in vivo can selectively inhibit low density lipoprotein degradation in macrophage-rich fatty streaks and slow the progression of atherosclerosis in the Watanabe heritable hyperlipidemic rabbit. *Proc. Natl. Acad. Sci. U.S.A.* 84:7725–7729.

46. Verlangieri, A. J., and Bush, M. J. 1992. Effects of d-alpha-tocopherol on experimentally induced primate atherosclerosis. *J. Am. Coll. Nutr.* 11:131–138.

47. Bjorkhem, I., Henriksson-Freyschuss, A., Breuer, O., Diczfalusy, U., Berglund, L., and Henriksson, P. 1991. The antioxidant butylated hydroxytoluene protects against atherosclerosis. *Arterioscler. Thromb.* 11:15–22.

48. Sparrow, C. P., Doebber, T. W., Olszewski, J., Wu, M. J., Ventre, J., Stevens, K. A., and Chao, Y. 1992. Low density lipoprotein is protected from oxidation and progression of atherosclerosis is slowed in cholesterol-fed rabbits by the antioxidant N,N'-diphenyl-phenylenediamine. *J. Clin. Invest.* 89: 1885–1891.

49. Morel, D. W., Hessler, J. R., and Chisholm, G. M. 1983. Low density lipoprotein cytotoxicity induced be free radical peroxidation of lipid. *J. Lipid. Res.* 24:1070–1076.

50. Smith, C., Mitchenson, M. J., Aruoma, O. I., and Halliwell, B. 1992. Stimulation of lipid peroxidation and hydroxyl-radical generation by the contents of human atherosclerotic lesions. *Biochem. J.* 286:901–905.

51. Jialal, I., Vega, G. L., and Grundy, S. M. 1990. Physiologic levels of ascorbate inhibit the oxidative modification of low density lipoprotein. *Atherosclerosis* 82:185–191.

52. Jialal, I., Norkus, E. P., Cristol, L., and Grundy, S. M. 1991. Beta-carotene inhibits the oxidative modification of low-density lipoprotein. *Biochim. Biophys. Acta* 1086:134–138.

53. Jessup, W., Rankin, S. M., DeWhalley, C. V., Hoult, R. S., Scott, J., and Leake, D. S. 1990. Alpha-tocopherol consumption during low-density-lipoprotein oxidation. *Biochem. J.* 265:399–405.

54. Esterbauer, H., Jurgens, G., Quehenberger, O., and Koller, E. 1987. Autoxidation of human low density lipoprotein: Loss of polyunsaturated fatty acids and vitamin E and generation of aldehydes. *J. Lipid. Res.* 28:495–509.

55. Stocker, R., Bowry, V. W., and Frei, B. 1991. Ubiquinol-10 protects human low density lipoprotein more efficiently against lipid peroxidation than does alpha-tocoperol. *Proc. Natl. Acad. Sci. U.S.A.* 88:1646–1650.

56. Esterbauer, H., Puhl, H., Waeg, G., Krebs, A., and Dieber-Rotheneder, M. 1992. The role of vitamin E in lipoprotein oxidation. In *Vitamin E: Biochemistry and Clinical Application*, eds. L. Packer and J. Fuchs, pp. 649–671. New York: Marcel Dekker.

57. Steinbrecher, U. P., Parthasarathy, S., Leake, D. S., Witztum, J. L., and Steinberg, D. 1984. Modification of low density lipoprotein by endothelial cells involves lipid peroxidation and degradation of low density lipoprotein phospholipids. *Proc. Natl. Acad. Sci. U.S.A.* 81:3883–3887.

58. Cathcart, M. K., McNally, A. K., Morel, D. W., and Chisholm, G. M. III. 1989. Superoxide anion participation in human monocyte-mediated oxidation of low-density lipoprotein and conversion of low-density lipoprotein to a cytotoxin. *J. Immunol.* 142:1963–1969.

59. Esterbauer, E., Dieber-Rotheneder, M., Striegl, G., and Waeg, G. 1991. Role of vitamin E in preventing the oxidation of low-density lipoprotein. *Am. J. Clin. Nutr.* 53:314S–321S.

60. Croft, K. D., Williams, P., Dimmitt, S., Abu-Amsha, R., and Beilin, L. 1995. Oxidation of low-density lipoproteins: Effect of antioxidant content, fatty acid composition and intrinsic phospholipase activity on susceptibility to metal ion-induced oxidation. *Biochim. Biophys. Acta* 1254:250–256.

61. Kontush, A., Hubner, C., Fincke, B., Kohlschutter, A., and Beisiegel, U. 1994. Low density lipoprotein oxidizability by copper relates to its initial ubiquinol-10 and polyunsaturated fatty acid content. *FEBS Lett.* 341:69–73.

62. Thomas, M. J., Thornburg, T., Manning, J., Hooper, K., and Rudel, L. L. 1994. Fatty acid composition of low-density lipoprotein influences its susceptibility to autoxidation. *Biochemistry* 33:1828–1834.
63. Schuster, B., Prassl, R., Nigon, F., Chapman, M. J., and Laggner, P. 1995. Core lipid structure is a major determinant of the oxidative resistance of low density lipoprotein. *Proc. Natl. Acad. Sci. U.S.A.* 92:2509–2513.
64. Tribble, D. L., Holl, L. G., Wood, P. D., and Krauss, R. M. 1992. Variations in oxidative susceptibility among 6 low density lipoprotein subfractions of differing density and particle size. *Atherosclerosis* 93:189–199.
65. Esterbauer, H., Gebicki, J., Puhl, H., and Jurgens, G. 1992. The role of lipid peroxidation and antioxidants in oxidative modification of LDL. *Free Radic. Biol. Med.* 13:341–390.
66. Dieber-Rotheneder, M., Puhl, H., Waeg, G., Streigl, G., and Esterbauer, H. 1991. Effect of oral supplementation with d-alpha-tocopherol on the vitamin E content of human low density lipoproteins and resistance to oxidation. *J. Lipid Res.* 32:1325–1332.
67. Jialal, I., and Grundy, S. M. 1992. Effect of dietary supplementation with alpha-tocopherol on the oxidative modification of low density lipoprotein. *J. Lipid. Res.* 33:899–906.
68. Jialal, I., Fuller, C. J., and Huet, B. A. 1995. The effect of alpha-tocopherol supplementation on LDL oxidation: A dose-response study. *Arterioscler. Thromb. Vasc. Biol.* 15:190–198.
69. Smith, D., O'Leary, V. J., and Darley-Usmar, D. M. 1993. The role of alpha-tocopherol as a peroxyl radical scavenger in human low density lipoprotein. *Biochem. Pharmacol.* 45:2195–2201.
70. Van Hinsbergh, V. W. M., Scheffer, M., Havekes, L., and Kempen, H. J. M. 1986. The role of endothelial cells and their products in the modification of low density lipoproteins. *Biochim. Biophys. Acta* 878:49–64.
71. Burton, G. W., and Ingold, K. U. 1984. Beta-carotene: An unusual type of lipid antioxidant. *Science* 224:569–573.
72. Naruszewicz, M., Selinger, E., and Davignon, J. 1992. Oxidative modification of lipoprotein (a) and the effect of beta-carotene. *Metabolism* 41:1215–1224.
73. Abbey, M., Nestel, P. J., and Baghurst, P. A. 1993. Antioxidant vitamins and low-density-lipoprotein oxidation. *Am. J. Clin. Nutr.* 58:525–532.
74. Reaven, P. D., Khow, A., Beltz, W., Parasarathy, S., and Witztum, J. L. 1993. Effect of dietary antioxidant combinations in humans. Protection of LDL by vitamin E but not by beta-carotene. *Arterioscler. Thromb.* 13:590–600.
75. Gaziano, J. M., Hatta, A., Flynn, M., Johnson, E. J., Krinsky, N., Ridker, P. M., Hennekens, C. H., and Frei, B. 1995. Supplementation with beta-carotene in vivo and in vitro does not inhibit low density lipoprotein oxidation. *Atherosclerosis* 112:187–195.
76. Jialal, I., and Grundy, S. M. 1993. Effect of combined supplementation with alpha-tocopherol, ascorbate, and beta carotene on low-density lipoprotein oxidation. *Circulation* 88:2780–2786.
77. Boscoboinik, D., Szewczyk, A., Hensey, C., and Azzi, A. 1991. Inhibition of cell proliferation by alpha-tocopherol: Role of protein kinase C. *J. Biol. Chem.* 266:6188–6194.
78. Ozer, N. K., Palozza, P., Boscoboinik, D., and Azzi, A. 1993. d-Alpha-tocopherol inhibits low density lipoprotein induced proliferation and protein kinase C activity in vascular smooth muscle cells. *FEBS Lett.* 322:307–310.
79. Ringer, T., DeLoof, M. J., Winterrowd, G. E., Francom, S. F., Gaylor, S. K., Ryan, J. A., Sanders, M. E., and Hughes, G. S. 1991. Beta-carotene's effects on serum lipoproteins and immunologic indices in humans. *Am. J. Clin. Nutr.* 53:688–694.
80. Parthasarathy, S., Barnett, J., and Fong, L. 1990. High-density lipoprotein inhibits the oxidative modification of low-density lipoprotein. *Biochim. Biophys. Acta* 1044:275–283.
81. Mackness, M., Abbott, C., Arrol, S., and Durrington, P. N. 1993. The role of high-density lipoprotein and lipid-soluble vitamins in inhibiting low-density lipoprotein oxidation. *Biochem. J.* 294:829–834.

Chapter Six

RESPONSE OF ANTIOXIDANT SYSTEM TO PHYSICAL AND CHEMICAL STRESS

Satu M. Somani, Kazim Husain, and Eric C. Schlorff

Exercise is not only a leisure activity, but also an effective preventive and therapeutic tool in medicine. Population, in general, irrespective of age and gender, perform regular exercise to maintain good health. Exercise is known to evoke numerous physiological changes in vital organ systems of the body. Among those changes, the most important is the enhanced respiration and utilization of oxygen in the body. Increased oxygen influx during exhaustive exercise may be potentially harmful to the body. During the last 10 years, much evidence has accumulated implicating generation of reactive oxygen species (ROS) and other free radicals during exercise in the muscle and heart (1, 2). Exposure to drugs and/or chemicals may also generate ROS. However, cells contain several antioxidant defense mechanisms to protect themselves from ROS injury. These include endogenous antioxidants (glutathione, vitamins C, A, E, uric acid and iron binding protein) and antioxidant enzymes (AOE) superoxide dismutase (SOD), catalase (CAT), and glutathione peroxidase (GSH-Px), referred to as the antioxidant system. The activity of AOE, which are well distributed in all organs of the body, are dependent on oxygen consumption rate, metabolic rate, and the amount of metal ions and fatty acids present. The AOE activity is prone to being altered by changes in oxygen consumption (oxidative stress). Oxidative stress (physical and/or chemical) can be described as a disturbance in the antioxidant system that is not able to adequately scavenge free radicals/ROS and arrest lipid peroxidation chain reactions. Responses of an antioxidant system are dependent on the type, mode, intensity, frequency, and duration of exercise, animal species, tissue specificity, and the extent of exposure to drugs and chemicals. The modulation of AOE activity primarily depends upon its substrate (ROS), cosubstrate production, nature of catalytic center, affinity of the enzyme to the substrate (K_m), and selectivity and specificity of the substrate. The activation of AOE activity occurs due to adenosine release during exercise. The induction of AOE activity may also occur through transcription/ translation or posttranslational processes as a result of cumulative/chronic physical stress (3).

These aspects of antioxidant system in relation to physical exercise and chemicals are discussed in this chapter.

PATHWAYS AND COMPONENTS OF THE ANTIOXIDANT SYSTEM

The substances that neutralize the potential harmful effects of ROS are generally known as the antioxidant defense system. The potential harm of oxygen derives from its ability to be converted to ROS such as superoxide anion ($O_2^{-\bullet}$), singlet oxygen, hydroxyl radicals, hydroperoxides, and lipid peroxides. These species are produced as a result of oxidation processes occurring in the mitochondria. Recent evidences suggest that there are two sources of $O_2^{-\bullet}$ within the electron transport chain in the mitochondria. The major source of $O_2^{-\bullet}$ is ubisemiquinone free radical generated by the reduction of ubiquinone during electron transport. Ubisemiquinone interacts with O_2 to form $O_2^{-\bullet}$

Ubiquinone $+ e^- \rightarrow$ Ubisemiquinone$^\bullet$

Ubisemiquinone$^\bullet + O_2 \rightarrow$ Ubiquinone $+ O_2^{-\bullet}$

A second source of mitochondrial $O_2^{-\bullet}$ formation has been reported to be NADH dehydrogenase. Quantitatively, NADH dehydrogenase generates 50% of the $O_2^\bullet$ produced by the ubisemiquinone (4). Ubisemiquinones are lipophilic and diffuse through the organelle and come in contact with oxygen to produce superoxides (5). The metabolism of certain drugs and chemicals in tissue microsomal complex may also result in the generation of ROS that interact with $O_2^\bullet$ to yield $O_2^{-\bullet}$ and H_2O_2 (6, 7). The superoxide anions are readily dismutated by Mn-superoxide dismutase in mitochondria and Cu,Zn-superoxide dismutase in the cytosol and extracellular fluid to produce hydrogen peroxide and singlet oxygen. This enzyme makes the reaction happen 10 million times faster than its spontaneous rate (8). It is estimated that hydrogen peroxide production is highest in endoplasmic reticulum (45%), followed by metal-catalyzed oxidations in peroxisomes (35%), oxidative phosphorylation in mitochondria (15%), and oxidation of xanthine in cytosol (5%) (5,9). Hydrogen peroxide generated after superoxide dismutase (SOD) catalyzed reaction is then quickly converted by either CAT or glutathione peroxidase (GSH-Px) to H_2O and oxygen.

$$2H_2O_2 \xrightarrow{\text{CAT}} 2H_2O + O_2$$

$$H_2O_2 + 2GSH \xrightarrow{\text{GSH-Px}} GSSG + 2H_2O$$

If hydrogen peroxide escapes this enzymatic onslaught, then it can react immediately with the transition metals, usually iron, to produce a very toxic and reactive hydroxyl radical (10). Lipid peroxides are converted to lipid alcohol by GSH-Px. Reduced glutathione (GSH) acts as a substrate in the GSH-Px reaction, producing oxidized glutathione (GSSG), which is recycled back to its reduced form by glutathione reductase (GR).

$$GSSG + NADPH + H^+ \xrightarrow{\text{GR}} 2\,GSH + NADP$$

To provide maximum protection, cells contain a variety of endogenous antioxidant substances capable of scavenging different ROS. These antioxidants and AOE are strategically compartmentalized in subcellular organelles within the cell to provide maximum protection. The components of antioxidant system are described next.

Endogenous Nonenzymatic Antioxidants

Vitamin E

Vitamin E is chemically referred to as α-tocopherol (11) and is the most widely distributed antioxidant. Because of the lipophilic property of the α-tocopherol molecule, it is the major free radical chain terminator in plasma lipoproteins. Intracellularly, vitamin E is associated with lipid-rich membranes such as mitochondria and endoplasmic reticulum.

Vitamin C

This is chemically known as ascorbic acid. It is hydrophilic and functions better in an aqueous environment. As a reducing and antioxidant agent, it directly reacts with $O_2^{-\bullet}$ and O$^\bullet$H. It can restore the antioxidant property of oxidized vitamin E, suggesting that it recycles vitamin E radical (12).

Vitamin A

This vitamin has long been considered an antioxidant because of its ability to scavenge ROS (13). Carotenoids protect lipids against oxidation by quenching free radicals (14). β-Carotene displays antioxidant activity through its inhibition of lipid peroxidation in the biological system (13).

Glutathione

Reduced glutathione (GSH), a tripeptide, is the most abundant thiol present in all mammalian cells. GSH concentration in the cell is in the millimolar range for most tissues, but there is variability in different organs depending on their function of oxidative capacity. The roles of GSH include: (1) to serve as a cosubstrate for GSH peroxidase (GPX), wherein GSH is used as a hydrogen donor to reduce hydrogen peroxide and organic peroxide to water and alcohol, respectively; (2) to conjugate exogenous and endogenous toxic compounds, catalyzed by GSH sulfur-transferase; (3) to reduce protein disulfide and GSH-protein mixed disulfide bonds, maintaining the sulfhydryl residues of certain proteins and enzymes in the reduced state; (4) to store cysteine in a nontoxic form; and (5) to assume a vital role in keeping α-tocopherol (vitamin E) and ascorbic acid (vitamin C) in the reduced states.

Uric Acid

Uric acid is an end product of purine metabolism in mammals. However, its antioxidant action was first reported by Howell and Wyngarden (15). Its antioxidant properties were confirmed by its protection against oxidative damage (16). Although the mechanism of the antioxidative action of uric acid is not well known, reports indicate that it may act by sparing ascorbate (17), probably by complexing iron and copper (18).

Endogenous Enzymatic Antioxidants

Superoxide Dismutase

The catalytic function of SOD was discovered by McCord and Fridovich (19). It exists in virtually all O_2-respiring organisms, and its major function is to catalyze the dismutative reactions.

$$O_2^{-\bullet} + O_2^{-\bullet} \xrightarrow{\text{SOD}} H_2O_2 + O_2$$

The rate of dismutation is $\sim 10^4$ times greater than that of chemical dismutation (20). Superoxide dismutases are classified into three distinct classes depending on the metal ion content: Cu,Zn-SOD, Mn-SOD, and Fe-SOD. In mammalian cells, there are two isozymes located in different subcellular compartments. The cytoplasm contains Cu,Zn-SOD (dimeric protein) and mitochondria contain Mn-SOD (tetrameric protein). Mammalian cells also have the third SOD isozyme, extracellular (EC) SOD, which is Cu,Zn-SOD (tetrameric protein) (21). The activity of the SOD varies among the tissues. The highest levels are seen in the liver, kidney, and red blood cells (RBC) of rat (Table 1) (22). The activity of SOD is regulated through biosynthesis, which is sensitive to tissue oxygenation (20); its biosynthesis was elevated in rats (23) subjected to high oxygen tensions.

Catalase

Catalase is a major component of the antioxidant system and catalyzes the decomposition of H_2O_2 to H_2O. It has also a peroxidic role in which the peroxide is utilized to oxidize H donors (AH_2).

$$\text{(i) } 2H_2O_2 \xrightarrow{\text{CAT}} 2H_2O + O_2$$

and

$$\text{(ii) } AH_2 + H_2O_2 \xrightarrow{\text{CAT}} A + 2H_2O$$

Although the tissue distribution of CAT is widespread, the level of activity varies not only between tissues but within the cell itself. Catalase is present predominantly in the peroxisomes (microbodies) in liver and kidney and also in the microperoxisomes of other tissues. The liver and kidney possess relatively high levels of CAT (Table 1). In hepatocytes, peroxisomes exhibit

Table 1. Levels of antioxidant enzyme activity in various body tissues of rat

	Enzyme activity		
Body tissue	Superoxide dismutase (SOD)	Catalase (CAT)	Glutathione peroxidase (GSH-Px)
Plasma	14.43 ± 1.63	29.05 ± 11.40	83.19 ± 10.25
RBC[a]	35.29 ± 3.49	0.62 ± 0.02	0.56 ± 0.11
WBC[a]	6.68 ± 1.02	0.48 ± 0.05	0.38 ± 0.04
Platelets[a]	16.92 ± 1.47	0.45 ± 0.07	0.58 ± 0.05
Brain	48.85 ± 2.54	22.25 ± 2.45	85.56 ± 4.44
Heart	63.34 ± 2.84	37.27 ± 2.14	95.36 ± 8.25
Kidney	78.14 ± 3.16	299.71 ± 13.88	162.15 ± 10.12
Liver	163.14 ± 10.33	525.44 ± 38.72	209.48 ± 22.13
Lung	58.88 ± 3.98	50.00 ± 2.85	105.55 ± 5.65
Muscle	30.33 ± 4.65	16.95 ± 1.88	71.54 ± 3.80
Testes	58.45 ± 3.15	28.12 ± 1.99	120.96 ± 8.87
Cochlea	38.13 ± 1.43	44.45 ± 5.42	68.82 ± 5.23

Note. Values are mean ± SEM. SOD, units/mg protein; CAT, mmol H_2O_2 degraded min/mg protein; GSH-Px, nmol NADPH oxidized/min/mg protein.

[a] Husain et al. (22).

an expectedly high CAT activity, although activity was also found in microsomes and in the cytosol (24).

Glutathione Peroxidase

This catalyzes the reduction of H_2O_2 and organic hydroperoxide and peroxides as follows:

$$2GSH + H_2O_2 \xrightarrow{\text{GSH-Px}} GSSG + 2H_2O$$

or

$$2GSH + ROOH \xrightarrow{\text{GSH-Px}} GSSG + H_2O + ROH$$

Both types of GSH-Px enzymes, selenium-dependent and selenium-independent, have been shown to catalyze these reactions and thus protect against radical damage by reducing peroxides. However, they possess different substrate specificities. Glutathione peroxidase is mostly present in the cytosol. The cytosolic and membrane-bound monomeric GSH-Px and the tetramer plasma GSH-Px are able to reduce phospholipid hydroperoxides. This enzyme has an absolute specificity for GSH as its electron-donating substrate: however, its specificity for peroxide is much less selective. Within the cell, the distribution of GSH-Px is almost complimentary to that of CAT. The tissue distribution of GSH-Px in rat is presented in Table 1. Once oxidized, GSH is regenerated from GSSG via the enzyme glutathione reductase (GR). Glutathione reductase is an ancillary enzyme to limit the amounts of ROS via its reduction of GSSG in the presence of an adequate supply of NADPH. Thus, the ratio of GSH/GSSG is maintained at a high level so that the cell maintains the capacity to combat oxidative stress.

$$GSSG + NADPH + H^+ \xrightarrow{\text{GR}} 2GSH + NADPH$$

Ratios involving the antioxidant enzymes could be considered as an index of oxidative stress. For instance, when the CAT/SOD or GSH-Px/SOD ratio increases, this might indicate more activation of the AOE due to ROS. When the ratio decreases, this may be indicative of an increase in ROS such enzymes are not scavenging ROS efficiently and oxidative damage may result. These two ratios can give insight into the adaptation and status of the AOE. Another important ratio is GR/GSH-Px. This ratio will show how well the system is recycling GSH. Reduced glutathione (GSH) is an important substrate that aids in reducing hydrogen peroxide and organic hydroperoxides to H_2O. When this ratio increases, then the system is able to keep the level of GSH high and provide protection to the cell. When this ratio decreases, this may be indicative of oxidative damage because the cell cannot produce enough GSH to get rid of the hydrogen peroxides and organic hydroperoxides or ROS. The question that is still unanswered is, "At which level do we consider that the oxidative damage is harmful?" in the short and long term. This question needs to be answered by correlating data with alterations in tissue structure and with pathological conditions where ROS or oxidative damage is presumed to be responsible.

ANTIOXIDANTS' RESPONSE TO EXERCISE

Endogenous Antioxidant Compounds

Ascorbic Acid (Vitamin C)

Ascorbate radicals play a pivotal role in the scavenging of ROS and functions as free radical chain-terminating agents by self-disproportionation (25). The ascorbate ion (A^-H) seems to have cardioprotective properties, and might play a critical role in ROS-mediated myocardial ischemic/reperfusion injury. A one-electron oxidation of ascorbate ion (A^-H) results in the production of the ascorbate free radical ($A^{\cdot-}sc$) (26). It seems that $A^{\cdot-}sc$ radicals are generated during exercise training and thus provides antioxidant protection to the heart muscle (2). The steady-state concentration of $A^{\cdot-}sc$ and the $A^{\cdot-}sc$ electron paramagnetic resonance (EPR) signal intensity can serve as a quantitative marker of oxidative stress. Because ascorbate free radicals have a much longer half-life than superoxide, singlet oxygen, hydroxyl radical, alkoxyl radical, peroxyl radical, and lipid peroxide, their stability will be greater, and more easily detected by EPR.

We have reported that the ascorbate radicals are generated in heart muscle during exercise training in aged rats (86 wk old) (2). The benefit of exercise could be attributed to the formation of ascorbate radical in the heart muscle of the aged rat. Exercise training can provide protection to the heart against oxidative damage via ascorbate ion and vitamin E.

Plasma levels of ascorbic acid have also been studied. An increase in ascorbic acid was present in the plasma after a 21-km race (27). Interestingly, acute swimming exercise significantly decreased plasma ascorbic acid concentration with 90 and 120 min of exercise (28). Dietary administration of vitamin C has also shown to be effective in preventing the oxidation of glutathione pool after exhaustive physical exercise (29).

Uric Acid

It has been shown that there is a decrease in accumulation of uric acid after exercise training (at 135% of VO_{2max} for 2 min) (30). However, no significant difference was seen at 106, 113, and 123% of VO_{2max} for 2 min. The authors believe that this decrease in uric acid in high exercise training may be due to an adaptation of the muscle, in which fewer adenine nucleotides are lost during highly intense exercise. Duthie et al. (27) observed the uric acid levels of trained male athletes after running a 21-km marathon race and showed a 24% increase of uric acid in plasma immediately following post race. Sjödin and Hellsten Westing (31) observed that after an 800-m run (at 110–115% VO_{2max}), plasma levels of hypoxanthine and uric acid rise sharply. The effect was not as remarkable at lower intensities of the exercise run.

Vitamin E

Several aspects of vitamin E's correlation to the antioxidant system have been investigated with respect to varying types of exercise: endogenous levels of vitamin E, vitamin E deficiency, and supplementation of dietary vitamin E. It has been shown that there is a decrease in the total level of vitamin E per unit of mitochondrial activity during exercise training, which may suggest higher rates of free radical formation (32). Also, exercise training decreased the level of vitamin E in the subepimyocardium of rat (33), suggesting a decrease in antioxidant status of the heart due to exercise. There was an increase in vitamin E levels in plasma after exercise training (34); however, no decrease in vitamin E level was noticed after acute submaximal exercise in untrained female rats (35). Vitamin E-deficient rats for 16 wk or in combination with an 8-wk endurance training program showed no effects on SOD, CAT, or GSH-Px activity in the heart, muscle, or liver tissues of females (35,36). Dietary vitamin E supplementation has also been shown to reduce lipid peroxidation after swimming exercise in rats (37).

In humans, dietary vitamin E supplementation has an effect on reducing lipid peroxidation. It has been shown that when supplementing vitamin E (dl-α-tocopherol) for 2 wk to adults, pentane breath levels were reduced (38). A similar finding was that in humans administered α-tocopherol, along with β-carotene, and ascorbate, there were significantly reduced pentane levels and serum MDA levels after moderate (60% of VO_{2max} for 30 min) or heavy exercise (100% VO_{2max} for 5 min) (39).

Glutathione

Exercise enhances the blood flow in most organs, but depletes it in the liver, which in turn will influence the compartmentalization of GSH. Exercise may alter the transport of glutathione (extracellular and intracellular), alter synthesis or degradation of glutathione due to an increased uptake of oxygen, or alter an increase or decrease in conjugation of glutathione with exogenous or endogenous compounds. The effect of exercise on the compartmentalization of GSH influences the oxidation of GSH and the onset of injury to the tissues (40).

Gohil et al. (41) reported that submaximal exercise influences blood glutathione status. A sharp increase (100%) in blood level of GSSG was observed within the first 15 min of exercising at 65% VO_{2peak}. In contrast to the finding of Gohil et al., a bout of exercise (70% VO_{2max}) that lasted for a mean duration of 134 min did not affect blood GSSG levels in healthy cyclists (42). Studies related to human blood glutathione oxidation during exercise are limited; however, a study did reveal that exhaustive exercising of rats remarkably increases oxidized glutathione level in the plasma (43). In moderately trained men (mean VO_{2peak}), a 50% decline in GSH level was observed in blood during the first 15 min of exercise. This effect was accompanied by an increase in GSSG level. Total glutathione level in the blood did not change significantly during the exercise. Reduced glutathione level in the blood returned to baseline after 15 min of postexercise recovery (43).

Acute exercise increased the levels of oxidized glutathione and glutathione in rat blood (44). Exercise training increased GSH level in human blood; however, there was no change in the GSSG level (42). In contrast, another finding in GSH level after acute exercise and exercise training showed decreased GSH level and increased GSSG level (45).

Exercise training significantly increased the GSH to GSSG ratio in the brainstem and in the cerebral cortex, whereas this ratio did not change in the corpus striatum. The benefit of exercise training seems to be evident as GSSG levels decreased and GSH levels remained unaltered. This study concluded that the exercise training altered the SOD activity and GSH to GSSG ratio differentially in various brain regions coping with oxidative stress (46).

Interestingly, in our study, GSSG levels were found to be almost twofold higher than GSH levels in the cytosolic fraction of the heart due to strenuous single exercise (47). We could not determine GSH or GSSG in the mitochondrial fraction of the heart. Mitochondrial GSH was found to originate from the cytosol and was then imported into mitochondria by a system that contains a high-affinity transporter (48, 49).

The glutathione metabolic profile of different types of skeletal muscle of rats and the effect of exercise intensity was reported (50). Total glutathione levels were observed to vary between different muscle types (soleus > deep vastus lateralis > superficial vastus lateralis). Exercise resulted in a 50% decrease in total glutathione in left soleus muscle, an effect that was interpreted as an index of oxidative stress. Recovery of muscle glutathione level was slow in the postexercise recovery period. A single bout of exercise resulted in glutathione loss from the skeletal muscle of mice (51, 52).

Hepatic total glutathione level was elevated in the trained rats; however, the single bout of exercise decreased muscle and liver total glutathione level (53).

Antioxidant Enzymes

Enzyme Activity

The status of the antioxidant enzymes varies after acute exercise and exercise training in different tissues in various animals and humans. In the blood, acute exercise activated the activity of SOD and GSH-Px (47), and GR (54). However, exercise training increased the activity of SOD,

CAT, and GSH-Px in rats (47). By contrast, in humans, no change in activity was observed in CAT and GSH-Px (42).

In the brain regions, the status of the antioxidant system varies after acute exercise and exercise training. SOD activity decreased in the cerebral cortex with no significant change in other brain regions due to acute exercise (55). Catalase activity increased due to acute exercise in the cerebral cortex and the corpus striatum, whereas GSH-Px activity increased profoundly in the cortex and decreased in the corpus striatum (55). After exercise training, different results appeared. The SOD activity increased in the brainstem and corpus striatum suggesting an adaptation to exercise (46). Glutathione reductase activity after exercise training did not change (46).

In the heart, the status of the antioxidant system was affected by acute exercise and exercise training. After acute exercise, the activities of Mn-SOD, CAT, GSH-Px, and GR increased (47). After exercise training, the activities of the enzymes did not change dramatically.

In the liver, after exercise training, Cu,Zn-SOD and CAT activity increased (56). GSH-Px activity in liver also increased with both acute exercise (57) and exercise training (56). In the lung, no significant change in CAT and GSH-Px activity due to exercise training was reported (58). There is a paucity of information on the status of the antioxidant system in the testes in response to exercise. Recently, we have reported an increase in (velocity) V_{max} of CAT, GSH-Px, and GR in the testes of exercise trained rats (59).

In the muscle, acute exercise increased the activities of SOD, CAT, GSH-Px, and GR (52, 60). Exercise training did not affect CAT or GSH-Px activity (52). The activity of SOD and GR after exercise training was not conclusive.

As can be seen, there is a profound difference in the status of the AOE in various tissues, and enzyme activity was enhanced differentially in different tissues due to exercise. There seems to be tissue specificity in the activation/induction of AOE activity due to acute exercise or exercise training.

Enzyme Kinetics

The increased AOE activity due to exercise in different tissues of animals has been discussed. However, the activation of AOE in response to exercise training using enzyme kinetics approach is described in this section. We determined the Michaelis-Menten constant (K_m) and maximum (V_{max}) of CAT, GSH-Px, and GR in different tissues of aged rats that were subjected to progressive exercise for 9 wk and sacrificed 23 h after the last bout of exercise (59). In the brain, liver, lung, muscle, and testes, the V_{max} and K_m values for CAT increased suggesting an activation of CAT. In the kidney, V_{max} and K_m values decreased for CAT, indicative of a higher affinity of CAT for H_2O_2 after exercise training. The V_{max} and K_m values for GSH-Px increased in the brain and liver, suggesting an activation of GSH-Px. In the lung and kidneys, V_{max} and K_m values for GSH-Px both decreased, suggesting a higher affinity for GSH. In the muscle, V_{max} for GSH-Px was not altered, but the K_m was decreased, suggesting a higher affinity for GSH. There was no alteration in V_{max} or K_m values in the testes for GSH-Px utilizing GSH as its substrate. When t-butyl hydroperoxide was used as a substrate, GSH-Px was activated in the testes. The V_{max} and K_m values for GR were both decreased in the liver, lung, and muscle, suggesting a higher affinity of GR for its substrate GSSG. In the kidney, the V_{max} increased and the K_m decreased, suggesting a slight activation of GR and a higher affinity for its substrate GSSG. In both the brain and testes, exercise training had no effect on the V_{max} and K_m values of GR. This study demonstrated the potential activation of AOE in specific tissues of exercise-trained aged rats.

Intracellular Genetic Regulation

The activation of AOE activity has been demonstrated using enzyme kinetics in different tissues of rat. However, the modulation of AOE activity may also be regulated intracellularly, which has been discussed in this section. Recent studies involving the proto-oncogene Bcl-2, heat shock protein, nuclear factor-κB (NF-κB), ACE1 activator protein, and nitric oxide have shown promising results in the identification and characterization of the intracellular regulation in the antioxidant enzymes during exercise.

Bcl-2, known to prevent apoptosis, has been shown to act on the antioxidant pathway after the generation of superoxide anion (61). Heat shock protein has been shown to increase after exercise (62). This increase in heat shock protein is probably due to hyperthermia, and possibly to the oxidative stress produced during exercise. Phorbol esters, known to induce oxidative stress, induce the activation of NF-κB (63). After treatment with the GR inhibitor BCNU (bischloroethylnitrosurea), the activation of NF-κB was remarkably greater (63). By inhibiting GR activity, an increase in GSSG levels occurs. During exercise, the level of GSSG also increases, which suggests that exercise may activate NF-κB leading to transcription and translation of proteins that protect the cell against ROS.

The ACE1 activator protein binds copper (I) ions, and forms a copper fist, which then can bind the ACE1 activator protein to its target DNA (64). The ACE1 activator protein, which expresses the level of cytosolic SOD, may actually protect the cell from oxidative stress because it is dependent on copper levels. Copper drives the Fenton-type reactions that lead to oxidative stress. It is likely that the activation of ACE1 activator protein may also occur during physical and chemical stress, thus regulating the expression of SOD.

Nitric oxide can interact with the superoxide anion, forming a more stable peroxynitrite. It has been shown that after chronic exercise, nitric oxide synthase gene expression is increased (65). Whether nitric oxide provides beneficial or harmful effects after exercise by regulating the antioxidant system is not known.

We have recently studied whether an increase in activity of (Mn-SOD and Cu,Zn-SOD), CAT, and GSH-Px during exercise training was associated with the increased levels of respective mRNAs (3). Although it is apparent that the increased enzyme activities of CAT and GR are not due to the transcriptional activation of respective genes, since such an activation should produce more copies of transcripts specific to the antioxidant enzymes. The increases in transcript levels are relatively low and are not concurrent with the increases in enzyme activity. This suggests other potential mechanisms by which exercise training can increase antioxidant enzyme activity. Our hypothesis is that the antioxidant enzyme induction might be regulated by both pretranslational (transcriptional) and posttranslation modification.

ADENOSINE GENERATION DURING EXERCISE

Exercise increases the intake of oxygen in various organs/tissues of the body, especially the skeletal muscles, heart, and brain, with increased breakdown of ATP. As shown in Figure 1, adenosine is formed intracellularly or extracellularly by dephosphorylation of adenosine monophosphate (AMP) via the ATP pathway catalyzed by the enzyme 5'-nucleotidase. In addition to the ATP pathway, adenosine can also be formed intracellularly by the degradation of S-adenosyl homocysteine (SAH), and catalyzed by the enzyme SAH pathway (Figure 1). The relative contribution of each metabolic pathway for the formation of adenosine varies according to the experimental conditions (66). Extracellular cyclic AMP can also lead to the formation of adenosine (67, 68). In human blood, the uptake of adenosine is such that the half-life is under 10 s (69, 70). Adenosine is removed through cellular uptake mechanisms in the tissues by either simple or facilitated diffusion, via a nucleoside transport system in the heart and by an Na^+-dependent active transport system in the brain. Adenosine is also degraded (deaminated) by the adenosine deaminase enzyme to inosine, which is mostly inactive, or adenosine is phosphorylated by adenosine kinase to AMP (71, 72). Treadmill exercise increased adenosine about threefold in cardiac muscle (73). The level of brain adenosine is elevated following increased energy consumption, such as during a seizure (68). Exercise increases the energy consumption that could increase the release of adenosine in the brain; this hypothesis has not been experimentally proved. Recently, it has been shown that the combination of exercise and adenosine infusion reduced the noncardiac side effects of vasodilation and major arrhythmias, indicating the clinical importance of exercise and adenosine (74).

Endogenous Activation of Antioxidant Enzymes by Adenosine

Adenosine plays a cytoprotective role in ischemic reperfusion injury where free radicals and ROS are generated. Our laboratory had already studied exercise-generated free radicals and ascorbate radicals in heart tissue, and also the increased activity of antioxidant enzymes in the heart mitochondria and brain regions. We thought that in both instances (i) free radicals are generated, (ii) an increased level of adenosine occurs, and (iii) increased activity of antioxidant

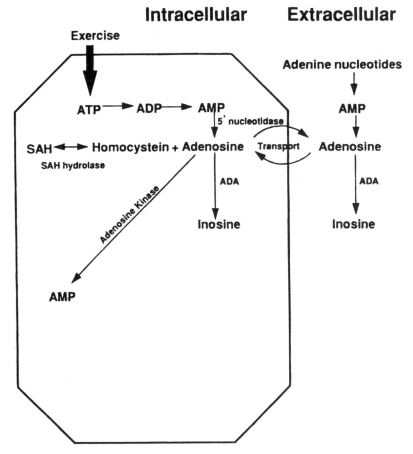

Figure 1. Exercise and generation of adenosine. ATP, adenosine triphosphate; ADP, adenosine diphosphate; AMP, adenosine monophosphate; SAH, S-adenocysteine; ADA, adenosine deaminase.

enzymes also occurs. We hypothesized, therefore, that there might be a relationship between adenosine and antioxidant enzyme activity.

During ischemia, the concentration of the nucleoside can increase almost eightfold, thereby playing a role of cytoprotection. Intracoronary infusion of adenosine during reperfusion, following a 90-min occlusion of the left anterior descending coronary artery, led to a significant decrease in infarct size (75). This protection was mimicked by adenosine receptor agonists (76) and abolished by antagonists (77). It seemed that adenosine played a beneficial role in reducing free radical formation during ischemia and in maintaining the integrity of lipid membranes by reducing lipid peroxidation (78).

Recently, we have shown that adenosine mediates cytoprotection by activating the antioxidant enzyme activity in mast cells, cardiac myocytes, and endothelial cells via the activation of the A_3 adenosine receptors (79). The role of adenosine in cytoprotection by activating antioxidant enzymes has been reviewed recently (80). This receptor is predominantly present in mast cells and rat leukocytes (81). The adenosine agonist R-phenyl isopropyl adenosine (R-PIA) also activated antioxidant enzyme activity in other cell types such as human and bovine endothelial cells, rat myocytes, DDT MF-2 smooth muscle cells, and P815 mastocytoma cells (79). CAT and GR activities showed maximum increase, whereas SOD and GSH-Px showed moderate increases. It seems that adenosine and its agonist could act as an endogenous activator of antioxidant enzyme activity.

Table 2. Effect of exercise, adenosine agonist R-PIA, and antagonist theophylline, and their combination on antioxidant enzyme activities in rat heart

	SOD	CAT	GSH-Px	GR
Sedentary control	100	100	100	100
AE (100% VO_{2max})	148	221	127	162
R-PIA	148	184	119	144
Th + AE	116	95	104	112
Th + R-PIA	115	115	103	106
Th	100	136	100	96

Note. Each value represents the percent of sedentary control. R-PIA, *R*-phenyl isopropyl adenosine; AE, acute exercise; Th, theophylline.

These studies suggest that the activation of antioxidant enzymes, GR, and malondialdehyde activity is mediated via the adenosine receptor G_i-protein complex. It is quite possible that the activation of the antioxidant enzymes and GR with exercise occurred through a similar pathway.

Adenosine and Antioxidant Enzyme in Exercise

Based on in vitro studies, our laboratory studied the effect of exercise and the administration of the agonist R-PIA and antagonist theophylline of adenosine A_3 receptor in rats. This study was the first to attempt to show that native antioxidant enzymes are activated via A_3 adenosine cell surface receptors. Antioxidant enzymes are also activated by single acute exercise and by exercise training (47, 77). The following questions then arise: Are cell surface receptors involved in the activation of antioxidant enzymes during exercise? Is the specific adenosine receptor activated by exercise, which, in turn, activates antioxidant enzymes and GR? To answer these questions, we extended the experiments to study the effect of the administration of adenosine receptor agonist R-PIA and antagonist theophylline on antioxidant enzymes, GR, and lipid peroxidation in the heart of the rat. We also studied whether these enzyme activities and lipid peroxidation are altered due to a combination of acute exercise and antagonist or agonist. Our preliminary study shows that the acute exercise (100% VO_{2max}) enhanced antioxidant enzyme activity in the rat heart (Table 2). This increase in enzyme activity did not occur when theophylline was administered to rats prior to exercise, indicating that theophylline blocked the activation of antioxidant enzyme and GR, and increased the lipid peroxidation.

On the other hand, the administration of an agonist of A_3 adenosine receptor, R-PIA, activated antioxidant enzyme activity similar to the administration of acute exercise, indicating that the adenosine receptor is possibly involved in the activation of antioxidant enzymes. We did not carry out an experiment on the combined effects of exercise and R-PIA. However, the concurrent administration of the agonist R-PIA and the antagonist (theophylline) reduced the AOE activity; the lipid peroxidation remained high in the rat heart (Table 3). The results of these experiments suggested that acute exercise stimulated the antioxidant enzyme activity in the rat heart, and that this was blocked by theophylline. *R*-Phenyl isopropyl adenosine increased antioxidant enzyme activities in the heart of the sedentary control rat. However, the concurrent administration of theophylline and R-PIA restored AOE activity to a normal level. Furthermore, the stimulation

Table 3. Effect of exercise, adenosine agonist R-PIA, and antagonist theophylline on lipid peroxidation (malondialdehyde) in rat heart

	Malondialdehyde (nmol/mg protein, % of control)
Sedentary control	100
AE (100% VO_{2max})	59
R-PIA	57
Theophylline + AE	94
Theophylline + R-PIA	87
Theophylline	97

Note. AE, acute exercise; R-PIA, *R*-phenyl isopropyl adenosine.

of AOE activity reduced the lipid peroxidation. Based on these preliminary results, we suggest that the stimulation of AOE activity, due to acute exercise, is mediated via the activation of the A_3 adenosine receptor. We have previously described the model scheme for adenosine receptor-mediated activation of antioxidant enzymes based on in vitro studies. It is possible that those steps are also followed during exercise; adenosine is released and activates adenosine receptor, thereby activating AOE. The activation of AOE can be blocked by the adenosine receptor antagonist theophylline during exercise.

INTERACTION OF EXERCISE AND CHEMICALS WITH ANTIOXIDANT SYSTEM

The biotransformation of chemicals can be both harmful and helpful to the body. Under certain conditions, a chemical can be safely biotransformed and excreted from the body. However, several metabolic pathways could form electrophilic intermediate metabolites and generate superoxide and free radicals. Chemicals that contain nitro, amine, iminium, or quinone functional groups usually form electrophilic intermediate metabolites. These metabolites can covalently bind to macromolecules and produce tissue-specific toxicity. Two specific pathways that produce free radicals or superoxide anions include NADPH-cytochrome P-450 reductase (82), and one-electron reduction under low oxygen tension (83). Once these reactive metabolites are generated, interaction with the ferrous ion could produce the hydroxyl radical, which could lead to toxic effects. Because exercise will increase oxygen consumption and the status of AOE, it may be a means of decreasing the toxicity that several chemicals produce. Exercise training was shown to prevent the toxicity of adriamycin, known to cause cardiac toxicity, by enhancing the activity of the AOE (56). Ethanol has been reported to generate α-hydroxyl ethyl radical and exert oxidative stress (84). We have recently demonstrated the interaction of acute exercise and ethanol on AOE activity in brain regions and liver of rats (85, 86). In the brain regions, the activity of GSH-Px increased in the hypothalamus and striatum. CAT and GR activity also increased in the striatum, and a decrease in SOD activity in the striatum and medulla was shown. Malondialdehyde levels increased in all brain regions after ethanol and acute exercise.

In the liver, where ethanol is mainly metabolized, the activity of SOD and GR significantly increased, whereas CAT and GSH-Px levels decreased after acute exercise and ethanol. Interestingly, malondialdehyde levels did not change.

Many drugs and chemicals produce tissue-specific toxicities (Table 4). For example, adriamycin produces cardiotoxicity, cisplatin produces nephrotoxicity and ototoxicity, vincristine causes neurotoxicity, and bleomycin, BCNU, and busulfan cause pulmonary toxicity. These toxicities may result from the free radical and oxidative stress that they generate. The question to be answered is whether exercise will decrease the free radicals and oxidative stress that these agents may generate. This is an area that has not been investigated.

FUTURE PERSPECTIVES

The role of antioxidant enzyme and glutathione in cellular antioxidant defense against toxic insults and various disease processes has been well established. Its significance in protecting against free radical-induced oxidative stress and tissue damage during physical exercise has not been well appreciated. Recent evidence has shown the activation/induction of antioxidant enzymes during exercise in specific tissues of the body. However, the mechanism of activation/induction of these enzymes needs to be further investigated. Since GSH plays a pivotal role in the maintenance of the intracellular redox status and AOE functions during acute exercise and exercise training, it is essential to study the exercise performance and intracellular interorgan homeostasis, especially in the brain. The recent work on A_3 receptors, through which adenosine enhances the activity of antioxidant enzymes, has opened a door for studying the interactions of agonists and antagonists of adenosine on antioxidant enzymes during exercise. The use of more selective agonists and antagonists of A_1, A_2 and A_3 receptors will provide a better understanding of the mechanisms responsible for the activation or inhibition of antioxidant enzyme during exercise. Exercise is known to enhance the production of nitric oxide (NO) in coronary vascular tissues. It is not known whether NO synthase inhibitors or activators would have antagonistic or synergistic

Table 4. Drugs and chemicals that produce toxicities mediated via free radicals/reactive intermediates metabolites

Drugs	Free radicals	Toxicity	Reference
Doxorubicin	Superoxide anion,	Cardiac	87
(anthracyclines)	hydroxyl radical		56
			88
Bleomycin	Superoxide anion	Pulmonary	89
Mytomycin	Free radical		90
Cisplatin	Possibly free radical	Nephrotoxicity	91
		Ototoxicity	92
BCNU (carmustine)	Methyl radical	Neurotoxicity	93
Procarbazine	Free radical	Neurotoxicity	94
Acetaminophen	Reactive intermediates metabolites	Hepatic	95
Isoniazid	Free radical	Hepatic	96
Ethanol	α-Hydroxyl ethyl radical	Hepatic and neurotoxicity	84
			55
Physostigmine	Eseroline to catechol to quinones	Neurotoxicity	97
Quinones	Reactive metabolites	Neurotoxicity	98
Morphine	Covalent binding	Neurotoxicity	99
Nitrofurantoin	Oxidant	Pulmonary	100
Paraquat	Oxidant	Pulmonary	101
Parathion	Reactive metabolites	Neurotoxicity	102
Carbon tetrachloride	Trichloromethyl radical	Hepatic	103
			104
Polycyclic aromatic hydrocarbons	Epoxides	Hepatic	105
Nitrofurazone	Free radical/ROS	Pulmonary	106
Metronidazole	Free radical/ROS	Pulmonary	106
6-Hydroxydopamine	Free radical/ROS	Neurotoxin	107
4-Hydroxyanisole	Free radical	—	107
Etoposide (VP-16)	Hydroxyl radical	—	107
Benzidine	Free radical	Bladder carcinogen	107
Aminopyrine	Free radical	Agranulocytosis	108
Clozaril	Free radical	Agranulocytosis	109
Phenylhydrazine	Free radical/ROS	Hemolytic anemia	110
3-methylindole	Free radical	Pulmonary	111
Probucol	Free radical	—	112
Ferrous sulfate	Hydroxyl radical	Iron overload	113
Methimazole	Free radical	—	107
Chloroprazine	Free radical	—	114
	Free radical	Phototoxicity, photoallergy	115
Salicylanilides	Free radical	Photoallergy	116
Mitoxantrone	Free radical	—	107
Daunomycin	Free radical/ROS	Cardiotoxicity	117

effects on adenosine receptor-induced responses during exercise. Both adenosine and NO are produced during exercise. Therefore, the actions and interactions of both agents, through modulatory receptors on AOE during exercise, would be an important study in the future. Moreover, it is likely that Ca^{2+}-dependent enzymes, such as protein kinase C and nitric oxide synthase, may contribute to the exercise-induced oxidative stress on cardiovascular and cerebrovascular systems. The use of calcium channel blockers, nucleoside transport blockers, protein kinase C, and NO synthase inhibitors will provide a better understanding of adenosine receptor-mediated antioxidant responses in specific tissues during exercise. Further studies of the role of Bcl-2, ACE1 activating protein, NF-κB, and heat shock protein may reveal the intracellular regulation that occurs in the antioxidant defense system during exercise.

SUMMARY

During exercise, an increase in oxygen consumption and an increase in adenosine triphosphate (ATP) utilization occur. Both of these aspects are instrumental in an increase of antioxidant enzyme activity and adenosine during exercise. The antioxidant system of the body comprises antioxidant enzymes (AOE) [superoxide dismutase (SOD), catalase (CAT), glutathione peroxidase (GSH-Px)], the ancillary enzyme glutathione reductase (GR), and antioxidants (vitamin C, vitamin E, and glutathione).

The generation of reactive oxygen species (ROS) and free radicals occurs due to an increase in oxygen consumption, and the antioxidant system scavenges the ROS and free radicals. The adenosine A_3 receptor appears to be correlated to the activation of AOE. Exercise increased the activity of SOD and GR, and also increased the level of oxidized glutathione (GSSG) and malondialdehyde in the blood irrespective of mode, intensity, and duration of exercise. Catalase and GSH-Px activity also increased moderately due to exercise. Antioxidant enzyme activity in various brain regions differed, and exercise differentially induced enzyme activity. The antioxidant enzyme activity in the heart is low when compared to the liver. Exercise specifically increased the activity of antioxidant enzyme in the mitochondria of the heart. The presence of ascorbate radicals in heart tissue 23 h after the last bout of exercise was detected, indicating that exercise training can provide protection to heart tissues against oxidative damage via ascorbate and vitamin E. The liver generates a very high amount of ROS and, correspondingly, there are high concentrations of antioxidants and high activity of the AOE to scavenge them. There is very little information about the AOE and exercise effects in kidney, lung, and testes tissues. The maximum velocity (V_{max}) of CAT, GSH-Px, and GR was favorably higher in most of the tissues of aged rats after exercise training. Similarly, the Michaelis-Menten constant (K_m) is generally increased due to exercise adaptation. Acute exercise increased membrane lipid peroxidation, whereas exercise training provided protection by depleting the level of malondialdehyde. With more knowledge of the intracellular regulation of the antioxidant system, pharmacological intervention may be plausible against oxidative tissue damage. Recent studies have demonstrated a beneficial role of adenosine in augmenting AOE activity and in reducing lipid peroxidation during exercise via the A_3 adenosine receptor. The mechanisms of interaction of drugs and exercise with respect to the antioxidant system would be an interesting study in the field of exercise research.

REFERENCES

1. Sjodin, B., Hellsten Westing, Y., and Apple, F. S. 1990. Biochemical mechanisms for oxygen free radical formation during exercise. *Sports Med.* 10:236–254.
2. Somani, S. M., and Arroyo, C. M. 1995. Endurance training generates ascorbate free radicals in rat heart. *Ind. J. Physiol. Pharmacol.* 39:323–329.
3. Somani, S. M. 1995. Exercise, drugs and tissue specific antioxidant enzymes. In *Pharmacology in Exercise and Sports*, ed. S. M. Somani. pp. 57–95. Boca Raton, FL: CRC Press.
4. Boveries, A., and Cadenas, E. 1982. Production of superoxide radicals and hydrogen peroxide in mitochondria. In *Superoxide Dismutase*, Vol. 2, ed. L. W. Oberley, p. 15. Boca Raton, FL: CRC Press.
5. Chance, B., Sies, H., and Boveris, A. 1979. Hydrogen peroxide metabolism in mammalian organs. *Physiol. Rev.* 59:72–77.
6. Gram, T. E., Okine, L. K., and Gram, R. A. 1986. The metabolism of xenobiotics by certain extrahepatic organs and its relation to toxicity. *Annu. Rev. Pharmacol. Toxicol.* 26:259–269.
7. Cadenas, E. 1985. Oxidative stress and formation of excited species. In *Oxidative Stress*, ed. H. Sies, pp. 311–326. London: Academic Press.
8. Halliwell, B., and Gutteridge, J. M. C., eds. 1989. *Free Radicals in Biology and Medicine*. Oxford: Clarendon Press.
9. Weindruch, R., Warner, H. R., Starke-Reed, P. E., and Yu, B. P., eds. 1993. Future directions of free radical research in aging. In *Free Radicals in Aging*, p. 269. Boca Raton, FL: CRC Press.
10. Cadenas, E. 1989. Biochemistry of oxygen toxicity. *Annu. Rev. Biochem.* 58:79–110.
11. Burton, G. W., Joyce, A., and Ingold, K. U. 1982. First proof that vitamin E is major lipid soluble, chain-breaking antioxidant in human blood plasma. *Lancet* 2:327.
12. Witting, L. A., and Horwitt, M. K. 1964. Effect of degree of fatty acid unsaturation in tocopherol deficiency-induced creatinuria. *J. Nutr.* 82:19–23.

13. Krinsky, N. I., and Deneke, S. M. 1982. Interaction of oxygen and oxy-radicals with carotenoid. *J. Natl. Cancer Inst.* 69:205–209.
14. Foote, C. S., and Denny, R. W. 1988. Chemistry of singlet oxygen. VIII. Quenching by β-carotene. *J. Am. Chem. Soc.* 90:6233–6235.
15. Howell, R. R., and Wyngarden, J. B. 1960. On the mechanisms of peroxidation of uric acid by hemoprotein. *J. Biol. Chem.* 235:3544–3550.
16. Ames, B. N., Cathcart, R., Schwiers, E. and Hochstein, P. 1981. Uric acid provides an antioxidant defense in humans against oxidant- and radical-caused aging and cancer: A hypothesis. *Proc. Natl. Acad. Sci. U.S.A.* 78:6858–6862.
17. Sevanian, A., Davies, K. J. A., and Hochstein, P. 1985. Conservation of vitamin C by uric acid in blood. *J. Free Radic. Biol. Med.* 1:117–124.
18. Davies, K. J., Sevanian, A., Muakkassah-Kelly, S. F., and Hochstein, P. 1986. Uric acid-iron ion complex: A new aspect of antioxidant function of uric acid. *Biochem. J.* 235:747–754.
19. McCord, J. M., and Fridovich, I. 1969. Superoxide dismutase: An enzymatic function for erythrocuprein (hemocuprein). *J. Biol. Chem.* 244:6049–6055.
20. Gregory, E. M. and Fridovich, I. 1973. Induction of superoxide dismutase by molecular oxygen. *J. Bacteriol.* 114:5443–5448.
21. Fridovich, I. 1995. Superoxide radical and superoxide dismutases. *Annu. Rev. Biochem.* 64:97–112.
22. Husain, K., Dube, S. N., Sugenolran, K., Singh, R., DasGupta, S., and Somani, S. M. 1996. Effect of topically applied sulphur mustard on antioxidant enzymes in blood cells and body tissues of rat. *J. Appl. Toxicol.*, 16:245–248.
23. Crapo, J. D., and Tierney, D. F. 1974. Superoxide dismutase and pulmonary oxygen toxicity. *Am. J. Physiol.* 226:1401–1407.
24. Thomas, C. E. and Aust, S. D. 1985. Rat liver microsomal NADPH-dependent release of iron from ferritin and lipid peroxidation. *Free Radic. Biol. Med.* 1:293–300.
25. Bielsi, B. H. J., Seib, P. A., and Tolbert, B. M., eds. 1984. *Ascorbic Acid: Chemistry, Metabolism and Uses*, pp. 81–100. Washington, DC: American Chemical Society.
26. Buettner, G. R., and Jurkiewicz, B. A. 1993. Ascorbate free radical as a marker of oxidative stress: An EPR study. *Free Radic. Biol. Med.* 14:49–55.
27. Duthie, G. C., Robertson, J. D., Maughan, R. J. and Morrice, P. C. 1990. Blood antioxidant status and erthrocyte lipid peroxidation following distance running. *Arch. Biochem. Biophys.* 282:78–83.
28. Koz, M., Erbas, D., Bilgihan, A., and Aricioğlu, A. 1992. Effects of acute swimming exercise on muscle and erythrocyte malondialdehye serum myoglobin, and plasma ascorbic acid concentrations. *Can. J. Physiol. Pharmacol.* 70:1392–1395.
29. Sastre, J., Asensi, M., Gasco, E., Pallardó, F. V., Ferrero, J. A., Furukawa, T., and Viña, J. 1992. Exhaustive physical exercise causes oxidation of glutathione status in blood: Prevention by antioxidant administration. *Am. J. Physiol.* 263:R992–R995.
30. Hellsten-Westing, Y., Balsom, P. D., Norman, B., and Sjödin, B. 1993. The effect of high-intensity training on purine metabolism in man. *Acta Physiol. Scand.* 149:405–412.
31. Sjödin, B., and Hellsten Westing, Y. 1990. Changes in plasma concentration of hypoxanthine and uric acid in man with short-distance running at various intensities. *Int. J. Sports. Med.* 11:493–495.
32. Gohil, K., Rothfuss, L., Lang, J., and Packer, L. 1987. Effect of exercise training on tissue vitamin E and ubiquinone content. *J. Appl. Physiol.* 63:1638–1641.
33. Kihlström, M. 1990. Protection of endurance training against reoxygenation-induced injuries in rat heart. *J. Appl. Physiol.* 68:1672–1678.
34. Meydani, M., Evans, W., Handelman, G., Fielding, R. A., Meydani, S. N., Fiatarone, M. A., Blumberg, J. B. and Cannon, J. G. 1992. Antioxidant response to exercise-induced oxidative stress and protection by vitamin E. *Ann. N.Y. Acad. Sci.* 669:363–364.
35. Tiidus, P. M., Behrens, W. A., Maderc, R. and Houston, M. E. 1993. Muscle vitamin E. levels following acute submaximal exercise in female rats. *Acta Physiol. Scand.* 147:249–250.
36. Tiidus, P. M., and Houston, M. E. 1993. Vitamin E status does not affect the responses to exercise training and acute exercise in female rats. *J. Nutr.* 123:834–840.
37. Brady, P. S., Brady, L. J. and Ullrey, D. E. 1979. Selenium, vitamin E and the response to swimming stress in the rat. *J. Nutr.* 109:1103–1109.
38. Dillard, C. J., Litov, R. E., Savin, W. M., Mumelin, E. E., and Tappel, A. L. 1978. Effects of exercise, vitamin E and ozone on pulmonary function and lipid peroxidation. *J. Appl. Physiol.* 45:927–932.
39. Kanter, M. M., Nolte, L. A., and Holoszy, J. O. 1993. Effects of an antioxidant vitamin mixture on lipid peroxidation at rest and postexercise. *J. Appl. Physiol.* 74:965–969.
40. Ji, L. L., Sen, C. K., Packer, L., and Hanninen, O., eds. 1994. *Exercise-Induced Oxidative Stress in the Heart, in Exercise and Oxygen Toxicity*, pp. 249–268. Amsterdam: Elsevier Science.
41. Gohil, K., Viguie, C., Stanley, W. C., Brooks, G. A., and Packer, L. 1988. Blood glutathione oxidation during human exercise. *J. Appl. Physiol.* 64:115–119.

42. Ji, L. L., Katz, A., Ronggen, F., Griffith, M., and Spencer, M. 1993. Blood glutathione status during exercise: Effect of carbohydrate supplementation. *J. Appl. Physiol.* 74:788–792.
43. Viguie, C. A., Frei, B., Shigenaga, M. K., Ames, B. N., Packer, L., and Brooks, G. A. 1993. Antioxidant status and indexes of oxidative stress during consecutive days of exercise. *J. Appl. Physiol.* 75:566–572.
44. Sen, C. K., Atalay, M., and Hänninen, O. 1994. Exercise-induced oxidative stress: Glutathione supplementation and deficiency. *J. Appl. Physiol.* 77:2177–2181.
45. Tessier, F., Margaritis, I., Richard, M., Moynot, C., and Marconnet, P. 1995. Selenium and training effects on the glutathione system and aerobic performance. *Med. Sci. Sports Exercise* 27:390–396.
46. Somani, S. M., Ravi, R. and Rybak, L. P. 1995. Effect of exercise training on antioxidant system in brain regions of rat. *Pharmacol. Biochem. Behav.* 50:635–637.
47. Somani, S. M., Frank, S., and Rybak, L. P. 1995. Responses of antioxidant system to acute and trained exercise in rat heart subcellular fractions. *Pharmacol. Biochem. Behav.* 51:627–634.
48. Griffith, O. W., and Meister, A. 1985. Origin and turnover of mitochondrial glutathione. *Proc. Natl. Acad. Sci. U.S.A.* 82:4668.
49. Martensson, J., Lai, J. C. K., and Meister, A. 1990. High affinity transport of glutathione is part of a multi-component system essential for mitochondrial function. *Proc. Natl. Acad. Sci. U.S.A.* 87:7185.
50. Ji, L. L., Fu, R., and Mitchell, E. W. 1992. Glutathione and antioxidant enzymes in skeletal muscle: Effects of fiber type and exercise intensity. *J. Appl. Physiol.* 73:1854–1859.
51. Duarte, J. A. R., Appell, H.-J., Carvalho, F., Bastos, M. L., and Soares, J. M. 1993. Endothelium-derived oxidative stress may contribute to exercise-induced muscle damage. *Int. J. Sports Med.* 14:440–443.
52. Salminen, A., and Vihko, V. 1983. Endurance training reduces the susceptibility of mouse skeletal muscle to lipid peroxidation in vitro. *Acta Physiol. Scand.* 117:109–113.
53. Sen, C. K., Marin, E., Kretzschmar, M., and Hänninen, O. 1992. Skeletal muscle and liver glutathione homeostasis in response to training, exercise, and immobilization. *J. Appl. Physiol.* 73:1265–1272.
54. Brady, P. S., Shelle, J. E., and Ullrey, D. E. 1977. Rapid changes in equine erythrocyte glutathione reductase with exercise. *Am. J. Vet. Res.* 38:1045–1047.
55. Somani, S. M., Husain, K., Diaz-Phillips, L., Lanzotti, D. G., Kareti, K., and Trammell, G. 1995. Effect of exercise and ethanol on antioxidant enzyme and lipid peroxidation in brain regions of rat. *Alcohol*, 13:603–616.
56. Kanter, M. M., Hamlin, R. L., Unverferth, D. V., Davis, H. W., and Merola, A. J. 1985. Effect of exercise training on antioxidant enzymes and cardiotoxicity of doxorubicin, *J. Appl. Physiol.* 59:1298–1303.
57. Ji, L. L. 1993. Antioxidant enzyme response to exercise and aging. *Med. Sci. Sports Exercise* 25:225–231.
58. Salminen, A., Kainulainen, H., and Vihko, V. 1984. Endurance training and antioxidants of Lung. *Experentia* 40:822–823.
59. Somani, S. M., and Husain, K. 1996. Exercise training alters kinetics of antioxidant enzymes in rat tissues. *Biochem. Mol. Biol. Int.*, 38:587–595.
60. Ji, L. L., and Fu, R. 1992. Responses of glutathione system and antioxidant enzymes to exhaustive exercise and hydroperoxide. *J. Appl. Physiol.* 72:1–6.
61. Hockenberry, D. M., Oltavi, Z. N., Yin, X. M., Milliman, C. L., and Korsmeyer, S. J. 1993. Bcl-2 functions in an antioxidant pathway to prevent apoptosis. *Cell* 75:241–250.
62. Thompson, H. S., Scordilis, S. P., and Clarkson, P. M. 1994. Muscle heat/stress protein changes after eccentric exercise. *Med. Sci. Exercise* 26:S134.
63. Galter, D., Mihm, S., and Droge, W. 1994. Distinct effects of glutathione disulphide on the transcription factors NF-κB and AP-1. *Eur. J. Biochem.* 221:639–648.
64. Fürst, P., Hu, S., Hackett, R., and Hamer, D. 1988. Copper activates metallothionein gene transcription by altering the conformation of a specific DNA binding protein. *Cell* 55:705–717.
65. Sessa, W. C., Pritchard, K., Seydi, N., Wang, J., and Hintze, T. H. 1994. Chronic exercise in dogs increases coronary vascular nitric oxide production and endothelial cell nitric oxide synthase gene expression. *Circ. Res.* 74:349–353.
66. Lloyd, H. G. E., Deussen, A., Wupperman, H., and Schrader, J. 1988. The transmethylation pathways as a source for adenosine in the isolated guinea pig heart. *Biochem. J.* 252:489–494.
67. Olsson, R. A., Snow, J. A., and Gentry, M. K. 1978. Adenosine metabolism in canine myocardial reactive hyperemia. *Circ. Res.* 42:358–362.
68. Pons, F., Bruns, R. F., and Daly, J. W. 1980. Depolarization-evoked accumulation of cyclic AMP in brain slices: The requisite intermediate adenosine is not derived from hydrolysis of released ATP. *J. Neurochem.* 34:1319–1323.
69. Klabunde, R. E. 1983. Dipyridamole inhibition of adenosine metabolism in human blood. *Eur. J. Pharmacol.* 93:21–26.
70. Ontyd, J., and Schrader, J. 1984. Measurement of adenosine, inosine, and hypoxanthine in human plasma. *J. Chromatogr.* 307:404.
71. Arch, J. R. S., and Newsholme, E. A. 1978. The control of the metabolism and the hormonal role of adenosine. *Essays Biochem.* 14:82–123.

72. Wu, P. H., and Phillis, J. W. 1984. Uptake by central nervous tissue as a mechanism for the regulation of extracellular adenosine concentrations. *Neurochem. Int.* 6:613–632.

73. Watkinson, W. P., Foley, D. H., Rubio, R., and Berne, R. M. 1979. Myocardial adenosine formation with increased cardiac performance in the dog. *Am. J. Physiol.* 236:H13-H21.

74. Pennel, D. J., Maurogeni, S. I., Forbat, S. M., Karwatowski, S. P., and Underwood, S. R. 1995. Adenosine combined with dynamic exercise for myocardial perfusion imaging. *J. Am. Coll. Cardiol.* 25:1300–1309.

75. Olafsson, B., Forman, M. B., Puett, D. W., Pou, A., Cates, C. U., Friesinger, G. C., and Virmani, R. 1987. Reduction of reperfusion injury in canine preparation by intracoronary adenosine: Importance of the endothelium and the no-reflow phenomenon. *Circulation* 76:1135–1145.

76. Liu, G. S., Thornton, J. D., Van Winkle, D. M., Stanley, A. W. H., Olsson, R. A., and Downey, J. M. 1991. Protection against infarction afforded by preconditioning is mediated by A_3 adenosine receptors in rabbit heart. *Circulation* 84:350–356.

77. Downey, J. M., Liu, G. S., and Thorton, J. D. 1993. Adenosine and the anti-infarct effects of preconditioning. *Cardiovasc. Res.* 27:3–8.

78. Fredholm, B. B., and Paton, D.M., eds. 1985. Methods used to study the involvement of adenosine in the regulation of lipolysis. In *Methods in Pharmacology*, p. 337. New York: Plenum Press.

79. Maggirwar, S. B., Dhanraj, D. N., Somani, S. M., and Ramkumar, V. 1994. Adenosine acts as an endogenous activator of the cellular antioxidant defense system. *Biochem. Biophys. Res. Commun.* 201:508–515.

80. Ramkumar, V., Zhongzhen, N., Rybak, L. P., and Maggirwar, S. B. 1995. Adenosine, antioxidant enzymes and cytoprotection. *TIPS* 16:283–285.

81. Ramkumar, V., Stiles, G. L., Beaven, M. A., and Ali, H. 1993. The A_3 adenosine receptor is the unique adenosine receptor which facilitates release of allergic mediators in mast cells. *J. Biol. Chem.* 268:16887–16880.

82. Ortiz de Montellano, P. R., ed. 1986. *Oxygen Activation and Transfer. Cytochrome P-450, Structure, Mechanism, and Biochemistry*, p. 217. New York: Plenum Press.

83. DeGroot, H., and Noll, T. 1983. Halothane hepatotoxicity: Relation between metabolic activation, hypoxia, covalent binding, lipid peroxidation and liver cell damages. *Hepatology* 3:601.

84. Knecht, K. T., Thurman, R. G., and Mason, R. P. 1993. Role of superoxide and trace metals in production of α-hydroxy ethyl radical from ethanol by microsomes from alcohol dehydrogenase deficient deer mice. *Arch. Biochem. Biophys.* 303:339–348.

85. Lanzotti, D. J., Husain, K., Kareti, K. B., Diaz-Phillips, L., and Trammel, G. L. 1995. Effect of exercise and ethanol on antioxidant enzymes and lipid peroxidation in brain regions of rat. *Med. Sci. Sport Exercise* 27:S121.

86. Rybak, L. P., Husain, K., and Somani, S. M. 1995. Influence of acute exercise and ethanol on hepatic antioxidant enzymes and lipid peroxidation in rats. *Med. Sci. Sports Exercise* 25:S122.

87. Pollakis, G., Goormaghtigh, E., Delmelle, M., Lion, Y., and Ruysschaert, J. B. 1984. Adriamycin and derivatives interaction with the mitochondrial membrane: O_2 consumption and free radical formation. *Res. Commun. Chem. Pathol. Pharmacol.* 44:445.

88. Ji, L. L., and Mitchell, E. W. 1994. Effects of adriamycin on heart mitochondrial function in rested and exercised rats. *Biochem. Pharmacol.* 47:877–885.

89. Petering, D. H., Byrnes, R. W., and Antholine, W. E. 1990. The role of redox-active metals in the mechanism of action of bleomycin. *Chem. Biol. Int.* 73:133.

90. Moore, H. W. 1977. Bioactivation as a model for drug design bioreductive alkylation. *Science* 197:527.

91. Somani, S. M., Ravi, R., and Rybak, L. P. 1995. Diethyldithiocarbamate protection against cisplatin nephrotoxicity: Antioxidant system. *Drug Chem. Toxicol.*, 18:151–170.

92. Rybak, L. P., Ravi, R., and Somani, S. M. 1995. Mechanism of protection by diethyldithiocarbamate against cisplatin ototoxicity: Antioxidant system. *Fundam. Appl. Toxicol.* 26:293–300.

93. Reed, D. J., May, H. E., Boorse, R. B., Gregory, K. M., and Beilstein, M. A. 1985. 2-Chloroethanol formation as evidence for a 2-chloroethyl alkylating intermediate during chemical degradation of 1-(2-chloroethyl)-3-cyclohexyl-1-nitrosourea and 1-(2- chloroethyl)-3-(*trans*-4-methylcyclohexyl)-1-nitrosourea. *Cancer Res.* 35:568.

94. Sinha, B. K. 1984. Metabolic activation of procarbazine: Evidence for carbon-centered free-radical intermediates. *Biochem. Pharmacol.* 33:2777.

95. Kaysen, G. A., Pond, S. M., and Roper, M. H. 1985. Combined hepatic and renal injury in alcoholics during therapeutic use of acetaminophen. *Arch. Int. Med.* 145:2019.

96. Mitchell, J. R., Zimmerman, H. J., Ishak, K. G., Thorgeirsson, U. P., Timbrell, J. A., Snodgrass, W. R., and Nelson, S. D. 1976. Isoniazid liver injury: Clinical spectrum, pathology, and probably pathogenesis. *Ann. Int. Med.* 84:181.

97. Somani, S. M., Kutty, R. K., and Krishna, G. 1990. Eseroline, a metabolite of physostigmine, induces neuronal cell death. *Toxicol. Appl. Pharmacol.* 106:28–37.

98. Kochli, H. W., Wermuth, B., and Von Wartburg, J. P. 1980. Characterization of a mitochondrial NADH-dependent nitro reductase from rat brain. *Biochem. Biophys. Acta* 616:133.

99. Nagamatsu, K., Kido, Y., Terao, T., Ishida, T., and Toki, S. 1983. Studies on the mechanism of covalent binding of morphine metabolites to proteins in mouse. *Drug Metab. Dispos.* 11:190.

100. Suntres, Z., and Shek, P. N. 1992. Nitrofurantoin-induced pulmonary toxicity: In vivo evidence for oxidative stress-mediated mechanisms. *Biochem. Pharmacol.* 43:1127.

101. Bus, J. S., and Gibson, J. E. 1984. Paraquat: Model for oxidant-initiated toxicity. *Environ. Health Perspect.* 55:37.

102. Chambers, J. E., Munson, J. R., and Chambers, H. W. 1989. Activation of the phosphothionate insecticide parathion by rat brain in situ. *Biochem. Biophys. Res. Commun.* 165:327.

103. Sipes, I. G., Krishna, G., and Gillette, J. R. 1977. Bioactivation of carbon tetrachloride, chloroform and bromotrichloromethane: Role of cytochrome P-450. *Life Sci.* 20:1541.

104. Mehendale, H. M. 1994. Amplified interactive toxicity of chemicals at nontoxic level: Mechanistic considerations and implications to public health. *Environ. Health Perspect.* 102:139.

105. Das, M., Seth, P. K., and Mukhtar, H. 1981. NADPH-dependent inducible aryl hydrocarbon hydroxylase activity in rat brain mitochondria. *Drug Metab. Dispos.* 9:69.

106. Rao, D. N. R., Harman, L., Motten, A., and Schreiber, J. 1987. Generation of radical anions of nitrofurantoin, isonidazole, and metronidazole by ascorbate. *Arch. Biochem. Biophys.* 255:419–427.

107. Aust, S. D., Chignell, C. F., Bray, T. M., Kalyaharaman, B., and Mason, R. P. 1993. Contemporary issues in toxicology: Free radicals in toxicology. *Toxicol. Appl. Pharmacol.* 120:168–178.

108. Griffin, B. W. 1977. Free radical intermediate in the *N*-demethylation of aminopyrine by horseradish peroxidase-hydrogen peroxide. *FEBS Lett.* 74:139.

109. Fischer, V., Haar, J. A., Greiner, L., Lloyd, R. V., and Mason, R. P. 1991. Possible role of free radical formation in clozapine (Clozaril)-induced agranulocytosis. *Mol. Pharmacol.* 40:846–853.

110. Smith, P., and Maples, K. R. 1985. EPR study of the oxidation of phenylhydrazine initiated by the titanous chloride/hydrogen peroxide reaction and by oxyhemoglobin. *J. Mag. Reson.* 65:491–496.

111. Bray, T. M., and Carlson, J. R. 1979. Role of mixed function oxidase in 3-methylindole-induced acute pulmonary edema in goats. *Am. J. Vet. Res.* 40:1268–1272.

112. Kalyanavaman, B., Darley-Usmar, V. M., Woods, J., Joseph, J., and Parthasarathy, S. 1992. Synergistic interaction between the probucal phenoxyl radical and ascorbic acid in inhibiting the oxidation of low density lipoprotein. *J. Biol. Chem.* 267:6789–6795.

113. Mason, R. P., and Holtzman, J. L. 1975. The role of catalytic superoxide formation in the O_2 inhibition of nitro reductase. *Biochem. Biophys. Res. Commun.* 67:1267.

114. Motten, A. G., and Chignell, C. F. 1985. ESR of cation radicals of phenothiazine derivatives. *Mag. Reson. Chem.* 23:834–841.

115. Motten, A. G., Buettner, G. R., and Chignell, C. F. 1985. Spectroscopic studies of autoneous photosensitizing agents. VIII. A spin-trapping study of light-induced radicals from chlorpromazine and promazine. *Photochem. Photobiol.* 42:9–15.

116. Chignell, C. F., and Sik, R. H. 1989. Spectroscopic studies of cutaneous photosensitizing agents-XIV. The spin trapping of free radicals formed during the photolysis of halogenated salicylanilide antibacterial agents. *Photochem. Photobiol.* 50:287–295.

117. Schreiber, J., Mottley, C., Sinha, B. K., Kalyanaraman, B., and Mason, B. P. 1987. One-electron reduction of daunomycin, daunomycinone, and 7-deoxydaunomycinone by the xanthine/xanthine oxidase system: Detection of semiquinone free radical by electron spin resonance. *J. Am. Chem. Soc.* 109:348–351.

Chapter Seven

ZINC AS A CARDIOPROTECTIVE ANTIOXIDANT*

Saul R. Powell

In the last 15–20 years extensive evidence has accumulated that links overproduction of reactive oxygen intermediates with postischemic organ dysfunction. The vast majority of evidence deals with cardiac injury (1–7); however, reactive oxygen intermediates have been implicated in damage to postischemic kidney (8–11), brain (12, 13), liver (14–16), pancreas (17, 18), skin (19, 20), and intestine (21–24). Much of the early evidence was indirect and involved demonstration of the protective effect of scavengers, such as superoxide dismutase and catalase, on postischemic injury. Many studies have demonstrated that prior administration of superoxide dismutase, or inclusion in perfusion buffers, decreases severity of injury observed in the ischemic heart (6, 25, 26), kidney (8, 11), skeletal muscle (27), skin flaps (19, 28), and intestine (22). From a kinetic standpoint, the use of antioxidants to scavenge reactive oxygen intermediates after they are formed would not be optimal, as large, often toxic concentrations at the site of formation are necessary. More recently, evidence has been obtained from observations that various reactive oxygen intermediates can be detected in blood or perfusates from postischemic organs. In general, these studies have used one of two techniques to chemically trap the short-lived radicals and then detect the reaction product: either spin trapping in combination with electron paramagnetic resonance spectroscopy, or salicylate in combination with electrochemical detection (29–38). These topics have been the subject of many extensive reviews and will not be discussed further (39–44). What will be described are the lines of evidence, ours and others, that support the role of redox-active transition metals as catalysts for production of reactive oxygen intermediates. Our research has concentrated on the role of redox-active metals as catalysts for formation of the more destructive secondary reactive species, such as ˙OH and carbon- and lipid-centered radicals, with the intent of interfering with this process. In this chapter evidence is presented supporting the role of redox-active transition metals as potential mediators of postischemic cardiac oxidative injury, and how zinc, a non-redox-active metal, may be used as an antagonist to preserve postischemic function.

A ROLE FOR THE REDOX-ACTIVE METALS IN POSTISCHEMIC OXIDATIVE CARDIAC INJURY

Redox-Active Transition Metals as Catalysts

Several redox-active metals, including iron and copper (20, 45–48), and possibly nickel (49) and cobalt (50), have been demonstrated to catalyze formation of reactive oxygen intermediates and thus have been hypothesized to be mediators of postischemic oxidative injury. There is a

*Some of the studies described in this chapter were supported by the National Institutes for Health (NHLBI) HL45534.

well-known requirement for trace amounts of iron or copper to catalyze formation of $^{\cdot}OH$ from H_2O_2 and O_2^- (51–54) through Fenton chemistry according to the following reactions.

$$2O_2^- + 2H_2 \rightarrow H_2O_2 + O_2 \qquad \text{(1) Haber-Weiss reaction}$$
$$O_2^- + M^n \rightarrow O_2 + M^{n-1} \qquad \text{(2)}$$
$$H_2O_2 + M^{n-1} \rightarrow OH^- + {}^{\cdot}OH + M^n \qquad \text{(3)}$$

$$3O_2^- + 2H^+ \rightarrow 2O_2 + OH^- + {}^{\cdot}OH \text{ (net)} \qquad \text{(4) Fenton reaction}$$

From a kinetic standpoint, reaction 1, the Haber-Weiss reaction, is relatively slow. Trace amounts of soluble redox-active metals (M^n), such as iron or copper, in the presence of reducing agents, such as O_2^-, can increase the rate of formation of $^{\cdot}OH$ by several orders of magnitude (reaction 4). Further, redox-active metals, and in particular copper and iron, may catalyze formation of reactive oxygen intermediates from several other endogenous substances (summarized in Table 1) (55–59).

Redox-Active Metals and Oxidative Tissue Injury

It has become clear that metal-catalyzed formation of reactive oxygen can initiate destructive processes such as lipid peroxidation and enzyme inactivation, leading to loss of cellular integrity and function.

Lipid Peroxidation

Numerous studies have demonstrated that the redox-active transition metals play a critical role in initiation and propagation of lipid peroxidation (Figure 1). Metal-catalyzed formation of $^{\cdot}OH$ can result in the abstraction of a hydrogen from an unsaturated fatty acid, leading to lipid radical formation, initiating the process. The lipid radical can then take part in a propagative cascade of cyclical reactions leading eventually to repetitive formation of short chain alkanes, such as ethane and pentane, and lipid acid aldehydes, such as malondialdehyde. The end result of this process is destruction of lipid bilayers (60).

Enzyme Inactivation

Two possible schemes illustrating how metal-catalyzed formation of reactive oxygen intermediates can result in enzyme inactivation are illustrated in Figure 2. In a series of papers, Stadtman and colleagues have demonstrated that the binding of a reduced redox-active transition metal to an enzyme to form a coordination complex can initiate the process of protein oxidation (61–65). They have hypothesized that H_2O_2 can react with the reduced metal, forming $^{\cdot}OH$,

Table 1.   Redox-active metal-catalyzed formation of reactive oxygen intermediates

$O_2^- + H_2O_2 \xrightarrow{\text{Cu/Fe}} {}^{\cdot}OH$ (metal-catalyzed Haber–Weiss reaction)

LOOH (lipid peroxide) $\xrightarrow{\text{Cu/Fe}}$ $LO^{\cdot}, LO_2^{\cdot}$ (alkoxyl and peroxyl radical)

Ascorbic acid + $O_2 \xrightarrow{\text{Cu/Fe}}$ ascorbyl radical, $^{\cdot}OH, H_2O_2$

NAD(P)H + $O_2 \xrightarrow{\text{Cu/Fe}}$ $NAD(P)^{\cdot}, O_2^{\cdot-}, {}^{\cdot}OH, H_2O_2$

Catecholamines + $O_2 \xrightarrow{\text{Cu/Fe}}$ $O_2^{\cdot-}, {}^{\cdot}OH, H_2O_2$

RSH (thiols) + $O_2 \xrightarrow{\text{Cu/Fe}}$ $O_2^{\cdot-}, {}^{\cdot}OH, H_2O_2, RS^{\cdot}$ (thiyl radical)

Source: From ref. 44, with permission.

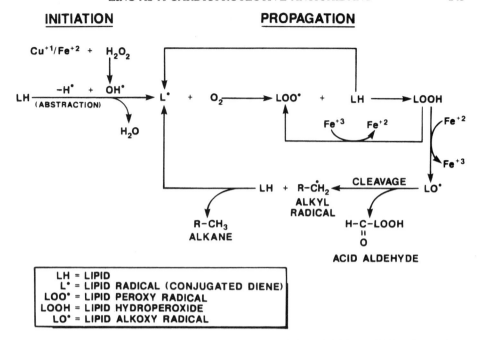

Figure 1. Role of redox-active metals in initiation and propagation of lipid peroxidation. From ref. 44 with permission from Academic Press, Inc.

which abstracts a hydrogen from the amino bearing carbon leading to the formation of a carbon-centered protein radical. An aldehyde or protein carbonyl is formed through a series of reactions that involves hydrolysis of the amino group. Once the amino group is hydrolyzed, the metal binding site is disrupted leading to dissociation of the metal. The enzyme has now been oxidatively modified and is now inactive and marked for selective degradation by proteases such as trypsin, chymotrypsin, calpain, cathepsin d, and pepsin. Because the oxidative modifications occur around the metal-binding site, the process of protein oxidation is said to be a "site-specific" reaction. An excellent review on the process of protein oxidation can be found in ref. 65. Another process that can lead to enzyme inactivation is protein sulfhydryl cross-linking. This process occurs when reactive oxygen intermediates oxidize constituent protein sulfhydryls to form disulfide linkages. Sulfhydryl oxidation along the same protein chain results in intraprotein disulfide cross-linking, which can alter the conformation of the enzyme. Sulfhydryl oxidation may not only inactivate enzymes, but can also activate enzymes, such as in the case of guanylate cyclase (66). Formation of a disulfide bond between sulfhydryls on two different chains may result in the cross-linking of proteins, generally leading to the formation of insoluble, inactive protein aggregates (67).

Redox-Active Metals and Postischemic Cardiac Oxidative Injury

There have been numerous studies implicating redox-active transition metals as catalysts during genesis of postischemic reperfusion injury. Most of these studies were indirect and involved the use of chelators. Since the use of chelators is more related to the concept of site specificity and the so-called "pull-technique," this aspect is discussed in the next section. Here I discuss the studies demonstrating one of the following two points: that preloading with a metal worsens postischemic injury, or that a myocardial preparation is susceptible to oxidative damage as a result of exposure to a metal-catalyzed reactive oxygen intermediate generating system. Since the majority of these studies deal with iron, this transition metal is discussed first. A series of

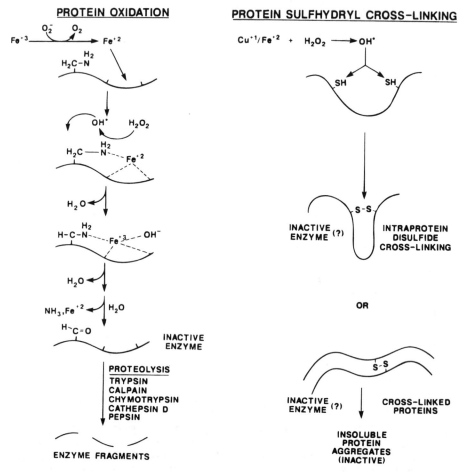

Figure 2. Role of redox-active metals in inactivation of enzymes. From ref. 44 with permission from Academic Press, Inc.

studies conducted by Pinson and colleagues in the cultured cardiomyocyte has established the deleterious effects of iron overload on cellular function. Pinson and co-workers have established a model of cultured ventricular myocytes, which allowed the study of iron overload at the cellular level (68). Among the parameters that have been shown to be affected by iron loading are: malondialdehyde formation (68), contractility and electrophysiology (69), lipid peroxidation (70), α-adrenergic stimulation (71), sarcolemmal structure and lysosomal fragility (72), and cellular SH-groups and activity of thiolic enzymes (73). In addition, Pinson and co-workers have published two studies that examined the effect of iron loading and ascorbate on cultured ventricular myocytes (74, 75). In these studies, it was demonstrated that incubation of iron-loaded ventricular myocytes with ascorbate significantly reduces iron uptake and increases malondialdehyde formation (74). Evidence for enhanced lipid peroxidation was derived from studies that demonstrated that addition of ascorbate changes the fatty acid composition of iron-loaded myocytes, resulting in decreased content of 20:4, 22:5, and 22:6 fatty acids (75). These studies confirm that the metal-loaded cell is more susceptible to oxidative injury. Other studies in whole heart preparations have provided some evidence to suggest that redox-active iron may increase or become delocalized during ischemia. For instance, it has been demonstrated that Fe^{+3} bound to low-molecular-weight substances increases in the ischemic dog heart (47). In rabbit kidney subjected to 24 h of cold ischemia, the pool of chelatable iron was shown to increase (76). In the Langendorff rat heart model, studies have demonstrated increases in intracellular low-molecular-weight iron (a pool of iron thought to be redox-active) during ischemia (77, 78) and increases in detectable iron in effluent from the heart during postischemic reperfusion (79). In regard to iron loading of the heart, the studies that have been performed, primarily in the Langendorff heart, all demonstrate increased susceptibility to postischemic oxidative injury, whether arrhythmia susceptibility (4) or cardiac function (80) was determined. Collectively, all of these studies provide a basis for a strong argument in favor of a mediatory role for iron in postischemic cardiac oxidative injury.

A role for copper is not as clear. Aside from our laboratory, copper has been virtually ignored as a potential mediator of cardiac oxidative injury. This is despite the fact that copper has been shown to be 10–25 times more reactive than iron in promoting cellular damage in single cell preparations in the presence of reducing agents (81, 82). In chemical systems copper is 60-fold more effective than iron in promoting $\cdot$OH formation from H_2O_2 (k_2(iron) for H_2O_2 = 76 mol^{-1} s^{-1}; k_2(copper) for H_2O_2 = 4.7 × 10^3 mol^{-1} s^{-1}) (83). Our initial studies of copper examined the effect of preloading on postischemic cardiac oxidative injury (84). Preloading with 20 μM copper bis-histidinate increased the severity of postischemic reperfusion injury and increased the release of lactate dehydrogenase, as well as $\cdot$OH formation, suggesting the importance of this metal as a mediator of oxidative injury to the postischemic heart (84). In a later study, we examined the effect of the copper-ascorbate reactive oxygen intermediate generating system on function of the isolated perfused rat heart (85). Perfusion of hearts with Krebs-Henseleit buffer containing this system resulted in loss of hemodynmamic function, the rate of which was dependent on the concentration of the copper, not the ascorbate. What was remarkable was that hemodynamic function was altered with as little as 0.5 μM copper. Further, we demonstrated that exposure to this generating system produced histologic changes similar to that seen in the postischemic heart. While these studies suggest that copper may be an important catalytic agent, they are not conclusive: thus our studies of this metal continue.

ZINC AND SITE-SPECIFIC FORMATION OF REACTIVE OXYGEN INTERMEDIATES ("PUSH VS. PULL")

Site-Specific Formation of Radicals

Since the late 1960s it has been known that transition metals, like iron and copper, tend to undergo hydrolytic polymerization and precipitation in aqueous media at neutral pH (86). Thus it has been inferred that these metals can remain in solution in physiological media only by association with some high- or low-molecular-weight cellular components, such as nucleotides, peptides, polypeptides, proteins, or DNA (52, 53, 57). Once the metal becomes complexed at one of these site, movement is hindered; thus the association site can then serve as a locus for repetitive free radical formation through repeated redox cycling of the metal. This has become known as "site-specific" formation of radicals. Because the metal is fixed in one place, any $\cdot$OH formed, by virtue of its high reactivity, would attack adjacent structures, resulting in severe localized damage (Figure 3). For this reason, the process of protein oxidation has been said to be a form of metal-catalyzed site-specific oxidative injury, since the modification to the protein occurs directly adjacent to the metal-binding site (65). An excellent review on the subject of site-specificity can be found in ref. 57.

"Push vs. Pull"

A basic assumption of the theory of site-specificity is that the pool of redox-active transition metals is associated with some cellular component, thus establishing a site for repetitive formation of $\cdot$OH. Accordingly, one can theorize two potential processes by which metal reactivity can be affected as a means of either decreasing $\cdot$OH formation, or possibly shifting the site the formation elsewhere to one less critical. From a kinetic standpoint, affecting metal reactivity, and thus preventing or inhibiting formation of the more reactive $\cdot$OH, should be more efficient than attempting to scavenge the radicals after they are formed. The first of these processes would be to remove or "pull" the metal off its binding site through the use of high-affinity chelators. In

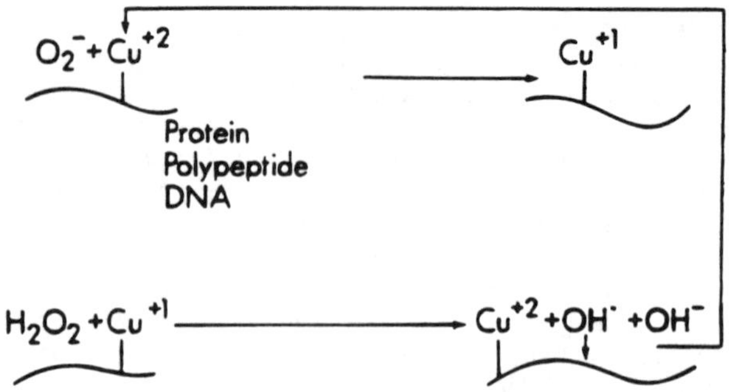

$$O_2^- + Cu^{+2} \longrightarrow Cu^{+1}$$

Protein
Polypeptide
DNA

$$H_2O_2 + Cu^{+1} \longrightarrow Cu^{+2} + OH^\cdot + OH^-$$

Figure 3. Proposed scheme for "site-specific" formation of ˙OH. From ref. 44 with permission from Academic Press, Inc.

the event that the chelator binds the metal so that it is no longer active, ˙OH formation would be interrupted. If the chelator simply removes the metal but does not inactivate it, ˙OH formation would be shifted into the cytosolic component, where it would be less likely to react with critical cellular components. The other potential means of affecting metal reactivity would be to force or "push" the metal off of its binding site through the use of some chemically similar, yet redox-inactive, agent. The net result would be to displace the metal into the cytosolic compartment, where it can undergo hydrolytic polymerization and precipitation as unreactive polynuclear structures (86, 87), or possibly redistribute to some other less critical site, thus shifting the site of formation of ˙OH (Figure 4).

Theoretically, if the metal were to bind to some other more critical site, oxidative injury could be enhanced. To this author's knowledge, the only example of an agent that can effectively compete for copper and/or iron binding sites is zinc. Zinc is a non-redox-active transition metal with similar coordination chemistry to Cu^{2+} and Fe^{3+} and has been demonstrated to compete effectively for Cu^{2+} site-specific binding in several heme proteins (88–90). By far, the use of chelators to "pull" the metal off of its binding site has been the process most extensively studied.

Effects of Chelators ("Pull")

There have been numerous studies examining the effect of chelators on postischemic reperfusion injury. The agent that has been tested most extensively is deferoxamine, an iron chelator. With several notable exceptions, deferoxamine has been demonstrated to have some beneficial (for

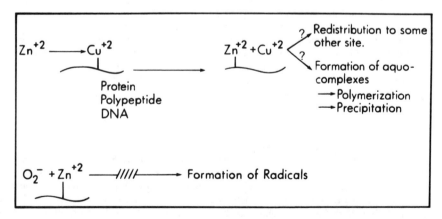

Figure 4. Rationale for the use of zinc to inhibit "site-specific" formation of ˙OH. From ref. 44 with permission from Academic Press, Inc.

the most part, minor) effect on postischemic reperfusion injury in many systems (20, 45–48, 91, 92). In one study, deferoxamine has been demonstrated to improve postischemic myocardial function and metabolism in the isolated rabbit heart model when infused at and during reflow (45). Of interest was the observation that the Fe^{3+}-deferoxamine complex was found in the effluent from the treated hearts. In light of a report that Fe^{3+} bound to low-molecular-weight substances increases in the ischemic dog heart (47), this is an important finding as it would suggest that this chelator can compete for delocalized iron. In vivo studies have demonstrated that pretreatment of dogs with deferoxamine was found to decrease the area of infarct following 2 h of coronary occlusion and 4 h of reperfusion (93). Like the previous in vitro study, urinary content of iron increased severalfold, indicating effective metal chelation and removal. In another in vivo dog study, intracoronary administration of deferoxamine was found to improve contractile function of the ischemic dog heart if infused 2 min prior to initiation of reperfusion (94). In this study, electron paramagnetic resonance (EPR) spin trapping experiments demonstrated apparent decreases in formation of free radicals during reperfusion. Largely because of concerns about toxicity and the short half-life of deferoxamine, other investigators have tested the hydroxyethyl starch conjugate in an in vivo dog model and have observed improved percent segment shortening during reperfusion (95). In a Langendorff rat heart model, deferoxamine has been observed to improve postischemic contractile function (37). Improvement was associated with decreased lactate dehydrogenase (LDH) leakage and membrane phospholipid breakdown, and decreased ·OH formation as detected using the salicylate method (37). The use of deferoxamine is not without some controversy, as there have been studies that were unable to find a beneficial effect of this chelator. For example, one in vivo study could not detect an effect of deferoxamine on infarct size in rabbit heart subjected to 45 min of LAD occlusion (96). Likewise, in dogs, both deferoxamine and deferoxamine-hydroxyethyl starch conjugate have been reported not to affect the area at risk (97). However, there is another issue with deferoxamine. Several studies have demonstrated that this chelator may be a direct scavenger of ·OH and O_2^-; thus reported beneficial effects might not be as a result of iron chelation (98–100). Finally, in regard to copper, neocuproine, a copper chelator, has been observed to decrease reperfusion arrhythmias in the isolated perfused rat heart (101).

Effect of Zinc ("Push")

That zinc possesses antioxidant properties has been known for some time. Abundant studies in various systems have demonstrated that zinc affects production of reactive oxygen intermediates or related pathologies (extensively reviewed in ref. 102). This has led to the generalization that zinc deficiency is deleterious and increases vulnerability to oxidative injury, while supplementation tends to be beneficial, and that actually one of the physiologic functions of zinc is as an endogenous antioxidant (102). What has been unclear is how zinc can effect alterations in reactive oxygen intermediates and related pathologies. We and others have hypothesized that by virtue of its similarities in coordination chemistry, zinc can enter the cell and displace or "push" metals from their site-specific binding sites.

The ability of zinc to compete with transition metals for site-specific binding sites in simple chemical and biological systems has been well documented and has been most clearly demonstrated in several heme proteins (89). Perhaps the earliest report of the ability of zinc to inhibit metal-mediated oxidative damage was the observation of antagonism of iron-mediated, xanthine/xanthine oxidase-induced peroxidation of erythrocyte membranes (103). Other studies have demonstrated that zinc nitrilotriacetate inhibits ascorbate and ferric fructose-mediated formation of methemoglobin in red blood cells and isolated hemoglobin (81, 104). That the inhibition may be of a competitive nature was demonstrated when zinc nitrilotriacetate was shown to competitively antagonize copper nitrilotriacetate-mediated paraquat-induced killing of *Escherichia coli* (105). In addition, zinc has been shown to antagonize copper-mediated free radical formation in several chemical systems including from hydrogen peroxide and ascorbate oxidation (106). These observations in chemical and biochemical systems have provided the rationale for the experimental use of zinc to decrease postischemic oxidative injury. Since our initial report in the heart (discussed later), zinc has been shown to decrease lipid peroxidation and area of erosion in the postischemic stomach of the rat following oral administration and to improve postischemic function of the rabbit kidney (107–109).

ZINC AS A CARDIOPROTECTIVE ANTIOXIDANT

The earliest reports to demonstrate possible antioxidant effects of zinc on oxidative cardiac damage were related to catecholamine-induced myocardial injury, a process thought to involve production of reactive oxygen intermediates (110, 111). In vitro and in vivo studies demonstrated that zinc has an inhibitory effect on isoproteronol-induced cardiac oxidative injury (110, 111). Virtually all of the data supporting the concept that zinc is cardioprotective in the postischemic heart is derived from studies conducted in our laboratory.

The Anti-Arrhythmic Effect of Zinc

Our initial report demonstrated the anti-arrhythmic effect of zinc in a Langendorff model of left anterior descending coronary artery occlusion (107). In this model the heart is subjected to 10 min of regional ischemia followed by 5 min of reperfusion. Regional ischemia of this duration renders the rat heart extremely vulnerable to onset of severe arrhythmias, such as ventricular fibrillation or tachycardia, during reperfusion (112). We demonstrated that when Zn-His$_2$ was added to the perfusion buffer, a concentration-dependent decrease in incidence and duration of reperfusion arrhythmias was observed (Table 2 and Figure 5). The effect of zinc on postischemic cardiac rhythm is best illustrated by the electrocardiograph (EKG) and pressure tracings in Figures 6 and 7. Figure 6 illustrates the onset of ventricular tachycardia, followed by rhythm degradation to ventricular fibrillation within seconds of the initiation of reperfusion. As is apparent in Figure 7, treatment with zinc resulted in immediate restoration of normal sinus rhythm at the start of reperfusion. In a subsequent study, we demonstrated that the anti-arrhythmic effect of zinc is dependent on its concentration and duration of preischemic perfusion and does not appear to be entirely dependent on the negative chronotropic activity of this metal (113). In relationship to inhibition of oxidative stress, the observation that zinc is anti-arrhythmic is important. Numerous studies have demonstrated that exposure of myocardial preparations to reactive oxygen intermediates appears to have a pro-arrhythmic effect, either resulting directly in rhythm disturbances or decreasing the vulnerability to onset of arrhythmias (114–117). Research demonstrating that zinc, and other more classical antioxidants, such as superoxide dismutase, glutathione and catalase, can be anti-arrhythmic in in vitro and in vivo models lends credence to the theory that production of reactive oxygen intermediates facilitates cardiac arrhythmogenesis (107, 113, 118–122).

Effect on Postischemic or Postarrest Cardiac Function

We conducted further studies on a Langendorff global ischemia model of "stunned" myocardium (123). In this model, the heart is subjected to 20 min of normothermic global ischemia followed

Table 2.   Effect of zinc on incidence of reperfusion arrhythmias

Concentration Zn/His (μM)	Incidence[a] of arrhythmia	
	VF	VT
Israeli rats		
0/0	16/16	13/16
12.5/25	7/9	9/9
25/50	12/12	12/12
50/100	8/11	9/11
100/200	6/11[b]	8/11
200/400	1/11[b]	5/11
0/400	11/11	11/11
United States rats		
0/0	7/7	7/7
37.5/75	1/6[b]	6/6

Source: From ref. 107, with permission.
[a] incidence = number of hearts demonstrating an arrhythmia/total number of hearts in group.
[b] $p < .01$ (Fisher's exact test) when compared with control.

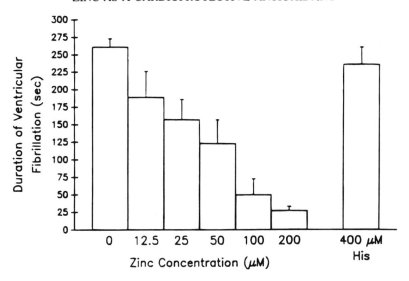

Figure 5. Effect of zinc on the duration of ventricular fibrillation during reperfusion of ischemic rat hearts. Isolated rat hearts were perfused with Zn-His$_2$ at the indicated concentrations or with histidine alone. During reperfusion, the duration of VF was significantly decreased by Zn-His$_2$ ($p < .01$, ANOVA) and was dose dependent. Each bar graph represents the mean ($\pm$SEM) of values obtained from 9–16 isolated hearts. From ref. 107 with permission from Elsevier Science.

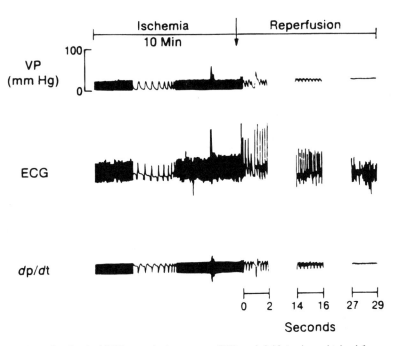

Figure 6. Example of typical ECG, ventricular pressure (VP), and *dp/dt* tracings obtained from a control heart during reperfusion. From ref. 107 with permission from Elsevier Science.

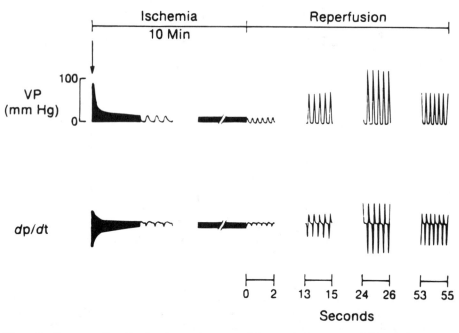

Figure 7. Recovery of cardiac function as demonstrated by ventricular pressure (VP) and dp/dt in the isolated rat heart perfused with 200 μM zinc/400 μM histidine. From ref. 107 with permission from Elsevier Science.

by reperfusion to produce an injury consistent with a severe stun. The myocardium is said to be "stunned" when a state of prolonged ventricular dysfunction exists in the postischemic period (124). Stunned tissue is still viable and may completely recover its function over an extended period of time. Numerous studies by Bolli and co-workers have indicated that stunning is associated with the production of reactive oxygen intermediates (33, 125, 126). In the stunned heart, we observed that addition of Zn-His$_2$ resulted in a concentration-dependent enhancement of postischemic left ventricular pressure development (Figure 8). Electron microscopic analysis of postischemic control hearts revealed grossly abnormal morphology. As shown in Figure 9a, mitochondria appeared to be grossly swollen with evidence of disorganized cristae and loss of inner matrix density (see insert). Furthermore, cardiac morphology was consistent with a heart in contracture as evidenced by an apparent decrease in Z-disc to Z-disc distance, with virtual loss of I-bands. Morphometric analysis of the electron micrographs confirmed these observations (Table 3). Morphology of this type is consistent with that previously observed in stunned myocardium (127). Quite to the contrary, the morphology of postischemic zinc-treated hearts was essentially normal (Figure 9b). There was virtually no evidence of mitochondrial swelling or loss of inner matrix density and little evidence of sarcomere contraction (Figure 9b and Table 3). These studies thus provided an ultrastructural correlate of the cardioprotective effect of zinc. Moreover, we observed that perfusion with zinc decreased LDH release in a dose-dependent manner (Figure 10), suggesting a sparing effect on cellular integrity. These studies clearly demonstrated a cardioprotective effect of zinc in the in vitro global ischemia stun model.

It then occurred to us that one possible clinical use of zinc might be as an additive to cardioplegic solutions. These are crystalloid or blood crystalloid solutions containing high concentrations of potassium to stop the heart, as well as possibly containing myocardial protectants to preserve function of the heart. The goal of this type of treatment is to allow the surgeon to perform a procedure on the heart in a nonmoving, bloodless field, thus facilitating the operative procedure (128). There was some question as to whether sufficient amounts of the metal would be available because of the low temperatures at which cardioplegic solutions are generally administered.

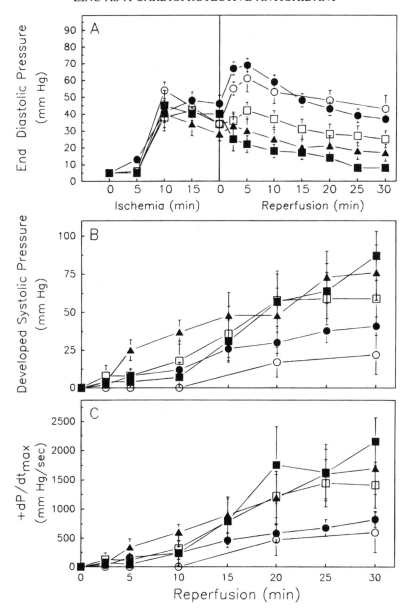

Figure 8. Effect of varying the concentration of zinc on postischemic functional recovery of the isolated rat heart. Hearts were equilibrated for 10 min with Krebs-Henseleit (KH) buffer and then perfused with KH containing 10, 20, or 30 μM zinc bis-histidinate (Zn-His$_2$) for 30 min. Control hearts ($\bullet$; 0 μM Zn-His$_2$) were perfused with Krebs-Henseleit buffer only throughout. All hearts were then subjected to 20 min of no-flow normothermic global ischemia followed by 30 min of reperfusion with identical buffer. $\square$, 10 μM Zn-His$_2$; $\blacktriangle$, 20 μM Zn-His$_2$; $\blacksquare$, 30 μM Zn-His$_2$; $\bigcirc$, 100 μM histidine. A, End diastolic pressure; B, systolic pressure developed; C, contractility or maximum rate of rise in pressure (+dP/dt$_{max}$). Values depicted represent the mean $\pm$SE of 8–14 observations. From ref. 123 with permission from the American Physiological Society.

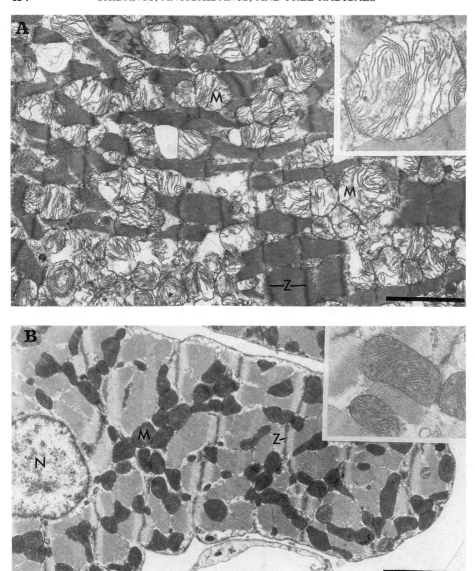

Figure 9. Representative postischemic ultramicroscopic changes in control (A) and Zn-treated (B) hearts. Hearts were equilibrated for 10 min and then perfused with Krebs-Henseleit (KH) buffer containing 20 μM Zn-His$_2$ for 40 min. Hearts were then subjected to 20 min of normothermic global ischemia followed by 15 min of reperfusion with like buffer. Control hearts were perfused with KH buffer only. At the end of the reperfusion period, hearts were perfusion-fixed and then prepared for electron microscopic study. A subset of three hearts were examined for each group. M, mitochondria; N, nucleus; Z, Z-disc. Bars, 3 μm. Magnification, $\times 7800$ (insets, $\times 22,000$). From ref. 123 with permission from the American Physiological Society.

Table 3. Effect of zinc on postischemic cardiac mitochondrial volume and sarcomere length

	Mitochondrial volume (μm^3/mitochondria)	Sarcomere length (μm)
Control	1.980 ± 0.361	1.65 ± 0.21
Zinc (20 μM)	0.705 ± 0.204^a	2.34 ± 0.40

Note. Values are means $\pm$ SE of multiple determinations in several samples (n = 3 hearts).
Source: From ref. 123, with permission.
$^a p < .05$ vs. control (Student's *t*-test).

Cellular uptake of zinc is known to be a facilitated process that has been demonstrated to be significantly impaired at hypothermic temperatures (129). Therefore, prior to initiating any preclinical or clinical studies, we felt it important to determine if zinc retained its efficacy in an in vitro model of hypothermic cardioplegic arrest (130). To demonstrate this, isolated rat hearts were perfused with standard Krebs-Henseleit buffer for 10 min at 37°C. After this time, the hearts were arrested by perfusion for either 5 min (Protocol 1) or 10 min (Protocol 2) with St. Thomas number 2 crystalloid cardioplegic solution, at 10°C, that had been supplemented with either 30 μM zinc (Protocol 1) or 40 μM zinc (Protocol 2). Hypothermic cardioplegic arrest was continued for 2 h, during which time the heart was maintained at 10°C. In Protocol 1 hearts, to simulate intermittent hypothermic arrest, cardioplegic solution ($\pm$30 μM Zn-His$_2$) was re-infused every 15 min for 5 min. In Protocol 2, hearts were arrested and then received no more cardioplegic solution until 5 min before rewarming and reperfusion. Reinfusion was with cardioplegic solution, not containing zinc. Rewarming and reperfusion was initiated by perfusion with standard Krebs-Henseleit buffer at 37°C. As illustrated in Figure 11, recovery of postarrest left ventricular systolic pressure development was significantly greater in zinc-treated hearts. Similar but less significant effects were observed on myocardial contractility ($+dP/dt_{max}$) (Figure 12) and isovolumetric rate of relaxation ($-dP/dt_{max}$) (Figure 13). These studies thus supported our original hypothesis and provided the rationale for studies in an in vivo swine model that are ongoing in our laboratory.

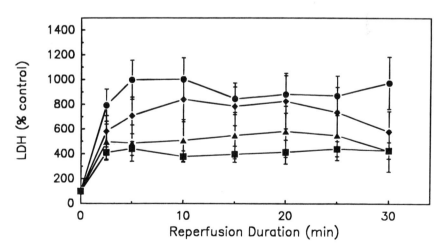

Figure 10. Effect of varying concentrations of zinc on postischemic lactate dehydrogenase (LDH) release (see Figure 7 for protocol). During reperfusion, LDH activity was determined from effluent from the pulmonary artery. Values are presented as the percent change from control (T_0), which is defined as the last preischemic determination. T_0 values: control (●), 1.7 ± 0.4; 10 μM Zn-His$_2$ (♦), 1.6 ± 0.2; 20 μM Zn-His$_2$ (▲), 5.0 ± 1.9; 30 μM Zn-His$_2$ (■), 3.0 ± 0.4 Racker units·min^{-1}·g tissue dry weight^{-1} × 10^3. Values are mean $\pm$SE of 7–12 observations. From ref. 123 with permission from the American Physiological Society.

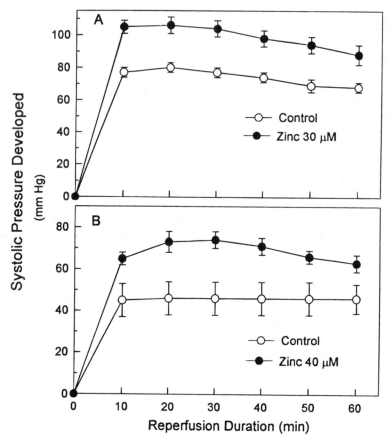

Figure 11. Effect of zinc-supplemented cardioplegic solution on postischemic left ventricular systolic pressure development. Isolated hearts were equilibrated with Krebs-Henseleit buffer at 37°C for 10 min and then subjected to 2 h of hypothermic (10°C) global ischemia. This was followed by 60 min of reperfusion with Krebs-Henseleit buffer at 37°C. (A) Intermittent hypothermic cardioplegia. Cardiac arrest was initiated with 5 min of perfusion with cardioplegic solution ±30 μM zinc, at 10°C, which was then reinfused for 5 min every 15 min during arrest (control, $n = 12$; zinc-treated, $n = 8$). (B) "No-flow" hypothermic arrest. Cardiac arrest was initiated with 10 min of perfusion with cardioplegic solution ±40 μM zinc, at 10°C. Just prior to reperfusion, all hearts were perfused with hypothermic cardioplegic solution (control, $n = 9$; zinc-treated, $n = 7$). From ref. 130 with permission from Mosby-Year Book, Inc.

Evidence of Inhibition of Transition Metal-Mediated Oxidative Stress

To determine if postischemic formation of reactive oxygen intermediates was affected by treatment with zinc, 100 μM salicylate as a chemical trap for ˙OH was included in the perfusate (recently reviewed in ref. 43). As illustrated in Figure 14, improvement in postischemic function was simultaneous with decreases in ˙OH formation as detected by dihydrobenzoate formation from salicylate. One posssible explanation for this observation deals with the use of the bis-histidinate complex of zinc. With respect to copper, this complex is rather unique, because the formation constant of Zn-His$_2$ is lower than that of the histidinate complex of copper (12.88 vs. 18.33, respectively) (131). This makes it possible for a ligand exchange reaction to occur at the tissue-specific site in which zinc exchanges with loosely bound copper, which then complexes with histidine, essentially a "push-pull" phenomena. The displaced copper can then be washed out of the cell, thus reducing the availability of the metal for participation in ˙OH formation.

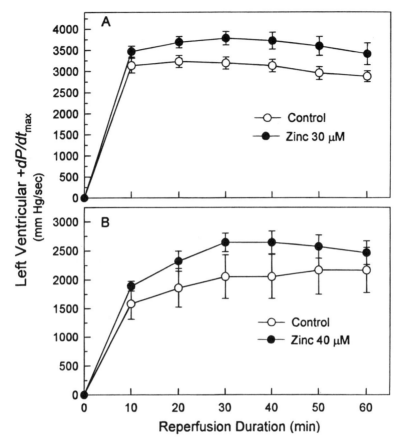

Figure 12. Effect of zinc-supplemented cardioplegic solution on postischemic left ventricular contractility ($+dP/dt_{max}$). Experimental conditions were as in Figure 11. From ref. 130 with permission from Mosby-Year Book, Inc.

That this process does indeed occur is illustrated in Figure 15, as perfusion with buffer containing Zn-His$_2$ resulted in decreased copper content. Although this observation suggests that zinc displaces endogenous copper, which is then washed out of the cell, it is still not clear if this due to a direct antagonistic effect. Nonetheless, the observations that improvements in postischemic function were associated with decreases in ˙OH and alterations in cardiac copper content provide a reasonable basis for the theory that zinc is cardioprotective as a result of inhibition of transition metal-mediated oxidative stress.

SUMMARY AND CONCLUSIONS

In this chapter, evidence supporting the role of redox-active transition metals as mediators of postischemic oxidative injury was presented. Many studies have been performed that collectively demonstrate how the redox-active metals iron and copper may become available during ischemia and participate in destructive processes during the postischemic period. Other evidence has been presented that demonstrates the ability of the non-redox-active transition metal zinc to act as an antagonist of site-specific processes catalyzed by copper and iron. Lastly, evidence from our laboratory indicates that zinc may be useful to decrease postischemic injury and thus act as a

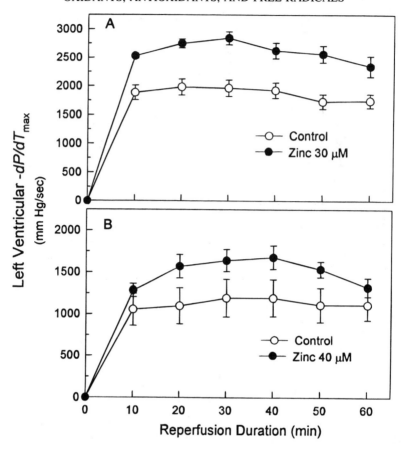

Figure 13. Effect of zinc-supplemented cardioplegic solution on postischemic left ventricular rate of relaxation ($-dP/dt_{max}$). Experimental conditions were as in Figure 11. From ref. 130 with permission from Mosby-Year Book, Inc.

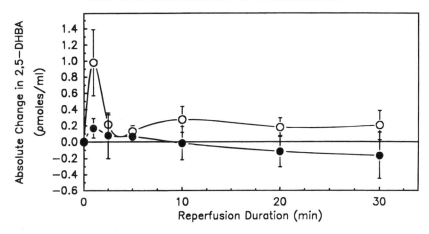

Figure 14. Effect of Zn on postischemic ˙OH formation. Hearts were equilibrated for 10 min with Krebs-Henseleit (KH) buffer and then perfused with KH buffer + 20 μM Zn-His$_2$ for 30 min (●). At this time 100 μM salicylate was added to the buffer containing Zn-His$_2$ and the hearts were perfused for an additional 10 min. Control hearts were perfused with KH buffer for 40 min before addition of 100 μM salicylate and further perfusion for 10 min. All hearts were then subjected to 20 min of "no flow" normothermic global ischemia followed by 30 min of reperfusion with identical buffer containing salicylate (○). During reperfusion the content of 2,5-dihydroxybenzoic acid (2,5-DHBA) was determined in the effluent. Data are presented as the absolute change in effluent 2,5-DHBA, which is the experimental value minus the value determined for the last preischemic sample. Values are the mean ±SE of 11 or 12 determinations. From ref. 123 with permission from the American Physiological Society.

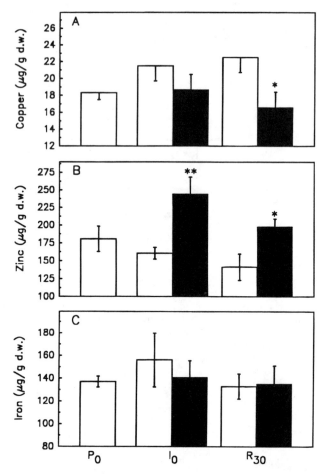

Figure 15. Effect of Zn on cardiac metal content. Hearts were equilibrated for 10 min with Krebs-Henseleit (KH) buffer and then perfused with KH plus 30 μM Zn-His$_2$ for 30 min (solid bars). Then, hearts were subjected to 20 min of "no flow" normothermic global ischemia followed by 30 min of reperfusion with identical buffer. Control hearts were perfused with KH buffer only (open bars). At indicated time points, myocardial copper (A), Zn (B), and iron (C) were determined. Time points: P$_0$, initial preischemic value at the end of the equilibration period; I$_0$, initial ischemic value at the end of the preischemic period; R$_{30}$, value at end of reperfusion. Values are mean $\pm$SE of 8–10 determinations (in μg/g dry weight). Double asterisk indicates significance at $p < .02$, and single asterisk at $p < .05$, compared with corresponding control (Student's t-test). From ref. 123 with permission from the American Physiological Society.

cardioprotective agent and possibly have utility in the clinical setting under the appropriate conditions.

REFERENCES

1. Hess, M. L., Manson, N. H., and Okabe, E. 1982. Involvement of free radicals in the pathophysiology of ischemic heart disease. *Can. J. Physiol. Pharmacol.* 60:1382–1389.
2. Rao, P. S., Cohen, M. V., and Mueller, H. S. 1983. Production of free radicals and lipid peroxides in early experimental ischemia. *J. Mol. Cell Cardiol.* 15:713–716.
3. Romson, J. L., Hook, B. G., Kunkel, S. L., Abrams, G. D., Schork, A., and Lucchesi, B. R. 1983. Reduction of the extent of ischemic myocardial injury by neutrophil depletion in the dog. *Circulation* 67:1016–1023.
4. Bernier, M., Hearse, D. J., and Manning, A. S. 1986. Reperfusion-induced arrhythmias and oxygen-derived free radicals. Studies with "anti-free radical" interventions and a free radical-generating system in the isolated perfused rat heart. *Circ. Res.* 58:331–340.
5. Jackson, C. V., Mickelson, J. K., Pope, T. K., Rao, P. S., and Lucchesi, B. R. 1986. O_2 free radical-mediated myocardial and vascular dysfunction. *Am. J. Physiol.* 251:H1225–H1231.
6. Przyklenk, K., and Kloner, R. A. 1986. Superoxide dismutase plus catalase improve contractile function in the canine model of the "stunned myocardium." *Circ. Res.* 58:148–156.
7. Zweier, J. L., Flaherty, J. T., and Weisfeldt, M. L. 1987. Direct measurement of free radical generation following reperfusion of ischemic myocardium. *Proc. Natl. Acad. Sci. U.S.A.* 84:1404–1407.
8. Koyama, I., Bulkley, G. B., Williams, G. M., and Im, M. J. 1985. The role of oxygen free radicals in mediating the reperfusion injury of cold-preserved ischemic kidneys. *Transplantation* 40:590–595.
9. Ratych, R. E., Bulkley, G. B., and Williams, G. M. 1986. Ischemia/reperfusion injury in the kidney. *Prog. Clin. Biol. Res.* 224:263–289.
10. Faedda, R., Satta, A., Branca, G. F., Turrini, F., Contu, B., and Bartoli, E. 1987. Superoxide radicals (SR) in the pathophysiology of ischemic acute renal failure (ARF). *Adv. Exp. Med. Biol.* 212:69–74.
11. Hansson, R., Johansson, S., Jonsson, O., Pettersson, S., Schersten, T., and Waldenstrom, J. 1986. Kidney protection by pretreatment with free radical scavengers and allopurinol: Renal function at recirculation after warm ischaemia in rabbits. *Clin. Sci.* 71:245–251.
12. Kogure, K., Arai, H., Abe, K., and Nakano, M. 1985. Free radical damage of the brain following ischemia. *Prog. Brain Res.* 63:237–259.
13. Itoh, T., Kawakami, M., Yamauchi, Y., Shimizu, S., and Nakamura, M. 1986. Effect of allopurinol on ischemia and reperfusion-induced cerebral injury in spontaneously hypertensive rats. *Stroke* 17:1284–1287.
14. Atalla, S. L., Toledo-Pereyra, L. H., MacKenzie, G. H., and Cederna, J. P. 1985. Influence of oxygen-derived free radical scavengers on ischemic livers. *Transplantation* 40:584–590.
15. Adkison, D., Hollwarth, M. E., Benoit, J. N., Parks, D. A., McCord, J. M., and Granger, D. N. 1986. Role of free radicals in ischemia-reperfusion injury to the liver. *Acta Physiol. Scand. (Suppl.)* 548:101–107.
16. McEnroe, C. S., Pearce, F. J., Ricotta, J. J., and Drucker, W. R. 1986. Failure of oxygen-free radical scavengers to improve postischemic liver function. *J. Trauma* 26:892–896.
17. Bounous, G. 1986. Pancreatic proteases and oxygen-derived free radicals in acute ischemic enteropathy. *Surgery* 99:92–94.
18. Sanfey, H., Sarr, M. G., Bulkley, G. B., and Cameron, J. L. 1986. Oxygen-derived free radicals and acute pancreatitis: A review. *Acta Physiol. Scand. (Suppl.)* 548:109–118.
19. Manson, P. N., Narayan, K. K., Im, M. J., Bulkley, G. B., and Hoopes, J. E. 1986. Improved survival in free skin flap transfers in rats. *Surgery* 99:211–215.
20. Angel, M. F., Narayanan, K., Swartz, W. M., Ramasastry, S. S., Kuhns, D. B., Basford, R. E., and Futrell, J. W. 1986. Deferoxamine increases skin flap survival: Additional evidence of free radical involvement in ischaemic flap surgery. *Br. J. Plast. Surg.* 39:469–472.
21. Schoenberg, M. H., Fredholm, B. B., Haglund, U., Jung, H., Sellin, D., Younes, M., and Schildberg, F. W. 1985. Studies on the oxygen radical mechanism involved in the small intestinal reperfusion damage. *Acta Physiol. Scand.* 124:581–589.
22. Granger, D. N., Rutili, G., and McCord, J. M. 1981. Superoxide radicals in feline intestinal ischemia. *Gastroenterology* 81:22–29.
23. Otamiri, T. 1989. Oxygen radicals, lipid peroxidation, and neutrophil infiltration after small-intestinal ischemia and reperfusion. *Surgery* 105:593–597.
24. Parks, D. A., Bulkley, G. B., Granger, D. N., Hamilton, S. R., and McCord, J. M. 1982. Ischemic injury in the cat small intestine: Role of superoxide radicals. *Gastroenterology* 82:9–15.
25. Gardner, T. J., Stewart, J. R., Casale, A. S., Downey, J. M., and Chambers, D. E. 1983. Reduction of myocardial ischemic injury with oxygen derived free radical scavengers. *Surgery* 94:423–427.

26. Gauduel, Y., and Duvelleroy, M. A. 1984. Role of oxygen radicals in cardiac injury due to reoxygenation. *J. Mol. Cell Cardiol.* 16:459–470.
27. Walker, P. M., Lindsay, T. F., Labbe, R., Mickle, D. A., and Romaschin, A. 1987. Salvage of skeletal muscle with free radical scavengers. *J. Vasc. Surg.* 5:68–75.
28. Im, M. J., Manson, P. N., Bulkley, G. B., and Hoopes, J. E. 1985. Effects of superoxide dismutase and allopurinol on the survival of acute island skin flaps. *Ann. Surg.* 201:357–359.
29. Arroyo, C. M., Kramer, J. H., Dickens, B. F., and Weglicki, W. B. 1987. Identification of free radicals in myocardial ischemia/reperfusion by spin trapping with nitrone DMPO. *FEBS Lett.* 221:101–104.
30. Kramer, J. H., Arroyo, C. M., Dickens, B. F., and Weglicki, W. B. 1987. Spin-trapping evidence that graded myocardial ischemia alters postischemic superoxide production. *Free Radic. Biol. Med.* 3:153–159.
31. Zweier, J. L., Kuppusamy, P., and Lutty, G. A. 1988. Measurement of endothelial cell free radical generation: evidence for a central mechanism of free radical injury in postischemic tissues. *Proc. Natl. Acad. Sci. U.S.A.* 85:4046–4050.
32. Garlick, P. B., Davies, M. J., Hearse, D. J., and Slater, T. F. 1987. Direct detection of free radicals in the reperfused rat heart using electron spin resonance spectroscopy. *Circ. Res.* 61:757–760.
33. Bolli, R., Patel, B., Jeroudi, M. O., Lai, E. K., and McCay, P. B. 1988. Demonstration of free radical generation in "stunned" myocardium of intact dogs with the use of the spin trap α-phenyl-*N*-tert-butyl nitrone. *J. Clin. Invest.* 82:476–485.
34. Bolli, R., Jeroudi, M. O., Patel, B. S., DuBose, C. M., Lai, E. K., Roberts, R., and McCay, P. B. 1989. Direct evidence that oxygen-derived free radicals contribute to postischemic myocardial dysfunction in the intact dog. *Proc. Natl. Acad. Sci. U.S.A.* 86:4695–4699.
35. Powell, S. R., and Hall, D. 1990. Use of salicylate as a probe for ˙OH formation in isolated ischemic rat hearts. *Free Radic. Biol. Med.* 9:133–141.
36. Das, D. K., Cordis, G. A., Rao, P. S., Liu, X., and Maity, S. 1991. High-performance liquid chromatographic detection of hydroxylated benzoic acids as an indirect measure of hydroxyl radical in heart: Its possible link with the myocardial reperfusion injury. *J. Chromatogr.* 536:273–282.
37. Liu, X., Prasad, M. R., Engelman, R. M., Jones, R. M., and Das, D. K. 1990. Role of iron on membrane phospholipid breakdown in ischemic-reperfused rat heart. *Am. J. Physiol. Heart Circ. Physiol.* 259:H1101–H1107.
38. Onodera, T., and Ashraf, M. 1991. Detection of hydroxyl radicals in the postischemic reperfused heart using salicylate as a trapping agent. *J. Mol. Cell Cardiol.* 23:365–370.
39. Bulkley, G. B. 1983. The role of oxygen free radicals in human disease processes. *Surgery* 94:407–411.
40. Parks, D. A., Bulkley, G. B., and Granger, D. N. 1983. Role of oxygen free radicals in shock, ischemia, and organ preservation. *Surgery* 94:428–432.
41. Sampson, P. J., and Lucchesi, B. R. 1987. Free radicals and myocardial ischemia and reperfusion injury. *J. Lab. Clin. Med.* 110:13–30.
42. Lucchesi, B. R., and Mullane, K. M. 1986. Leukocytes and ischemia-induced myocardial injury. *Annu. Rev. Pharmacol. Toxicol.* 26:201–24.
43. Powell, S.R. 1994. Salicylate trapping of ˙OH as a tool for studying postischemic oxidative injury in the isolated rat heart. *Free Radic. Res. Commun.* 21:355–370.
44. Powell, S. R., and Tortolani, A. J. 1992. Recent advances in the role of reactive oxygen intermediates in ischemic injury. *J. Surg. Res.* 53:417–429.
45. Ambrosio, G., Zweier, J. L., Jacobus, W. E., Weisfeldt, M. L., and Flaherty, J. T. 1987. Improvement of postischemic myocardial function and metabolism induced by administration of deferoxamine at the time of reflow: The role of iron in the pathogenesis of reperfusion injury. *Circulation* 76:906–915.
46. Fuller, B. J., Lunec, J., Healing, G., Simpkin, S., and Green, C. J. 1987. Reduction of susceptibility to lipid peroxidation by desferrioxamine in rabbit kidneys subjected to 24 hour cold ischemia and reperfusion. *Transplantation* 43:604–606.
47. Holt, S., Gunderson, M., Joyce, K., Nayini, N. R., Eyster, G. F., Garitano, A. M., Zonia, C., Krause, G. S., Aust, S. D., and White, B. C. 1986. Myocardial tissue iron delocalization and evidence for lipid peroxidation after two hours of ischemia. 15:1155–1159.
48. Angel, M. F., Narayanan, K., Swartz, W. M., Ramasastry, S. S., Basford, R. E., Kuhns, D. B., and Futrell, J. W. 1986. The etiologic role of free radicals in hematoma-induced flap necrosis. *Plast. Reconstr. Surg.* 77:795–803.
49. Torreilles, J., and Guérin, M.-C. 1990. Nickel(II) as a temporary catalyst for hydroxyl radical generation. *FEBS Lett.* 272:58–60.
50. Moorhouse, C. P., Halliwell, B., Grootveld, M., and Gutteridge, J. M. 1985. Cobalt(II) ion as a promoter of hydroxyl radical and possible "crypto-hydroxyl" radical formation under physiological conditions. Differential effects of hydroxyl radical scavengers. *Biochim. Biophys. Acta* 843:261–268.
51. Haber, F., and Weiss, J. 1934. The catalytic decomposition of H_2O_2 by iron salts. *Proc. R. Soc. Lond. Ser. A* 147:332–351.

52. Czapski, G., Aronovitch, J., Godinger, D., Samuni, A., and Chevion, M. 1984. On the mechanisms of cytotoxicity induced by superoxide. In *Oxygen Radicals in Chemistry and Biology*, eds. W. Bors, M. Saian, and D. Tait, pp. 225–228. Berlin: Walter de Gruyter.
53. Chvapil, M., Ryan, J. N., Elias, S. L., and Peng, Y. M. 1973. Protective effect of zinc on carbon tetrachloride-induced liver injury in rats. *Exp. Mol. Pathol.* 19:186–196.
54. Kohen, R., and Chevion, M. 1985. Paraquat toxicity is mediated by transition metals. In *Biochemical and Inorganic Aspects of Copper Coordination Chemistry*, eds. K. D. Karlin and J. Zubeita, pp. 159–172. New York: Adenine Press.
55. Kalyanaraman, B., Felix, C. C., and Sealy, R. C. 1985. Semiquinone anion radicals of catechol(amine)s, catechol estrogens, and their metal ion complexes. *Environ. Health Perspect.* 64:185–198.
56. Aust, S. D., Morehouse, L. A., and Thomas, C. E. 1985. Role of metals in oxygen radical reactions. *J. Free Radic. Biol. Med.* 1:3–25.
57. Chevion, M. 1988. A site-specific mechanism for free radical induced biological damage: The essential role of redox-active transition metals. *Free Radic. Biol. Med.* 5:27–37.
58. Jewett, S. L., Eddy, L. J., and Hochstein, P. 1989. Is the autoxidation of catecholamines involved in ischemia-reperfusion injury. *Free Radic. Biol. Med.* 6:185–188.
59. Halliwell, B., and Gutteridge, J. M. C. 1990. Role of free radicals and catalytic metal ions in human disease: An overview. *Methods Enzymol.* Vol:1–85.
60. Kappus, H. 1985. Lipid peroxidation: Mechanisms, analysis, enzymology and biological relevance. In *Oxidative Stress*, ed. H. Sies, pp. 273–310. New York: Academic Press.
61. Rivett, A. J. 1985. Preferential degradation of the oxidatively modified form of glutamine synthetase by intracellular mammalian proteases. *J. Biol. Chem.* 260:300–305.
62. Fucci, L., Oliver, C. N., Coon, M. J., and Stadtman, E. R. 1983. Inactivation of key metabolic enzymes by mixed-function oxidation reactions: Possible implication in protein turnover and aging. *Proc. Natl. Acad. Sci. U.S.A.* 80:1521–1525.
63. Starke-Reed, P. E., and Oliver, C. N. 1989. Protein oxidation and proteolysis during aging and oxidative stress. *Arch. Biochem. Biophys.* 275:559–567.
64. Oliver, C. N., Starke-Reed, P. E., Stadtman, E. R., Liu, G. J., Carney, J. M., and Floyd, R. A. 1990. Oxidative damage to brain proteins, loss of glutamine synthetase activity, and production of free radicals during ischemia/reperfusion-induced injury to gerbil brain. *Proc. Natl. Acad. Sci. U.S.A.* 87:5144–5147.
65. Stadtman, E. R. 1990. Metal-ion catalyzed oxidation of proteins: Biochemical mechanism and consequences. *Free Radic. Biol. Med.* 9:315–325.
66. Niroomand, F., Rössle, R., Mülsch, A., and Böhme, E. 1989. Under anaerobic conditions, soluble guanylate cyclase is specifically stimulated by glutathione. *Biochem. Biophys. Res. Commun.* 161:75–80.
67. Davies, K. J. A. 1988. Proteolytic systems as secondary antioxidant defenses. In *Cellular Antioxidant Defense Mechanisms*, ed. C. K. Chow, Vol. II, pp. 25–67. Boca Raton, FL: CRC Press.
68. Link, G., Pinson, A., and Hershko, C. 1985. Heart cells in culture: A model of myocardial iron overload and chelation. *J. Lab. Clin. Med.* 106:147–153.
69. Link, G., Athias, P., Grynberg, A., Pinson, A., and Hershko, C. 1989. Effect of iron loading on transmembrane potential, contraction, and automaticity of rat ventricular muscle cells in culture. *J. Lab. Clin. Med.* 113:103–111.
70. Link, G., Pinson, A., and Hershko, C. 1993. Iron loading of cultured myocytes modifies sarcolemmal structure and increases lysosomal fragility. *J. Lab. Clin. Med.* 121:127–134.
71. Link, G., Athias, P., Grynberg, A., Hershko, C., and Pinson, A. 1991. Iron loading modifies β-adrenergic responsiveness in cultured ventricular myocytes. *Cardioscience* 2:185–188.
72. Link, G., Pinson, A., and Hershko, C. 1993. Effect of iron loading on sarcolemmal membrane structure and lysosomal fragility. *J. Lab. Clin. Med.* 121:127–134.
73. Link, G., Pinson, A., and Hershko, C. 1994. The ability of orally effective iron chelators dimethyl- and dimethyl-hydroxypyridine-4-one and of deferoxamine to restore sarcolemmal thiolic enzyme activity in iron loaded cells. *Blood* 83:2697.
74. Hershko, C., Link, G., and Pinson, A. 1987. Modification of iron uptake and lipid peroxidation by hypoxia, ascorbic acid, and α-tocopherol in iron-loaded rat myocardial cell cultures. *J. Lab. Clin. Med.* 110:355–361.
75. Link, G., Pinson, A., Kahane, I., and Hershko, C. 1989. Iron loading modifies the fatty acid composition of cultured rat myocardial cells and liposomal vesicles: Effect of ascorbate and α-tocopherol on myocardial lipid peroxidation. *J. Lab. Clin. Med.* 114:243–249.
76. Sohal, R. S., and Brunk, U. T. 1990. Lipofuscin as an indicator of oxidative stress and aging. *Adv. Exp. Med. Biol.* 266:17–30.
77. Van Jaarsveld, H., Potgieter, G. M., Barnard, S. P., and Potgieter, S. 1990. Improvement of ischemic and postischemic mitochondrial function by deferrioxamine: The role of iron. *Adv. Exp. Med. Biol.* 264:361–366.

78. Voogd, A., Sluiter, W., van Eijk, H. G., and Koster, J. F. 1992. Low molecular weight iron and the oxygen paradox in isolated rat hearts. *J. Clin. Invest.* 90:2050–2055.

79. Chevion, M., Jiang, Y., Har-el, R., Berenshtein, E., Uretzky, G., and Kitrossky, N. 1993. Copper and iron are mobilized following myocardial ischemia: Possible predictive criteria for tissue injury. *Proc. Natl. Acad. Sci. U.S.A.* 90:1102–1106.

80. Van der Kraaij, A. M. M., Mostert, L. J., van Eijk, H. G., and Koster, J. F. 1988. Iron load increases the susceptibility of rat hearts toward reperfusion damage. Protection by the anti oxidant (+)-cyanidanol-3 and deferoxamine. *Circulation* 80:442–449.

81. Shinar, E., Rachmilewitz, E. A., Shifter, A., Rahamim, E., and Saltman, P. 1989. Oxidative damage to human red blood cells induced by copper and iron complexes in the presence of ascorbate. *Biochim. Biophys. Acta* 1014:66–72.

82. Korbashi, P., Kohen, R., Katzhandler, J., and Chevion, M. 1985. Iron mediates paraquat toxicity in *Escherichia coli. J. Biol. Chem.* 261:12472–12476.

83. Halliwell, B., and Gutteridge, J. M. C. 1989. *Free Radicals in Biology and Medicine.* Oxford: Clarendon Press.

84. Powell, S. R., Hall, D., and Shih, A. 1991. Copper loading of hearts increases postischemic reperfusion injury. *Circ. Res.* 69:881–885.

85. Powell, S. R., Hyacinthe, L., Teichberg, S., and Tortolani, A. J. 1992. Mediatory role of copper in reactive oxygen intermediate-induced cardiac injury. *J. Mol. Cell Cardiol.* 24:1371–1386.

86. Spiro, T. G., Pape, L., and Saltman, P. 1967. The hydrolytic polymerization of ferric citrate. *J. Am. Chem. Soc.* 89:5555–5559.

87. Eguchi, L. A., and Saltman, P. 1984. The aerobic reduction of Fe(III) complexes by hemoglobin and myoglobin. *J. Biol. Chem.* 259:14337–14338.

88. Hegetschweiler, K., Saltman, P., Dalvit, C., and Wright, P. E. 1987. Kinetics and mechanisms of the oxidation of myoglobin by Fe(III) and Cu(II) complexes. *Biochim. Biophys. Acta* 912:384–397.

89. Reid, L. S., Gray, H. B., Dalvit, C., Wright, P. E., and Saltman, P. 1987. Electron transfer from cytochrome b5 to iron and copper complexes. *Biochemistry* 26:7102–7107.

90. Cotton, F. A., and Wilkinson, G. 1972. *Advanced Inorganic Chemistry.* London: John Wiley and Sons.

91. Smith, S. M., Grisham, M. B., Manci, E. A., Granger, D. N., and Kvietys, P. R. 1987. Gastric mucosal injury in the rat. Role of iron and xanthine oxidase. *Gastroenterology* 92:950–956.

92. Smith, J. K., Carden, D. L., Grisham, M. B., Granger, D. N., and Korthuis, R. J. 1989. Role of iron in postischemic microvascular injury. *Am. J. Physiol.* 256:H1472–H1477.

93. Reddy, B. R., Kloner, R. A., and Przyklenk, K. 1989. Early treatment with deferoxamine limits myocardial ischemic/reperfusion injury. *Free Radic. Biol. Med.* 7:45–52.

94. Bolli, R., Patel, B. S., Jeroudi, M. O., Li, X.-Y., Triana, J. F., Lai, E. K., and McCay, P. B. 1990. Iron-mediated radical reactions upon reperfusion contribute to myocardial "stunning." *Am. J. Physiol. Heart Circ. Physiol.* 259:H1901–H1911.

95. Maruyama, M., Pieper, G. M., Kalyanaraman, B., Hallaway, P. E., Hedlund, B. E., and Gross, G. J. 1991. Effects of hydroxyethyl starch conjugated deferoxamine on myocardial functional recovery following coronary occlusion and reperfusion in dogs. *J. Cardiovasc. Pharmacol.* 17:166–175.

96. Maxwell, M. P., Hearse, D. J., and Yellon, D. M. 1989. Inability of desferrioxamine to limit tissue injury in the ischaemic and reperfused rabbit heart. *J. Cardiovasc. Pharmacol.* 13:608–615.

97. Lesnefsky, E. J., Hedlund, B. E., Hallaway, P. E., and Horwitz, L. D. 1990. High-dose iron-chelator therapy during reperfusion with deferoxamine-hydroxyethyl starch conjugate fails to reduce canine infarct size. *J. Cardiovasc. Pharmacol.* 16:523–528.

98. Sinaceur, J., Ribère, C., Nordmann, J., and Nordmann, R. 1984. Desferrioxamine: A scavenger of superoxide radicals. *Biochem. Pharmacol.* 33:1693–1694.

99. Halliwell, B. 1985. Use of desferrioxamine as a "probe" for iron-dependent formation of hydroxyl radicals. Evidence for a direct reaction between desferal and the superoxide radical. *Biochem. Pharmacol.* 34:229–233.

100. Hoe, S., Rowley, D. A., and Halliwell, B. 1982. Reactions of ferrioxamine and desferrioxamine with the hydroxyl radical. *Chem. Biol. Interact.* 41:75–81.

101. Appelbaum, Y. J., Kuvin, J., Borman, J. B., Uretzky, G., and Chevion, M. 1990. The protective role of neocuproine against cardiac damage in isolated perfused rat hearts. *Free Radic. Biol. Med.* 8:133–143.

102. Bray, T. M., and Bettger, W. J. 1990. The physiologic role of zinc as an antioxidant. *Free Radic. Biol. Med.* 8:281–291.

103. Girotti, A. W., Thomas, J. P., and Jordan, J. E. 1985. Inhibitory effect of zinc(II) on free radical lipid peroxidation in erythrocyte membranes. *J. Free Radic. Biol. Med.* 1:395–401.

104. Eguchi, L., and Saltman, P. 1987. Kinetics and mechanisms of metal reduction by hemoglobin I. Reduction of Fe(III) complexes. *Inorg. Chem.* 26:3665–3669.

105. Korbashi, P., Katzhandler, J., Saltman, P., and Chevion, M. 1989. Zinc protects E. coli against copper-mediated paraquat-induced damage. *J. Biol. Chem.* 264:8479–8482.

106. Lovering, K. E., and Dean, R. T. 1992. Restriction of the participation of copper in radical-generating systems by zinc. *Free Radic. Res. Commun.* 14:217–225.
107. Powell, S. R., Saltman, P., Uretzky, G., and Chevion, M. 1990. The effect of zinc on reperfusion arrhythmias in the isolated perfused rat heart. *Free Radic. Biol. Med.* 8:33–46.
108. Yoshikawa, T., Naito, Y., Tanigawa, T., Yoneta, T., Yasuda, M., Ueda, S., Oyamada, H., and Kondo, M. 1992. Effect of zinc-carnosine chelate compound (Z-103), a novel antioxidant, on acute gastric mucosal injury induced by ischemia-reperfusion in rats. *Free Radic. Res. Commun.* 14:289–296.
109. Hegenauer, J., Saltman, P., Fairchild, R., and Halasz, N. A. 1991. Improved function of reperfused rabbit kidney following administration of zinc histidine. *J. Trace Elem. Exp. Med.* 4:103–107.
110. Persoon-Rothert, M., van der Valk-Kokshoorn, E. J. M., Egas-Kenniphaas, J. M., Mauve, I., and van der Laarse, A. 1989. Isoproterenol-induced cytotoxicity in neonatal rat heart cell cultures is mediated by free radical formation. *J. Mol. Cell Cardiol.* 21:1285–1291.
111. Singal, P. K., Kapur, N., Dhillon, K. S., Beamish, R. E., and Dhalia, N. S. 1982. Role of free radicals in catecholamine-induced cardiomyopathy. *Can. J. Physiol. Pharmacol.* 60:1390–1397.
112. Lubbe, W. F., Daries, P. S., and Opie, L. H. 1978. Ventricular arrhythmias associated with coronary artery occlusion and reperfusion in the isolated perfused rat heart: A model for assessment of antifibrillatory action of antiarrhythmic agents. *Cardiovasc. Res.* 12:212–220.
113. Aiuto, L. T., and Powell, S. R. (1995). Characterization of the anti-arrhythmic effect of the trace element, zinc, and its potential relationship to inhibition of oxidative stress. *J. Trace Elem. Exptl. Med.* 8:173–182.
114. Barrington, P. L. 1990. Effects of free radicals on the electrophysiological function of cardiac membranes. *Free Radic. Biol. Med.* 9:355–365.
115. Bernier, M., Kusama, Y., Borgers, M., Ver Donck, L., Valdes-Aguilera, O., Neckers, D. C., and Hearse, D. J. 1991. Pharmacological studies of arrhythmias induced by rose bengal photoactivation. *Free Radic. Biol. Med.* 10:287–296.
116. Cerbai, E., Ambrosio, G., Porciatti, F., Chiariello, M., Giotti, A., and Mugelli, A. 1991. Cellular electrophysiological basis for oxygen radical-induced arrhythmias: A patch-clamp study in guinea pig ventricular myocytes. *Circulation* 84:1773–1782.
117. Lesnefsky, E. J., Williams, G. R., Rubinstein, J. D., Hogue, T. S., Horwitz, L. D., and Reiter, M. J. 1991. Hydrogen peroxide decreases effective refractory period in the isolated heart. *Free Radic. Biol. Med.* 11:529–535.
118. Woodward, B., and Zakaria, M. N. M. 1985. Effects of some free radical scavengers on reperfusion induced arrhythmias in the isolated rat heart. *J. Mol. Cell Cardiol.* 17:485–493.
119. Nishinaka, Y., Kitahara, S., Sugiyama, S., Yokota, M., Saito, H., and Ozawa, T. 1991. The cardioprotective effect of gamma-glutamylcysteine ethyl ester during coronary reperfusion in canine hearts. *Br. J. Pharmacol.* 104:805–810.
120. Nejima, J., Knight, D. R., Fallon, J. T., Uemura, N., Manders, T., Canfield, D. R., Cohen, M. V., and Vatner, S. F. 1989. Superoxide dismutase reduces reperfusion arrhythmias but fails to salvage regional function or myocardium at risk in conscious dogs. *Circulation* 79:143–153.
121. Bernier, M., Manning, A. S., and Hearse, D. J. 1989. Reperfusion arrhythmias: Dose-related protection by anti-free radical interventions. *Am. J. Physiol.* 256:H1344–H1352.
122. Hatori, N., Miyazaki, A., Tadokoro, H., Rydén, L., Moll, J., Rajagopalan, R. E., Fishbein, M. C., Meerbaum, S., Corday, E., and Drury, J. K. 1989. Beneficial effects of coronary venous retroinfusion of superoxide dismutase and catalase on reperfusion arrhythmias, myocardial function, and infarct size in dogs. *J. Cardiovasc. Pharmacol.* 14:396–404.
123. Powell, S. R., Hall, D., Aiuto, L., Wapnir, R. A., Teichberg, S., and Tortolani, A. J. 1994. Zinc improves postischemic recovery of the isolated rat heart through inhibition of oxidative stress. *Am. J. Physiol. Heart Circ. Physiol.* 266:H2497–H2507.
124. Braunwald, E., and Kloner, R. A. 1982. The stunned myocardium: Prolonged, postischemic ventricular dysfunction. *Circulation* 66:1146–1149.
125. Li, X.-Y., McCay, P. B., Zughaib, M., Jeroudi, M. O., Triana, J. F., and Bolli, R. 1993. Demonstration of free radical generation in the "stunned" myocardium in the conscious dog and identification of major differences between conscious and open-chest dogs. *J. Clin. Invest.* 92:1025–1041.
126. Sekili, S., McCay, P. B., Li, X.-Y., Zughaib, M., Sun, J.-Z., Tang, L., Thornby, J. I., and Bolli, R. 1993. Direct evidence that the hydroxyl radical plays a pathogenetic role in myocardial "stunning" in the conscious dog and demonstration that stunning can be markedly attenuated without subsequent adverse effects. *Circ. Res.* 73:705–723.
127. Kloner, R. A., Ellis, S. G., Lange, R., and Braunwald, E. 1983. Studies of experimental artery reperfusion. Effects on infarct size, myocardial function, biochemistry, ultrastructure and microvascular damage. *Circulation* 68(Suppl. I):I-8–I-15.
128. Gravlee, G. P., Davis, R. F., and Utley, J. R. 1993. *Cardiopulmonary Bypass. Principles and Practice.* Philadelphia: Williams & Wilkins.

129. Bobilya, D. J., Briske-Anderson, M., and Reeves, P. G. 1992. Zinc transport into endothelial cells is a facilitated process. *J. Cell. Physiol.* 151:1–7.
130. Powell, S, Aiuto, L., Hall, D., and Tortolani, A. J. 1995. Zinc supplementation enhances the effectiveness of St. Thomas, Hospital No. 2 cardioplegic solution in an *in vitro* model of hypothermic cardiac arrest. *J. Thorac. Cardiovasc. Surg.* 110:1642–1648.
131. Ashmead, H. D., Graff, D. J., and Ashmead, H. H. 1985. *Intestinal Absorption of Metal Ions and Chelates.* Springfield, IL: Charles C. Thomas.

Chapter Eight

ASCORBIC ACID, MELATONIN AND THE ADRENAL GLAND: A COMMENTARY

Madeline M. Hall, Harry Van Keulen, and James M. Willard

Albert Szent-Gyorgyi is credited with the discovery of ascorbic acid. Recent reviews on the biological, chemical, and physiological properties of ascorbic acid suggest it acts mainly as an antioxidant in enzymatic reactions, resulting in the formation of a variety of substances, for example, carnitine, cholic acid, hydroxyproline, hydroxylysine, homogentisic acid, norepinephrine, and serotonin (1–4). Pirani's earlier review of 1952 (5) examines the relationship of ascorbic acid and adrenalcortical function—an area we feel has been overlooked by recent reviews. We would like to comment on selected biochemical and physiological issues, such as glucocorticoid synthesis/storage/release, ascorbic acid release by adrenals, and the effect of melatonin on adrenal function, all of which should be revisited.

Next to the pituitary, the adrenals contain the highest amounts of ascorbic acid (55 mg%) in mammals. It is now well known that activation of the pituitary-adrenal axis by stressful stimuli releases the trophic hormone adrenocorticotropin (ACTH) from the anterior pituitary, which in turn activates the adrenal release of first ascorbic acid and then glucocorticoid, cholesterol, and other lipids. Initially, this axis was monitored by measuring the total ascorbic acid content of the adrenal (5–9). Adrenal stimulation by ACTH brought about adrenal ascorbic acid depletion. The measurement of such depletion served for years as an assay method for ACTH activity. In 1960, Lipscomb and Nelson (10) showed that following ACTH injection, ascorbic acid and corticosterone concentrations could be measured in adrenal venous blood (see Figure 1). Administration of ACTH in the rat model prompted in the adrenal vein a rapid rise of first ascorbic acid and then glucocorticoid with a subsequent decline in ascorbic acid levels. A continuous, sustained rise in corticosterone occurred following the decline in the secretion of ascorbic acid. Since the measurements were in adrenal venous blood, this indicates release, not depletion..

Scorbutic guinea pigs, deficient in ascorbate, can still release glucocorticoids (5). Such evidence suggests that there may be two independent biochemical events. Both ascorbate release and glucocorticoid release occur due to stimulation by ACTH; however, ascorbate may not be a needed precursor for glucocorticoid release as indicated with the data on guinea pigs. Release of ascorbic acid occurs throughout the entire gland, both cortex and medulla. Hypertrophy of the adrenal and ascorbate depletion can occur when the stimulus or stress is too prolonged. In situations of continuous stress, death can ensue (5), indicating a need for ascorbate in adrenal function. Replenishing vitamin C by infusion enhances cortisol production activated by ACTH. Kodama et al. (11) showed the use of vitamin C infusion in the clinical treatment of autoimmune disease and allergy, disease conditions that are states of stress. Intravenous vitamin C enhances endogenous cortisol production and allows clinical control of immune disorders. Kodama et al. (11) showed that vitamin C infusion produces a surge of plasma ACTH and cortisol. The mechanism is unknown. Implied in the experiment of Kodama et al. is that exogenous administra-

167

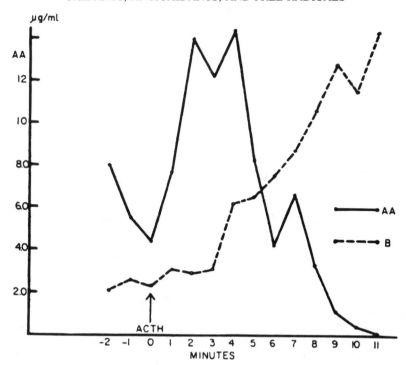

Figure 1.   Secretion of ascorbic acid (AA) and corticosterone (B) in the adrenal vein of a hypophysectomized rat following ACTH administration. From ref. 10 with permission of the authors and *Endrocrinology*.

tion of vitamin C can enhance the release of adrenal steroids, making it a necessary cofactor in steroid secretion.

Today individuals in stressful situations are taking vitamin C as a supplement (12). Since vitamin C rapidly clears the body as a water-soluble vitamin, time-release tablets that are now available can maintain blood levels of vitamin C useful in periods of stressful states. If ascorbic acid is released with glucocorticoids during stress, how fast is it replenished and, further, is there a need to take supplements to replace the ascorbate lost in order to maintain the integrity of adrenal function?

Ascorbic acid depletion served as a useful measure of the pituitary-adrenal axis until the 1960s with the advent of the acid fluorescence technique for measuring adrenal steroids (13). What should be of current interest is the link between ascorbate and the cortex's release of glucocorticoids and the medulla's release of norepinephrine and epinephrine. Is ascorbate functioning in both the synthesis and secretion of these hormones? Is the release of ascorbic acid independent of the release of the glucocorticoids? Insulin release from scorbutic guinea pig pancreatic islets appears to depend on ascorbic acid (14). In this study the proposed mechanism is not the redox cycle in protein disulfide formation. Rather, ascorbic acid, coupled to glucose-induced insulin release, modulates calcium channels in pancreatic beta cells by inactivating the slow-deactivating calcium channels. Is the ascorbate mechanism one of oxidation-reduction or an effect on calcium channels? If it is acting on calcium channels, does this have an effect on glucocorticoid release as it does on insulin release?

There appears to be no evidence regarding ascorbate's role in glucocorticoid synthesis or release. However, in the enzymatic conversion of cholesterol to glucocorticoids, hydroxylation of C-20, C-22, and C-11 depends on the mitochondrial cytochrome P-450. Could these or other hydroxylation reactions involve ascorbate? Could the mitochondrial cytochrome P-450 be similar to the microsomal P-450 of mammalian liver, which is involved in the detoxification of xenobiotics and is a mixed function oxidase? In ascorbate-deficient guinea pigs, the level of microsomal cytochrome P-450 is dramatically decreased. Alternatively, could the hydroxlyation reactions leading to the enzymatic conversion of cholesterol to glucocorticoids be in any way similar to the enzymatic conversion of cholesterol to cholic acid, for which it is known there is an O_2 and ascorbate requirement?

Information on catecholamine release is also lacking. However, the requirement of ascorbate for catecholamine synthesis is well documented. In the metabolic sequence phenylalanine → tyrosine → DOPA → dopamine → norepinephrine → epinephrine, the enzymatic conversions of phenylalanine → tyrosine and that of dopamine → norepinephrine by monooxygenases require O_2, the folic acid derivative, tetrahydrobiopterin, and ascorbate. Tetrahydrobiopterin serves as a proton and electron donor, thus being oxidized to dihydrobiopterin. Ascorbate appears to reduce dihydrobiopterin back to tetrahydrobiopterin. Davies, Austin, and Partridge, in their 1991 review on vitamin C (2), pointed out that these monooxygenase reactions result in the formation of a highly reactive and potentially dangerous oxygen atom. Is the fate of this oxygen known? (Most investigators merely show the formation of water from this oxygen.) The question prevails also for the resultant oxygen atom formed with the hydroxylation of tryptophan to 5-hydroxytrptophan in the biosynthetic pathway leading to serotonin and melatonin formation in the pineal gland.

Although melatonin formation is associated with the pineal gland and its function is more closely associated with circadian rhythms, its ability to act as a potent, ubiquitous hydroxyl and peroxyl radical scavenger has recently been demonstrated (15, 16). The synthesis of melatonin is obviously ascorbate dependent (discussed earlier). The rate-limiting enzyme in melatonin formation, N-acetyltransferase, converting 5-hydroxytryptamine (serotonin) to N-acetyl-5-hydroxytryptamine, is regulated by norepinephrine (itself ascorbate dependent) from the fibers of the sympathetic nervous system innervating the pineal. Melatonin has been shown to affect the hypothalamic-pituitary-adrenal axis in several ways. At the level of the hypothalamus, release of the corticotrophin-releasing hormone is inhibited, possibly through inhibition of cAMP formation (17). At the level of the adrenal, melatonin enhances glucocorticoid release (18). Furthermore, Poeggeler et al. (19) have demonstrated that melatonin acts synergistically with ascorbate to enhance by almost threefold the scavenging activity of melatonin alone. Could it be that melatonin is present in the adrenals and is scavenging the oxygen released during monooxygenase activity?

Melatonin exerts many of its known effects by signaling through G-protein-coupled receptors, all of which thus far are known to inhibit cAMP formation (20). The amine serotonin also has such receptors. Melatonin is well established as an antioxidant and has receptor mechanisms. Concurrently, are the effects of ascorbic acid due to reduction, antioxidant activity, or receptor mediated? Can melatonin substitute for ascorbic acid?

A feasible approach to dissecting this biochemical puzzle is to observe the pharmacodynamic profile of plasma glucocorticoid in the presence of melatonin alone, ascorbic acid alone, and in the presence of both, measuring the dependence and interdependence of various doses of melatonin and ascorbic acid on the secretion of adrenal glucocorticoids.

Many questions arise as to the role of antioxidants and adrenal secretion. Is the mechanism of secretion due to oxidation-reduction or mediated by a calcium channel? What dose of ascorbic acid, if any, should be taken in the stress state? Should plasma levels of ascorbic acid be maintained with time-release vitamin C or multiple dosing?

As yet, the importance of megadoses of vitamin C or doses of time-release vitamin C in the adrenals is still clouded in mystery and requires further research and controlled scientific investigation.

REFERENCES

1. Englard, S., and Seifter, S. 1986. The biochemical functions of ascorbic acid. *Annu. Rev. Nutr.* 6:365–406.
2. Davies, M. B., Austin, J., and Partridge, D. A. 1991. *Vitamin C: Its Chemistry and Biochemistry.* Cambridge: Royal Society of Chemistry.
3. Brown, L. A. S., and Jones, D. P. 1996. The biology of ascorbic acid. In *Handbook of Antioxidants*, eds. E. Cadenas and L. Packer, pp. 117–154. New York: Marcel Dekker.
4. Buettner, G. R., and B. A. Jurkiewicz, B. A. 1996. Chemistry and biochemistry of ascorbic acid. In *Handbook of Antioxidants*, eds. E. Cadenas and L. Packer, pp. 91–115. New York: Marcel Dekker.

5. Pirani, C. L. 1952. Review: Relation of vitamin C to adrenalcortical function and stress phenomena. *Metab. Clin. Exp.* 1:197–222.
6. Munson, P. L., and Toepel, W. 1958. Detection of minute amounts of adrenocorticotropic hormone by the effect of adrenal venous ascorbic acid. *Endocrinology* 63:785–793.
7. Brodish, A., and Long, C. N. H. 1960. Characteristics of the adrenal ascorbic acid response to adrenocorticotrophic hormone (ACTH) in the rat. *Endocrinology* 66:151–159.
8. Schwartz, N. B., and Kling, A. 1960. Stress-induced adrenal ascorbic acid depletion in the cat. *Endocrinology* 66:308–310.
9. Vernikos-Danellis, J. 1964. Estimation of corticotropin-releasing activity of rat hypothalamus and neurohypophysis before and after stress. *Endrocrinology* 75:514–520.
10. Lipscomb, H. S., and Nelson, D. H. 1960. Dynamic changes in ascorbic acid and corticosteroids in adrenal vein after ACTH. *Endocrinology* 66:144–146.
11. Kodama, M., Kodama, T., Murakami, M., and Kodama, M. 1994. Vitamin C infusion treatment enhances cortisol production of the adrenal via the pituitary route. *In Vivo* 8:1079–1085.
12. Vernikos-Danellis, J., and Hall, M. M. 1965. Inhibition of adrenocortical responsiveness to ACTH by Actinomycin D in vivo. *Nature* 207:766–768.
13. Wells, W. W., Dou, C.-Z., Dybas, L. N., Jung, C.-H., Kalbach, H. L., and Xu, D. P. 1995. Ascorbic acid is essential for the release of insulin from scorbutic guinea pig pancreatic islets. *Proc. Natl. Acad. Sci. U.S.A.* 92:11869–11873.
14. Tan, D.-X., Chen, L.-D., Poeggeler, B., Manchester, L. C., and Reiter, R. J. 1993. Melatonin: A potent, endogenous hydroxyl radical scavenger. *Endocr. J.* 1:57–60.
15. Reiter, R. J., Melchiorri, D., Sewerynek, E., Poeggeler, B., Barlow-Walden, L. R., Chuang, J., Ortiz, G. G., and Acuna-Castroviejo, D. 1995. A review of the evidence supporting melatonin's role as an antioxidant. *J. Pineal Res.* 18:1–11.
16. Chamberlain, R. S., and Herman, B. H. 1990. A novel biochemical model linking dysfunctions in brain melatonin, proopiomelanocortin peptides, and serotonin in autism. *Biol. Psychiatry* 28:773–793.
17. Weidenfeld, Y., Schmidt, U., and Nir, I. 1993. The effect of exogenous melatonin on the hypothalamic-pituitary-adrenal axis in intact and pinealectomized rats under basal and stressed conditions. *J. Pineal Res.* 14:60–66.
18. Poeggeler, B., Reiter, R., Hardeland, J. R., Sewerynek, E., Melchiorri, D., and Barlow-Walden, L. R. 1995. *Neuroendocrinol. Lett.* 17:87–92.
19. Reppert, S. M., and Weaver, D. R. 1995. Melatonin madness—Minirevew. *Cell* 83:1059–1062.

Chapter Nine

ANTIOXIDANT PROPERTIES OF GLUTATHIONE AND ITS ROLE IN TISSUE PROTECTION

Alfred M. Sciuto

Over the past 20 years, much has been written about the protective properties of glutathione. The importance of glutathione in biochemical and biomedical research is reflected by the number of citations that include glutathione either in the title or as a keyword. Since 1992, there have been over 3400 citations concerning glutathione, which includes close to 600 citations just for the first 10 months of 1995. It is apparent, therefore, that scientific and medical interest in glutathione and its role in metabolism has exploded in an attempt to link its presence to a protective role against free-radical-mediated injury in cells and tissues.

HISTORICAL BACKGROUND

The first reference to glutathione (GSH) was made by the French scientist J. de Rey-Pailhade in 1888 when he described experiments that involved reactions of "philothion" (1). GSH was first isolated in yeast and rat liver by Sir F. Gowland Hopkins, an English biochemist at Cambridge, in the early 1920s (2). Hopkins initially believed that GSH was comprised of two amino acids, glutamic acid and cysteine; however, work by Hunter and Eagles in 1927 (3), to whom Hopkins later gave appropriate credit, caused him to reconsider the dipeptide status of GSH by reclassifying it as a tripeptide with the inclusion of a third amino acid, glycine (4). In 1929, Pirie and Pinhey, also at Cambridge, presented data tentatively outlining the structure of the tripeptide (5). In 1935, Harrington and Meade presented data on the first method to synthesize GSH, thereby firmly establishing that its chemical structure was γ-glutamylcysteinylglycine (6). The physical properties of GSH are listed in Table 1 (7).

BIOSYNTHESIS OF GSH

Glutathione (GSH) is a nonprotein tripeptide found in several prokaryotes and in most eukaryotic cells in concentrations ranging from 10^{-6} to 10^{-3} M. GSH is part of the naturally occurring antioxidant defense strategy employed by cells and tissues to combat oxidative stress caused by the presence of free radicals. The concentration of GSH, especially in tissues such as the lung, can vary widely among species (8). GSH and other endogenous intracellular/cytosolic antioxidants such as catalase, superoxide dismutase, GSH-reductases/peroxidases, and vitamin E are all part of the cell's natural capacity to protect itself from oxidant damage (Figure 1). GSH is synthesized from three constituent amino acids by the sequential action of two ATP-dependent enzymes:

$$\text{L-glutamate} + \text{L-cysteine} + \text{ATP} \rightarrow \text{L-}\gamma\text{-glutamyl-L-cysteine} + \text{ADP} + \text{P}_i \qquad (1)$$

The opinions or assertions contained herein are the private views of the author and are not to be construed as official or as reflecting the views of the Army or the Department of Defense.

Table 1. Physical properties of glutathione

Structural formulas	$$\overset{\displaystyle O \qquad\quad\ O}{\overset{\displaystyle \|\qquad\quad\ \|}{^+H_3NCHCH_2CH_2CNHCHCNHCH_2COO^-}}$$
	$$\underset{\text{reduced }\gamma\text{-Glu-Cys-Gly (GHS)}}{\overset{\displaystyle |\qquad\qquad\quad |}{COO^-\qquad\qquad CH_2SH}}$$

γ-Glu-Cys-Gly
     |
     S
     |
     S
     |
γ-Glu-Cys-Gly
oxidized (GSSG)

White, crystalline solid	mp 192–195°C
Molecular weight	
GSH	307.3
GSSG	612.6
Solubility	Freely in water, ethanol, liquid ammonia
pK_{a_1} of -COOH	2.12
pK_{a_2} of -SH	9.2
pK_{a_3} of -NH$_2$	8.66
pK_{a_4} of -COOH	3.53

Source: Data from Budavari (7).

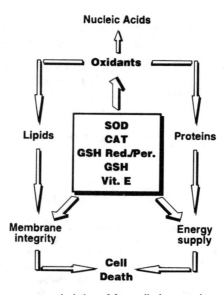

Figure 1. Oxidative injury can cause depletion of free radical scavenging antioxidants and antioxidant enzymes listed within the block. Significant loss of activity of any one of these, and others not included, can lead eventually to compromised cell/tissue metabolic dysfunction and ultimately cell death. However, supplementation with compounds that can maintain or up-regulate antioxidant activity can lead to enhanced defense capabilities.

$$\text{L-}\gamma\text{-glutamyl-L-cysteine} + \text{glycine} + \text{ATP} \rightarrow \text{glutathione} + \text{ADP} + \text{P}_i \qquad (2)$$

Reaction 1 is catalyzed by γ-glutamylcysteine synthetase, whereas reaction 2 is catalyzed by γ-glutathione synthetase and is the rate-limiting step in GSH formation. GSH is protected from proteolytic breakdown from peptidases and proteases by the γ-glutamyl bond and from the γ-glutamyl cyclotransferases by the C-terminal glycine moiety (9).

The primary site for de novo GSH synthesis is the liver, which supplies approximately 90% of the circulating plasma GSH. Hepatic GSH metabolism has been extensively investigated (10, 11). GSH is utilized by most of the major organs such as the lung, kidney, intestines, and brain. The amino acid cysteine is required for the formation of GSH. Since cysteine levels in the plasma fraction of the blood are low and GSH is poorly taken up directly by the cell, cysteine required for Eq. 1 can be furnished by the cleavage of the S–S (disulfide) bond of cystine. Cystine, the oxidized form of cysteine, forms disulfide cross-linkages in proteins. Cotgreave and Schuppe-Koistinen presented data indicating that cystine may be transported into cells via a carrier-mediated mechanism in the presence of γ-glutamyl transpeptidase to increase the availability of -SH (sulfhydryls) groups for the up-regulation of GSH (12). GSH can also be synthesized by way of the cystathionase pathway, which utilizes methionine. The required amount of cysteine for de novo synthesis in other organs comes from the liver. Cells possess the ability to make more GSH in the presence of excess -SH from sulfur proteins. Additional GSH production is then shut off. This appears to be related to the negative feedback on γ-glutamylcysteine synthetase that is required for GSH formation (Eq. 1). Once GSH is formed, cells acquire the ability to protect themselves from free-radical-induced injury by detoxifying highly reactive electrophilic species, toxic endogenous cellular metabolites, and exogenous xenobiotics.

The primary function of GSH is to form conjugates with reactive electrophiles and render them nontoxic, inactive, and more easily excretable. The nucleophilic -SH moiety of GSH combines with toxic compounds consisting of electrophilic carbon atoms. This reaction, with the -SH thiolate anion (GS$^-$), produces a thioether bond between the carbon atoms and the -SH group of GSH. It should be stated, however, that nucleophilic attack by GSH can occur on other atoms such as electrophilic nitrogen and sulfur and is not limited solely to carbon. The formation of these mercapturic acid conjugates is made possible by the availability of a variety of catalytic isoenzymic forms of glutathione (GSH) S-transferases, which are highly abundant in the livers of animals (13–15). Glutathione S-transferases are multifunctional proteins, with molecular mass in the 40–50 kD range. They are present in many tissues and serve not only as detoxifying enzymes but as intracellular binding proteins as well. GSH conjugates are eventually cleaved in the kidney by N-acetyltransferases to mercapturic acid derivatives and excreted. It stands to reason that the amount of GSH available for detoxification is a function of the net balance between its synthesis and release to the tissues. The signal responsible for increased hepatic synthesis and release of GSH into the plasma is related to the amount of oxidative stress "sensed" by the liver. This sensing mechanism may be regulated by circulating hormones. Stress hormones such as vasopressin and angiotensin II have been shown to increase the delivery of GSH to the systemic circulation, possibly by a carrier-mediated mechanism (16).

GSH also serves as a storage site and transport form of cysteine and can be directly or indirectly utilized in many important metabolic functions such as protein synthesis, enzyme activity, DNA synthesis, leukotriene and prostaglandin (PGE$_2$) formation, and amino acid transport (γ-glutamyl cycle). Glutathione (GSH) is required as a substrate for both the GSH-transferases and the GSH-peroxidases, which are necessary for cell detoxification processes. GSH exists primarily in two forms, reduced glutathione, GSH, and the oxidized form, glutathione disulfide (GSSG). There are other forms of GSH such as protein mixed disulfides and thioethers, which contribute to the GSH pool; however, these are not discussed in this chapter. The reader is referred to a review by Brigelius on mixed disulfides (17). The sum of reduced and the oxidized forms is referred to as "total" GSH. To measure the amount of reduced GSH, it is necessary to analyze total GSH, GSSG, and calculate the reduced GSH concentration (red [GSH]) by total[GSH] = red[GSH] + 2[GSSG].

Cells need to be in a balanced reduction–oxidation (redox) state. The redox status of cell and tissues can be quantitatively determined by calculating the GSH/GSSG ratio. Additional assessment of the cellular redox status can also be determined by measuring the NADPH/NADP$^+$

ratio, which supplies reducing equivalents for the formation of reduced GSH in the presence of GSH reductase (Figure 2). During conditions of optimal metabolism, the in vivo ratio of reduced GSH to GSSG can exceed 100:1 (10). However, as cells become more oxidatively stressed this ratio can decrease as GSH becomes oxidized to produce GSSG. This reaction is driven by the enzymes present in the GSH redox cycle (Figure 2). Increased oxidative stress may overwhelm a cell's natural ability to detoxify electrophilic species via GSH synthesis. Fasting and starvation as well as oxidative biochemical stress can deplete tissue GSH to less than 30% of normal values. This can cause alterations in metabolism of xenobiotics and increase the toxicity of reactive metabolites (18–20). Depletion of GSH, especially in the liver, can cause irreversible damage and cell death (21, 22). An indication of in vivo oxidative stress can be quantitatively determined by measuring plasma GSSG concentration (23, 24). As outlined later, GSH plays an important role in protecting cells against reactive O_2 intermediates and free radicals.

There are documented cases in which GSH deficiency seriously compromised the health status of the organism. Genetic errors in coding for GSH synthetase, the enzyme required for GSH synthesis, are associated with 5-oxoprolinemia and 5-oxoprolinuria (25). Also, there are patients who have an inborn GSSG reductase deficiency and are prone to hemolytic anemia and cataracts. Hemolytic anemia can be exacerbated through yet another defect in GSH metabolism. GSH is supplied reducing equivalents from NADPH, which in turn receives them from the hexose monophosphate shunt pathway (Figure 2). This reaction is catalyzed by glucose-6-phosphate dehydrogenase (G6PD) (26). G6PD deficiency is a common genetic disorder that can be characterized by the loss of NADPH and can result in increased red blood cell oxidant sensitivity and hemolysis (27).

The specialized intracellular properties of GSH may also be involved in apoptosis or what has been generally referred to as "programmed" cell death. Apoptosis can be characterized by cell shrinkage, nuclear fragmentation, condensation of cellular debris into membrane-bound apoptotic bodies, and their ingestion by phagocytic macrophages. This process allows the elimination of certain selective cells during growth, development, or homeostatic processes. An interesting study by Fernandes and Cotter has shown that GSH may play a role in the cell's pathway to death either by apoptosis or necrosis when challenged with alkylating agents such as melphalan and chlorambucil (28). By gradually depleting human cell lines, such as HL-60, of GSH with buthionine sulfoximine these cells have the time to activate an apoptotic death sequence when toxin challenge is low; that is, decreasing GSH levels increases the cytotoxicity of melphalan. In contrast, if toxic challenge is too high, necrosis is the route to cell death. The authors conclude that necrosis is the physiological route to cell death when GSH levels are reduced. However, recent work by Slater et al. has demonstrated that thymocyte apoptotic events could be regulated by the availability of GSH (29). When GSH was depleted, apoptosis was increased, and when

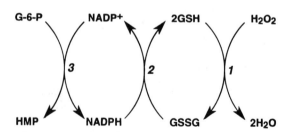

Figure 2. The glutathione redox cycle is the single most important antioxidant pathway used for cell defense. Under conditions of oxidative stress caused by hydrogen peroxide (H_2O_2), reduced GSH becomes oxidized, catalyzed by GSH peroxidase (**1**), and forms GSSG. To maintain an effective cellular GSH levels, GSSG must be reduced back to reduced GSH, a process catalyzed by glutathione reductase (**2**). This reaction sets into motion the transfer of reducing equivalents in the form of hydrogen atoms (transhydrogenation) from the hexose monophosphate shunt pathway of glucose metabolism to NADP$^+$. This reaction is catalyzed by glucose-6-phosphate dehydrogenase (**3**). Oxidative injury progresses when the cell's ability to reform reduced GSH becomes overwhelmed because of loss of some or all of the components of the GSH redox cycle.

GSH levels were increased, apoptosis was inhibited. It is apparent from this study that the normal life–death cycle of cellular metabolism depends on the capacity of GSH to maintain a certain degree of metabolic order.

Experimentally, it has been shown that the addition of cysteine-containing compounds to oxidatively stressed preparations can replenish intracellular cysteine stores, thereby increasing GSH levels. Cells can manufacture GSH when supplied with its amino acid precursor, cysteine. In the liver, GSH can be reformed through the *trans*-sulfuration pathway by the addition of methionine, which can convert, through a series of enzymatic steps utilizing cystathionase, methionine into cysteine (30). In contrast, other cells, such as those present in the lung, lack the *trans*-sulfuration pathway and can only reform GSH via cysteine addition (31). Generally, the liver is the organ that has the greatest capacity to detoxify xenobiotics, but the lung is exposed to inhaled as well as bloodborne toxins and also can play a vital role in protecting itself and other tissues against toxin challenge.

ANTIOXIDANT PROPERTIES

It is agreed that at least two basic conditions must be met for a substance to be classified as an antioxidant: (1) When present in appropriate concentrations antioxidants can postpone, subdue, retard, or prevent autooxidation of free-radical-mediated oxidation of the substrate and (2) produce a stable free radical. The availability of GSH to the cell intracellularly or extracellularly could be the difference between life or death. Depletion of GSH would result in compromising the cell's ability to detoxify highly reactive metabolic intermediates formed during oxidatively stressful conditions. Oxidative stress can be damaging to cells and tissues because it is caused by free radicals. Free radicals, by definition, are atoms or molecules comprised of one or more unpaired electrons in their outer orbital shells and are capable of independent existence. Free radical species consist of many molecular compounds such as the superoxide anion ($O_2^{\cdot-}$) and hyroxyl ion ($^{\cdot}OH$). Cells and tissues can be challenged with oxidative stress from endogenous sources such as normal aerobic respiration and mitochondrial oxygen consumption, oxidative bursts from inflammatory cell phagocytic metabolism, and hydrogen peroxide (H_2O_2) generated from peroxisome action on fatty acids and other substances. Oxidative stress can be caused also by direct exogenous chemical challenge with compounds such as H_2O_2, *tert*-butyl hydroperoxide, or paraquat. Furthermore, inhalation of oxidant gases, such as ozone, phosgene, or nitrogen dioxide, normal cellular metabolism, ingestion of toxic food additives, drugs, pesticides/herbicides, and even exercise are all capable of causing oxidative stress-induced injury. Reactive-centered chemical intermediates responsible for oxidative stress are particularly damaging if they are formed in the presence of catalytic metal cofactors such as iron or copper. These reactions can be summarized by the Haber–Weiss (Eqs. 3 and 4) and Fenton (Eq. 5) reactions, both of which have been heavily investigated over the past 25 years:

$$Fe^{3+} + O_2^{\cdot-} \rightarrow Fe^{2+} + O_2 \tag{3}$$

$$O_2^{\cdot-} + O_2^{\cdot-} + 2H^+ \xrightarrow{\text{superoxide dismutase}} H_2O_2 + O_2 \tag{4}$$

$$H_2O_2 + Fe^{2+} \rightarrow {}^{\cdot}OH + OH^- + Fe^{3+} \tag{5}$$

An event correlated with GSH depletion is lipid peroxidation, a free-radical-mediated process linked to oxidative stress. Lipid peroxidation is an autocatalytic mechanism in which cell membrane polyenoic fatty acids are degraded by means of a chain reaction. This is characterized by the abstraction of an H atom from an unsaturated fatty acid with the subsequent formation of lipid peroxides, aldehydes, alkenes, alkanes, alkanals, and alkenals. The overall general schema for this chain reaction can be represented by:

Initiation $$L(lipid)H + {}^{\cdot}OH \rightarrow L^{\cdot} + H_2O \tag{6}$$

$$L^{\cdot} + O_2 \rightarrow LOO^{\cdot} \tag{7}$$

Chain propagation $LOO^{\bullet} + LH \rightarrow LOOH + L^{\bullet}$ (8)

Glutathione's role as a chain breaker and detoxifier may involve these general reactions (32):

$$^{\bullet}OH + GSH \rightarrow H_2O + GS^{\bullet} \tag{9}$$

$$LO^{\bullet} + GSH \rightarrow LOH + GS^{\bullet} \tag{10}$$

$$LOO^{\bullet} + GSH \rightarrow LOOH + GS^{\bullet} \tag{11}$$

$$2GS^{\bullet} \rightarrow GSSG \tag{12}$$

GSH can participate directly in the detoxification of H_2O_2 by:

$$2GSH + H_2O_2 \xrightarrow{\text{GSH peroxidase}} GSSG + 2H_2O \tag{13}$$

$GS^{\bullet}$ (glutathione thiyl) radical is the reactive by-product of GSH chemical reactions.

The extent of lipid peroxidation is affected by many factors that include not only the availability of GSH but the lipid/protein ratio, the membrane-bound cholesterol, the lipid content of the membrane, and the redox state of membrane-bound vitamin E. In spite of these factors, lipid membrane susceptibility for the formation of lipid peroxides could be related to the species under investigation. Singh and co-workers found varying amounts of vitamin E in the lipid membranes from several species, which may make comparison and, hence, interpretation of lipid peroxidation results from different species somewhat tenuous (33). Nevertheless, all of these elements work in collaboration with GSH to help maintain the functional integrity of lipid membranes, which, in the case of some organs such as the lung and liver, can be the last line in detoxification defense. Formation of reactive lipid mediators such as phospholipases can induce release of arachidonic acid by-products. These metabolites can set into motion additional deleterious phenomena. For example, they may act as signals for leukocyte macrophage sequestation at the site of injury. These inflammatory cells are then capable of enhancing the injurious response by releasing toxic cellular mediators such as proteases and lysosomal enzymes. As this process continues, the ability to scavenge free radical species diminishes as protective thiol levels become overwhelmed and further membrane and cellular damage occurs (Figure 1).

Thus, it is clear that injury to certain cell types can beget further injury. The rate of GSH replacement is for all intents and purposes dependent upon cysteine stores as a precursor for GSH synthesis. However, this is organ dependent since the liver, as mentioned earlier, is the only organ that possesses the *trans*-sulfuration machinery capable of synthesis of GSH from either cysteine or methionine amino acids. Most other organs are not so fortunate and therefore must rely on the availability of free cysteine or exogenously supplied sulfhydryl compounds such as *N*-acetylcysteine. GSH delivery compounds are reviewed later.

As mentioned earlier, continued oxidant stress can and does lead to the formation of GSSG by the enzyme GSH peroxidase (Figure 2). This has been shown in the isolated, perfused, ischemic rat liver model following challenges with the highly oxidizing chemical *tert*-butyl hydroperoxide. Mathews et al. found that plasma levels of GSSG increased by 500–1000% after ischemic challenge. This was highly correlated with increased lipid peroxidation (34). In the same study, infusion of *tert*-butyl hydroperoxide also increased both lipid peroxidation and plasma levels of GSSG. However, even with perfusion in Krebs–Ringer solution with or without blood (hypoxic O_2 conditions), rat livers could boost export of increased amounts of GSSG to the plasma via the bile (35). These results indicate that under conditions of metabolic stress, the liver is extremely important in GSH metabolism. Continued oxidative stress from many sources can compromise the liver's capacity to produce sufficient antioxidants such as GSH even for self-protection.

There are two main pools of GSH, one located in the cytosolic fraction of the cell and the other in the mitochondrial matrix. It is the cytosolic form that is used mostly in the overall tissue response to oxidative stress. It is widely believed that the cytosolic form is required to maintain the functional integrity of mitochondrial respiration, which is crucial for cell metabolism even while cytosolic GSH levels continue to be depleted because of oxidative stress. The hepatic turnover ($t_{1/2}$) of the cytosolic pool is about 2 h, compared to the mitochondrial pool of approximately 30 h. This implies that mitochondria synthesize their own pool of GSH (11). Recent work done by Kristal and co-workers demonstrated that the mitochondrial transcription system is very sensitive to an oxidizing environment consisting of peroxyl radical ($RO_2^{\cdot}$) species (36). Addition of exogenous GSH to liver mitochondrial extracts only attenuated slightly the transcriptional response to oxidative stress. Therefore, the function of GSH may be limited in terms of its capacity to reach certain parts of the cell in the concentrations that are necessary to maintain some degree of protection.

TISSUE PROTECTION AND RESPONSE TO STRESS

Environmental Stress

Strenuous Environments

GSH levels have been found to be important during conditions of environmental stress. Siems et al. recently completed work showing that oxidative stress can occur when humans are exposed to harsh environmental conditions (37). They measured blood levels of GSH in cold-water swimmers who swam in 2–3°C water for 5 min. GSH levels in erythrocytes were higher 1 h after hypothermic exposure. This increase could indicate the red blood cells' capacity to boost GSH production needed to increase their tolerance to cold-induced stress. Therefore, increased levels of GSH may be important in environmentally stressful conditions.

Human serum exposed to pulsed ultrasound normally will induce DNA clastogenic mechanisms when added to isolated cells in response to this stressful condition. Pinamonti et al. effectively demonstrated that exogenous addition of 20 mM GSH to the serum either before or after sonication of lymphocytes inhibited DNA damage by 90% (38). The attenuation of sonication-induced DNA damage by GSH indicates that ultrasound may cause damage by a free-radical-mediated process.

Exercise

Strenous exercise also can produce oxidant stress. Most of the oxygen consumed during exercise is utilized in the mitochondria. However, up to about 5% of this available oxygen may be converted to toxic oxygen species (39). Under conditions of prolonged exercise, the liver can export GSH into the plasma. Decreased liver GSH and increased plasma GSH concentrations have been found in several species of rodents during strenuous exercise (40–42). In humans, blood GSH levels were elevated after intensive exercise (43). In contrast to these findings, Duthie et al. found that during strenuous long-distance running blood levels of GSH actually decreased (44). The implication of these results is that hepatic stores of GSH are depleted and the capacity of the liver to release GSH to the periphery is jeopardized by the excessive and continued buildup of reactive oxygen species. A study by Leeuwenburgh and Ji demonstrated that when GSH in mice was moderately depleted with buthionine sulfoximine, it did not affect their swimming duration time (42). However, when further GSH depletion occurred, there was a direct correlation with increased levels of lipid peroxidation as measured by malondialdehyde formation.

Malnutrition

Protein energy malnutrition, which is commonly referred to in nutrition as marasmus and kwashiorkor diseases, is characterized by body wasting, edema, skin lesions, and fatty livers, all of which could result from oxidative stress. Erythrocyte GSH levels measured in children afflicted with these diseases are depleted. It has been suggested that to treat the physical condition associated with these diseases, supplementation with GSH may be therapeutically beneficial (45). The efficacy of GSH supplementation may be useful in treating other disorders where protein energy malnutrition is a factor. Protein energy malnutrition not only arises from nutrition-deficient disease conditions mentioned earlier, but is also secondary to other nutrition-related disorders such as acquired immune deficiency syndrome (AIDS), cancer, burn injury, and alcoholism. In AIDS-infected patients, GSH levels are decreased, which may increase inflammatory stress and further compromise the immune system (46).

Metals

Not only are we overburdened by numerous and varied organic pesticides, solvents, vapors, and food additives, etc., but we are also constantly being challenged with toxic metals that have been linked to neuro-, nephro-, and hepatotoxic diseases (47).

The literature is replete with many studies showing that transition metals such as iron, lead, copper, zinc, chromium, mercury, and vanadium play an important role in the formation of toxic oxygen free radical species. Transition metals (M) are capable of forming reactive oxygen free radical species; thus (48)

$$M^{(x)} + O_2^{\cdot-} \rightarrow M^{(x-1)} + O_2 \tag{14}$$

$$2O_2^{\cdot-} + 2H^+ \rightarrow H_2O_2 + O_2 \tag{15}$$

$$M^{(x-1)} + H_2O_2 \rightarrow M^{(x)} + {}^\cdot OH + OH^- \tag{16}$$

The presence of GSH as an antioxidant can serve a double purpose in that it can act not only as a chain-breaking lipid peroxidation inhibitor, but may also directly chelate potential dangerous Fenton reaction-mediating metals in a toxic environment. From Table 1, it can be seen that GSH possesses many potential metal ligand binding sites: a thiol group, the amino group, two peptide linkages, and two carboxylic acid groups. In fact, GSH carboxylate groups can form ligands with Fe^{3+} (49). Multiple complexes can be formed also with Ag^+ and Cu^{2+}. Work done by Perrin and Watt indicated that there is a 1:1 complex formation between GSH and Zn^{2+} at 37°C in the physiological pH range of 6.1–7.7 (50). Interestingly, when DNA damage was investigated in the copper–phenanthroline complex, it was found that worse damage resulted when GSH was present. This occurred even though in the presence of H_2O_2, Cu^{2+}, and AA, GSH appeared to inhibit free radical formation. This apparent pro-oxidant/antioxidant nature of GSH is thought to be solely a function of the redox cycling capability of GSH in the presence of a phenanthroline–Cu^{2+} complex (51). Resistance to cadmium-induced oxidative stress in V79 Chinese hamster fibroblasts was increased in the presence of GSH (52). Metals such as arsenic and cadmium have toxic effects on cytoskeletal elements such as microfilaments, microtubules, and intermediate filaments, which are required to maintain cell architecture and act as conduits for the transport of nutrients, especially in neuronal axons. For example, in an organ such as the lung, cytoskeletal integrity could make the difference between a dry alveolar compartment and one that is flooded due to failure of cytoskeletal elements to withhold fluid movement from the capillaries into the alveolar compartment. GSH prevented cadmium-induced alteration of cytoskeletal sulfhydryls in cultured 3T3 cells (53). This is a condition often observed during oxidant-induced lung injury. Also in Swiss 3T3 mouse cells, As^{3+} challenge actually increased the amounts of GSH, which could be a possible cellular response to help maintain enhanced protection (54). The deleterious effects of reduced Na^+,K^+-ATPase activity and its hydrolysis in rat liver mitochondrial cells and HeLa cells exposed in vitro to the gasoline antiknocking agent triethyllead were reversed by the addition of 1 mM GSH (55).

On the other hand, there are trace metals that are extremely crucial in detoxification processes regulated by GSH. Selenium (Se) is required along with GSH as the cosubstrate for the formation of glutathione peroxidase. The functional role of glutathione peroxidase is attributed to a seleno-cysteine group at each of the four catalytic sites on the enzyme. In the isolated perfused rat lung, the capacity of the lung to detoxify infused tert-butyl hydroperoxide was decreased due to a Se-deficient diet (56). Selenium deficiency not only leads to the reduced protective function of glutathione peroxidase, but also to various diseases in animals as well as to cardiomyopathy in humans, Keshan disease, and white muscle dystrophy in Se-deficient lambs and calves (57). Conversely, too much selenium can be detrimental. Ingestion by animals of plants high in selenium can cause paralysis and death (58). A review of selenium toxicity has been published by Spallholz (59). Supplementation with selenium can enhance protective mechanisms against peroxidative processes by being accessible for increased synthesis of glutathione peroxidase. This may have been the case in a study in which the addition of 6 ppm Na selenite to the drinking water for 6 wk significantly diminished aflatoxin-induced lipid peroxidation in rat liver

(60). It should be mentioned, however, that not all glutathione peroxidase enzymes are selenium dependent (61).

Physiological Effects

Kidney

As briefly alluded to earlier, the kidney is extremely important in maintaining metabolic homeostasis by regulating the composition of body fluids and purging the body of toxic wastes generated from homeostatic metabolic functions, food additives, drugs and xenobiotic compounds, etc. The kidney, along with the liver, is important for the interorgan circulation of GSH. Moreover, it is the site of N-acetyltransferase enzymes that when combined with GSH-conjugated toxins enable them to become more water soluble to facilitate excretion. However, the kidney can also succumb to the lack of GSH protection. GSH depletion in the rat can be responsible for decreased glomerular filtration rates and increased Na^+ and K^+ excretion, and may cause a urinary acidification defect. Indeed, chronic analgesic nephropathy has been linked to acetaminophen and phenacetin overdosage, both of which cause GSH depletion (62). Intermediates of lipid peroxidation have been shown to increase in the kidneys following reduced blood flow and ischemia (63). Ischemic processes can lead to increased free radical production, which is a significant challenge for the maintenance of homeostatic mechanisms. Clearly, reduction in blood flow would have profound implications regarding urinary-regulated clearance mechanisms, especially since these mechanisms are dependent upon the functional integrity of both the renal medulla and cortex. It has been shown that target cells for GSH depletion are located in the ascending limb of the Loop of Henle (64). Torres et al. have shown that renal function, as measured by glomerular filtration rates, was altered when GSH levels were depleted. Rats treated with diethyl maleate had a glomerular filtration rate that was 43% lower than control rats when measured after 1 h. Renal tissue GSH concentration was lowered by 75%. In comparison to reduced glomerular filtration rates at 1 h, Torres et al. found that renal function recovered completely within 48 h after diethyl maleate challenge (64). These data indicate that the kidney, owing to its ability to synthesize GSH within 30 min, is able to replenish GSH to restore homeostatic processes. Studies completed by Scaduto et al. showed that in rats renal GSH is depleted under conditions of ischemia. However, when GSH was increased by exogenous pretreatment with GSH monoethyl ester prior to ischemia, renal dysfunction was enhanced (65). These results may suggest that under conditions of ischemic renal damage, the kidney may become sensitized and that the use of such therapeutic compounds may actually enhance injury.

Lung

The lung, along with the skin, is the only organ connected directly to the external environment. As a consequence, it would be logical that the lung should possess sufficient amounts of antioxidants to combat direct inhalational challenges. We inhale many toxic compounds on a daily basis, such as cigarette smoke, nitrogen dioxide, sulfur dioxide, ozone, and occupational exposures consisting of mists, fumes, and dusts. Ozone, nitrogen dioxide, hyperbaric O_2, and the industrial gas phosgene have been shown to cause tissue injury with the subsequent formation of lipid peroxidation by-products following acute or chronic exposure. The capacity of the lung to detoxify inhaled compounds is functionally dependent upon the cellular site of injury. There are over 40 cell types in the lung that possess varying metabolic capabilities required to reduce lung injury. GSH is present in many of these cells, albeit synthesized at different rates under varying stressful situations. Bend et al. have measured GSH synthesis and uptake rates of exogenously administered cysteine in type II cells, macrophages, and Clara cells isolated from rabbits. In addition, they studied the cell selectivity of the free-radical-forming herbicide paraquat. GSH synthesis was fastest in macrophages and slowest in type II cells, and paraquat was selectively accumulated in type II cells but not macrophages. These data clearly indicate that under in vitro experimental conditions different cell types coexisting within the same organ have varying capabilities for the regulation and synthesis of GSH and the inactivation of electrophiles, and are selectively sensitive to toxic insults (66). This multifunctional response of the lung to toxic insult presents a severe challenge to the pulmonary toxicologist. Not only are there varying responses to toxic insult throughout the tracheo–bronchio–alveolar complex, but effective drug therapy may also follow a regiospecific pattern as well.

The mucus lining, or extracellular lining fluid, of the lung protects the underlying epithelium from airborne toxins and dehydration. Since it is the mucus layer that receives the direct hit of inhaled toxins, it has become a valuable source of information regarding lung defense mechanisms and can be used to evaluate lung injury. GSH levels have been shown to be depleted in both lung tissue and in the epithelial lining fluid of the lung in pulmonary diseases such as acute respiratory distress syndrome and idiopathic pulmonary fibrosis (67, 68). In contrast, GSH levels in asthmatic patients have been shown to be higher when measured in their bronchoalveolar lavage fluid (69). Under normal circumstances the epithelial lining fluid contains very high levels of GSH, averaging about 140-fold the plasma levels in nonsmoking individuals compared to about 80% in smokers (70). These concentrations are believed to be sufficient to protect against H_2O_2 that would be released from macrophages during phagocytizing events in the lower airway that are triggered by oxidative lung injury.

Brain

The brain is a highly metabolic organ. It occupies about 2–3% of the body weight, requires up to 20% of the cardiac output, and demands over 20% of the total body oxygen consumption. During the course of a 24-h day, a 1400-g brain requires 1000 L blood, 71 L oxygen, and 100 g glucose. It also releases 71 L carbon dioxide and 7 g lactic acid (71). From this physiological portrait, it is readily apparent that the brain, which may have limited capability to oppose free-radical-induced stress, is a potentially "high risk" organ for toxic insult. The free radical metal cofactor iron is present in some areas of the brain in high concentrations (72). Damage caused by traumatic or ischemic events can modify the GSH redox status of the affected cell, tissue, and organ involved. This is especially the case for the brain, which requires an environment that is "free" from potentially harmful reactive chemical intermediates. It has been found that oxygen radicals can cause parenchymal and vascular injury in the central nervous system (73). Under conditions of ischemia, metal ions such as iron can become available and accelerate the injury process by acting as a Fenton reagent (Eq. 5). Free-radical-induced injury can lead to membrane disruption, causing an increase in free fatty acids via lipid peroxide formation and propagation (Eqs. 6–8). The brain is very rich in polyunsaturated fatty acid side chains. Generation of reactive oxygen species catalyzed in part by the presence of iron can kill neurons (74). Although GSH is found in the brain, its role in the protection of neurons from free radical damage is not completely understood (75). Makar et al. have shown, in vitro, that chick astrocytes possess significantly more GSH and γ-glutamylcysteine synthetase, the enzyme required for GSH synthesis, than cultured forebrain neurons (76). It is believed that astrocytes, which form structural supports between the capillaries and neurons in the central nervous system and are part of the blood–brain barrier, may be more resistant to oxidative damage than are other brain cells. Astrocytes may also play an active role in protecting the brain from potentially toxic free radical injury. The importance of GSH in the amelioration of neurotoxicity was observed by Shivakumar and Ravindranath when they investigated acrylamide-induced injury in mice. It was found that acrylamide injury was enhanced when GSH was depleted using buthionine sulfoximine or diethylmaleate. GSH pretreatment replacement therapy with L-2-oxothiazolidine-4-carboxy-late attenuated acrylamide-induced neurotoxicity (77). In experiments using rat brains, ligation of the right middle cerebral and carotid arteries caused brain infarcts associated with cerebral ischemia. GSH levels on the ischemic cortex side decreased over time. Further depletion of GSH using buthionine sulfoximine exacerbated cortical function and edema, suggesting that brain ischemia is a free-radical-induced injury and that endogenous GSH may be important in the defense against tissue damage caused by reactive oxygen species (78).

Oxidative damage may be a causative factor in several neural-related disorders such as Parkinson's disease. Sofic and co-workers found reduced concentrations of GSH in the substantia nigra in post-mortem brain tissue in patients confirmed to have Parkinson's disease (79). There may also be a connection between iron and Alzheimer's disease. McLachlan et al. demonstrated that treatment of Alzheimer's patients with the iron chelator deferroximine improved their "daily living skills" (80). For a discussion on the effect of free radicals on the central nervous system, the reader is referred a recent review by Halliwell (39).

Gastrointestinal Tract

The gastrointestinal (GI) system, a collection of interconnected organs, is also susceptible to constant and potentially injurious free radical chemical attack. The GI system is subject to ingestion of xenobiotics in the form of food additives, pesticides, herbicides, drugs, and diet-bound fatty acids. An increased intake of these compounds has been linked to heart disease, cancer, aging, and digestion disorders. The role of GSH in protection against such toxic challenges has been under investigation. Mårtensson et al. have demonstrated that GSH is essential for intestinal function (81). Jejunal and colonic epithelial cells are dependent upon the availability of GSH, since cellular degeneration was observed when GSH was depleted. It is believed that biliary GSH, which is in the millimolar range under normal conditions, is necessary for the protection of intestinal epithelia. In the intestinal mucosa, GSSG reductase and GSH peroxidase enzymes are present to detoxify lipid hydroperoxides that may be present in the lumen (82). Furthermore, intestinal lumen thiols, which consist of cysteine and GSH, have been found to be secreted from the intestinal mucosa into the lumen (83). Therefore, the mucosa may represent a source of protective thiols needed to maintain intestinal function. This protective factor may be significantly enhanced by the bioavailability of GSH following oral administration. Hagen et al. have established that dietary GSH can be absorbed intact and can result in increased plasma concentration (84). Plasma GSH increased nearly threefold within 90 min after gavage administration of 90 μmol GSH to rats. Exogenous pretreatment with 1 mM GSH protected isolated rat jejunum epithelial cells, as measured by cell viability, against oxidative stress following challenge with *tert*-butyl hydroperoxide and menadione (85). Owing to the rapid turnover of intestinal epithelial cells and relatively scant biosynthesis of GSH, these cells may need to have a continuous supply of GSH available to protect against chemical injury. Moreover, it has been shown through compartmentation-kinetic studies in rats that the plasma GSH is the primary source of cysteinyl compounds. It was also observed that the capacity of tissue to take up intact GSH from the plasma may be related to the developmental stage of the animal (86). Could oral GSH treatment be therapeutically useful to enhance tissue stores of glutathione? These data may be important in certain nutritional diseases, as mentioned earlier, where protein energy malnutrition may have an effect on the uptake, synthesis, or transport of GSH, especially for children in whose early development certain extrahepatic tissues may have special requirements for protection against an oxidizing environment. Additional information regarding intestinal transport mechanisms and uptake of GSH can be found in an excellent review by Vincenzini et al. (87).

GSH and Coprotection with Vitamins C and E

The protective properties of GSH cannot be ascribed solely to the presence of the cysteinyl moiety. GSH, the substrate for the GSH *S*-transfersases and GSH reductases, works in collaboration with many other antioxidants such as vitamins C and E to form part of the cellular defense mechanism. The primary function of vitamin E is to prevent free-radical lipid peroxidation in the cell membrane. Vitamin E, a lipid-soluble phenolic antioxidant, is known as a lipid peroxidation chain breaker because it can interfere with xanthine oxidase synthesis an initiator of free-radical processes (88). Organs that are most resistant to lipid peroxidation generally have the highest vitamin E content (89). The cell's overall antioxidant protective machinery largely depends upon the simultaneous availability of both vitamin E and GSH. In a study investigating lipid peroxidation in rat liver microsomes, it was demonstrated that the protective capacity of GSH to prevent lipid peroxidation was lost in animals that were vitamin E deficient (90).

The use of therapeutic compounds such as doxorubicin for the treatment of cancer can result in cardiotoxicity associated with free-radical-induced tissue damage through lipid peroxidation. Since doxorubicin is an anthracyclic antibiotic, free-radical damage may occur by the intracellular redox cycling (electron shuttling) of the anthraquinone compound. Powell and McKay have demonstrated this in the rat liver and in bovine cardiac microsomes (91). GSH in the presence of vitamin E protected against the toxicity of doxorubicin by inhibiting lipid peroxidation. In the presence of GSH, oxidized vitamin E can be reduced back to its native form as described by Pryor (92) and determined by Niki et al. (93) according to the following sequence:

$$\text{Vit-E} + \text{LOO}^\bullet \rightarrow \text{Vit-E}^\bullet + \text{LOOH} \tag{17}$$

$$\text{Vit-E}^{\cdot} + \text{GSH} \rightarrow \text{Vit-E} + \text{GS}^{\cdot} \tag{18}$$

$$2\text{GS}^{\cdot} \rightarrow \text{GSSG} \tag{19}$$

$$\text{H}^+ + \text{GSSG} + \text{NADPH}_2 \xrightarrow{\text{GSH reductase}} 2\text{GSH} + \text{NADP}^+ \tag{20}$$

Although there has been considerable evidence for the synergistic effects of combined antioxidant therapy involving vitamin E and GSH, there appears to be a growing amount of anecdotal evidence that vitamin C, AA, and GSH can also function in a similar manner. The synergistic link between AA and GSH antioxidant activity was first noted in plants by Hopkins and Morgan, followed by Borsook et al. (94), and more recently by Dalton et al. (95) in soybean root nodules (96).

Humans and guinea pigs require AA but are unable to synthesize it on their own and must rely on dietary sources to increase tissue levels. AA has a tendency to accumulate in tissues while plasma levels remain low, which may help to explain its role as an antioxidant. Clinically, AA deficiency can lead to scurvy, which, if not corrected by dietary supplementation, can have life-threatening consequences, especially in the context of protein energy malnutrition described earlier. Winkler has shown, in vitro, that there is a stoichiometric relationship between the formation of GSSG and AA from GSH and dehydroascorbic acid (DHA) (97):

$$2\text{GSH} + \text{DHA}^- \rightarrow \text{GSSG} + \text{AA} \tag{21}$$

For a more extensive review of AA and GSH physical and chemical perspectives, the reader is referred to Winkler et al. (98).

Mårtensson et al. have shown in newborn rats, made GSH deficient by the administration of buthionine sulfoximine, that AA concentration was also depleted in liver, lung, brain, and kidney (99). Furthermore, mortality was significantly decreased when rats were administered both GSH monoethyl ester, which enters the cell much more readily than does native GSH, and AA (100). In guinea pigs made AA deficient by treatment with scorbutic acid, the addition of GSH monoethyl ester to the diet significantly delayed the onset of scurvy (101). Eiserich et al. found that depletion of AA and GSH directly coincided with the formation of 3-nitrotyrosine in a study investigating cigarette smoke-induced nitration of the amino acid tyrosine (102). These data imply that inhalation of cigarette smoke can modify proteins and that the addition of GSH and AA can attenuate protein oxidation.

Cataract opacities are thought to be formed by protein aggregates within the lens of the eye. In studies from the Linus Pauling Institute, thermal protein aggregate formation in the eye lens was shown to be significantly higher in the AA-deficient group and lower in the AA-supplemented group (103). This protective effect, measured by less protein aggregate formation, was the direct result of increased AA levels rather than GSH, which showed no change in any of the AA dietary regimentations. A similar AA/GSH relationship was found in guinea pigs fed atherogenic diets supplemented with AA (104). It appears, therefore, that the protective and synergistic action between AA and GSH remains open for further investigation. This is in light of the fact that there does appear to be a direct connection of AA-deficient diets being responsible for decreased tissue levels of vitamin E (105). The links between vitamin E, AA, and GSH and their redox cycling interaction can be summarized in a familiar schematic (Figure 3).

Aging

It is well known that cataract formation in the eye lens is associated with tissue damage and that it is a common condition found in the aged. However, cataracts can be formed in individuals exposed to hyperbaric oxygen therapy (106). This is due primarily to the overwhelming of the natural antioxidant system comprised of GSH, catalase, GSH peroxidase, GSH reductase, and superoxide dismutase by chemically reactive oxygen species. Supplementation with lipoic acid was found to decrease the amount of cataract formation by helping to maintain protective levels of ascorbic acid thereby preventing -SH groups from becoming oxidized (107). It is conceivable from these data that boosting GSH or thiol levels could have important clinical potential for the preventive or therapeutic treatment of cataract formation.

Another metabolic dysfunction generally found in all aging cells concerns brown pigment

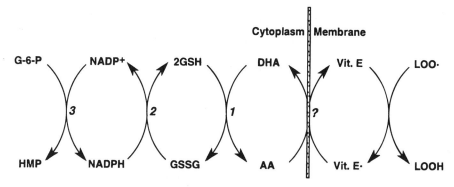

Figure 3. Glutathione redox cycle with the inclusion of the hypothetical transmembrane coupling of the interaction of AA and vitamin E. The intramembrane presence and function of vitamin E has been established, especially as a chain breaker in the detoxification of lipid peroxides, LOO·. Reactions between ascorbic acid and vitamin E do occur in vivo. However, enzymatic control between ascorbic acid and vitamin E (?) remains to be proven. This figure was modified from Winkler et al. (98). Enzymic regulation designations are the same as for Figure 2: (**1**) glutathione peroxidase, (**2**) glutathione reductase, and (**3**) glucose-6-phosphate dehydrogenase.

formation caused by lipofuscin. Brunk et al. have put forth a hypothesis regarding possible mechanisms involved in the formation of age-related pigmentation or lipofuscinogenesis (108). These mechanisms are centered around an oxidant-stress theory that lipofuscin pigments are the result of an imbalance in the cells antioxidant defense system regulated by the availability of GSH. The theory states that lipofuscinogenesis is brought about by an increase in the production of reduced oxygen species and the degradation of bioorganic macromolecules by secondary lysosomes and that pigment accumulation is the result of lipid/protein oxidation undergoing autophagocytic degradation. The amount and rate of accumulation of these brown-pigment macromolecules are dependent upon the availability of antioxidants and antioxidant enzymes needed to degrade this lipid accumulation. Gao et al. demonstrated in cultured neonatal rat cardiac myocytes that lipofuscin pigment accumulation was increased following treatment with buthionine sulfoximine, a specific inhibitor of γ-glutamylcysteine synthetase, the enzyme required for GSH synthesis. The authors concluded that the increased production of H_2O_2 and accumulation in the lysosomal interior enhanced the formation of ·OH and increased pigment formation (109). These results may suggest that treatment with antioxidant compounds such as GSH may inhibit or retard the formation of age-related pigmentation. For the purposes of this review, time and space do not allow ample discussion on the role of oxidants and antioxidants in the degenerative disease processes observed in aging. The reader is referred to an excellent review on these events recently published by Ames et al. (110).

GSH and Tumor Cell Resistance

We have seen throughout this review that GSH plays an important role in the overall antioxidant defense strategy utilized by cells to defend against free-radical-induced oxidative stress. In reality, however, the presence of GSH can be indirectly, yet potentially, damaging to the organism. It is the killing power of free radicals and alkylating agents that is mostly responsible for the reduction in tumor spread and increased cure rates seen in certain forms of cancers. In other words, they increase cytotoxicity at the tumor site, which is crucial for the death of cancer cells. Excess GSH has been observed in some forms of cancer, which may enhance their resistance to chemo- and radiotherapy. For example, there is resistance to chemotherapy in patients afflicted with acute lymphoblastic leukemia. In patients with leukemic blast formation, upon relapse, a twofold increase in GSH concentration was observed (111). It was concluded that elevated GSH levels may be responsible for increased drug therapy resistance in patients with acute lymphoblastic leukemia. Indeed, studies done by Zaman et al. demonstrate that part of the biochemical puzzle behind drug resistance may be related to the presence of a plasma membrane glycoprotein that can confer multidrug resistance by increasing GSH–drug-conjugated efflux from target cells (112). As mentioned previously, buthionine sulfoximine, which interferes with GSH synthesis by blocking the effects of γ-glutamylcysteine synthetase, is now in clinical trials in cancer patients so that they may benefit more directly and efficiently from tumor treatment therapy.

PROPHYLACTIC AND THERAPEUTIC UP-REGULATION OF GSH

Naturally, it would seem that if we could bolster and sustain the amount of GSH made available to the cell or tissue, we could gain some degree of enhanced or "mega-protection." It is generally believed that by furnishing additional -SH groups, such as cysteine, one of the amino acid components of GSH, increased synthesis may occur. N-Acetylcysteine presumably works in this manner. N-Acetylcysteine has been used clinically and safely as a mucolytic agent. It liquefies mucus and DNA by the mucolytic action of its free sulfhydryl group, which opens the disulfide bonds of mucoproteins, thereby lowering viscosity (113). It can be given as a nebulized solution or as oral therapy to treat asthmatic patients. Oral treatment may not be preferred, since it has been shown in rats that N-acetylcysteine is deacetylated in the intestinal mucosa, which may decrease its bioavailability (114). In the lung, N-acetylcysteine reversibly binds to proteins by disulfide bridges, probably allowing free cysteine to become readily available (115).

The addition of N-acetylcysteine has been found to protect nonenzymatically against endotoxin-induced damage (116) and radiation-induced cellular damage (117) by either replenishing depleted cysteine or acting as a hydrogen donor. N-Acetylcysteine has also been found to protect against liver toxicity caused by acetaminophen overdose (118), against chronic bronchitis (119), and is effective against hyperbaric O_2- and bleomycin-induced lung damage in mice (120). Meyer et al. have also shown that intravenous N-acetylcysteine administration increased the reduced form of GSH, and may possibly enhance protection in lung tissue in patients with idiopathic pulmonary fibrosis (121). This could prove to be an effective clinical treatment for idiopathic pulmonary fibrosis, since Cantin et al. have shown that in these patients the epithelial fluid lining of their lungs is GSH deficient (68). Oral administration of N-acetylcysteine caused an increase in plasma GSH concentration in patients with AIDS-related wasting syndrome (122). In contrast, Bridgemen and co-workers observed that high oral doses of 600 mg N-acetylcysteine 1–3 times a day for 5 days did not produce sustained blood levels of GSH in patients with chronic obstructive pulmonary disease (123).

Because of its unique role and direct link to potentially oxidizing environments, the lung is one of the major organs in which research with compounds that up-regulate GSH or cysteine concentration has been extensively investigated. The lung cannot synthesize GSH from methionine since it lacks the cystathionase pathway (124). Exogenous administration of 20 μM GSH to cultured rat alveolar type II cells provided significant protection from paraquat-induced cellular injury as measured by cell viability assays (125). In contrast, a later study by Mårtensson and colleagues demonstrated that native GSH markedly increased the plasma levels of GSH, but they did not observe a similar increase in either the lung or the liver. However, treatment with GSH monoethyl ester substantially increased GSH in these two organs because it is more readily "taken up" by the tissues (126).

In inhalation exposure studies, differences in the effects of N-acetylcysteine used therapeutically to prevent lung injury have also been observed. Postexposure treatment with a loading dose of 140 mg/kg N-acetylcysteine ameliorated pulmonary edema formation in rabbits exposed to a high concentration of the toxic industrial gas phosgene. The protective quality of N-acetylcysteine was not necessarily a function of dramatically increased lung tissue GSH, but was rather a function of N-acetylcysteine acting directly as an antioxidant. This was evidenced by a significant decrease in the lipid peroxidation byproduct malondialdehyde (127). The protective action of N-acetylcysteine may be associated with its -SH group, which may act as a nucleophile, thereby trapping electrophilic intermediates (128). There have been studies showing that pretreatment with a loading dose of 170 mg/kg/h of N-acetylcysteine failed to protect rats exposed to NO_2 as determined by protein concentration, angiotensin II and alkaline phosphatase levels in bronchoalveolar lavage fluid (129). Drost et al. found that oral administration of N-acetylcysteine failed to prevent H_2O_2 and O_2^- generation from activated lung phagocytes (130). Work done by Critchley and colleagues also failed to indicate protection against hyperbaric O_2 following ip injection of *N*-acetylcysteine (131). Although space does not permit an extensive comparison of studies using *N*-acetylcysteine, it is apparent that the mode and site of delivery could determine the efficacy of the treatment. In spite of contradictory studies, the widespread impression is that *N*-acetylcysteine has been efficacious and remains under intensive investigation in many experimental injury models.

Compounds such as L-2-oxothiazolidine-4-carboxylate (OTC) have been used to up-regulate GSH. Williamson and Meister found that OTC is converted to L-cysteine by oxoprolinase and used for GSH synthesis (132). In a later study, Williamson and co-workers found that OTC protected against acetaminophen-induced hepatic toxicity (133). OTC has been found to be very effective in protecting calf pulmonary endothelial cells from H_2O_2 injury (134). Likewise, Kuzuya and co-workers found reduced toxicity of oxidized lipoprotein in cultured bovine endothelial cells pretreated with 2 m*M* OTC (135). Rat heart cardiac function improved following ischemia/ reperfusion injury after the addition of 250 and 500 μM OTC (136). Recovery was enhanced when OTC was used in combination with various amino acids. In studies with mice, OTC pretreatment attenuated the toxic effects of acrylamide-induced neurotoxicity (77). The importance of the thiols, such as *N*-acetylcysteine, OTC, or GSH, is readily apparent from these studies in cell/tissue protection in a variety of injury models. However, can cell/tissue protection occur even if GSH cannot be increased? Successful treatment using thiol derivatives in various toxicity models has been shown to ameliorate toxic effects. Indeed, research completed at Porton Down in the United Kingdom showed that several cysteinyl esters, when administered as pretreatments, increased survival rates in rats exposed to the toxic hydrophobic gas perfluoroisobutylene (PFIB). Additionally, these esters protected cultured rat lung slices from sulfur mustard injury as well (137, 138). While cysteine ester treatment increased plasma and tissue cysteine concentrations, GSH levels were unchanged. This indicates that GSH itself is not necessarily important in this injury model, since pretreatment with buthionine sulfoximine, an inhibitor of GSH synthesis, followed by treatment with cysteine esters also reduced the toxic effects of PFIB.

As mentioned earlier, GSH monoethyl esters have been used successfully in several injury models. Because GSH is poorly transported into cells, there is a need to have high protective concentrations available at the intracellular level. Apparently, GSH monoethyl esters can fill this need. Injection and oral dosing with GSH monoethyl esters increased GSH in the liver, kidney, spleen, and heart in mice (139). In a later study, mice given toxic levels of $CdCl_2$ had an increased survival rate of 80% following pretreatment with GSH monoethyl ester (140). In contrast to the protective qualities of GSH monoethyl esters, there is some evidence that increases in GSH utilizing this treatment could exacerbate injury especially in the kidney (65). Possible therapeutic benefits of GSH monoethyl ester treatments have been published (141).

In addition to the thiol compounds that have exhibited protection, there are other nonthiol-like substances that may also help to up-regulate GSH or at least prevent its depletion. The endogenous intracellular compound nitric oxide (NO) has displayed a protective capacity in cell injury, especially against reactive oxygen intermediates (142). The administration of NO and NO donors to rat lung fibroblasts, bovine pulmonary artery endothelial cells, and bovine pulmonary artery smooth muscle cells significantly increased GSH (143). In yet another oxidant injury model, dibutyryl cAMP prevented the oxidation of reduced GSH, thereby inhibiting lipid peroxidation, leukotriene production, and decreased pulmonary edema formation in rabbits exposed to a high concentration of phosgene (144).

TOXICITY OF GSH METABOLISM

Although there is no disagreement about the importance of the presence or absence of GSH as part of the defense strategy of cells and tissue to combat potentially disabling oxidative stress, GSH can produce toxic intermediates via its own metabolism. From Eq. 9–12 and 18, it is seen that GS· (glutathione thiyl radical) can become a potentially highly reactive electrophilic species in its own right, capable of initiating lipid peroxidation (Eqs. 6–8). There is evidence that this could occur. It is generally thought that the formation of the thiyl radical is controlled by the eventual formation of GSSG (Eqs. 12 and 19). However, using linoleic, linolenic, and arachidonic acid in pulsed radiolysis studies, Schöneich and colleagues demonstrated that thiyl radicals are capable of abstracting a biallylic H from polyunsaturated fatty acids (145). This prooxidant feature is readily apparent from data gathered by Sudhar and Armstrong, who have measured the redox potential of sulfur radicals and determined that $E°$ for GS˙ is +0.84V (146). A more detailed description of GSH chemical reactions has been presented by Beuttner (32).

CONCLUSION

This review is intended to raise the level of awareness of the importance of the role of glutathione in tissue protection against a variety of toxic challenges. It certainly is not an exhaustive summary of recent work. Instead, it provides an assortment of citations as a means to increase interest in this vitally important endogenous scavenger. A considerable amount of effort was put into treatment with GSH delivery systems such as N-acetylcysteine, GSH ester, and L-2-oxothiazolidine-4-carboxylate because many of the models described involved what are believed to be injuries mediated by free-radical attack. The positive results of these treatments in the cell/tissue/animal models may encourage additional investigation into the use of thiol therapy in human disease.

REFERENCES

1. De Rey-Pailhade, J. 1888. Sur un corps d'origine organique hydroenant le soufre a froid. *C. R. Hebd. Seances Acad. Sci.* 106:1683–1684.
2. Hopkins, F. G. 1921. On an autoxidisable constituent of the cell. *Biochem. J.* 15:286–305.
3. Hunter, G., and Eagles, B. A. 1927. Glutathione: A critical study. *J. Biol. Chem.* 72:147–166.
4. Hopkins, F. G. 1929. On glutathione, a reinvestigation. *J. Biol. Chem.* 84:269–320.
5. Pirie, N. W., and Pinhey, K. G. 1929. The titration curve of glutathione. *J. Biol Chem.* 84:321–333.
6. Harrington, C. R., and Meade, T. H. 1935. Synthesis of glutathione. *Biochem J.* 29:1602–1611.
7. Budavari, S., ed. 1989. *The Merck Index. An Encyclopedia of Chemical, Drugs, and Biologicals.* 11th ed. Rahway, NJ: Merck and Co.
8. Slade, R., Crissman, K., Norwood, J., and Hatch, G. 1993. Comparison of antioxidant substances in bronchoalveolar lavage cells and fluids from humans, guinea pigs and rats. *Exp. Lung Res.* 19:469–484.
9. Meister, A., and Anderson, M. E. 1983. Glutathione. *Annu. Rev. Biochem.* 52:711–760.
10. Sies, H., Brigelius, R., and Akerboom, T. P. M. 1983. Intrahepatic glutathione status. In *Function of Glutathione: Biochemical, Physiological, Toxicological, and Clinical Aspects*, ed. A. Larrson, pp. 51–65. New York: Raven Press.
11. Kaplowitz, N., M., Aw, T. Y., and Ookhtens, M. 1985. The regulation of hepatic glutathione. *Annu. Rev. Pharmacol. Toxicol.* 25:715–749.
12. Cotgreave, I., and Schuppe-Koistinen, I. 1994. A role for γ-glutamyl transpeptidase in the transport of cystine into human endothelial cells: Relationship to intracellular glutathione. *Biochim. Biophys. Acta* 1222:375–382.
13. Grover, P. 1982. Glutathione S-transferases in detoxification. *Biochem. Soc. Trans.* 10:80–82.
14. Lee, M. J., and Dinsdale, D. Immunolocalization of glutathione S-transferases isoenzymes in the bronchiolar epithelium of rats and mice. *Am. J. Physiol. (Lung Cell. Mol. Physiol. 11):* L766–774.
15. Boyer, T. D. 1989. The glutathione S-transferases: An update. *Hepatology.* 9:486–496.
16. Raiford, D. S., Sciuto, A. M., and Mitchell, M. C. 1991. Effects of vasopressor hormones and modulators of protein kinase C on glutathione efflux from the perfused rat liver. *Am J. Physiol. (Gastrointest. Liver Physiol. 24)* 261:G578–G584.
17. Brigelius, R. 1985. Mixed disulfides: Biological function and increase in oxidative stress. In *Oxidative Stress*, ed. H. Sies, pp. 243–272. London: Academic Press.
18. Plummer, J. L. Smith, B. R., Sies, H., and Bend, J. R. 1981. Chemical depletion of glutathione in vivo. *Methods Enzymol.* 77:50–59.
19. Jaeger, R. J., Connolly, R. B., and Murphy S. D. 1973. Diurnal variation of hepatic glutathione concentration and its correlation with 1,1-dichloroethylene inhalaiton toxicity. *Res. Commun. Chem. Pathol. Pharmacol.* 6:465–471.
20. Bray, T. M. and Taylor, C. G. 1994. Tissue glutathione, nutrition, and oxidative stress. *Can. J. Physiol. Pharmacol.* 71:746–751.
21. Sies, H. 1985. Hydroperoxides and the thiol oxidants in the study of oxidative stress in intact cells and organs. In *Oxidative Stress*, ed. H. Sies, pp. 73–90. London: Academic Press.
22. Reiter, R., and Wendel, A. 1982. Chemically-induced glutathione depletion and lipid peroxidation. *Chem. Biol. Interact.* 40:365–374.

23. Adams, J. D., Lauterburg, B. H., and Mitchell, J. R. 1983. Plasma glutathione and glutathione disulfide in the rat: Regulation and response to oxidative stress. *J. Pharmacol. Exp. Ther.* 227:749–754.

24. Abdalla, E., Caty, M., Guice, K., Hinshaw, D. B., and Oldham, K. T. 1990. Arterial levels of oxidized glutathione (GSSG) reflect oxidant stress in vivo. *J. Surg. Res.* 48:291–296.

25. Wellner, V. P., Sekura, R. , Meister, A., and Larrson, A. 1974. Glutathione synthetase deficiency: A inborn error of metabolism involving the γ-glutamyl cycle in patients with 5-oxoprolinuria (pyroglutamic aciduria). *Proc. Natl. Acad. Sci. U.S.A.* 71:2969–2972.

26. Meister, A., and Larrson, A. 1989. Glutathione synthetase deficiency and other disorders of the γ-glutamyl cycle. In *The Metabolic Basis of Inherited Diseases*, eds. C. R. Scriver, A. L. Beaudet, W. S. Sly, and D. Valle, 6th ed., pp. 855–868. New York: McGraw Hill.

27. Scott, M. D., Zuo, L., Lubin, B. H., and Chiu, D. T.-Y., 1991. NADPH not glutathione status modulates oxidant sensitivity in normal and glucose-6-phosphate dehydrogenase deficient erythrocytes. *Blood* 77:2059–2064.

28. Fernandes, R. S., and Cotter, T. G. 1994. Apoptosis or necrosis: Intracellular levels of glutathione influence mode of cell death. *Biochem. Pharmacol.* 48:675–681.

29. Slater, A. F. G., Nobel, C. S. I., Maellaro, E , Bustamente, J., Kimland, M., and Orrenius, S. 1995. Nitrone spin traps and a nitroxide antioxidant inhibits a common pathway of thymocyte apoptosis. *Biochem. J.* 306:771–778.

30. Beatty, P. W., and Reed, D. J. 1980. Involvement of the cystathionine pathway in the biosynthesis of glutathione by isolated rat hepatocytes. *Arch. Biochem. Biophys.* 204:80–87.

31. Berggren, M., Dawson, J., and Moldéus, P. 1984. Glutathione biosynthesis in the isolated perfused rat lung: Utilization of extracellular glutathione. *FEBS* 176:189–192.

32. Beuttner, G. R. 1993. The pecking order of free radicals and antioxidants: Lipid peroxidation, α-tocopherol, and ascorbate. *Arch. Biochem. Biophys.* 300:535–543.

33. Singh, Y., Hall, G. L., and Miller, M. G. 1992. Species differences in membrane susceptibility to lipid peroxidation. *J. Biochem. Toxicol.* 7:97–105.

34. Mathews, W. R., Guido, D. M., Fisher, M. A., and Jaeschke, H. 1994. Lipid peroxidation as molecular mechanism of liver cell injury during reperfusion after ischemia. *Free Radic. Biol. Med.* 16:763–770.

35. Ballatori, N., Truon, A. T., Ma, A. K., and Boyer, J. L. 1989. Determinants of glutathione efflux and biliary GSH/GSSG ratio in the perfused rat liver. *Am. J. Physiol. (Gastrointest. Liver Physiol. 19)* 256:G482–G490.

36. Kristal, B. S., Park, B. J., and Yu, B. P. 1994. Antioxidants reduce peroxyl-mediated inhibition of mitochondrial transcription. *Free Radic. Biol. Med.* 16:653–660.

37. Siems, W. G., van Kuijk, F. J. G. M., Maass, R., and Brenke, R. 1994. Uric acid and glutathione levels during short-term whole body cold exposure. *Free Radic. Biol. Med.* 16:299–305.

38. Pinamonti, S., Chicca, M. C., Muzzoli, M., Papi, A., Fabbri, L. M., and Ciaccia, A. 1994. Oxygen radical scavengers inhibit clastogenic activity induced by sonication of human serum. *Free Radic. Biol. Med.* 16:363–371.

39. Halliwell, B. reactive oxygen species and the central nervous system. 1992. *J. Neurochem.* 59:1609–1623.

40. Lew, H., Pyke, S., and Quintanilha, A. 1985. Changes in the glutathione status of plasma, liver, and muscle following exhaustive exercise. *FEBS Lett.* 185:262–266.

41. Sen, C. K., Marin, E., Kretzschmer, M., and Hanninen, O. 1992. Skeletal muscle, and liver glutathione homeostasis in response to training, exercise, and immoblization. *J. Appl. Physiol.* 73:1265–1272.

42. Leeuwenburgh, C., and Ji, L. L. 1995. Glutathione depletion in rested and exercised mice: Biochemical consequence and adaptation. *Arch. Biochem. Biophys.* 316:941–949.

43. Ji, L. L., Katz, A., Fu, R. G., Griffiths, M., and Spence, M. 1993. Blood glutathione status during exercise: Effect of carbohydrate supplementation. *J. Appl. Physiol.* 74:788–792.

44. Duthie, G. G., Robertson, J. D., Maughan, R. J., and Morrice, D. C. 1990. Blood antioxidant status and erythrocyte lipid peroxidation following long distance running. *Arch. Biochem. Biophys.* 282:78–83.

45. Becker, K., Leichsenring, M., Gana, L., Bremer, H. J., and Schirmer, R. H. 1995. Glutathione and associated antioxidant systems in protein energy malnutrition: Results of a study in Nigeria. *Free Radic. Biol. Med.* 18(2):257–263.

46. Anderson, M. T., Staal, F. J. T., Roederer, M., Gitler, C., Herzenberg, L. A., and Herzenberg, L A. 1994. Decreased glutathione in those HIV infected may account for decreased proliferative response and increased inflammatory stress. *Therapeutic Potential of Biological Antioxidants Symp.*, Tiburon, CA, September 29-October 1, abstr. 39.

47. Goyer, R. A. 1991. Toxic effects of metals. In *Casarett and Doull's Toxicology: The Basic Science of Poisons*, 4th ed., eds. M. O. Amdur, J. Doull and C. D. Klaassen, pp. 623–680. New York: Pergamon Press.

48. Stohs, S. J., and Bagchi, D. 1995. Oxidative mechanisms in the toxicity of metal ions. *Free Radic. Biol. Med.* 18:321–336.

49. Spear, N., and Aust, S. D. 1994. Thiol-mediated NTA-Fe(III) reduction and lipid peroxidation. *Arch. Biochem Biophys.* 312:198–202.

50. Perrin, D. D., and Watt, A. E. 1971. Complex formation of zinc and cadmium with glutathione. *Biochim. Biophys. Acta* 230:96–104.

51. Milne, L., Nicotera, P., Orrenius, S., and Burkitt, M. J. 1993. Effects of glutathione and chelating agents on copper-mediated DNA oxidation: Prooxidant and antioxidant properties of glutathione. *Arch. Biochem. Biophys.* 304:102–109.

52. Chubatsu, L. S., Gennari, M., and Meneghini, R. 1992. Glutathione is the antioxidant responsible for resistance to oxidative stress in V79 Chinese hamster fibroblasts rendered resistant to cadmium. *Chem. Biol. Interact.* 82:99–110.

53. Li, W., Yinzhi, Z., and Chou, I.-N. 1993. Alterations in cytoskeletal protein sulfhydryls and cellular glutathione in cultured cells exposed to cadmium and nickel ions. *Toxicology* 77:65–79.

54. Li, W. and Chou, I.-N. 1992. Effects of sodium arsenite on the cytoskeleton and cellular glutathione levels in cultured cells. *Toxicol. Appl. Pharmacol.* 114:132–139.

55. Münter, K., Athanasiou, M., and Stournaras, C. 1989. Inhibition of cellular activities by triethyllead: Role of glutathione and accumulation of triethylead in vitro. *Biochem. Pharmacol.* 38(22):3941–3945.

56. Jenkinson, S. G., Spence, T. H., Lawrence, R. A., Hill, K. E., Duncan, C. A., and Johnson, K. H. 1987. Rat lung glutathione release: Response to oxidative stress and selenium deficiency. *J. Appl. Physiol.* 62(1):55–60.

57. Yang, G. Q., Wang, S. Z., Zhou, R. H., and Sun, S. Z. 1983. Endemic selenium intoxication of humans in China. *Am. J. Clin. Nutr.* 37:872–881.

58. Hogberg, J., and Alexander, J. 1986. Selenium. In *Handbook on the Toxicology of Metals.* Vol II, 2nd ed, *Specific Metals,* eds. L. Friberg, G. F. Nordberg, and V. B. Vouk, pp. 482–512. Amsterdam: Elsevier.

59. Spallholz, J. E. 1994. On the nature of selenium toxicity and carcinostatic activity. *Free Radic. Biol. Med.* 17:45–64.

60. Shen, H.-M., Shi, C.-Y, Lee, H.-P, and Ong, C.-N. 1994. Aflatoxin B₁-induced lipid peroxidation in the rat liver. *Toxicol. Appl. Pharmacol.* 127:145–150.

61. Jenkinson, S. G., Lawrence, R. A., Burk, R., and Gregory, P. 1983. Non-selenium dependent glutathione peroxidase activity in rat lung: Association with lung glutathione S-transferase activity and effects of hyperoxia. *Toxicol. Appl. Pharmacol.* 68:399–404.

62. Duggin, G. G. 1980. Mechanisms in the development of analgesic nephropathy. *Kidney Int.* 18:553–561.

63. Green, C. J., Healing, G., Simpkin, S., Lunee, J., and Fuller, B. J. 1986. Increased susceptibility to lipid peroxidation in rabbit kidneys: A consequence of warm ischemia and subsequent reperfusion. *Comp. Biochem. Physiol. B. Biol. Sci.* 83:603–606.

64. Torres, A., Rodriguez, J. V., Ochoa, J. E., and Elias, M. M. 1986. Rat kidney function related to tissue glutathione levels. *Biochem. Pharmacol.* 35:3355–3358.

65. Scaduto, R. C., Gattone, V. H., Grotyohan, L.W., Wertz, J., and Martin, L. F. 1988. Effect of an altered glutathione content on renal ischemic injury. *Am. J. Physiol. (Renal Fluid Electrolyte Physiol. 24)* 255:F911–F921.

66. Bend, J. D., Horton, J. K., Brigelius, R., Dostal, L.A., Mason, R.P., and Serabjit-Singh, C. J. 1989. Cell selective toxicity in the lung: Role of metabolism. *Proc. Int. Cong. Toxicology,* eds. C. J. Volans, J. Sims, F. M. Sullivan, and P. Turner, July 16–21, pp. 220–232. London: Taylor & Francis.

67. Bunnel, E., and Pacht, E. R. 1993. Oxidized glutathione is increased in the alveolar fluid of patients with adult respiratory distress syndrome. *Am. Rev. Respir. Dis.* 148:1174–1178.

68. Cantin, A. M., Hubbard, R. C., and Crystal, R.G. 1989. Glutathione deficiency in the epithelial lining fluid of the lung of the lower respiratory tract in idiopathic pulmonary fibrosis. *Am. Rev. Respir. Dis.* 139:370–372.

69. Smith, L. J., Houston, M., and Anderson, J. 1993. Increased levels of glutathione in bronchoalveolar fluid from patients with asthma. *Am. Rev. Respir. Dis.* 147:1461–1464.

70. Cantin, A. M., North, S. L., Hubbard, R. C., and Crystal, R. G. 1987. Normal alveolar epithelial lining fluid contains high levels of glutathione. *J. Appl. Physiol.* 63:145–157.

71. Müller, R., and Lehrach, F. 1981. Haemorheology and cerebrovascular disease: Multifunctional approach with pentoxifylline. *Curr. Med. Res. Opin.* 7:253–263.

72. Youdim, M. B. H. 1988. *Brain Iron. Neurochemical and Behavioural Aspects.* New York: Taylor and Francis.

73. Kontos, H. 1989. Oxygen radicals in CNS damage. *Chem. Biol. Interact.* 72:229–255.

74. Liu, D., Yang, R., Yan, X., and McAdoo, D. J. 1994. Hydroxyl radicals generated in vivo kill neurons in the rat spinal cord: Electrophysiological, histological, and neurochemical results. *J. Neurochem.* 62:37–44.

75. Slivka, A., Mytilineou, C., and Cohen, G. 1987. Histochemical evaluation of glutathione in brain. *Brain Res.* 409:275–284.

76. Makar, T., Nedergaard, M., Preuss, A., Gelbard, A. S., Perumal, A. S., and Cooper, A. J. L. 1994.

Vitamin E, ascorbate, glutathione, glutathione disulfide, and enzymes of glutathione metabolism in cultures of chick astrocytes and neurons: Evidence that astrocytes play an imporatnt role in antioxidative processess in the brain. *J. Neurochem.* 62:45–53.

77. Shivakumar, B. aand Ravindrananth, V. 1992. Selective modulation of glutathione in mouse brain regions and its effect on acrylamide-induced neurotoxicity. *Biochem Pharmacol.* 43:263–269.

78. Mizui, T., Kinouchi, H., and Chan, P. 1992. Depletion of brain glutathione by buthionine sulfoximine enhances cerebral ischemic injury in rats. *Am. J. Physiol. (Heart Circ. Physiol. 31)* 262:H313–H317.

79. Sofic, E., Lange, K. W., Jellinger, K., and Riederer, P. 1992. Reduced and oxidized glutathione in the substantia nigra of patients with Parkinsons's disease. *Neurosci. Lett.* 142:128–130.

80. McLachlan, D. R. C., Dalton, A. J., Kruck, T. P, A., Bell, M. Y., Smith, W. L., Kalow, W., and Andrews, D. F. 1991. Intramuscular desferrioxamine in patients with Alzheimer's disease. *Lancet* 337:1304–1308.

81. Mårtensson, J., Jain, A., and Meister, A. 1990. Glutathione is required for intestinal function. *Proc. Natl. Acad. Sci. U.S.A.* 87:1715–1719.

82. Aw, T. Y., Williams, M. W., and Gray, L. 1992. Absorption and lymphatic transport of peroxidized lipids by rat small intestine in vivo: Role of mucosal GSH. *Am. J. Physiol. (Gastrointest. Liver Physiol. 25)* 262:G99–G106.

83. Dahm, L. J., and Jones, D. P. 1994. Secretion of cysteine and glutathione from mucosa to lumen in rat small intestine. *Am. J. Physiol. (Gastrointest. Liver Physiol. 30)* 267:G292–G300.

84. Hagen, T. M., Wierzbicka, G. T., Sillau, A. H., Bowman, B. B., and Jones, D. P. 1990. Bioavailability of dietary glutathione: Effect on plasma concentration. *Am. J. Physiol. (Gastrointest. Liver Physiol. 22)* 259:G524–G529.

85. Lash, L. H., Hagen, T. M., and Jones, D P. 1986. Exogenous glutathione protects intestinal epithelial cells from oxidative injury. *Proc. Natl. Acad. Sci. U.S.A.* 83:4641–4645.

86. Ookhtens, M. and Mittur, A. V. 1994. Developmental changes in plasma thiol-disulfide turnover in rats: A multicompartmental approach. *Am. J. Physiol. (Regulatory Integrative Comp. Physiol. 36)* 267:R415–R425.

87. Vincenzini, M. T., Favilli, F., and Iantomasi, T. 1992. Intestinal uptake and transmembrane transport system of intact GSH; characteristics and possible biological role. *Biochim. Biophys. Acta* 1113:13–23.

88. McCord, J. M. 1985. Oxygen-derived free radicals in post-ischemic tissue injury. *N. Engl. J. Med.* 312:159–163.

89. Kornbrust, D. J. and Mavis, R. D. 1980. Relative susceptibility of microsomes from lung, heart, liver, kidney, brain, and testes to lipid peroxidation: Correlation with vitamin E content. *Lipids* 15:315–322.

90. Reddy, C. C., Scholz, R. W., Thomas, C. E., and Massaro, E. J. 1982. Vitamin E dependent reduced glutathione inhibition of rat liver microsomal lipid peroxidation. *Life Sci.* 31:571–576.

91. Powell, S., and McCay, P. B. 1995. Inhibition of doxorubicin-induced membrane damage by thiol compounds: toxicological implications of a glutathione-dependent microsomal factor. *Free Radic. Biol. Med.* 18:159–168.

92. Pryor, W. A. 1976. The role of free radical reactions in biological systems. In *Free Radic. in Biology,* Vol. 1, ed. W. A. Pryor, pp. 1–47. New York: Academic Press.

93. Niki, E., Tsuchiya, J., Tanimura, R., and Kamiya, Y. 1982. Regeneration of vitamin E from α-chromanoxyl radical by glutathione and vitamin C. *Chem. Lett.* 789–792.

94. Hopkins, F. G., and Morgan J. M. C. 1936. Some relations between ascorbic acid and glutathione. *Biochem. J.* 30:1446–1462.

95. Borsook, H., Davenport, H. W., Jefferys, C. E. P., and Warner, R. C. 1937. The oxidation of ascorbic acid and its reduction in vitro and in vivo. *J. Biol. Chem.* 117:237–279.

96. Dalton, D. A., Russell, S., Hanus, F. J., Pascoe, G. A., and Evans, H. J. 1986. Enzymatic reactions of ascorbate and glutathione that prevent peroxide damage in soybean root nodules. *Proc. Natl. Acad. Sci. U.S.A.* 83:3811–3815.

97. Winkler, B. S. 1992. Unequivocal evidence in support of the nonenzymatic redox coupling between glutathione, glutathione disulfide and ascorbic acid/dehydroascorbic acid. *Biochim. Biophys. Acta* 1117:287–290.

98. Winkler, B. S., Orselli, S. M., and Rex, T. S. 1994. The redox couple between glutathione and ascorbic acid: A chemical and physiological perspective. *Free Radic. Biol. Med.* 17:333–349.

99. Mårtensson, J. and Meister, A. 1991. Glutathione deficiency decreases tissue ascorbate levels in newborn rats: Ascorbate spares glutathione and protects. *Proc. Natl. Acad. Sci. U.S.A.* 88:4656–4660.

100. Mårtensson, J., Jain, A., Stole, E., Frayer, W., Auld, P. A. M., and Meister, A. 1991. Inhibition of glutathione synthesis in the newborn rat: A model for endogenously produced oxidative stress. *Proc. Natl. Acad. Sci. U.S.A.* 88:9360–9364.

101. Han, J., Mårtensson, J., Meister, A., and Griffith, O. W. 1992. Glutathione ester but not glutathione delays onset of scurvy in guinea pigs fed vitamin C-deficient diet. *FASEB J.* 6:5631.

102. Eiserich, J. P., Vossen, V., van der Vliet, A., O'Neill, C. A., Halliwell, B., and Cross, C. E. 1994.

Cigarette smoke-induced nitration of tyrosine and its inhibition by ascorbate, urate, and glutathione. *Therapeutic Potential of Biological Antioxidants Symp.*, Tiburon, CA, September 29-October 1, abstr. 46.

103. Roomi, M. W., and Tsao, C. 1994. Antioxidants and thermal aggregation studies in guinea pig eye lens protein and their modulation by dietary vitamin C. *Therapeutic Potential of Biological Antioxidants Symp.*, Tiburon, CA, September 29-October 1, abstr. 78.

104. Netke, S. P., and Niedzwiecki, A. 1994. Liver antioxidants in guinea pigs fed atherogenic diet with varying intakes of ascobic acid. *Therapeutic Potential of Biological Antioxidants Symp.*, Tiburon, CA, September 29–October 1, abstr. 68.

105. Hrubá, F., Nováková, V., and Ginter, E. 1982. The effect of chronic marginal vitamin C deficiency on the α-tocopherol content of the organs and plasma of guinea pigs. *Experientia* 38:1454–1455.

106. Palmquist, B. M., Philipson, B., and Barr, P. O. 1984. Nuclear cataract and myopia during hyperbaric oxygen therapy. *Br. J. Opthalmol.* 68:113.

107. Maitra, I., Seebinova, E., Trischler, H., and Packer, L. 1995. α-Lipoic acid prevents buthionine sulfoximine-induced cataract formation in newborn rat. *Free Radic. Biol. Med.* 18:823–829.

108. Brunk, U. T., Jones, C. B., and Sohal, R. S. 1992. A novel hypothesis of lipofuscinogenesis and cellular ageing based on interactions between oxidative stress and autophagocytosis. *Mutat. Res.* 275:395–403.

109. Gao, G., Öllinger, K., and Brunk, U. T. 1994. Influence of intracellular glutathione concentration of lipofuscin accumulation in cultured neonatal rat cardiac myocytes. *Free Radic. Med. Biol.* 16:187–194.

110. Ames, B. N., Shigenaga, M. K., and Hagen, T. M. 1993. Oxidants, antioxidants and the degenerative diseases of aging. *Proc. Natl. Acad. Sci. U.S.A.* 90:7915–7922.

111. Maung, Z. T., Hogarth, L., Reid, M. M., Proctor, S.J., Hamilton, P. J., and Hall, A. G. 1994. Raised intracellular glutathione levels correlate with in vitro resistance to cytotoxic drugs in leukaemic cells from patients with acute lymphoblastic leukaemia. *Leukemia* 8:1487–1491.

112. Zaman, G. J. R., Lankelma, J., van Tellingen, O., Beijnen, J., Dekker, H., Paulusma, C. Oude Elferink, R. P. J., Baas, F., and Borst, P. 1995. Role of glutathione in the export of compounds from cells by the multidrug-resistant-associated protein. *Proc. Natl. Acad. Sci. U.S.A.* 92:7690–7694.

113. Sheffner, A. L., medler, E. M., Jacobs, L. W., and Sarett, H. P. 1964. The in vitro reduction in viscosity of human tracheo-bronchial secretions by acetylcysteine. *Am. Rev. Respir. Dis.* 90:721–729.

114. Sjödin, K., Nilsson, E., Hallberg, A., and Tunek, A. 1989. Metabolism of *N*-acetylcysteine: Some structural requirements for the deacylation and consequences for oral bioavailability. *Biochem. Pharmacol.* 38:3981–3985.

115. Bonanomi, L., and Gazzaniga, A. 1980. Toxicological, pharmacological, and metabolic studies on acetylcysteine. *Eur. J. Respr. Dis.* 61(suppl. 111):45–51.

116. Bernard, G., Lucht, W. D., Niedermeyer, M. E., Snapper, J. R., Ogletree, M. L., and Brigham, K. L. 1984. Effect of *N*-acetylcysteine on pulmonary response to endotoxin in the awake sheep and upon in vitro granulocyte function. *J. Clin. Invest.* 73:1772–1784.

117. Revesz, L., and Malaise, E. P. 1983. Significance of cellular glutathione in radioprotection and repair of radiation damage. In *Functions of Glutathione: Biochemical, Physiological, Toxicological, and Clinical Aspects*, ed. A. Larsson, pp. 163–173. New York: Raven Press.

118. Prescott, L. F., Illingworth, R. N., Critchley, J. A., Stewart, M. J., Adam, R. D., and Proudfoot, A. T. 1979. Intravenous *N*-acetylcysteine: The treatment of choice for paracetamol poisoning. *Br. Med. J.* 2:1097–1110.

119. Bowman, G, Bäcker, U., Larsson, S., Melander, B., and Wahlinder, L. 1983. Oral acetylcysteine reduces exacerbation rates in chronic bronchitis: Report of a trial organized by the Swedish Society for Pulmonary Diseases. *Eur. J. Respir. Dis.* 64:405–415.

120. Jamieson, D. D., Kerr, D. R., and Unsworth, I. 1987. Interaction of I-acetylcysteine and bleomycin on hyperbaric oxygen-induced lung damage in mice. *Lung* 165:239–247.

121. Meyer, A., Buhl, R., Kampf, S., and Magnussen, H. 1995. Intravenous *N*-acetylcysteine and lung glutathione of patients with pulmonary fibrosis and normals. *Am. J. Respir. Crit. Care Med.* 152:1055–1060.

122. de Quay, B., Malinverni, R., and Lauterburg, B. H. 1992. Glutathione depletion in HIV-infected patients: Role of cysteine deficiency and effect of oral *N*-acetylcysteine. *AIDS* 6:815–819.

123. Bridgeman, M. M. E., Marsden, M., Selby, C., Morrison, D., and MacNee W. 1994. Effect of *N*-acetylcysteine on the concentrations of thiols in plasma, bronchoalveolar lavage fluids and lung tissue. *Thorax* 49:670–675.

124. Horton, J. K., Meredith, M. J., and Bend, J. R. 1987. Glutathione biosynthesis from sulfur-containing amino acids in enriched populations of Clara and type II cells and macrophages freshly isolated from rabbit lung. *J. Pharmacol. Exp. Ther.* 240:376–380.

125. Hagen, T. M., Brown, L. A., and Jones, D. P. 1986. Protection against paraquat-induced injury by exogenous GSH in pulmonary alveolar type II cells. *Biochem. Pharmacol.* 35:4537–4542.

126. Mårtensson, J., Jain, A., Frayer, W., and Meister, A. 1989. Glutathione metabolism in the lung:

Inhibition of its synthesis leads to lamellar body and mitochondrial defects. *Proc. Natl. Acad. Sci. U.S.A.* 86:5296–5300.

127. Sciuto, ·A. M., Strickland, P. T., Kennedy, T. P., and Gurtner, G. H. 1995. Protective effects of N-acetylcysteine treatment after phosgene exposure in rabbits. *Am J. Repir. Crit. Care Med.* 151:768–772.
128. Moldéus, P., Cotgreave, I., and Berggren, M. 1986. Lung protection by a thiol-containing antioxidant: *N*-acetylcysteine. *Respiration* 50(suppl. 11):31–42.
129. Meulenbelt, J., van Bree, L., Dormans, J. A. M. A., and Sangster, B. 1994. No beneficial effect of N-acetylcysteine treatment on broncho-alveolar lavage fluid variables in acute nitrogen dioxide intoxicated rats. *Hum. Exp. Toxicol.* 13:472–477.
130. Drost, E., Lannan, S., Bridgeman, M. M. E., Brown, D., Selby, C., Donalson, K., and MacNee, W. 1991. Lack of effect of N-acetylcysteine on the release of oxygen radicals from neutrophils and alveolar macrophages. *Eur. Respir. J.* 4:723–729.
131. Critchley, J. A. J. H., Beely, J. M., Clark, R. J., Summerfield, M., Bell, S., Spurlock, M. S., Edgington, J. A. G., and Buchanan, J. D. 1990. Evaluation of *N*-acetylcysteine and methylprednisolone as therapies for oxygen and acrolein-induced lung damage. *Environ. Health Perspect.* 85:89–94.
132. Williamson, J., and Meister, A. 1981. Stimulation of hepatic glutathione formation by administration of L-2-oxothiazolidine-4-carboxylate, a 5-oxoprolinase substrate. *Proc. Natl. Acad. Sci. U.S.A.* 78:936–939.
133. Williamson, J. M., Boettcher, B., and Meister, A. 1982. Intracellular cysteine delivery system that protects against toxicity by promoting glutathione synthesis. *Proc Natl. Acad. Sci. U.S.A.* 79:6246–6249.
134. Tsan, M.-F., Danis, E. N., Del Vecchio, P. J., and Rosano, C. L. 1985. Enhancement of intracellular glutathione protects endothelial cells against oxidant damage. *Biochem. Biophys. Res. Comm.* 127:270–276.
135. Kuzuya, M., Naito, M., Funaki, C., Hayashi, T., Asai, K., and Kuzuya, F. 1989. Protective role of intracellular glutathione against oxidized low density lipoproteins in cultured endothelial cells. *Biochem. Biophys. Res. Commun.* 163:1466–1472.
136. Shug, A. L., and Madsen, D. C. 1994. Protection of the ischemic rat heart by procysteine and amino acids. *J. Nutr. Biochem.* 5:356–359.
137. Lailey, A. F., Hill, L., Lawston, I. W., Stanton, D., and Upshall. D. 1991. Protection by cysteine esters against chemically-induced pulmonary oedema. *Biochem. Pharmacol.* 42(suppl):S47–S54.
138. Wilde, P. E., and Upshall, D. 1994. Cysteine esters protect cultured rodent lung slices from sulfur mustard. *Human Exp. Toxicol.* 13:743–748.
139. Anderson, M. E., Powrie, F., Puri, R. N., and Meister, A. 1985. Glutathione monoethyl ester: Preparation, uptake by tissues, and conversion to glutathione. *Arch. Biochem. Biophys.* 239(2):538–548.
140. Singhal, R. K., Anderson, M. E., and Meister, A. 1987. Glutathione, a first line of defense against cadmium toxicity. *FASEB J.* 1:220–233.
141. Meister, A. 1991. Glutathione deficiency produced by inhibition of its synthesis, and its reversal: Applications in research and therapy. *Pharmacol. Ther.* 51:155–194.
142. Beckman, J. S., Beckman, T. W., Chen, J., Marshall, P. A., and Freeman, B. A., 1990. Apparent hydroxyl radical production by peroxynitrite: Implications for endothelial injury from nitric oxide and superoxide. *Proc. Natl. Acad. Sci. U.S.A.* 87:1620–1624.
143. White, A. C., Maloney, E. K., Boustani, M. R., Hassoun, P. M., and Fanburg, B. L. 1995. Nitric oxide increases cellular glutathione levels in rat lung fibroblasts. *Am. J. Respir. Cell. Mol. Biol.* 13:442–448.
144. Sciuto, A. M., Strickland, P. T., Kennedy, T. P., Guo, Y.-L., and Gurtner, G. H. 1996. Intratracheal administration of dibutyryl cAMP attenuates edema formation in phosgene-induced acute lung injury. *J. Appl. Physiol.* 80:149–157.
145. Schöneich, C., Asmus, K.-D, Dillinger, U., and von Bruchhasuen, F. 1989. Thiyl radical attack on fatty acids: A possible route to lipid peroxidation. *Biochem. Biophys. Res. Commun.* 161:113–120.
146. Surdhar, P. S., and Armstrong, D. A. 1986. Redox potentials of some sulfur-containing radicals. *J. Phys. Chem.* 90:5915–5917.

Chapter Ten

ANTIOXIDANT EFFECTS OF HYPOTAURINE AND TAURINE

Steven I. Baskin, Kokiku Wakayama, Melanie A. Banks,
Dale W. Porter, and Harry Salem

Hypotaurine (2-aminosulfinic acid) is metabolized from cysteamine (2-aminoethanethiol) in the heart of rats, oxen, horses, and pigs but not sheep (1) and from cysteine sulfinic acid (β,β'-dithiodialanine sulfinic acid) in brain, liver and perhaps the prostate in rats (2). Several studies have shown that hypotaurine (2-aminosulfinic acid) is further oxidized to taurine (2-aminoethanesulfonic acid) through a disulfone intermediate (3) or to thiotaurine (2-aminoethanethiolsulfonic acid) (4). Loading rats with hypotaurine leads to greater excretion of taurine in urine and some formation of sulfate (5). A number of studies in several systems have shown that hypotaurine can inhibit the uptake of taurine and beta-alanine (6, 7).

In addition to being a metabolic intermediate, hypotaurine has been suggested to be an antioxidant in a number of biological systems, including the lung, neutrophils, and sexual tissue such as sperm. Taurine and hypotaurine are necessary compounds for sperm capacitation, fertilization, and embryo development (8) and should be carefully considered in in vitro fertilization (9). Earlier studies in horses (10) and later studies in humans (11) demonstrated hypotaurine in sperm serves as an antioxidant to protect against infertility.

It has been indicated that hypotaurine acts to protect against toxic hydroxyl free radical (3). In addition to scavenging the initiator ˙OH, hypotaurine affects other reactions relevant to the initiation, propagation, and termination phases of lipid peroxidation. Hypotaurine decreases ferrous auto-oxidation, leading to generation of perferryl iron, decreases ferrous oxidation by cumene hydroperoxide that forms alkoxy radical, and inhibits lipid peroxidation (12).

A number of studies have suggested that there is an interaction between hypotaurine and taurine biochemically and possibly functionally as antioxidants. Thus, the antioxidant relationship may be determined by some mathematical formula related to the concentrations and turnover of both chemical species.

Many of the studies in which tissue hypotaurine and/or taurine levels were measured utilized a combination of oxidizing acids (e.g., picric acid, sulfosalicylic acid) or salts of acids (e.g., dansyl chloride, dabsyl chloride) to extract amino acids from tissues and derivatize to provide quantitative analysis, respectively. It appears that while these compounds are excellent for some amino acids, they appear to oxidize sulfinic acids to sulfonic acids. Thus, certain chromophores, such as dansyl cholide and dabsyl chloride, may convert hypotaurine to taurine. In the study of amino acids in the pineal gland (13) and in a model of leukemic cells (14) using dabsylation to provide derivatize amino acids, good separation of the sulfonic amino acids was found. In

The opinions or assertions contained in this article are the private views of the authors and are not to be construed as official or as reflecting the views of the Army or Department of Defense.

the Krusz study, for example, pineal homotaurine (or that which had the same retention time as authentic homotaurine) was separated from a much larger taurine peak by high-performance liquid chromatography (HPLC). In general, it is difficult to distinguish these two substances in a biological matrix. Figure 1 illustrates an example of colorimetric dabsyl chloride derivatives from a HPLC amino acid analysis. The chromatographic conditions are indicated on the figure. This method appeared to provide a better separation of the acidic amino acids than orthophthalde-hyde (OPA) derivatized amino acids. Figure 2 illustrates an example of fluorometric amino acid analysis using OPA. This analysis is good for the neutral amino acids, but as can be observed from Figure 2, it is more difficult to separate the acidic amino acids derivitized with OPA.

To examine if the derivatizing agent could oxidize sulfinic acids, we injected a known quantity of sulfinic acid alone and observed the results. Injecting either hypotaurine (1.5 mM) or cysteine sulfinic acid (0.625 mM) under the dabsylating conditions described (13), there was conversion of either hypotaurine to taurine or cysteine sulfinic acid to cysteic acid (Figure 3). If one changes the concentration of hypotaurine in the analysis sample, the conversion of hypotaurine to taurine changes (Figure 4). Additionally, altering the concentration of the dabsyl reagent can also change the ratio of taurine to hypotaurine using the excess dabsyl chloride reagent (Figure 5). These data suggest that these derivatizing reagents, although useful for separating amino acids by HPLC, may increase measured taurine levels by oxidizing hypotaurine to taurine. Several reported assays suggested taurine concentrations in tissues that were higher than the solubility of taurine in solution. These values may be explained by conversion of hypotaurine to taurine during HPLC analysis. Conversely, the concentration of hypotaurine may be much larger than previous studies have determined. If this is true, then hypotaurine concentration may be of similar magnitude as that of other antioxidants such as glutathione, which has been reported to be in the range of 15 mM (15). Hypotaurine, together with taurine, and its related enzyme systems may serve as another fundamental antioxidant regenerating system.

The antioxidant properties of taurine seem to be related not to its ability to prevent oxidation, but to ameliorate the damage of the oxidation. More specifically, the results of studies conducted

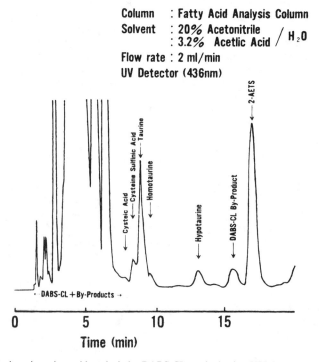

Column   : Fatty Acid Analysis Column
Solvent : 20% Acetonitrile
         : 3.2% Acetlic Acid / H$_2$O
Flow rate : 2 ml/min
UV Detector (436nm)

Figure 1.   Colorimetric amino acid analysis by DABS-CL method using HPLC.

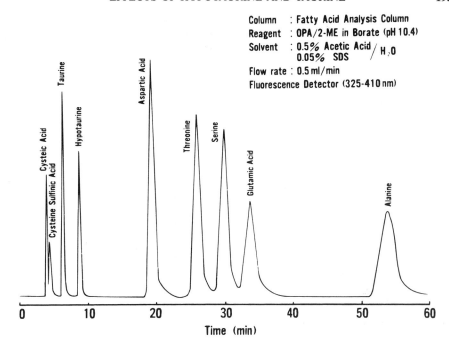

Figure 2. Fluorometric amino acid analysis using HPLC.

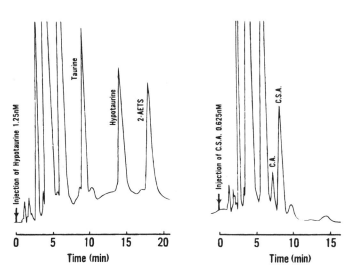

Figure 3. (left) Conversion from hypotaurine to taurine by DABS-CL (500 μl). (right) Conversion from CSA to CA by DABS-CL (500 μl).

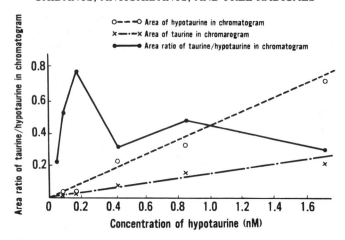

Figure 4. Conversion ratio from hypotaurine to taurine under different concentrations of hypotaurine (DABS-CL method).

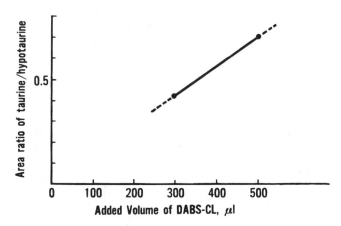

Figure 5. Conversion ratio from hypotaurine to taurine different concentrations of DABS-CL (area ratio of taurine/hypotaurine in chromatogram).

with various lung cell types suggest that taurine increases membrane integrity, decreasing the loss of cellular ions and other important cellular antioxidant defenses, such as glutathione. It becomes obvious then that this mechanism assumes that taurine is able to interact with the plasma membrane or membrane constituents. Is there evidence to suggest that taurine is capable of interaction with membranes?

Taurine has ionic and structural similarities to the head groups of phosphatidylcholine and phosphatidylethanolamine, which are neutral phospholipids. On the basis of these similar properties, a low-energy interaction between taurine and the neutral phospholipids was postulated (16). Equilibrium dialysis studies of sarcolemma membranes have confirmed that taurine does form a low-affinity interaction with membranes, with a $V_{max} = 2677$ nmol taurine/mg phospholipid and $K_d = 19.2$ mM (17). Positive cooperativity was indicated by the determination of the Hill coefficient as 1.90.

The results of the equilibrium dialysis studies were extended using phospholipid vesicles composed of individual phospholipids (18). Neutral phospholipids had a molar ratio of taurine:lipid from 0.5 to 1.0:1.0 and also had the highest binding affinity. Phosphatidylinositol, an acidic phospholipid, had no taurine bound. The inclusion of 50% cholesterol, which would result in increased membrane fluidity, increased membrane affinity for taurine. Thus, membrane binding of taurine occurs specifically with neutral phospholipids, probably due to ion-pair or dipole–dipole interactions, and is effected by membrane fluidity.

Cations, particularly Ca^{2+}, have an antagonistic effect on membrane-bound taurine (19). This observation and the finding of positive cooperativity of taurine binding to membrane phospholipids suggested that taurine binding to neutral phospholipids may alter Ca^{2+} binding to acidic phospholipids and vice versa (16). This proposal was investigated in a study of Ca^{2+} binding to cardiac sarcolemma membranes (19). High-affinity ($K_d = 0.03$ mM) and low-affinity ($K_d = 3.94$ mM) Ca^{2+} binding sites were determined to exist in the sarcolemma membranes. Taurine did not affect the low-affinity binding component, but resulted in a significant stimulation of Ca^{2+} binding at the physiological Ca_i^{2+}, 10^{-6} M.

The stimulatory effect of taurine on Ca^{2+} binding was more thoroughly investigated using mixed phospholipid vesicles. The stimulatory effect of taurine on Ca^{2+} binding was much higher in the mixed phospholipid vesicles than that predicted using vesicles containing only one type of phospholipid. This indicated that the stimulatory effect of taurine on Ca^{2+} binding is not additive (18). Therefore, it was proposed that taurine interaction with the neutral phospholipids leads to modification of Ca^{2+} binding to the acidic phospholipids.

Taurine can also modify intrinsic membrane protein activities by directly modifying membrane composition. In a study using guinea pigs, taurine has been reported to effect metabolic interconversions of phospholipids (20). A more recent study using rats reported that taurine inhibits phospholipid N-methylation and consequently the conversion of phosphatidylethanolamine to phosphatidylcholine. In addition, taurine also prevented a fourfold inhibition of the rat heart Na^+–Ca^{2+} exchanger activity produced by perfusion with 300 μM methionine, and this action of taurine is postulated to be achieved by its inhibitory effect on phospholipid N-methylation (21). However, studies using cats fed three levels of dietary taurine found no evidence of an effect on metabolic interconversions of phospholipids, but did find that taurine supplementation decreased free fatty acids and increased the proportion of all other lipid classes, particularly the neutral lipids, in cat hepatocytes (22).

The interaction of taurine with membrane phospholipids may also allow taurine to indirectly modify other lipid-dependent cell activities, such as the function of intrinsic membrane proteins. The operation of the erythrocyte Ca^{2+} channel is dependent on membrane fluidity, being stimulated by membrane enrichment with cholesterol and inhibited by cholesterol depletion (23). Taurine stimulates an ATPase activity of the human erythrocyte, which was thought to be the Ca^{2+},Mg^{2+}-ATPase (24). These researchers suggested that an interaction between taurine and phospholipids in the plasma membrane is responsible for the stimulatory action of taurine on the Ca^{2+},Mg^{2+}-ATPase. Additional support for this view comes from the observation that Ca^{2+}-ATPase activity is dependent on membrane fluidity (25).

Although taurine is known to function in bile acid conjugation, as a neurotransmitter, as a modifier of cardiac function and as a possible measure in trauma (hypotaurinemia) patients (26), recent evidence also suggests that it functions as an antioxidant in tissues that are susceptible to oxidant injury, such as the lung.

Exposure to oxidant gases such as hyperbaric oxygen, nitrogen dioxide, and ozone results in pulmonary injury that is characterized by destruction of capillary endothelial cells, edema, hypertrophy and hyperplasia of the bronchiolar epithelium, bronchiolization of the alveolar duct epithelium, and an influx of macrophages and polymorphonuclear leukocytes into the alveolar air spaces (27–30). Lung oxidant injury is accompanied by metabolic changes including lipid peroxidation, membrane leakage, and the mobilization of antioxidant compounds such as glutathione, ascorbic acid, and vitamin E (31–39). The distribution of the injury in the lung varies with the type of oxidant insult: With NO_2 and ozone exposures, the epithelium of the conducting airways and the alveoli near the terminal bronchioles (proximal alveoli) are primarily involved (40). Alveolar pneumocytes exhibit a wide range of sensitivity to oxidant injury, with type II epithelial cells and alveolar macrophages being the most resistant, while type I epithelial cells are the most susceptible (41–46).

Dietary taurine supplementation (0.5% in the drinking water for 14 days) has been shown to protect against bronchiolar damage induced by in vivo exposure of hamsters to both 7 ppm and

30 ppm NO_2 for 24 h (47). Taurine was effective in preventing NO_2-induced damage to Clara cells and ciliated cells in the bronchiolar epithelium and the accompanying infiltration of neutrophils and macrophages into the bronchiolar and alveolar duct regions. In a later study, Gordon et al. also noted that taurine brought about the formation of gap junction-like aggregates among the tight junctions between type I pneumocytes in both air- and NO_2-exposed lungs of hamsters, and suggested that this might confer a protective effect upon the alveolar epithelium (48). On the other hand, Lehnert et al. reported that upon exposure of rats to higher concentrations of NO_2 (100 ppm), even for brief periods of time (15 min), 0.5% taurine administered in the drinking water over a 7-day or a 14-day period failed to prevent an increase in lung wet weight, a crude indicator of pulmonary edema (49). However, more sensitive indices of lung injury were apparently not measured in this trial, and it is possible that at a cellular level taurine did confer some protection against the oxidant injury.

Taurine formed via the glutathione-dependent metabolism of cysteamine that is taken up by the lung is retained by rat lung slices and does not undergo further metabolism in isolated perfused rat lung but is effused into the medium (50, 51). Thus it seems likely that taurine has an endogenous role in lung tissue. Exogenously supplied taurine accumulates in rat lungs (52). Taurine uptake by isolated rat alveolar macrophages and type II epithelial cells has been demonstrated by Banks and her co-workers (53). Furthermore, the taurine content of alveolar macrophages and type I and type II epithelial cells has been shown to inversely correlate with their susceptibility to oxidant injury (54). In in vitro studies, isolated alveolar macrophages exposed to ozone (0.45 ppm for 15–60 min) exhibited decreased viability, increased lipid peroxidation, increased membrane leakage of protein, glutathione, and potassium ion, and decreased Na^+,K^+-ATPase activity. These metabolic changes were accompanied by an increase in intracellular taurine levels despite increased leakage of taurine into the medium, suggesting that taurine is mobilized from bound intracellular stores in response to oxidant injury in these cells (55). In studies with exogenously added taurine (0.45 ppm ozone, 15 min; 0–500 μM taurine), the effects of ozone on cell viability, lipid peroxidation, and membrane leakage were mitigated, but taurine supplementation had no effect on the activity of the Na^+,K^+-ATPase except at 100 μM, which is its normal plasma concentration in the rat. Interestingly, the degree of protection from oxidant injury also was maximal at 100 μM taurine and declined at 250 and 500 μM concentrations of exogenous taurine, despite demonstrated increases in the intracellular taurine concentration at these levels of exogenous taurine after ozone exposure. These results suggest that the protective effect of taurine against ozone-induced damage to alveolar macrophages is concentration dependent (56, 57).

A protective effect of taurine against acute paraquat (40 mg/kg) intoxication has been observed in renal but not lung tissue in rats. Through infusion of 2.5% or 5% taurine into the tail vein for 1 h after sc injection of paraquat significantly decreased the lung/blood ratio of taurine, longer infusions (3 or 5 h) significantly elevated the paraquat concentration in the lung. Paraquat is known to induce lung lesions through generation of superoxide anion and lipid peroxidation (58).

In certain types of oxidant-induced pulmonary injury, unless a proliferation of alveolar type II cells occurs to replace damaged type I cells, intra-alveolar fibrosis results (59). Bleomycin, like paraquat, is thought to undergo continuous cycling from the oxidized to the reduced state, generating a superoxide ion in the process (60). Bleomycin-induced lung fibrosis is attenuated by taurine. Administration of 500 mg/kg ip taurine to hamsters for 1 wk prior to intratracheal instillation of bleomycin (7.5 units/kg), followed by continuation of taurine administration in the drinking water (1%) and ip (250 mg/kg), resulted in inhibition of increases in lung collagen and a decrease in the number of neutrophils in lung lavage fluid. Less than half as many lung lesions, consisting of diffuse nuclear alveolitits and multifocal fibroplasia, and smaller consolidated foci were seen in the taurine-treated, bleomycin-exposed animals. When taurine was administered to saline-treated animals in the same protocol, elevations of plasma taurine and bronchoalveolar lavage fluid taurine concentrations resulted, indicating that administration of exoginous taurine can elevate circulating taurine levels (61). Wang and co-workers have also shown that combined treatment with taurine and niacin results in even greater protection against bleomycin-induced lung injury than taurine alone. When taurine and niacin were administered either (1) in the drinking water (taurine 1% w/v) and ip (250 mg/5 ml/kg) or (2) in the diet (both 2.5% w/w), reductions in lung malonaldehyde level, superoxide dismutase activity (therefore reducing the amount of hydrogen peroxide generated), calcium level, hydroxyproline level, prolyl hydroxylase activity, and poly(ADP-ribose) polymerase activity were observed versus the group treated with bleomycin only (60). The effects of taurine and niacin in reducing lung collagen formation were postulated to take place through (1) their combined inhibition of bleomycin-induced increases in collagen reactivity, (2) the individual effects of taurine to scavenge hypochlorous acid produced by neutrophils and to stabilize the pulmonary membranes, and (3) niacin to replenish the bleomycin-induced depletion of NAD and ATP (62). Interestingly, the combination of taurine and niacin (each 2.5% w/w in the feed, administered for 6 days before and continued after treatment) was also found to attenuate the induction of lung fibrosis and phospholipidosis by amiodarone (1.5 μmol/5 ml/100 g body weight, administered by intratracheal instillation to hamsters), an antiarrhythmic drug for which the mechanism of pulmonary toxicity is yet unknown. In this case, taurine and niacin individually and the combination thereof were equally effective in reducing amiodarone-induced increases in lipid peroxidation and superoxide dismutase activity (though not to the levels of the saline-treated controls) and collagen accumulation, but only niacin and the niacin–taurine combination were (equally) effective in preventing the increase in total phospholipid content of the hamster lungs (63).

Similarly, Venkatesan and Chandrakasan (64) have shown taurine plus niacin (50 mg/100 g each in the drinking water for 7 days prior to and 2 days following treatment) protect against lung fibrosis induced by cyclophosphamide (single ip dose of 20 mg/100 g in rats), a cytotoxic and immunosuppressive drug that is thought to produce pulmonary injury through the generation of reactive oxygen metabolites. The combined agent was successful in inhibiting cyclophospha-mide-induced lipid peroxidation in lung tissue, preventing increases in protein, lactate dehydrogenase, and angiotensin-converting enzyme leakage into lung fluid, and preventing depletion of the antioxidants glutathione and ascorbic acid from lung lavage fluid (65). These same authors have reported that taurine plus niacin also ameliorates paraquat-induced early lung injury in rats (64).

Though taurine is generally considered to be protective against lung injury in its capacity as an antioxidant and/or membrane stabilizer, it cannot be ignored that taurine may also contribute to pulmonary injury by reacting with hypochlorous acid formed in neutrophils to form taurine chloramines, which are relatively long-lived oxidants. The accumulation of polymorphonuclear leukocytes in the lung accompanies lung oxidant injury, and high concentrations of taurine have been found in the airway secretions of patients with cystic fibrosis and other inflammatory lung diseases (66). When the vascular surface of an isolated perfused rabbit lung was exposed to taurine chloramine, albumin flux across the endothelium was increased in the absence of cytolysis of the cells making up the endothelial barrier (67). Similar results were observed in cocultures of dedifferentiated type II cells and fibroblasts from rat lung that were exposed to increasing concentrations of taurine chloramine; albumin flux increased accordingly (68). On the other hand, taurine may indeed play a protective role against further damage to lung endothelium in the presence of these inflammatory cells. Uptake of taurine chloramine through anion transporters has been shown to induce cytotoxicity in a cat epithelial cell line. However, a molar excess of taurine relative to hypochlorous acid reduced the cytotoxicity by decreasing the formation of dichloramines and increasing the formation of monochloramines. Thus addition of taurine increased the amount of hypochlorous acid necessary to induce lung epithelial cell cytotoxicity (66). Also, the migration of neutrophils through the endothelium at sites of inflammation may be facilitated by oxidant-mediated disruption of cellular junctions, and Tatsumi and Fliss have shown that hypochlorous acid increased the microvascular permeability in isolated perfused rabbit lungs, but that taurine chloramine did not have a similar effect (69).

In summary, taurine appears to play a role in modulating lung oxidant injury. Its precise mode of action at the cellular level still remains unclear. Because it seems to be involved in stabilizing the membranes of certain cell types within the lung (i.e., alveolar epithelial cells and macrophages) yet potentiates the action of oxidants released from recruited inflammatory cells (i.e. neutrophils), the protective effects of taurine in the lung may well be concentration dependent relative to not only the amount and duration of exposure but also the particular cell types affected by a particular oxidant agent.

REFERENCES

1. Dupré, S. and De Marco, C. Activity osome anima tissues on the oxidation of cysteamine to hypotaurine in the presence of sulfide. *J. Bio. Chemistry* 13:386–390, 1964.
2. Chatagner, F., Lefauconnier, J.-M., and Portemer, C. 1976. On the formation of hypotaurine in various tissues of different species. In *Taurine*, eds. R. Huxtable and A. Barbeau, pp. 67–72. New York: Raven Press.
3. Fellman, J. H., Green, T. R., and Eicher, A. L. 1987. The oxidation of hypotaurine to taurine: Bis-aminoethyl α-disulfone, a metabolic intermediate in mammalian tissue. In *The Biology of Taurine— Methods and Mechanisms*, eds. R. J. Huxtable, F. Franconi, and A. Giotti, pp. 39–48. New York: Plenum Press. 1987.
4. Cavallini, D., DeMarco, C., Mondovi, B., and Mori, B. G. 1960. The cleavage of cystine by cystathionase and the transulfuration of hypotaurine. *Enzymologia* 22:161–173.
5. Fujiwara, M., Ubuka, T., Abe, T., Yukihiro, K., and Tomaozawa, M. 1995. Increased excretion of taurine, hypotaurine and sulfate after hypotaurine loading and capacity of hypotaurine metabolism in rats. *Physiol. Chem. Med NMR* 27:131–137.
6. Baskin, S. I., Zaydon, P. T., Kendrick, Z. V., Katz, T. C., and Orr, P. L. 1980. Pharmacokinetic studies of taurine in bovine Purkinje fibers. *Circ. Res.* 47:763–769.
7. Saransaari, P., and Oja, S. S. 1993. Uptake and release of beta-alanine in cerebellar granule cells in primary culture: Regulation of release by glutamatergic and GABAergic receptors. *Neuroscience* 53:475–481.
8. Barnett, D. K., and Bavister, B. D. 1992. Hypotaurine requirement for in vitro development of golden hamster one-cell embryos into morulae and blastocysts, and production of term offspring from in vitro-fertilized ova. *Biol. Reprod.* 47:297–304.
9. Guerin, P., Guillaud, J., and Menezo, Y. 1995. Hypotaurine in spermatozoa and genital secretions and its production by oviduct epithelial cells in vitro. *Hum. Reprod.* 10:866–872.
10. Kochakian, C. D. 1976. Influence of testosterone on the concentration of hypotaurine and taurine in the reproductive tract of the male guinea pig, rat and mouse. In *Taurine*, eds. R. Huxtable and A. Barbeau, pp. 327–334. New York: Raven Press.
11. Holmes, R. P. Goodman, H. O. Shihabi, Z. K., and Jarow, J. P. 1992. The taurine and hypotaurine content of human semen. *J. Androl.* 13:289–292.
12. Tadolini, B., Pintus, G., Pinna, G. G., Bennardini, F., and Franconi, F. 1995. Effects of taurine and hypotaurine on lipid peroxidation. *Biochem Biophys. Res. Commun.* 213:820–826.
13. Krusz, J. C. Kendrick, Z. V. and Baskin, S. I. 1979. Pineal taurine content. In *Aging in Nonhuman Primates*, ed. D. M. Bowden, pp. 106–111. New York: Van Nostrand Reinhold.
14. Baskin, S. I., Wakayama, K., Knight, T., Jepson, J. H., and Besa, E. C. 1982. The sulfur-containing amino acid pathway in normal and malignant cell growth. In *Taurine in Nutrition and Neurology*, eds. R. J. Huxtable and H. Pasantes-Morales, pp. 127–141. *Adv. Exp. Med. Biol.* New York: Plenum Press.
15. Sciuto, A. M. 1997. Glutathione. In *Antioxidants, oxidants and free radicals*, eds. S. I. Baskin and H. Salem. Lancaster PA: Techomic Press.
16. Huxtable, R. J. 1976. Metabolism and function of taurine in the heart. In *Taurine*, eds. R. J. Huxtable and A. Barbeau, pp. 99–120. New York: Raven Press.
17. Sebring, L., and Huxtable, R. J. 1985. Taurine modulation of calcium binding to cardiac sarcolemma. *J. Pharmacol. Exp. Ther.* 232:445–451.
18. Huxtable, R. J., and Sebring, L. A. 1986. Towards a unifying theory for the action of taurine. *Trends Pharmacol. Sci.* 7:481–485.
19. Sebring, L., and Huxtable, R. J. 1986. Low affinity binding of taurine to phospholipids and cardiac sarcolemma. *Biochem. Biophys. Acta* 884:559–566.
20. Cantafora, A., Mantovani, A., Masella, M., and Alvaro, D. 1986. Effect of taurine administration on liver lipids in guinea pig. *Experientia* 42:407–408.
21. Schaffer, S. W., Azuma, J., and Hamaguchi, T. 1991. Inhibition of phospholipid N-methylation by taurine. *FASEB J.* 5:A591.
22. Cantafora, A., Blotta, I., Hoffmann, A. F., and Sturman, J. A. 1991. Dietary taurine content changes liver lipids in cats. *J. Nutr.* 121:1522–1528.
23. Neyes, L., Locher, R., Stimpel, Streuli, R., and Vetter, W. 1985. Stereospecific modulation of the calcium channel in human erythrocytes by cholesterol and its oxidized derivatives. *Biochem. J.* 227:105–112.
24. Igisu, H., Izumi, K., Goto, I., and Kina, K. 1976. Effects of taurine on the ATPase activity in the human erythrocyte membrane. *Pharmacology* 14:362–366.
25. Warren, G. B., Houslay, M. D., and Metcalfe, J. C. 1975. Cholesterol is excluded from the phospholipid annulus surrounding an active calcium transport protein. *Nature* 255:684–687.
26. Paauw, J. D., and Davis, A. T. 1994. Taurine supplementation at three dosages and its effect on trauma patients. *Am J. Clin. Nutr.* 60:203–206.
27. Chow, C. K., Hussian, M. Z., Cross, C. E., Dungsworth, D. L., and Mustafa, M. G. 1976. Effects of

low levels of ozone on rat lungs. I. Biochemical responses during recovery and reexposure. *Exp. Mol. Pathol.* 25:182–188.

28. Crapo, J. D., Barry, B. E., Foscue, H. A., and Shelburne, J. 1980. Structural and biochemical changes in rat lungs occurring during exposures to lethal and adaptive doses of oxygen. *Am. Rev. Respir. Dis.* 122:123–143.

29. Mustafa, M. G., and Tierney, D. F. 1978. Biochemical and metabolic changes in the lung with oxygen, ozone and nitrogen dioxide toxicity. *Am. Rev. Respir. Dis.* 118:1061–1090.

30. Plopper, C. C., Chow, C. K., Dungworth, D. L., Brummer, M., and Nemeth, T. J. 1978. Effect of low level of ozone on rat lungs. II. Morphological responses during recovery and reexposure. *Exp. Mol. Pathol.* 29:400–411.

31. Chow, C. K., and Tappel, A. L. 1972. An enzymatic protective mechanism against lipid peroxidation damage to lungs of ozone-exposed rats. *Lipids* 1:518–524.

32. Sagai, M., Ichinose, T., Oda, H., and Kubota, K. 1982. Studies on biochemical effects of nitrogen dioxide. II. Changes of the protective systems in rat lungs and of lipid peroxidation by acute exposure. *J. Toxicol. Environ. Health* 9:153–164.

33. DeLucia, A. J., Mustafa, M. G., Hussain, M. Z., and Cross, C. E. 1975. Ozone interaction with rodent lung. III. Oxidation of reduced glutathione and formation of mixed disulfides between protein and nonprotein sulfhydryls. *J. Clin. Invest.* 55:794–802.

34. Rietjens, I. M. C. M., Alink, G. M., and Vos, R. M. E. 1985. The role of glutathione and changes in thiol homeostasis in cultured lung cells exposed to ozone. *Toxicology* 35:207–217.

35. Kratzing, C. C., and Willis, R. J. 1980. Decreased levels of ascorbic acid in lung following exposure to ozone. *Chem. Biol. Interract.* 30:53–56.

36. Sutherland, M. W., Glass, M., Nelson, J., Lyrn, Y., and Forman, H. J. 1985. Oxygen toxicity: Loss of lung macrophage function without metabolite depletion. *J. Free Radic. Biol. Med.* 1:209–214.

37. Elsayed, N. M. 1987. Influence of vitamin E on polyamine metabolism in ozone-exposed rat lungs. *Arch. Biochem. Biophys.* 255:392–399.

38. Goldstein, B. D., Buckley, R. M., Cardenas, R., and Balchum, O. J. 1970. Ozone and vitamin E. *Science* 18:631–632.

39. Rietjens, I. M., Poelen, M. C., Hempeniu, R. A., Gijbels, M. J., and Alink, G. M. 1986. Toxicity of ozone and nitrogen dioxide to alveolar macrophage: Comparative study revealing differences in their mechanism of toxic action. *J. Toxicol. Environ. Health* 19:555–568.

40. Crapo, J. D., Marsh-Salin, J., Ingram, P., and Pratt, P. C. 1978. Pulmonary morphology and morphometry. *J. Appl. Physiol.* 44:370–379.

41. Crapo, J. D., Barry, B. E., Chang, L. Y., and Mercer, R. R. 1984. Alterations in lung structure caused by inhalation of oxidants. *J. Toxicol. Environ. Health* 13:301–321.

42. Adamson, I. Y. R., and Bowden, D. H. 1974. The type 2 cell as progenitor of alveolar epithelial regeneration. A cytodynamic study in mice after exposure to oxygen. *Lab. Invest.* 30:35–42.

43. Evans, M. J., Cabral, L. J., Stephens, R. J., and Freeman, G. 1973. Renewal of alveolar epithelium in the rat following exposure to NO₂. *Am. J. Pathol.* 70:175–198.

44. Evans, M. J., Cabral, L. J., Stephens, R. J., and Freeman, G. 1975. Transformation of alveolar type 2 cells to type I cells following exposure to NO₂. *Exp. Mol. Pathol.* 22:142–150.

45. Kleinerman, J., Ip, M. P. C., and Sorensen, J. 1982. Nitrogen dioxide exposure and alveolar macrophage elastase in hamsters. *Ann. Rev. Respir. Dis.* 125:203–207.

46. Sturrock, J. E., Nunn, J. R., and Jones, A. J. 1980. Effects of oxygen on pulmonary macrophages and alveolar epithelial type II cells in culture. *Respir. Physiol.* 41:381–390.

47. Gordon, R. E., Shaked, A. A., and Solano, D. F. 1986. Taurine protects hamster bronchioles from acute NO₂-induced alterations. A histological, ultrastructural and freeze-fracture study. *Am. J. Pathol.* 125:585–600.

48. Gordon, R. E., Heller, R. F., DelValle, J. R., and Heller, R. F. 1989. Membrane perturbations and mediation of gap junction formation in response to taurine treatment in normal and injured alveolar epithelia. *Exp. Lung Res.* 15:895–908.

49. Lehnert, B. E., Archuleta, D. C., Ellis, T., Session, W. S., Lehnert, N. M., Gurley, L. R., and Stavert, D. M. 1994. Lung injury following exposure of rats to relatively high mass concentrations of nitrogen dioxide. *Toxicology* 89:239–277.

50. Lewis, C. P. L., Cohen, G. M., and Smith, L. L. 1990. The identification and characterization of an uptake system for taurine into rat lung slices. *Biochem. Pharmacol.* 39:431–437.

51. Sharma, R., Kodavanti, U. P., Smith, L. L., and Mehendale, H. M. 1995. The uptake and metabolism of cysteamine and taurine by isolated perfused rat and rabbit lungs. *Int. J. Biochem. Cell. Biol.* 27:655–664.

52. Lewis, C. P., Haschek, W. M., Wyatt, I., Cohen, G. M., and Smith, L. L. 1989. The accumulation of cysteamine and its metabolism to taurine in rat lung slices. *Biochem. Pharmacol.* 38:481–488.

53. Banks, M. A., Martin, W. G., Pailes, W. H., and Castranova, V. 1989. Taurine uptake by isolated alveolar macrophages and type II cells. *J. Appl. Physiol.* 66:1079–1086.

54. Banks, M. A., Porter, D. W., Martin, W. G., and Castranova, C. 1991. Ozone-induced lipid peroxidation and membrane leakage in isolated rat alveolar macrophages: Protective effects of taurine. *J. Nutr. Biochem.* 2:308–313.
55. Banks, M. A., Porter, D. W., Martin, W. G., and Castranova, V. 1990. Effects of in vitro ozone exposure on peroxidative damage, membrane leakage, and taurine content of rat alveolar macrophages. *Toxicol. Appl. Pharmacol.* 105:55–65.
56. Banks, M. A., Porter, D. W., Martin, W. G., and Castranova, V. 1992. Taurine protects against oxidant injury to rat alveolar pneumocytes. *Adv. Exp. Med. Biol.* 315:341–354.
57. Banks, M. A., Porter, D. W., Pailes, W. H., Schwegler-Berry, D., Martin, W. G., and Castranova, V. 1991. Taurine content of isolated rat alveolar type I cells. *Comp. Biochem. Physiol.* 100B:795–799.
58. Nagata, T., Masaoka, T., and Akahori, F. 1991. Protective effect of taurine against acute paraquat intoxication in rats. *J. Toxicol. Sci.* 16:11–27.
59. Haschek, W. M., and Witschi, H. P. 1979. Pulmonary fibrosis: A possible mechanism. *Toxicol. Appl. Pharmacol.* 51:475–481.
60. Wang, Q., Giri, S. N., Hyde, D. M., and Li, C. 1991. Amelioration of bleomycin-induced pulmonary fibrosis in hamsters by combined treatment with taurine and niacin. *Biochem. Pharmacol.* 42:1115–1122.
61. Wang, Q., Giri, S. N., Hyde, D. M., and Nakashima, J. M. 1989. Effects of taurine on bleomycin-induced lung fibrosis in hamsters. *Proc. Soc. Exp. Biol. Med.* 190:330–338.
62. Giri, S. N., Blaisdell, R., Rucker, R. B., Wang, Q., and Hyde, D. M. 1994. Amelioration of bleomycin-induced lung fibrosis in hamsters by dietary supplementation with taurine and niacin: Biochemical mechanisms. *Environ. Health Perspect.* 102:137–148.
63. Wang, Q., Hollinger, M. A., and Giri, S. N. 1992. Attenuation of amiodarone-induced lung fibrosis and phospholipidosis in hamsters by taurine and/or niacin treatment. *J. Pharmacol. Exp. Ther.* 262:127–132.
64. Venkatesan, N., and Chandrakasan, G. 1993. Ameliorative effects of taurine and niacin on paraquat induced early lung injury in rats. *J. Clin. Biochem. Nutr.* 15:127–134.
65. Venkatesan, N., and Chandrakasan, G. 1994. In vivo administration of taurine and niacin modulate cyclophosphamide-induced lung injury. *Eur. J. Pharm.* 292:75–80.
66. Cantin, A. M. 1994. Taurine modulation of hypochlorous acid-induced lung epithelial cell injury in vitro. Role of anion transport. *J. Clin. Invest.* 93:606–614.
67. Shasby, D. M., and Hampson, F. 1989. Effects of chlorinated amines on endothelial and epithelial barriers in vitro and ex vivo. *Exp. Lung Res.* 15:345–357.
68. Mangum, J. B., Everitt, J. I., Bonner, J. C., Moore, L. R., and Brody, A. R. 1990. Co-culture of primary pulmonary cells to model alveolar injury and translocation of proteins. *In Vitro Cell. Dev. Biol.* 26:1135–1143.
69. Tatsumi, T., and Fliss, H. 1994. Hypochlorous acid and chloramines increase endothelial permeability: Possible involvement of cellular zinc. *Am. J. Physiol.* 267:H1597–H1607.

Chapter Eleven

ANTIOXIDATIVE ACTIVITY OF ERGOTHIONEINE AND OVOTHIOL

Robert D. Short, Jr. and Steven I. Baskin

Active species of oxygen have been implicated in a variety of adverse health effects (1). These effects include aging, circulatory diseases, cataract formation, and the induction of cancer. Two important biomolecules, ergothioneine and ovothiol C, which protect some biological systems against oxidative damage, are discussed in this chapter.

ERGOTHIONEINE

Background

Ergothioneine (ET) is the trimethylbetaine of 2-thiolhistidine (2-thiol-L-histidine betaine) and it exists in a variety of biological systems. The early literature was summarized by Melville (2) and Stowell (3) and more recently by Hartman (4). ET was initially isolated from the ergot fungus *Claviceps purpurea* and characterized as $C_9H_{15}N_3O_2S$ in 1909. ET was subsequently identified in mammalian erythrocytes, and for a number of years these were thought to be its only sources in biological systems. While ET was a popular focus of investigations following its identification in erythrocytes, it has been called a "neglected molecule" because there was a limited amount of published research following these initial efforts (5). More recent studies, however, have proposed a variety of biological roles for ET (4, 6, 7).

Occurrence

As revived by the early investigators, ET was synthesized only by fungi and mycobacterium and from 1909 to 1925 ET was thought to be a peculiarity of the ergot fungus. After improvements were made in the analytical methods for measuring ET, it was found in the erythrocytes of every mature species investigated. In addition to erythrocytes, ET has been found in the liver, kidneys, and seminal fluids at nearly millimolar levels following continuous ingestion. ET is stored in the free form. The actual content varies between tissues and depends on dietary concentration as well as the animal species.

The measurement of ET levels in plants is complicated by the presence of interfering substances. For example, 17 mg/kg of ET was measured in rolled oats while only 2.8 mg/kg of pure ET was isolated. While ET maybe present in food sources as a result of microbial contamination, plants take up ET from the soil through their roots.

Since mammals ingest ET in their diet and conserve it with minimal metabolism, humans may receive ET in their diets from plant and animal sources. After reviewing literature on the distribution of ET, Hartman (4) concluded that while most mammals are exposed to ET, humans may be consuming lower levels of what may be regarded as a "universally treasured molecule" since we consume less mold-contaminated foods and resulting fungal toxins than animals in native environments. However, we do not know the sources or the amount of ET consumed by humans.

Biological Activity in Whole Animals

The reviews of ET just cited did not identify any harmful effects. There are some limited studies that have attempted to identify a biological role for ET. For example, as soon as ET was identified in erythrocytes, Tainter (8) attempted to identify its pharmacological role. After rabbits were treated subcutaneously with 100–200 mg/kg, no changes were observed in blood sugar (4 h after treatment), respiration, pulse rate, or pupillary diameter. There was no change in blood pressure, pulse rate, respiration, or biliary secretion in cats treated intravenously with 1–60 mg/kg. In addition, excised strips of the cats duodenum showed no change in tonus, rate, or amplitude of signals with ET concentrations in the bath up to 0.06%.

Rabbits that consumed diets containing 6.7 mg% ET for 3 mo produced erythrocytes that formed methemoglobin at a slower rate than controls fed diets without additional ET (9). The erythrocytes of treated rabbits contained 23 mgET/100 ml of erythrocytes, while erythrocytes from control rabbits contained no measurable ET.

Rats treated with 16 mg/kg (ip) of ^{3}H-ET for 1 wk exhibited limited metabolism of ET, and the ET content of erythrocytes and liver tissue was increased by 12-fold and 7-fold (10).

Pregnant mice received 80 mg/kg of ET by the intraperitoneal route on days 5–9 of gestation (11). No treatment-related effects were observed on gross, soft tissue, or skeletal defects. Treatment with ET, however, reduced the incidence of exencephaly, open eye, and skeletal defects produced by 2 mg/kg of cadmium administered by the same route on day 7 of gestation. This protective effect apparently was not due to changes in the maternal distribution or placental transfer of cadmium.

Biological Activity in Organ, Tissue, or Culture Systems

The oxidative state of myoglobin may be a critical event in the tissue damage associated with cardiac ischemia/reperfusion states (12). The higher valence states of myoglobin have to oxidize cellular components through their peroxidative activity. On the other hand, the reduction of higher valence states may permit this heme protein to serve as an electron sink and provide a protective mechanism against oxidative injury to the heart. In support of a protective role, ET reduced ferrylmyoglobin, Mb(IV), to Mb(III) (13). In addition, the addition of 100 μM ET to the perfusion bath of Langendorff rat heart preparations exposed to brief periods of ischemia prevented myocardial damage, as monitored by lactate dehydrogenase (LDH) release, following reperfusion. The isolated rat diaphragm was also used to monitor the oxidation state of myoglobin (14). The addition of 2 mM ET either prevented the formation or dissipated ferrylmyoglobin, depending on the time of addition. Similar results were also reported for sperm whale ferrylmyoglobin (13). Additional experiments by other investigators failed to demonstrate a protective effect of 10 and 100 μM ET in isolated rabbit hearts (15).

ET also protects cells in culture against reactive molecules with genetic activity. For example, ET blocked the mutagenicity of nitrosation products of spermidine, cumene, and t-butyl hydroperoxides in *Salmonella* strain TA1950 (5). ET also has a strong protective effect against the inactivation of bacteriophages P22 and T4 by gamma irradiation (4) and prevented the alkylation of RNA by tris-(2-chloroethyl)-amine (16).

Dahl et al. (17) investigated the ability of several prevalent biological molecules, including ET, to protect a target substrate against oxidation at physiological concentrations and neutral pH. Singlet oxygen, which is the lowest energy electronically excited state of molecular oxygen, is important because it has a long lifetime and can react with a variety of biological substrates. While ET did not quench singlet oxygen at physiological relevant concentrations of 1 mM in their system, ET prevented its formation.

What Are the Roles of ET in Biological Systems?

While various roles have been proposed for ET in biological systems (6,7), the recent literature supports a defensive role for ET as a scavenger for hydrogen peroxide and some radical species and electrophilic molecules (4). The following examples demonstrate the role of ET as a defensive molecule:

1. ET protects tissues and organs against injury by agents that produce oxidative damage. (a) ET may protect cardiac tissue against oxidative damage associated with ischemia/reperfusion states by altering the oxidative state of myoglobin, as discussed earlier. (b) Ram semen is protected against damage by hydrogen peroxide by the addition of ET (18). (c) ET protected against the inactivation of bacteriophages T4 and P22 by gamma radiation (4). In this system, inactivation is dependent on hydrogen peroxide or radicals derived from hydrogen peroxide. (d) ET (80 mg/kg, subcutaneously for 7 days) reduced the levels of lipid peroxidation in rats treated with ethionine (19). Similar protective effects against lipid peroxidation were also observed when ET was added to liver homogenate from mice at levels of 5–250 mM (20).
2. ET blocks the mutagenicity of the nitrosation products of spermidine in *Salmonella* strain TA1950 (5). ET also reduces the mutagenicity of cumene and *t*-butyl hydroperoxides. ET, however, produced only minimal protection against DNA methylase inhibition by direct-acting carcinogens in an in vitro system (4).

Potential Clinical Applications

Since ET is a relatively nontoxic naturally occurring molecule that protects biological systems against damage by certain agents, several potential clinical applications have been suggested for it (4). For example, ET may be useful to (1) protect long-term cultures of human cells against hydrogen peroxide-mediated toxicity, (2) protect tissue against damage associated with the oxidative burst following reperfusion, (3) treat erythrocyte disorders that are thought to be associated with oxidative stress (eg., sickle-cell anemia, thalassemia, deficiencies in glucose-6-phosphate dehydrogenase activity, etc.). As discussed earlier, active oxygen species have been implicated in a wide variety of health effects, such as aging, circulatory disease, cataract formation, and the induction of cancer (1). ET may have a beneficial role in these disease processes, since it provided protection against the formation of active oxygen species (17).

OVOTHIOLS

Background

Ovothiols are a family of mercaptohistidine compounds that have been isolated from marine invertebrate eggs and posses redox activity (21). The family currently consists of three members: ovothiol A, ovothiol B, and ovothiol C. The redox activity of this family of aromatic thiols, 1-methyl-4-mercaptohistidines, was discovered while studying the biology of fertilization using sea urchin gametes as the experimental system.

Occurrence

The striking feature of the ovothiols is their high concentration in the eggs of marine invertebrates (21). For example, ovothiol C is present at levels equivalent to about 4.3 mM in eggs of the sea urchin *Stronglocentrotus purpuratus*, which only has ovothiol C in the eggs and ovary with none in somatic cells. The bay scallop *Clamys hastata* has primarily ovothiol B in the ovary but not the testes or somatic tissues. The starfish *Evasteria troschelii* has ovothiol A primarily in the ovary but with smaller amounts in the testes and gut and none in somatic cells.

Biological Activity

These heterothiols are present at millimolar concentrations in sea urchin eggs and they may protect the fertilized egg from free-radical-induced damage (22). After sperm entry a sea urchin egg develops a protective envelope that is formed by cross-linking adjacent polypeptide chains through tyrosine dimerization. The cross-linking is catalyzed by oviperoxidase, which utilizes hydrogen peroxide generated by the "respiratory burst" oxidase. While the use of hydrogen peroxide as an extracellular oxidant presents significant problems for the early embryo, studies suggest that systems have evolved both for generating hydrogen peroxide as well as controlling its levels during development (22).

REFERENCES

1. Ames, B. N., Cathcart, R., Schwiers, E., and Hochstein, P. 1981. Uric acid provides an antioxidant defense in human against oxidant- and radical-caused aging and cancer; A hypothesis. *Proc. Nat. Acad. Sci. U.S.A.* 78:6858–6862.
2. Melville, D. B. 1959. Ergothionine. *Vitamins Hormones* 17:155–203.
3. Stowell, E. C. 1961. Ergothionine. In *Organic Sulfur Compounds*, ed. N. Kharasch, vol. 1, pp. 488–642. New York: Pergamon Press.
4. Hartman, P. E. 1990. Ergothionine as antioxidant. *Methods Enzymol.* 186:310–318.
5. Hartman, Z., and Hartman, P. E. 1987. Interception of some direct-acting mutagens by ergothionine. *Environ. Mol. Mutagen.* 10:3–15.
6. Brumel, M. C. 1985. In search of a physiological function for L-ergothionine. *Med.Hypoth.* 18:351–370.
7. Brumel, M. C. 1989. In search of a physiological function for L-ergothionine-II. *Med.Hypoth.* 30:39–48.
8. Tainter, M. L. 1926. Note on the pharmacology of ergothionine. *Proc. Soc. Exp. Biol. Med.* 24:261.
9. Spicer, S. S. 1951. Ergothionine depletion in rabbit erythrocytes and its effects on methemoglobin formation and reversion. *Proc. Soc. Expl. Biol. Med.* 77:418–420.
10. Mayumi, T., Kawano, H., Sakamoto, Y., Suehisa, E., Kawai, Y., and Hama, T. 1978. Studies on ergothionine. V. Determination by high performance liquid chromatography and application to metabolic research. *Chem. Pharm. Bull.* 26:3772–3778.
11. Mayumi, T., Okamoto, K., Yoshida, K., Kawai, Y., Kawano, H., Hama, T., and Tanaka, K. 1982. Studies on ergothionine. VIII. Preventive effects of ergothionine on cadmium induced teratogenesis. *Chem. Pharm. Bull.* 30:2141–2146.
12. Galaris, D., Eddy, L., Arduini, A., Cadenas, E., and Hochstein, P. 1989. Mechanism of reoxygenation injury in myocardial infarction: Implications of a myoglobin redox cycle. *Biochem. Biophys. Res. Commun.* 160:1162–1168.
13. Arduini, A., Eddy, L., and Hochstein, P., The reduction of ferryl myoglobin by ergothionine: A novel function of ergothionine. Arch. Biochem. Biophys. 281:41–43, 1990.
14. Eddy, L., Arduini, A., and Hochstein, P. 1990. Reduction of ferrylmyoglobin in rat diaphragm. *Am. J. Physiol.* 259(*Cell Physiol.* 28):C995–C997.
15. Cargnoni, A., Bernocchi, P., and Ceconi, C. 1994. In vitro ergothionine administration failed to protect isolated ischaemic and reperfused rabbit heart. *Adv. Exp. Med. Biol.* 366:448–449.
16. Sziniez, I., Albrecht, G. J., and Weger, N. 1981. Effects of various compounds on the reaction of tris-(2-chloroethyl)amine with ribonucleic acid in vitro and on its toxicity in mice. *Arzneim. Forsch./Drug Res.* 31:1713–1717.
17. Dahl, T. A., Midden, R. W., and Hartman, P. E. 1988. Some prevalent biomolecules as defenses against singlet oxygen damage. *Photochem. Photobiol.* 47:367–362.
18. Mann, T., and Leone, E. 1953. Studies on the metabolism of semen. *Biochem. J.* 53:140–147.
19. Kawano, H., Cho, K., Haruna, Y., Kawai, Y., Mayumi, T., and Hama, T. 1983. Studies on ergothionine. X. Effects of ergothionine on the hepatic drug metabolizing enzyme system and on experimental hepatic injury in rats. *Chem. Pharm. Bull.* 31:1676–1681.
20. Kawano, H., Murata, H., Iriguchi, S., Mayumi, T., and Hama, T. 1983. Studies on ergothionine. XI. Inhibitory effects on lipid peroxide formation in mouse liver. *Chem. Pharm. Bull.* 31:1682–1687.
21. Turner, E., Klevit, R., Hager, L. J., and Shapiro, B. M. 1987. Ovothiols, a family of redox-active mercaptohistidine compounds from marine invertebrate eggs. *Biochem. J.* 26:4028–4036.
22. Shapiro, B. M., and Hopkins, P. B. 1991. Ovothiols: Biological perspective. In *Advances in Enzymology and Related Areas of Moleculer Biology*, ed. A. Meister, vol. 64, pp. 291–315. New York: John Wiley.

Chapter Twelve

THE TOXICOLOGY OF ANTIOXIDANTS

Harry Salem and Steven I. Baskin

Antioxidants have been reported to improve the health of those using them and to prolong the shelf-life of foods. They are not, however, without adverse effects, especially when overused. This chapter reviews the adverse effects or toxicology of some of the more prominent antioxidants. These include beta-carotene, vitamins A, C, and E, as well as copper, manganese, selenium, and zinc, which are considered essential nutrients.

The naturally occurring antioxidants include retinoids (vitamin A), tocopherols (vitamin E), and ascorbic acid (vitamin C). The synthetic antioxidants include butylated hydroxytoluene (BHT), butylated hydroxyanisole (BHA), and propylgallate. Both natural and synthetic antioxidants are added to food to prevent deterioration. Foods preserved with antioxidants include vegetable oils, bread, and cheese.

The billions of cells in our body require energy to carry out their biological functions. This energy is generated via the oxidation process. The metabolic by-products of oxidation include free radicals, also known as reactive oxidants. Free radicals may also be generated by the inhalation of air pollutants and cigarette smoke. These highly reactive and unstable molecules with one or more unpaired electrons can cause damage to cell structures and are believed to play a major role in contributing to disease and strokes, acquired immune deficiency syndrome (AIDS), ozone-induced pulmonary disorders, cataracts, and other diseases associated with aging. They do this by stealing electrons from other molecules within the cell, disrupting normal chemical processes (1).

Repeated exposure to free radicals can damage cells in the following ways: by inhibiting enzyme activity, by lipid peroxidation, and by membrane damage (2), as well as by strand breaks or base alterations of DNA (3).

Antioxidants neutralize free radicals and other reactive chemicals, and are being studied individually and in various combinations to determine their activity in preventing or reducing the risks of the above mentioned diseases and conditions.

Some foods are rich sources of antioxidants. Antioxidants are also available commercially in tablet and capsule form. Beta-carotene, which is converted in the body to vitamin A, is found in yellow-orange fruits and vegetables, as well as in dark green vegetables. Good natural sources of vitamin E include nuts, vegetable oils, and green leafy vegetables, while vitamin C is found in many fruits and vegetables and green leafy vegetables.

Table 1 lists the common food sources of antioxidants.

TOXICOLOGY

Vitamins continue to be the world's most widely used pharmaceutical products, based partially on the misconception that more is better (6). Indiscriminate use and self-medication with vitamins

---

The views of the authors do not purport to reflect the position of the Department of Defense. The use of trade names does not constitute an official endorsement or approval of the use of such commercial hardware or software.

Table 1. Common food sources of antioxidants

Beta carotene:	Sweet potatoes	
	Winter squash	
	Pumpkins	
	Carrots	
	Cantaloupes	
	Peaches	
	Spinach	
	Collard greens	
	Apricots	
	Papayas	
Vitamin E:	Nuts (almond, peanut, walnut)	
	Olive oil	
	Soybean oil	
	Corn oil	
	Sunflower oil	
	Green leafy vegetables	
	Sunflower seeds	
	Wheat germ	
	Whole grains	
Vitamin C:	Grapefruit	Squash
	Pineapples	Green peppers
	Lemons	Cantaloupe
	Limes	Broccoli
	Tangerines	Kale
	Oranges	Potatoes
	Strawberries	Tomatoes
	Asparagus	Green leafy vegetables
	Brussel Sprouts	
	Cauliflower	
	Cabbage	
	Green beans	
	Okra	
Zinc:	Meat (liver)	
	Eggs	
	Seafoods (oysters)	
	Cereals	
Copper:	Organ meats (liver)	
	Seafood	
	Nuts	
	Seeds	
Selinium:	Seafood	
	Meat (liver, kidney)	
Manganese:	Cereal	
	Whole grains	

and dietary supplements may produce toxicoses, even though research reported by Pryor (7) implies that high levels of antioxidants may be beneficial. The safety and efficacy issues related to megavitamin therapy have been reviewed by Omaye (8) and Ovesen (9). Vitamins are essential for normal metabolism but cannot be synthesized in the cells of the body. Therefore, when they are lacking in the diet or supplement intake, specific metabolic deficits can occur. The U.S. Recommended Dietary Allowances (U.S. RDA) for the antioxidants are presented in Table 2.

Salem (10) discussed the different shapes of toxicity dose-response curves for different pharmacologically active drugs. These shapes include V, U, linear, and angled. The graded dose response for vitamins and essential nutrients may also demonstrate a U-shaped toxicity dose response. At very low doses, there is a high level of adverse effect from deficiency, which decreases with increasing doses. At doses where there is no longer deficiency, no adverse effect occurs and the organism is in a state of homeostasis. When doses become abnormally high, the adverse effects

Table 2. Dietary requirements for antioxidants

Antioxidant	U.S. RDA
Vitamin A	5000 IU[a] (men)
	4000 IU (women)
Beta-carotene	3 mg[b]
Vitamin C	60 mg
Vitamin E	30 IU
Zinc	15 mg
Copper	2 mg
Selenium	[c]
Manganese	[c]

Source: Ref. 4

[a] IU, international units.

[b] U.S. RDA for vitamin A, 3 mg = 5000 IU.

[c] Essential but no U.S. RDA established. However, safe and adequate daily dietary intakes (SADDI) for adults have been estimated by the Food and Nutrition Board of the National Research Council to be 0.05–0.2 mg for selenium and 2.5–5.0 mg for manganese.

of vitamins and essential nutrients appear to increase with increasing doses. However, the adverse effects at these high doses are usually qualitatively different from those observed at deficiency doses (11).

VITAMIN A

Vitamin A, the first fat-soluble vitamin to be discovered, cannot be synthesized in the cells of the body. However, its precursors occur in abundance in many foods as carotenoids (beta-carotene), which have structures similar to that of vitamin A and can be converted into vitamin A by the body. Vitamin A exists in the body mainly as retinol. It is stored in and regulated by the liver. The circulating levels of vitamin A increase with age through the production of retinol-binding proteins (RBP) in the liver. Usually vitamin A circulates bound to RBP. In hypervitaminosis A, there is insufficient RBP to bind all of the circulating vitamin A. This may result in significantly higher circulating levels of free retinol and retinyl esters (12). Although all of the metabolic functions of vitamin A are not known, it is involved in the formation of retinol photochemicals utilized in vision. In addition, vitamin A is necessary for normal growth of most cells, especially for normal growth and proliferation of different types of epithelial cells. The lack of vitamin A results in stratification and keratinization of epithelial structures. This is manifest as scaliness of the skin and sometimes acne, failure of normal growth and reproduction, atrophy of the germinal epithelium of the testes, and sometimes interruption of the female sexual cycle. In addition, keratinization of the cornea can occur, resulting in corneal opacity and blindness. Thus vitamin A is indicated therapeutically for deficiency states such as night blindness, steatorrhea, hyperkeratinosis, xerophthalmia, possibly for acne vulgaris, certain immune disorders, and cancer.

Vitamin toxicity may not be detected since many of the signs and symptoms of poisoning are similar to those that characterize the underlying condition for which the vitamin is taken.

Acute Human Toxicity

In 1857 the first case of vitamin A toxicity was reported by Artic explorers who consumed large quantities of polar bear liver (13). Usually, poisoning from vitamin A is due to overambitious prophylactic use, food fadism, and parents giving inadvertant overdoses to children. Doses of 75,000–300,000 IU can cause acute toxicosis in children, while 2–5 million IU can cause acute toxicosis in adults. Doses of 500,000 IU can cause acute reversible side effects (14). The signs and symptoms of acute intoxication in children include anorexia, bulging fontanelles, increased intracranial pressure, drowsiness, irritability, and vomiting, while in adults the signs and symptoms include anorexia, blurred vision, headache, drowsiness, hypercalcemia, irritability, abdominal pain, muscle weakness, nausea, vomiting, peripheral neuritis, and skin desquamation (15). Ingestion of from 25,000 to 50,000 IU of retinol daily for 30 days induced signs of increased intracranial pressure in infants (16). The signs and symptoms of chronic vitamin A toxicity in children include alopecia, anorexia, bone pain and tenderness, bulging of fontanelles, craniotabes, fissuring at lip corners, hepatomegaly, hyperostosis, premature epiphyseal closure, photophobia, pruritis, pseudotumor cerebri, skin desquamation, and skin erythema (15), while in adults they include alopecia, anemia, anorexia, ataxia, bone pain, bone abnormalities, brittle nails, cheilitis, conjuncti-vitis, diarrhea, diplopia, dryness of mucus membranes, dysuria, edema, elevated cerebral spinal fluid pressure, epistaxis, exanthema, facial dermatitis, fatigue, fever, headache, hepatomegaly, hepatotoxicity, hyperostosis, insomnia, irritability, menstrual abnormalities, muscular stiffness and pain, nausea, negative nitrogen balance, nervous abnormalities, papilledema, petechiae, pruritis, pseudomator cerebri, skin desquamation, skin erythema, skin rash, skin scaliness, spleno-megaly, vomiting, and weight loss (15).

Chronic Human Toxicity

Doses of 18,000–60,000 IU per day can cause chronic toxicity in infants, while from 100,000 up to 600,000 IU for a few months to several years can cause chronic toxicity in adults (17).

In a cross-sectional epidemological study of 562 elderly (60–98 yr) healthy individuals (181 males; 381 females) who received from 5000 to 10,000 IU vitamin A daily, Krasinsky et al. (18) found that 5 had an elevation of 1 of 2 liver enzymes, and all 5 had a modest elevation in serum retinyl ester levels. However, these ester levels were noted as 10-fold lower than those associated with hypervitaminosis A. Based on these limited data, it was suggested that the elderly should limit their intake of vitamin A supplementation, particularly over the long term. Although these abnormalities in clinical indices may have been due to vitamin A toxicity, it was suggested by Hathcock et al. (15) that it was also possible that preexisting liver damage led to elevated serum levels. In 2 other prospective longitudinal 5-yr studies, no evidence of liver toxicity, as defined by elevated liver enzymes, was seen following up to 5 yr of vitamin A supplementation. In one of the these studies, 284 healthy elderly postmenopausal women (40–70 yr) received 5000 IU (1250–25,000) vitamin A as part of a multivitamin preparation. In the other study, 116 healthy elderly volunteers (47 male; 69 females), aged 64–88 yr, received vitamin A supplementation in doses of less than or greater than 10,000 IU as part of a multivitamin and mineral preparation. Results in these studies showed that serum retinyl ester levels did increase with dose, but not with the duration of supplement use. However, serum liver enzyme levels did not increase (19, 20).

The potential for human teratology does not appear associated with supplementation of 5000 IU vitamin A in women of childbearing potential and 8000 IU in pregnant women (12). However, in their evaluation of vitamin A toxicity, Hathcock et al. (15) stated that extrapolation of the unequivocal data from several species of animals very strongly suggests that at sufficiently high doses, vitamin A would produce teratogenic effects in humans. In addition, they reported that there is a temporal association of high intakes of vitamin A by pregnant women with babies born having defects characteristic of effects seen in animals given excessive doses of vitamin A. Although the evidence is convincing, there is uncertainty about the dose of vitamin A required to produce these effects.

DiPalma and Ritchie (13) reviewed the toxicology of vitamins. They reported the suggestion that clinical toxicity results in hypervitaminosis A when the amount of retinol binding protein is insufficient to bind the vitamin and the cell membrane is exposed to unprotected vitamins. The Registry of Toxic Effects of Chemical Substances (RTECS) reports that in women taking vitamin A from wk 1 through wk 39 of pregnancy, the toxic dose low (TDLo) or the lowest dose that produced specific developmental abnormalities in the urogenital system of the offspring was 68 mg/kg. The TDLo for specific developmental abnormalities for eyes and ears was 200 mg/kg when vitamin A was administered through 8 wk of pregnancy.

In a study based on 22,748 women who became pregnant in the mid 1980s, Rothman et al. (21) suggested that 1 of every 57 babies born to women who take more than 10,000 units of vitamin A daily from early pregnancy on would have a birth defect. These defects include malformations of the face, head, heart, and nervous system. In this study of over 22,000 pregnancies, it was estimated that only 5 or 6 babies were harmed by vitamin A. The authors also suggested that vitamin intake should not exceed the U.S. RDA.

In addition, they cautioned that pregnant women be careful combining vitamin supplements with large servings of liver or vitamin-enriched cereals that contain high levels of vitamin A. They further noted that beta-carotene, which appears to be completely safe, can be substituted for vitamin A. Vitamin A plays a critical role in regulating organ formation during early growth of the embryo. Synthetic forms of vitamin A, such as the acne medication 13-*cis*-retinoic acid, are known to cause animal and human birth defects. Although vitamin A has been suspected of causing human birth defects, the Rothman et al. (21) study was an attempt to determine the maximum safe dose. The Teratology Society has recommended that doses of vitamin A not exceed 6000 IU per day during pregnancy (13).

Ingestion of excessive amounts of carrots, other yellow vegetables, and some green vegetables can cause a condition called carotenemia (22). This condition is characterized by a yellow-orange discoloration of the skin (xanthodermia). Localization of the pigment to the palms and soles is consistent clinically with the diagnosis of hypercarotenemia (23). The absence of yellow pigment in the sclera and oral cavity distinguishes carotenemia from jaundice. Carotenemia was previously described under the names of aurantiasis and carotenosis cutis (24). This may occur in infants when mothers give their babies large amounts of carrots in their food. Carotenemia is considered a benign condition. Despite massive doses of carotene, vitamin A poisoning does not occur because the conversion of carotene to vitamin A is slow. Hypothyroidism, diabetes mellitus, hepatic diseases and renal disease may be associated with carotenemia, but are not caused by ingestion of carotene (22).

Hypercarotenemia, in addition to discoloration of the skin, is characterized by high serum concentrations of beta-carotene and may occur following consumption of 30 mg beta-carotene per day. Amenorrhea was reported to be associated with weight reduction, long-distance running, stress, and/or high intakes of fruits and vegetables resulting in carotenemia (25, 26).

A similar condition lycopenemia results from the ingestion of large amounts of tomatoes (22). The natural carotenoids including lycopene (from tomatoes) and the tocopherols are the major natural protective agents against free-radical-mediated liver damage (27). The carotenoids and lycopene may also play a protective role in cervical carcinogenesis (28).

The radioprotective activity in mice from lycopene is greater than that from beta-carotene, which is consistent with their relative antioxidant activity. Lycopene also exhibits antimutagenic activity (29). Zhang et al. (30) proposed that in their system of chemically induced neoplastic transformation in 10T1/2 cells, carotenoid-enhanced intracellular gap junctional communication and inhibition of lipid peroxidation provide a mechanistic basis for the cancer chemoprotective action of carotenoids.

In January 1996, the National Cancer Institute (NCI) announced that their two clinical trials yielded negative results. These studies, in which volunteers took beta-carotene to prevent cancer and heart disease, were terminated 21 mo early because of a possible cancer risk. These results were consistent with those of the Finnish study that suggested that beta-carotene increased the risk of lung cancer in smokers. The current study, Beta Carotene and Retinol Efficacy Trial (CARET), included 18,314 participants who were smokers and people exposed to high levels of asbestos. For 8 yr they received daily doses of beta-carotene, vitamin A, or placebo. Those taking beta-carotene had 28% more lung cancers and 17% more deaths than those taking the placebo. The second beta-carotene study, the Physicians' Health Study, included 22,071 physicians. The subjects, of whom 11% were smokers, took beta-carotene or a placebo every other day for 12 yr. The results of the study were disappointing since cancer and cardiovascular disease were not prevented, but no harm was done (31).

Several human studies have observed a direct association between retinol (vitamin A) intake and risk of prostate cancer. Other studies have reported either an inverse association or no association of intake of the major provitamin A, beta-carotene, with risk of prostate cancer. Giovannucci et al. (32), using a validated, semiquantitative food frequency questionnaire, calculated the relative risk (RR) for prostate cancer. Intake of the carotenoids beta-carotene, alpha-carotene, lutein, and beta-cryptoxanthin was not associated with risk of prostate cancer. However, lycopene intake was related to lower risk of prostate cancer. Of the 46 vegetables, fruits, or related products, 4 were significantly associated with lower prostate cancer risk. These were tomatoes, tomato sauce, pizza, and strawberries. Except for strawberries, the other three, which are primary sources of lycopene, account for 82% of the lycopene intake. These findings suggest that tomato-based foods may be especially beneficial regarding prostate cancer risk.

Le Marchand et al. (33) reexamined the data of a previous study that demonstrates a positive association of prostate cancer with beta-carotene intake. In evaluating the consistency among the main food sources of beta-carotene as well as the other phytochemicals in fruits and vegetables thought to be cancer inhibitors, they found the following. Papaya was positively associated with prostate cancer risk among men 70 years and older, while consumption of other yellow-orange fruits and vegetables, tomatoes, dark green vegetables, and cruciferous vegetables was not associated with prostate cancer risk. The chemicals in these fruits and vegetables not associated with prostate cancer risk include beta-carotene, lycopene, lutein, indoles, phenols, and other phytochemicals.

Animal Toxicology: Acute Toxicity

The Registry of Toxic Effects of Chemical Substances (RTECS) (34) lists the vitamin A oral LD50 in rats as 2 g/kg and in mice as 1.51 g/kg. When administered intraperitoneally in mice the LD50 is also reported as 1.51 g/kg. Hathcock et al. (15) list LD50 values for mice and rats shown in Table 3.

Following a single intramuscular injection in young monkeys of the equivalent of 300 mg/kg of retinol as retinyl acetate, the monkeys became progressively weak, developed dyspnea, lapsed into coma, and died. None of the monkeys who received the equivalent of 100 mg/kg retinol died. The intramuscular LD50 in monkeys for retinol was estimated to be 168 mg/kg (15).

Animal Toxicity: Repeat Dose

The toxicology of vitamin A was reviewed by Hathcock et al. (15). Following administration for 2 wk to 3 mo the toxic effects of hypervitaminosis A resulted in hair loss, localized erythema, and thickened epithelium. Fatty infiltration of the liver and fatty changes in the heart and kidney were observed, as well as testicular hypertrophy in adult rats. Weanling rats, but not adults, exhibited degenerative testicular changes following prolonged treatment. Treatment for 3 mo with the equivalent of 3 or 6 mg/kg/day caused extensive foci of degenerative myocardial fibers in rats. Blood effects following retinol administration included decreased hemoglobin concentration and transient increases in total circulating lipids and serum cholesterol. Serum triglycerides were also elevated. At daily doses 10,000 times the amount required to maintain growth, a limping gait in mice and rats was produced. Examination by x-rays revealed bone fractures. Since bone mineralization was normal as indicated by bone-ash determinations, the toxicity was associated with the formation or destruction of bone matrix.

Excess retinol elicited changes in bone whose longitudinal growth was greater than the circumferential growth, resulting in a thin fragile cortex that ultimately fractured (35,36). The effects of hypervitaminosis A on bones have also been demonstrated in other species including dogs (37). Reduced formation of dentin and atrophy of lingal odontoblasts were also observed.

Rats and dogs on chronic studies for 10 mo who received vitamin A palmitate did not show any adverse affect in body growth and hematologic indices. The rats received doses of 0, 5.5,

Table 3.   Animal toxicity of vitamin A

	LD50	
	Orally in mice (mg/kg)	Oral intubation in rat (mg/kg)
Retinol	2570	7900 (retinyl palmitate)
All-*trans*-retinoic acid	1100–4000	2000
13-*cis*-Retinoic acid	3389–26000	>4000
Etretinate	>4000	>4000

13.8, and 27.5 mg/kg, equivalent to 0, 10,000, 25,000, and 50,000 IU, and the dogs received doses of 0, 0.6, 2.8, and 13.8 mg/kg, equivalent to 0, 1000, 5100, and 25,000 IU (38).

Animal Reproductive Studies

The results of the reproductive studies shown in Table 4 have been reported in the Registry of Toxic Effects of Chemical Substances (34).

Table 4. Animal reproductive studies

Route of administration	Days of pregnancy	TDLo (mg/kg)	Adverse effect
			Rat
Oral	8–10	76.6	Growth statistics, behavioral
Oral	10	113	Musculoskeletal system
Oral	7–10	150	Central nervous system, craniofacial
Oral	4–16	683	Postimplantation mortality, fertility
Oral	1–22	22	Weaning/lactation index
			Growth statistics
Oral	9	240	Postimplantation mortality
			Extra embryonic structures
Intraperitoneal	10	600.6	Musculoskeletal system
			Developmental abnormalities
Intraperitoneal	10	60	Eyes, ears
Intraperitoneal	10	60	Craniofacial
Intraperitoneal	11	30	Musculoskeletal system
			Mouse
Oral	3	30	Skin and skin appendages
Oral	12	120	Craniofacial
Oral	8–12	9610	Live birth index
			Viability index
			Growth statistics
Oral	1 Post	30	Skin and skin appendages
Oral	7–11	375	Postimplantation mortality
Intraperitoneal	8	45	Eyes, ears
	8	145	Postimplantation mortality
			Litter size
			Fetotoxicity
Intraperitoneal	9	145	Central nervous system
			Eyes, ears
			Craniofacial
Intraperitoneal	10	145	Fetal death
			Musculosketal system
Intraperitoneal	10	180	Body wall
			Gastrointestinal system
Subcutaneous	11–14	96	Craniofacial
			Musculoskeletal
Subcutaneous	11–14	480	Fetal death
Intravenous	8	12	Central nervous system
			Musculoskeletal system
			Mouse
Intramuscular	10–13	480	Craniofacial
			Musculoskeletal system
Intramuscular	8	120	Musculoskeletal system
Intramuscular	14–17	240	Hepatobiliary system
			Rabbit
Oral	12	45	Fetotoxicity
			Craniofacial
			Musculoskeletal system

Hathcock et. al. (15) reviewed the teratogenicity studies in animals and concluded that hypervitaminosis A results in congenital malformations in most of the organ systems. The major factors include the dose, and the stages of gestation when vitamin A is administered. They summarized the structural effects in animals that are analogous to some of the malformations observed in humans. These are in the brain (anencephaly), spinal cord (spina bifida), face (cleft lip, cleft palate, microgonathia), eye (microphthalmia), malformations of the ear, teeth, salivary glands, aortic arch, heart (ventricular septal defects), lungs, gastrointestinal tract (imperforate anus, omphalocele), liver and gallbladder, urinary system (renal agenesis, polycystic kidney, hydronephrosis), genitalia, pituitary, thyroid, thymus, skull, vertebrae, ribs, extremities (phocomelia, digit malformations), muscles, and situs inversus (39). In addition, permanent learning disabilities in animals can occur at doses much lower than those that produce gross teratology. Early damage resulting in permanent behavioral deficits in adulthood occurs despite apparent cytochemical repair (40).

Mutagenicity

In the Ames test, the retinoids (vitamin A, tretinoin, and isotretinoin) were reported as nonmutagenic (38). RTECS (34) lists vitamin A in the EPA Genotox Program (1988) as negative in the mouse sperm morphology test. Although the toxicity of hypervitaminosis A has been demonstrated in animals and humans, no adverse effects have been reported following daily supplementation of beta-carotene at 120 times the normal intake for over 15 yr. The only side effect of large doses appears to be hypercarotenemia, which reverses when supplementation is discontinued. The amenorrhea and leukopenia associated with hypercarotenemia may be due to other factors or components in the carotene-containing foods, since these effects have not been observed in people ingesting large amounts of beta-carotene.

Benign intracranial hypertension was reported in people taking tetracycline and vitamin A in doses of 40,000–150,000 IU per day. Since both vitamin A and tetracycline can cause this condition, the combination may increase the severity (41–43).

In animals the toxic effects of hypervitaminosis A can be antagonized by vitamin D and the growth retardation can be prevented or counteracted by high dietary doses of vitamin E (44–46).

VITAMIN C

Vitamin C, one of the water-soluble vitamins, is not synthesized by the body and thus humans must be dependent on dietary sources. Ascorbic acid is the reduced form of vitamin C. Vast quantities of vitamin C are used in the food and pharmaceutical industries. Although the physiological role of vitamin C remains to be completely defined, the failure to synthesize collagen in skin and bones, which leads to a weakening and failure of repair processes in the extracellular matrix, is associated with vitamin C deficiency. This condition is known as scurvy. In normal physiology, the main electron-transfer role of vitamin C appears to be to reduce metals so that the associated enzyme systems can act in the transport of molecular oxygen, such as in the hydroxylation of proline during collagen synthesis and in the formation of noradrenaline (47). Ascorbic acid is essential for activating the enzyme prolyl hydroxylase, which promotes the hydroxylation step in the formation of hydroxyproline, an integral constituent of collagen. Without ascorbic acid, the collagen that is formed is defective and weak. Therefore, this vitamin is essential for growth of subcutaneous tissue, cartilage, bone, and teeth. A 20- to 30-wk deficiency of ascorbic acid occurred frequently during long sailing voyages in the olden days. When this was recognized, British sailors were issued a lime per day to prevent scurvy. Thus the nickname for the British "limey" was born.

The effects of scurvy or ascorbic acid deficiency include the failure of wounds to heal because the cells fail to deposit collagen fibrils and intracellular cement substances. This results in longer times for wounds to heal, that is, several months rather than days. The lack of ascorbic acid also causes the cessation of bone growth. The cells of the growing epiphyses continue to proliferate, but no new matrix is laid down between the cells, and thus bones tend to fracture easily at the point of growth, because of the failure to ossify. When already ossified bones fracture in people with ascorbic acid deficiency, osteoblasts cannot secrete a new matrix for deposition of new bone. Consequently, the fractured bone does not heal.

The blood vessel walls become extremely fragile in scurvy, because of the failure of endothelial cells to cement together properly to form collagen fibrils, normally present in the vessel walls. Capillaries especially are likely to rupture and result in many small petechial hemorrhages throughout the body. Hemorrhages beneath the skin cause purpuric blotches, sometimes over the entire body. In cases of extreme scurvy, the muscle cells sometimes fragment. Lesions of the gums with loosening of the teeth, mouth infections, bloody vomit, and bloody stools as well as cerebral hemorrhages may also result. Finally, high fever may develop prior to death (48). Vitamin C prevents the formation of nitrosamines and nitrosamides, thus reducing or eliminating cancer risk at various target organs such as the liver, upper gastrointestinal tract, and respiratory tract (49).

Human Toxicity

Behavioral changes, reduced drug metabolism, and reduced immunocompetence have been reported following marginal inadequacy of vitamin C (50).

Although serious toxicity to vitamin C is uncommon, numerous side effects may occur with overdosing or in susceptible individuals. Vitamin C can induce serious renal toxicity in a small number of susceptible individuals (17). Large doses of ascorbic acid may cause diarrhea and acidification of the urine, which can induce precipitation of cysteine or oxalate stones in the urinary tract (51). This increased renal excretion of oxalate, uric acid, and calcium by ascorbic acid in doses of 1 g or more daily increases the potential for stone formation in the kidney and bladder (52). Current consensus (16) is that doses up to 6 g per day and even up to 12 g per day will not significantly alter urine pH.

Large doses of ascorbic acid taken during pregnancy may cause scurvy in newborns after they are removed from the high ascorbic acid environment (51). This effect has been reported to occur in newborns of some women taking as little as 400 mg ascorbic acid daily throughout their pregnancy. It has been suggested that vitamin C enhances the development of fetal liver microsomal enzymes, which then causes scurvy because of greater destruction of the vitamin after birth. Another possibility is that the fetus recognizes the danger of increased vitamin C concentrations and increases its own metabolic rate to destroy the excess amount. This increased metabolic rate continues after birth so that the signs and symptoms are seen shortly after delivery. A similar but less dangerous rebound scurvy has been reported in adults who suddenly withdraw gradually from megadoses of vitamin C (17). Thus it has been suggested to withdraw gradually from megadoses by 10% to 20% daily (53).

Megadoses of vitamin C over a period of years have been reported to destroy substantial amounts of vitamin B12 and produce symptoms of megaloblastic anemia (54, 55).

Ascorbic acid in patients maintained on heparin or warfarin has been shown to shorten prothrombin time (56). Doses of grams per day for 3–7 days caused sustained uricosuria with a fall in uric acid that could cause precipitation of gouty arthritis or renal calculi in susceptible individuals (57).

Gossel and Bricker (17) have listed the common side effects of vitamin C administration as follows: interference with urine and stool testing (negates occult blood tests; causes false negative reactions with glucose oxidase tests); decreased absorption of vitamin B12; rebound scurry following prolonged administration of megadoses; increased urinary oxalate, cysteine, and uric acid; acidification of urine; diarrhea and occasional nausea.

The Registry of Toxic Effects of Chemical Substances lists the toxic dose low (TDLo), the lowest dose reported to produce any toxic effect in humans, to be 2300 mg/kg intravenously in men over 2 days to produce oxidant-related anemia and 900 mg/kg intravenously in women to produce changes in the kidney tubules.

Toxicology—Animals

The Registry of Toxic Effects of Chemical Substances (34) lists the ascorbic acid/vitamin C LD50 values shown in Table 5.

Table 5. Ascorbic acid/vitamin C LD50 values

Species	Route	LD50 (mg/kg)	Effect
Rat	Oral	11,900	Lacrimation Somnolence Gastrointestinal hypermotility, diarrhea
Rat	Subcutaneous	>10,000	—
Rat	Subcutaneous	>4000	Altered sleep time Somnolence
Mouse	Oral	3367	—
Mouse	Intraperitoneal	643	—
Mouse	Intravenous	518	—

The reproductive studies reported in RTECS are given in Table 6.

In the EPA Genotox Program (1988); vitamin C/ascorbic acid was positive in the following tests: in vitro sister chromatid exchange (SCE) in human lymphocytes and in vitro SCE, nonhuman. It was negative in mouse sperm morphology and inconclusive in the mammalian micronucleus and in the histidine reversion—Ames test. In the National Toxicology Program (NTP) Carcinogenesis Bioassay feeding study in mice and rats there was no evidence of carcinogenicity (NTP-TR-247, 83).

VITAMIN E

There are eight naturally occurring tocopherols with vitamin E activity. Alpha-tocopherol (5, 7, 8-trimethyl tocol) is considered the most important since it comprises about 90% of the tocopherols in animal tissue. The tocopherols deteriorate slowly when exposed to air or ultraviolet light. Vitamin E is the collective term for all ecologically active tocopherols. As an antioxidant, the fat-soluble vitamin E presumably prevents the oxidation of essential constituents such as coenzyme Q (ubiquinone) and prevents the formation of toxic oxidation products such as peroxidation products from unsaturated fatty acids (16). Chemically unrelated substances such as synthetic antioxidants, selenium, sulfur-containing amino acids, and the coenzyme Q groups can prevent or reverse some of the symptoms of vitamin E deficiency (58). Vitamin E enhances the intestinal absorption of vitamin A, which may be related to its antioxidant activity. It also appears to protect against the effects of hypervitaminosis A (59).

During normal respiration, oxidant intermediates may be formed, which because of their high chemical reactivity can cause damage to biological membranes. The particular function of vitamin E is to protect membranes from oxidative damage. In vitamin E deficiency, membrane turnover may accelerate so that the normal order and compartmentalization of cells is destroyed. Because free radical reactions may be responsible for health disorders of major concern worldwide (cardiovascular disease and cancer), control of free radical generation, their proliferation, and their effects are of great importance. Although direct evidence implies that vitamin E, selenium,

Table 6. Reproductive studies for vitamin C in animals

Species	Route	TDLo (mg/kg)	Days of pregnancy	Effects
Rat	Oral	2500	1–22	Postimplantation mortality
Mouse	Intraperitoneal	6680	11	Fetal death
Mouse	Intravenous	800	8	Central nervous system abnormalities Musculoskeletal system abnormalities
Guinea pig	Oral	19,500	30–58	Biochemical/metabolic in newborns
Guinea pig	Oral	5800	1–58	Stillbirth, viability index
Guinea pig	Oral	2471	Multigeneration	Growth statistics
Mouse	Oral	546,000	13 wk intermittently	Lethality

vitamin C, and beta-carotene act as modulators or moderators of the biochemical charges leading to disease, direct evidence is lacking for their activity in disease prevention. However, there is a large and steadily growing body of indirect evidence that implies the importance of vitamin E in preventing a number of diseases, in particular cardiovascular and certain cancers (60).

Vitamin E is considered the most important fat-soluble antioxidant, while vitamin C is the most important water-soluble antioxidant. This functions to break free-radical-initiated chain reactions among the polyunsaturated fatty acids of biological membranes. Vitamin E is transported in the lymph and in blood as free tocopherol, bound to lipoproteins. Thus all body tissues, that is, cell membranes and subcellular organelles, are supplied with vitamin E and thereby stabilized. There is also a functional connection of tocopherols and membrane-associated enzymes. Initially, vitamin E was considered to be responsible for the maintenance of pregnancy and fertility in rats (61,62). Currently its role is considered as a lipid-soluble antioxidant and protective component of membranes of cells and organelles (63), as well as an inhibitor of lipid peroxidation (64), and it plays a special role in porphyrin metabolism (65, 66). Vitamin E also affects the metabolism of arachidonic acid, inhibition of thromboxane, leukotriene biosynthesis, and increased prostacyclin formation have been reported. These may be involved in vitamin E's effect on the inhibition of inflammation and thrombotic disease (62). Diplock et. al. (67), however, suggested a biphasic effect of vitamin E on the arachidonic acid cascade in which small amounts of vitamin E stimulate the biosynthesis of prostaglandins and larger amounts inhibit it.

Vitamin E deficiency leads to premature cell aging and to the binding of immunoglobulin G to erythrocytes (68). Although other deficiency symptoms occur in animals, vitamin E deficiency in humans is rare.

In lower animals, lack of vitamin E causes degeneration of the germinal epithelium in the testes and therefore causes male sterility. In females it causes resorption of the fetus after conception. Because of this, vitamin E has been called the antisterility vitamin.

Vitamin E deficiency in animals also causes paralysis of the hind quarters with pathological changes in muscles similar to those in muscular dystrophy patients. However, vitamin E administration did not benefit these patients. As with most of the other vitamins, vitamin E deficiency prevents normal growth. Vitamin E appears to function mainly in relation to unsaturated fatty acids, preventing oxidation of unsaturated fat in the cells is diminished, causing abnormal structure and function of such cellular organelles as the mitochondria, lysosomes, and even the cell membrane (69).

Human Toxicity

The lack of toxicity of high doses of vitamin E has been reported by many authors (13, 16, 70, 71). However, undesirable side effects of megadoses of vitamin E have been reported to include headache, nausea, fatigue, dizziness, and blurred vision. These may be related to the fact that large doses of vitamin E may antagonize the action of vitamin A. In addition, inflammation of the mouth, chafing of the lips, gastrointestinal disturbance, muscle weakness, hypoglycemia, increased bleeding tendencies, degenerative changes, and emotional disorders have also been reported (72). Gossel and Bricker (17) list other reported side effects from megadoses of vitamin E including disturbance in growth, thyroid function, mitochondrial respiration rate, bone calcification, and decreased hematocrit. Aeling et al. (73) reported that humans can tolerate doses of vitamin E up to 1000 IU per day without developing toxicity. Roberts (74) reported on 7 women with thrombophlebitis who complained of sore breasts while taking 400 to 800 IU vitamin E per day for over a year. In all seven, breast pain disappeared when vitamin E administration was discontinued. Since no reference to breast pain was made in the many double-blind studies with vitamin E, and other investigators reported on favorable effects of vitamin E on benign breast disease (75), the relationship between vitamin E and breast pain was considered questionable (60).

In humans as well as in experimental animals, vitamin E may interfere with vitamin K metabolism, resulting in a prolonged prothrombin time. Excessive vitamin E intake in experimental animals decreases the rate of wound healing, while in humans it induces gastrointestinal symptoms and creatinuria (17).

Among the other drug interactions reported for vitamin E are the increased requirements for vitamins A and D (76) and that the maintenance dose of digitalis should be reduced by 50% in the presence of high doses of vitamin E (77).

COPPER

Copper, an essential element, is widely distributed in nature. It is a required component of some proteins including superoxide dismutase, amine oxidases, uricase, catalase, peroxidase, cytochrome oxidase, and other oxidative enzymes, and is essential for the incorporation of iron into hemoglobin. It is also an essential component of the enzyme tyrosinase, which is involved in the formation of melanin pigment (78, 79). Copper deficiency is characterized by hypochromic, microcytic anemia resulting from defective hemoglobin synthesis. Iron deficiency anemia in infancy is sometimes accompanied by copper deficiency (80) as well as neutropenia and impaired growth, especially in children (81). Mildly copper-deficient diets may increase cardiovascular risk factors such as cholesterol levels and glucose tolerance, although this is inconclusive (81).

The estimated safe and adequate daily dietary intake (ESADDI) of copper is 1.5–3.0 mg (81), while the adult RDA is 2.0–3.0 mg (82). Humans ingest copper from foods and water in amounts that vary from less than 10 to more than 25,000 µg/100 calories (79). Rich sources of copper include oysters, nuts, liver, kidney, and dried legumes (82).

Copper is actively absorbed from the stomach and duodenum. Usually half of the ingested dose is absorbed, which can be decreased by competition with zinc and binding by ascorbic acid (79).

Absorbed copper is bound by ceruloplasmin and transported in the plasma to be stored in the liver, brain and muscles. An average blood level of copper is 1 mg/L, and a 70-kg man stores from 70 to 120 mg (78, 79).

The U.S. Environmental Protection Agency (U.S. EPA, 1985) maximum contaminant level proposed for copper in drinking water adopted in 1981 was 1.3 mg/L.

Copper, like iron, can participate in oxidation–reduction reactions, and free copper ions can be toxic by causing oxidation of lipids, proteins, DNA and other biomolecules (83). Acute overdoses by ingestion cause an immediate metallic taste, epigastric burning, nausea, vomiting (blue-green color), and, in more severe cases, diarrhea. Ulcers and other local damage to the gastrointestinal tract, jaundice (with liver necrosis and biliary stasis), and suppression of urine may also occur. Fatal cases include secondary effects such as hypotension, melena, coma, and jaundice (80,84).

Acute copper poisoning can also cause hemolytic anemia (85).

Except for patients with inborn metabolic disorders, there are few reports of chronic toxicity. Sheep appear to be highly susceptible to copper toxicity. Excessive copper ingestion results in lethal hemolytic anemia accompanied by severe degeneration of the liver, kidney, and spleen (78, 84).

MANGANESE

Manganese is an essential element and a cofactor in enzymatic reactions, particularly those involved in phosphorylation, cholesterol, and fatty acids synthesis. It is a component of mitochondrial enzymes such as pyruvate carboxylase and superoxide dismutase and activates a wide variety of enzymes. Manganese can substitute for magnesium in many enzymes (85, 86). The enzyme systems in which manganese is essential are involved with protein and energy metabolism and in mucopolysaccharide formation (82). Although no cases of human manganese deficiencies have been recorded, the recommended daily allowance for humans is 2.5–5 mg (82).

Manganese is widely distributed in foods and water. It makes up 0.1% of the earth's crust. Good sources of manganese include nuts, whole cereals, and teas, while meats, fish, and dairy products are poor sources. Oral absorption of manganese from the gastrointestinal tract is slow and incomplete, and ranges from 1 to 4%. Manganese is transported in plasma bound to a B_1-globulin, thought to be transferrin or transmang, and is widely distributed in the body (80), with the highest concentrations in the bone, liver, kidney, pancreas and intestines and within cells. Concentrations are higher in the mitochondria than in the cytosol and other organelles. The biological half-life of manganese in the body is 37 days and longer in the brain, even though it crosses the blood–brain barrier readily (80, 87).

Human manganese toxicity has only been seen following exposure to high levels of manganese in the air (82).

It is uncertain whether oral ingestion of manganese can be neurotoxic as it is by inhalation. Case reports and animal studies suggest that high levels of at least 1 g/day for adults and 10 mg/day for infants may produce neurotoxicity (88). The U.S. EPA has derived a chronic oral RfD for manganese of 0.1 mg/kg/day.

In animals, manganese deficiency can cause impaired growth, skeletal abnormalities, and altered metabolism of carbohydrates and lipids (88).

SELENIUM

Oxidative stress may result in cellular selenium dysfunction and damage. Quantifiable endpoints for assessing oxidative stress include antioxidant enzyme activity, redox status, and lipid peroxidation. Reduced glutathione (GSH) and associated antioxidant enzymes are major combatants of oxidative stress that influence redox status. Glutathione, the most important nonprotein thiol in living systems, plays a vital role in protecting nucleophile sites in liver and other tissues from electrophile attack by xenobiotics and their metabolites, including polycyclic aromatic hydrocarbons, heavy metals, selenium, and certain industrial solvents. Selenium salts and vitamin E are cofactors in glutathione peroxidase, which may be responsible for the activation and detoxification of the metabolites of carcinogens (49).

Historically, the biological roles of selenium have been associated with antioxidant activity (89). In addition, selenium is a requirement for the enzymatic activity of glutathione peroxidase (90) and is considered an essential nutrient (5). Dietary selenium in excess of 4 ppm is generally considered to be toxic to mammals (91), and high environmental concentrations have been reported to adversely affect reproduction, survival, and development of wild aquatic birds (92, 93).

Hoffman et al. (94) studied the effects of excess dietary selenium in mallard ducklings. They found that excess dietary selenium as selenomethionine had a more pronounced effect on hepatic glutathione metabolism and lipid peroxidation in ducklings then selenite, which may be related to the patterns of accumulation. The effects of selenomethionine appeared to be less pronounced in ducklings than reported in laboratory rodents. Selenomethionine, which occurs in vegetation, is of particular interest for the health of wild aquatic birds.

It was proposed that selenium toxicity is related to changes in intracellular concentrations of reduced glutathione (GSH), and excess selenium has been shown to interact with cellular sulfhydryls (95). Selenium in the oxidant selenite undergoes reductive metabolism, utilizing GSH and NADPH as a source of reducing equivalents (96).

Although the estimated intake in the United States is 132 μg/day (97), the diet may actually contribute from 0.7 to 7.0 mg selenium for a 70-kg man (98). Seafoods, kidney, liver, and meat provide a good dietary source of selenium (82). Reports of selenium's potential anticarcinogenic activity may be responsible in part for the overuse of selenium supplements (87).

Selenium is an essential nutrient with a recommended dose of 50–200 μg/day considered to be adequate and safe for adults (82). More selenium is required if diets are also deficient in vitamin E. Specifically, selenium is an essential component of glutathione peroxidase, which destroys hydrogen peroxide and hydroperoxides, thus protecting cell membranes from oxidative damage. Vitamin E is also implicated in this system in which its role is to prevent the formation of liquid hydroperoxides (99).

In addition to its anticarcinogenic activity, selenium also protects against heavy metal toxicity, that is, from cadmium, mercury, and silver. Dietary deficiencies of selenium have been associated with increased incidence of heart disease (100).

In the Keshan county of the northern Chinese province of Herlungjiang, an endemic cardiomyopathy among children and young women was attributed to a dietary selenium deficiency resulting from the consumption of vegetables grown in soil of low selenium content (101). This Keshan disease was reported to have affected over 8000 people in each of the 3 peak years of 1959, 1964, and 1970. The victims of Keshan disease showed depressed selenium levels in body fluids and tissues (102), which were selenium responsive, and dietary sodium selenite (Na_2SeO_3) has become the established prophylactic procedure where the disease is most prevalent (103).

The clinical effects of selenium have been reported as follows: Elemental selenium is low in toxicity; however, all selenium salts may produce toxicity by ingestion, inhalation, and percutaneous absorption. They can produce hypotension, pulmonary edema, and cardiopulmonary arrest, as well as dizziness, decreased reflexes, CNS depression, and coma. In addition, vomiting, salivation, a burning sensation in the nostrils, and a garlic-like odor on the breath have also been reported; as have dermatitis, nasal irritation, mild tubular degeneration, and fatty degeneration and cirrhosis of the liver.

The adequate dietary intake of selenium suggested by the Food and Nutrition Board of the National Research Council is 50–200 µg/day. Observations in Japanese fishermen suggested that selenium intake from 10 to 200 times normal have not produced toxic effects. However, ingestion of 31 mg/day for 11 days produced toxicity, while individuals who ingested 312–617 mg/day chronically reported toxicity.

The Hazardous Substances Data Base (HSDB) reports that elemental selenium is nontoxic in the therapeutic dose range of 50–200 µg/day. The estimated toxic doses derived from toxic cases ranged from 27 to 2310 mg and exhibited the following signs and symptoms: nausea, vomiting, nail changes, fatigue, and irritability. In addition to dietary, occupational, and self-medication exposure, ingestion of selenium may also occur from drinking water. For human health, the ambient water quality criterion for selenium in the Clean Water Act is identical to the existing water standard, which is 10 µg/L. This is the same as the FDA requirement for bottled water. Because selenium constitutes about 0.09 ppm in the earth's crust and is present in the major oceans and inland waters, selenium may also be available from plants.

Schroeder et al. (104) reported that in humans, the signs and symptoms of chronic, sublethal selenium intoxication included discolored and decayed teeth, pallor or yellow skin color, skin eruptions, chronic arthritis, atrophic brittle nails, edema and gastrointestinal disorders, and, in some cases, lassitude and partial loss of hair and nails.

Animal Toxicology

Grazing animals require a minimum concentration of 0.04 ppm selenium in their diet, and it is beneficial up to 0.1 ppm. At greater than 4 ppm it becomes toxic. Years ago, cowboys of the American West observed that cattle grazing on milk-vetch (legumes of the *Astragalus* family) grown in the selenium-rich soils of Wyoming and South Dakota became disoriented and suffered weakness, lassitude, visual impairment, loss of appetite, and paralysis with respiratory failure. It was recognized in 1934 that this vegetation concentrated large amounts of selenium from the alkaline soil, and that the animals consuming this "loco-weed" developed the "blind staggers" or "alkali disease" (105). At the other extreme, calves, lambs, and foals of animals grazing in pastures with low concentrations of selenium in the soil become afflicted with "white muscle disease," a degeneration of the striated muscles. Selenium deficiency in livestock has been reported in the United States, Canada, New Zealand, Australia, Scotland, Finland, Sweden, Denmark, France, Germany, Greece, and Russia (101).

The Registry of Toxic Effects of Chemical Substances (34) lists toxicity for selenium as shown in Table 7.

Selenium dusts produce respiratory-tract irritation manifested by nasal discharge, loss of smell, epistaxis, and cough.

ZINC

Zinc is ubiquitous in the environment and ranges from 5 to 200 ppm in the earth's crust. It is present in most foodstuffs, water, and air. Zinc is a component of and essential for the activity of the following enzymes: alcohol dehydrogenase, carboxypeptidase, leucine aminopeptidase, alkaline phosphatase, carbonic anhydrase, RNA-polymerase, and DNA-polymerase. Thymidine kinase may also be zinc dependent (106).

Meats, fish, and poultry contain an average of 24.5 mg zinc/kg while grains contain 8 and

Table 7. Selenium toxicity values

Species	Route	LD50 (mg/kg)	Effects
Rat	Oral	6700	Somnolence
			Dyspnea
			Nutritional
			Gross metabolic
Rat	Inhalation	33 (8 hr LDLo)	Hemmorhage
			Emphysema
			Pulmonary edema
Rat	iv	6	
Rabbit	iv	2.5 (LDLo)	
Mouse	Oral	480 (TDLo), 60 days	Tumors on skin
Mouse	Oral	134 (TDLo)	Fetal death (Multigeneration)

potatoes 6. The recommended daily allowances (RDAs) for zinc are 15, 20, and 25 mg/day for adults, pregnant women, and lactating women, respectively (84).

The bioavailability of zinc from food ranges from 10 to 40% (107), while Snyder et al. (98) consider the best value to be 35%. Absorption is from the small intestine and is affected by age, content of zinc, protein, phytic acid, and calcium fiber in the diet. Zinc from animal sources is more available for human absorption than zinc from plants, which may be due to zinc complexing with phytate (inositol hexaphosphate) and other plant constituents (108).

The absorption mechanism is considered to be homeostatically controlled and probably a carrier-mediated process (109). It is also influenced by protaglandins E_2 and F_2 and is chelated by picolinic acid, a tryptophane derivative. In the blood, zinc is bound to albumin and B_2-macroglobulin. Zinc enters the gastrointestinal tract as a component of metallothionein secreted by the salivary glands, intestinal mucosa, pancreas, and liver (109).

The highest concentrations of zinc in the body are found in the male reproductive system, with the prostate gland having the greatest. It is also found in muscle, bone, liver, kidney, pancreas, thyroid, and some other endocrine glands. Zinc content in the kidneys and liver is dependent on cadmium concentrations. Zinc is stored in metallothionein (106).

Since more than 200 metalloenzymes require zinc as a cofactor, deficiency results in a wide spectrum of clinical effects depending on age, stage of development, and deficiencies of related metals. Reported effects of zinc deficiency include growth failure and delayed sexual maturity in boys, accompanied by protein-caloric malnutrition, pellagra, and iron and folate deficiency. In the newborn, dermatitis, loss of hair, impaired healing, susceptibility to infections, and neuropsychological abnormalities have been reported. Dietary inadequacies in people with liver disease from chronic alcoholism may be associated with dermatitis, night blindness, testicular atrophy, impotence, and poor wound healing. In addition, clinical disorders such as ulcerative colitis and the maladsorption syndrome, chronic renal disease, and hemolytic anemia have been reported. Zinc deficiency may also exacerbate impaired copper nutrition, and zinc interaction with cadmium and lead may modify the toxicity of these metals (80).

One of the major clinical abnormalities associated with zinc deficiency is acrodermatitis enteropathica, which is an inherited autosomal recessive trait. Infants are the most frequent victims of this disease and exhibit chronic diarrhea and seborrheic skin lesions, usually located at body orifices (101). A decline in the zinc status of Dutch children with acute diarrhea was reported by Van Wouwe et al. (110), while zinc supplementation cleared up the skin lesions and normalized bowel function (111, 112). The genetic defect may affect the synthesis of picolinic acid or facilitation of zinc absorption, and thus the deficiency may be a result of malabsorption rather than dietary deficiency. Acrodermatitis enteropathica has been reviewed by Van Wouwe (113).

Dietary zinc deficiency has also been reported as responsible for retarded growth and delayed sexual maturation in Iranian and Egyptian males. Dietary supplementation of 100 mg zinc sulfate per day restored sexual function and increased growth an average of 10.5 cm during a 5-day confinement period. Subjects treated with the same diet, but without zinc supplementation, required 224 days to sexually mature, and only grew an average of 4.2 cm during the 59-day

Table 8.   Toxicity values for zinc sulfate

Species	Route	LD50 (mg/kg)	Effects
Rat	Oral	2150	
Rat	Intraperitoneal	200	
Rat	Subcutaneous	330 (LDLo)	
Rat	Intravenous	49 (LDLo)	
Mouse	Oral	2200	
Mouse	Intraperitoneal	75	
Dog	Subcutaneous	78 (LDLo)	
Dog	Intravenous	66 (LDLo)	
Rabbit	Oral	1914 (LDLo)	
Rabbit	Intravenous	44 (LDLo)	
Guinea pig	Subcutaneous	590 (LDLo)	Somnolence
			Coma
			Hypermotility
			Diarrhea
Multiple dose TDLo (mg/kg)			
Rat	Oral	226,226 (13 wk)	Changes in structure or function of exocrine pancreas
			Weight loss or decreased weight gain
			Transaminase changes
Mouse	Oral	443,898 (13 wk)	Changes in structure or function of exocrine pancreas
			Changes in kidney, ureter, bladder
			Changes in thymus weight
Mutagenicity dose (mg/kg)			
Mouse	Intraperitoneal	20,000	DNA inhibition

confinement period (114). The relationship between zinc status and recurrent respiratory infections in children (115), recovery of human adult burn victims (116), and bottle versus breast feeding in infants (117) have been investigated. Another consequence of zinc deficiency responsive to zinc supplementation is hypogeusia (101).

Toxicity from excessive ingestion of zinc is uncommon, but gastrointestinal distress and diarrhea have been reported following ingestion of beverages standing in galvanized cans or from use of galvanized utensils. However, no evidence of hematologic, hepatic, or renal toxicity has been observed in individuals who ingested as much as 12 g of elemental zinc over a 2-day period (80).

Patients who took zinc at doses 10 times the RDA for months and years did not demonstrate any adverse reactions. However, excessive zinc intakes may inhibit copper absorption and thus aggravate marginal copper deficiency (82). However, Carson et al. (84) reported that ingestion of 2 g or more of zinc produces toxic symptoms in humans.

The Registry of Toxic Effects of Chemical Substances (RTECS) lists toxicity for zinc sulfate as shown in Table 8.

REFERENCES

1. Diplock, A. T. 1991. Antioxidant nutrients and disease prevention: An overview. *Am. J. Clin. Nutr.* 53:189S–193S.
2. Slater, T. F. 1991. *Am. J. Clin. Nutr.* 53:394S–396S.
3. Imaly, J. A., and Linn, S. 1988. DNA damage and oxygen radical toxicity. *Science* 240:1302–1309.
4. *Federal Register.* 1973. 38:20708, 38:20730, 2 August.
5. National Research Council Subcommittee on Selenium 1983. *Selenium Nutrition*, pp. 1–136. Washington, DC: National Academy Press.
6. Cummings, F., Briggs, M., and Brigg, M. 1981. Clinical toxicology of vitamin supplements. In *Vitamins in Human Biology and Medicine*, ed. M. H. Briggs, pp. 187–243. Boca Raton, FL: CRC Press.

7. Pryor, W. A. 1991. The antioxidant nutrients and disease prevention—What do we know and what do we need to find out? *Am. J. Clin. Nutr.* 53:391S–393S.
8. Omaye, S. T. 1988. Safety of megavitamin therapy. In *Nutritional and Toxicological Aspects of Food Safety*, ed. M. Friedman, pp. 169–203. New York: Plenum Press.
9. Ovesen, L. 1984. Vitamin therapy in the absence of obvious therapy. What is the evidence? *Drugs* 27:148–170.
10. Salem, H. 1987. Factors influencing toxicity. In *Inhalation Toxicology*, ed. H. Salem, pp. 35–57. New York: Marcel Dekker.
11. Eaton, D. L., and Klausen, C. D. 1996. Principles of toxicology. In *Casarett & Doull's Toxicology*, pp 13–33. New York: McGraw-Hill.
12. Bendich, A. 1992. Safety issues regarding the use of vitamin supplements. *Ann. N.Y. Acad. Sci.* 669:300–312.
13. DiPalma, J. R., and Ritchie, D. M. 1977. Vitamin toxicity. *Annu. Rev. Pharmacol. Toxicol.* 17:133–148.
14. Bendich, A., and Langseth, L. 1989. Safety of vitamin A. *Am. J. Clin. Nutr.* 49:358–371.
15. Hathcock, J. N., Hattan, D. G., Jenkins, M. Y., McDonald, J. T., Ramnathansundaresan, P., and Wilkening, V. L. 1990. *Am. J. Clin. Nutr.* 52:183–202.
16. Mandel, H. G., and Cohn, V. H. 1985. Fat soluble vitamins. In *The Pharmacological Basis of Therapeutics*, 7th ed., eds. A. G. Gilman, L. S. Goodman, and T. W. Rall, pp. 1573–1591. New York: Macmillan.
17. Gossel, T. A., and Bricker, J. D. 1994. *Principles of Clinical Toxicology*, 3rd ed., pp. 403–417. New York: Raven Press.
18. Krasinski, S. D., Russell, R. M., Otradovec, C. L., Sadowski, J. A., Hartz, S. C., Jacob, R. A., and McGandy, R. B. 1989. *Am. J. Clin. Nutr.* 49:112–120.
19. Johnson, E. J., Krall, E. A., Dawson-Hughes, B., Dallal, G. E., and Russell, R. M. 1992. Lack of an effect of multivitamins containing vitamin A on serum retinyl esters and liver function tests in healthy women. *J. Am. Coll. Nutr.* 11:682–686.
20. Stauber, P. M., Sherry, M. B., Vanderjagt, D. J., Bhagavan, H. N., and Garry, P. J. 1991. *Am. J. Clin. Nutr.* 54:878–883.
21. Rothman 1995
22. Lascari, A. D. 1981. Carotenemia: A review. *Clin. Pediatr.* 20:25–29.
23. Stack, K. M., Churchwell, M. A., and Skinner, R. B. 1988. Xanthoderma: A case report and differential diagnosis. *Cutis* 41:100–102.
24. Aloy Pantin, M., and Torres Peris, V. 1985. Clinical features of carotinemia. *Med. Cutan. Ibero. Lat. Am.* 13:31–34.
25. Frumar, A. M., Meldrum, D. R., and Judd, H. L. 1979. Hypercarotenemia in hypothalamic amenorrhea. *Fertil. Steril.* 32:261–264.
26. Deuster, P. A., Kyle, S. B., and Moser, P. B. 1986. Nutritional intakes and status of highly trained amenorrheic and eumenorrheic women runners. *Fertil. Steril.* 46:636–643.
27. Leo, M. A., Rosman, A. S., and Lieber, C. S. 1993. Differential depletion of carotenoids and tocopherol in liver disease. *Hepatology* 17:977–986.
28. Batieka, A. M., Armenian, H. K., Norkus, E. P., Morris, J. S., Spate, V. E., and Comstock, G. W. 1993. Serum micronutrients and the subsequent risk of cervical cancer in a population-based nested case control study. *Cancer Epidemiol. Biomarkers Prev.* 2:335–339.
29. Kapitanov, A. B., Pimenov, A. M., Obukhova, L. K., and Izmailov, D. M. 1994. Radiation-protective effectiveness of lycoperal. *Radiat. Biol. Radioecol.* 34:439–445.
30. Zhang, L. X., Cooney, R. V., and Bertram, J. S. 1991. Carotenoids enhance gap functional communication and inhibit lipid peroxidation in C3H/10T1/2 cells: Relationship to their cancer chemoprotective action. *Carcinogenesis* 12:2109–2114.
31. Peterson, K. 1996. Natural cancer prevention trial halted. *Science* 271:441.
32. Giovannucci, E. 1995. Intake of carotenoids and retinol in relation to risk of prostate cancer. *J. Natl. Cancer Inst.* 87:1767–1776.
33. Le Marchand, L. 1991. Vegetable and fruit consumption in relation to prostate cancer risk in Hawaii: A reevaluation of the effect of dietary beta-carotene. *Am. J. Epidemiol.* 133:215–219.
34. Registry of Toxic Effects of Chemical Substance (RTECS)
35. Nieman, C., and Obbink, H. J. K. 1954. The biochemistry and pathology of hypervitaminosis A. *Vitamin Horm.* 12:69–99.
36. Irving, J. T. 1949. The effects of avitaminosis and hypervitaminosis A upon the incisor teeth and incisal alveolar bone of rats. *J. Physiol.* 108:92–101.
37. Maddock, C. L., Wolbach, S. B., and Maddock, S. 1949. Hypervitaminosis A in the dog. *J. Nutr.* 39:117–137.
38. Kamm, J. J., Ashenfelter, K. O., and Ehmann, C. W. 1984. Preclinical and clinical toxicology of selected retinoids. In *The Retinoids*, vol. 2, eds M. B. Sporn, A. B. Roberts, and D. S. Goodman. New York: Academic Press.

39. Shenefelt, R. E. 1972. Gross congenital malformation. Animal model: Treatment of various species with a large dose of vitamin A at known stages in pregnancy. *Am. J. Pathol.* 66:589–592.
40. Vacca, L., and Hutchings, D. E. 1977. Effect of maternal vitamin A excess on 5–100 in neonatal rat cerebellum: A preliminary study. *Dev. Psychobiol.* 10:171–176.
41. Walters, B. N., and Gubbay, S. S. 1981. Tetracycline and benign intracranial hypertension. *Br. Med. J.* 282:19–20.
42. Pearson, M. G., Littlewood, S. M., and Bowden, A. N. 1981. Tetracycline and benign intracranial hypertension. *Br. Med. J.* 282:568–569.
43. Krausz, M. M., Feinsod, M., and Beller, A. J. 1978. Bilateral transverse sinus obstruction in benign intracranial hypertension due to hypervitaminosis A. *Isr. J. Med. Sci.* 14:858–861.
44. Metz, A. L., Walser, M. M., and Olson, W. G. 1985. The interaction of dietary vitamin A and vitamin D related to skeletal development in the turkey poult. *J. Nutr.* 115:929–935.
45. McCuaig, L. W., and Motzok, I. 1970. Excessive dietary vitamin E: Its alleviation of hypervitaminosis A and lack of toxicity. *Poult. Sci.* 49:1050–1051.
46. Jenkins, M. Y., and Mitchell, G. V. 1975. Influence of excess vitamin E on vitamin A toxicity in rats. *J. Nutr.* 105:1600–1606.
47. Marks, J. 1988. *The Vitamins: Their Role in Medical Practice.* Lancaster, PA: MTP Press.
48. Aruoma, O. I. 1991. Pro-oxidant properties: An important consideration for food additives and/or nutrient components? In *Free Radicals and Food Additives*, eds. O. I. Aruoma and B. Halliwell. New York: Taylor and Francis.
49. Mirvish, S. S. 1986. Effects of vitamin C and E on *N*-nitroso compound formation of carcinogenesis and cancer. *Cancer* 58:1842–1850.
50. Brimm, M. 1982. In *Ascorbic Acid, Chemistry, Metabolism and Uses*, eds. P. A. Seib and B. M. Tolbert, pp. 369–379. Washington, DC: American Chemical Society.
51. Cochrane, W. A. 1965. Overnutrition in prenatal and neonatal life: A problem? *Can. Med. Assoc. J.* 93:893–899.
52. Sestili, M. A. 1983. Adverse health effects of vitamin C and ascorbic acid. *Semin. Oncol.* 10:299–304.
53. Herbert, V. D. 1979. Megavitamin therapy. *NY J. Med.* February:278.
54. DiPalma, J. 1978. Vitamin toxicity. *Am. Family Physician* 10:106.
55. Herbert, V. D. and Jacob, E. 1974. Destruction of vitamin B12 by ascorbic acid. *J. Am. Med. Assoc.* 230:241–242.
56. Rosenthal, G. 1971. Interaction of ascorbic acid and warfarin. *J. Am. Med. Assoc.* 215:1671.
57. Stein, H. B., Hasan, A., and Fox, I. R. 1976. Ascorbic acid-induced uricosuria. *Ann. Intern. Med.* 84:385–388.
58. Wasserman, R. H., and Taylor, A. N. 1972. Metabolic roles of fat-soluble vitamins D, E, and K. *Annu. Rev. Biochem.* 41:179–201.
59. Underwood, B. A. 1984. Vitamin A in animal and human nutrition. In *The Retinoids*, Vol. 1, eds. M. B. Sporn, A. B. Roberts, and D. W. S. Goodman, pp. 263–374. New York: Academic Press.
60. Kappus, H., and Diplock, A. J. 1992. Tolerance and safety of vitamin E: A toxicological position report. *Free Radic. Biol. Med.* 13:55–74.
61. Diplock, A. T. 1985. Vitamin E. In *Fat-Soluble Vitamins. Their Biochemistry and Application*, pp. 154–224. Lancaster: Technomic.
62. Machlin, L. J. 1984. *Handbook of Vitamins: Nutritional, Biochemical and Clinical Aspects*, pp 99–145. New York: Marcel Dekker.
63. Lucy, J. A. 1972. Functional and structural aspects of biological membranes: A suggested structural role for vitamin E in the control of membrane permeability and stability. *Ann. N.Y. Acad. Sci.* 203:29–45.
64. Green, J. 1972. Vitamin E and the biological antioxidant theory. *Ann. N.Y. Acad. Sci.* 203:29–45.
65. Pinelli, A., Pozzo, G., Formento, M. L., Favalli, L., and Coglio, G. 1972. Effect of vitamin E in urine porphyrin and steroid profiles in porphyria cutanea tarda, Report of four cases. *Eur. J. Clin. Pharmacol.* 5:100–103.
66. Briggs, M., and Briggs, M. 1974. Are vitamin E supplements beneficial? *Med. J. Aust.* 1:434–437.
67. Diplock, A. T., Xu, G., Yeow, C., and Okikiola, M. 1989. Relationship of tocopherol structure to biological activity, tissue uptake and prostaglandin biosynthesis. *Ann. N.Y. Acad. Sci.* 570:72–84.
68. Kay, M. M., Bosman, G. J., Shapiro, S. S., Bendich, A., and Bassel, P. S. 1986. Oxidation as a possible mechanism of cellular aging: Vitamin E deficiency causes premature aging and IgG binding to erythrocytes. *Proc. Natl. Acad. Sci. U.S.A.* 83:2463–2467.
69. Guyton, A. C. 1987. *Human Physiology and Mechanisms of Diseases*, 4th ed. Philadelphia: W. B. Saunders.
70. Murphy, B. F. 1974. Hypervitaminosis E. *J. Am. Med. Assoc.* 227:1381.
71. Roels, O. A. 1967. Present knowledge of vitamin E. *Nutr. Rev.* 25:33–37.
72. Kingman, A. M. 1982. Vitamin E toxicity. *Arch. Dermatol.* 118:289.

73. Aeling, J. L., Panagotacos, P. J., and Andreozzi, R. J. 1973. Allergic contact dermatitis to vitamin E aerosol deodorant. *Arch. Dermatol.* 108:579–580.
74. Roberts 1978
75. Sundaram, G. S., London, F., Manimekalai, S., Nair, P. P., and Goldstein, P. 1981. Alpha tocopherol and serum lipoproteins. *Lipids* 16:223–227.
76. Bieri, J. G. 1975. Vitamin E. *Nutr. Rev.* 33:161–167.
77. Vogelsang, A. 1970. Twenty-four years using alpha-tocopherol in degenerative cardiovascular disease. *Angiology* 21:275–279.
78. Piscator, M. 1977. Copper. In *Toxicology of Metals*, Vol. 2, pp. 206–221. Springfield, VA: National Technical Information Service. PB 268–324.
79. U.S. Environmental Protection Agency. 1980. *Ambient Water Quality Criteria for Copper.* Springfield, VA: National Technical Information Service. PB 81-117475.
80. Goyer, R. A. 1996. Toxic effects of metals. In *Casarett and Doull's Toxicology*, 5th ed., pp. 691–736. New York: McGraw-Hill.
81. National Research Council. 1980. *Recommended Dietary Allowances*, 10th ed. Washington, DC: National Academy Press.
82. National Academy of Sciences. 1980. *Recommended Dietary Allowances*, 9th ed. Washington, DC: National Academy of Sciences.
83. Bast, A., Timmerman, H., Van Der Goot, H., and Wijker, J. E. 1986. Superoxide dismutase activity of copper-II-1,10-phenananthroline and copper-II-2,9-dimethyl-1,10-phenthroline. *Br. J. Pharmacol.* 89(proc. suppl):689.
84. Carson, B. L., Ellis, H. V., and McCann, J. L. 1986. Copper. In *Toxicology and Biological Monitoring*, pp. 93–98.
85. Finelli, V. N., Boscolo, P., Salimei, E., Messineo, A., and Carelli, G. 1981. Anemia in men occupationally exposed to low levels of copper. Heavy Metals Environ. Int. Conf. 3rd, pp. 475–478. Edinburgh, UK: CEP Consulting Ltd.
85. National Academy of Sciences. 1973.
86. National Research Council. 1989.
87. Crounse, R. G., Pories, W. J., Bray, J. T., and Mauger, R. L. 1983. Geochemistry and man: Health and disease. 1. Essential elements. In *Applied Environmental Geochemistry*, ed. I. Thornton, pp. 309–333. London: Academic Press.
88. Agency for Toxic Substances and Disease Registry. 1991. Toxicological Profile for Manganese. NTIS/PB93-110781. Atlanta GA: ATSDR.
89. Schwarz, K., and Foltz, C. M. 1957. Selenium as an integral part of factor 3 against dietary liver degeneration. *J. Am. Chem. Soc.* 79:3292.
90. Rotruck, J. T., Pope, A. L., Ganther, H. E., Hafeman, D. G., Swanson, A. B., and Hockstra, W. G. 1973. Selenium: Biochemical role as a component of glutathione peroxidase. *Science* 197:588–590.
91. Olson, O. E. 1986. Selenium toxicity in animals with emphasis on man. *J. Am. Coll. Toxicol.* 5:45–70.
92. Ohlendorf, H. M., Hoffman, D. J., Saiki, M. K., and Aldrich, T. W. 1986. Embryonic mortality and abnormalities of aquatic birds: Apparent impacts of selenium from irrigation drain water. *Sci. Total Environ.* 52:49–63.
93. Ohlendorf, H. M., Kilness, A., Simmons, J. L., Stroud, R. K., Hoffman, D. J., and Moore, J. F. 1988. Selenium toxicosis in wild aquatic birds. *J. Toxicol. Environ. Health* 24:67–92.
94. Hoffman, D. J., Heinz, G. H., and Krynitsky, A. J. 1989. Hepatic glutathione metabolism and lipid peroxidation in response to excess dietary selenomethionine and selenite in mallard ducklings. *J. Toxicol. Environ. Health* 27:263–271.
95. Combs, G. F., Jr., and Combs, S. B. 1986. *The Role of Selenium in Nutrition.* San Diego: Academic Press.
96. Ganther, H. E. 1986. Pathways of selenium metabolism including respiratory excretory products. *J. Am. Coll. Toxicol.* 5:1–5
97. U.S. Environmental Protection Agency. 1980. Ambient Water Quality Criteria for Selenium. Springfield, VA: National Technical Information Service. PB 81-117814.
98. Snyder, W. S., Cook, M. J., Nasset, E. S., Karhausen, L. R., Howells, G. P. and Tipton, I. H. 1975. International Commission on Radiological Protection. Report of the Task Group on Reference Man. ICRP Publication 23. New York: ICRP.
99. Holkstra, W. G. 1975. Biochemical function of selenium and its relations to vitamin E. *Fed. Proc.* 34:2083–2088.
100. Schnell R. C., and Angle C. R. 1983. Selenium—Toxin or panacea. *Fundam. Appl. Toxicol.* 3:409–410.
101. Katz, S. A. 1995. The toxicity/essentiality of dietary minerals. *Arh. Hig. Rad. Toksikol.* 46:333–345.
102. Katz and Chatt 1988.
103. Iyengar, I. Y., and Gopal-Ayengar A. R. 1988. Human health and trace elements including effects on high-altitude populations. *Ambio* 17:31–35.

104. Schroeder, H. A., Frost, D. V., and Balassa J. J. 1970. Essential trace elements in man. Selenium. *J. Chron. Dis.* 23:227–243.
105. Rosenfeld, I., and Beath, O. H. 1964. *Selenium.* New York: Academic Press.
106. U.S. Environmental Protection Agency. 1980. Ambient Water Quality Criteria for Zinc. Springfield, VA: National Technical Information Service. PB 81-117897.
107. Solomons, N. W. 1982. Biological availability of zinc in humans. *Am. J. Clin. Nutr.* 35:1048–1075.
108. Taylor, M. C., De Mayo, A., and Taylor, K. W. 1982. Effects of zinc on humans, laboratory and farm animals, terrestrial plants, and freshwater aquatic life. *Crit. Rev. Environ. Control* 12:113–181.
109. Davies 1980.
110. Van Wouwe, J. P., Van Gelderen, H. H., Enschede, F. A. J., and Van De Velde, E. A. 1988. Acute diarrhea nonresponsive to dietary restriction, zinc deficiency and subclinical growth retardation in preschool children. *Trace Elem. Med.* 5:90–92.
111. Moynahan, E. J. 1974. Acrodermatitis enteropathica: A lethal inherited zinc deficiency disorder. *Lancet* 2:399.
112. Nelder, K. H., and Hambidge, K. M. 1975. Zinc therapy of acrodermatitis enteropathica. *N. Engl. J. Med.* 292:879–881.
113. Van Wouwe, J. P. 1989. Clinical and laboratory diagnosis of acrodermatitis enteropathica. *Eur. J. Pediatr.* 149:2–8.
114. Prasad, A. S. 1966. *Zinc Metabolism.* Springfield, IL: Charles C. Thomas.
115. Van Wouwe, J. P., Van Gelderen, H. H., and Bos J. H. 1987. Subacute zinc deficiency in children with recurrent upper respiratory tract infection. *Eur. J. Pediatr.* 146:293–295.
116. De Haan, K. E. C., De Gaeji, J. J. M., Van Den Hamer, C. J. A., Boxma, H., and De Groot, C. J. 1992. Changes in zinc metabolism after burns: Observations, explanations, clinical interpretations. *J. Trace Elem Electrolytes Health Dis.* 6:195–201.
117. Van Wouwe, J. P., Van Den Hamen, C. J. A., and Van Tricht, J. B. 1986. Serum zinc concentrations in exclusively breast-fed infants and in infants fed on adapted formula. *Eur. J. Pediatr.* 144:598–599.
118. Bast, A., Haenen, G. R., and Doelman C. J. 1991. Oxidants and antioxidants: State of the art. *Am. J. Med.* 91(ISS 3C):2S–13S.
119. Woolliscrift, J. A. 1983. Megavitamins: Fact and fancy. *DM* 29:1–56

Chapter Thirteen

TOXICITY OF OXYGEN AND OZONE

Harry Salem and Steven I. Baskin

This chapter, although it recognizes the toxicology of oxidants and free radicals, focuses primarily on oxygen and ozone. Other aspects of this and similar topics are found in recent excellent reviews (Chiueh, Emerit, Nriagu; Simmons, Sies, Foote) and because of space limitations and to reduce redundancy will not be addressed in this chapter. Many of the reviews provide a historical perspective reaching back to data of Joseph Priestly (Chieuh) or the dawn of "oxidant toxicology" (Parke) from the individual author's vantage point.

The oxidation process in biological systems is necessary for the billions of cells in our body to generate energy in order to carry out their biological functions. The metabolic by-products of oxidation include free radicals also known as reactive oxidants or reactive oxidant species (ROS) (superoxide anion, O_2^-; hydrogen peroxide, H_2O_2; hydroxyl radical, ˙OH; singlet oxygen, O_2^*). Reactive oxygen species result from normal functioning of cells and are released in small amounts (1, Klein et al.). Antioxidant defenses including vitamin E, vitamin C, the glutathione redox system, catalase, and superoxide dismutase (SOD) readily intercept the ROS that escape from the cells; however, if the released ROS exhaust the antioxidant defenses, they may be associated with and may play a major role in contributing to cardiovascular disease and stroke, acquired immunodeficiency syndrome (AIDS), ozone-induced pulmonary disorders, cataracts, and other diseases associated with aging. Free radicals may also be generated by the inhalation of air pollutants and cigarette smoke. These highly reactive and unstable molecules, with one or more unpaired electrons, can cause damage to cell structures by stealing electrons from other molecules within the cell, disrupting normal chemical processes (Diplock, 1991).

Details of oxidant injury from inhaled particulate matter are described in Chapter 16 by McGrath in this book. He also describes the release of ROS by the macrophages in maintaining the integrity and sterility of the respiratory tract and the initiation of and participation of metals in the oxidative reactions.

The oxides produced by photoxidation of air pollutants demonstrate adverse effects not only on the respiratory system, but also on the cardiovascular system.

OXYGEN

Oxygen is a life-supporting gas that has also been called the breath of life and is a chemical element. All living things need oxygen to sustain life (except anaerobic bacteria). Oxygen combines with other chemicals in plant and animal cells to produce energy for life processes and the release of heat. At one time in early history, it was thought that breathing pure oxygen would cause death. It was incorrectly thought that when breathing pure oxygen, the body cells would utilize the oxygen so fast that the person would die of exhaustion. In some cases, breathing

The opinions or assertions contained herein are the private views of the authors and are not to be construed as official or as reflecting the views of the Army or the Department of Defense.

pure oxygen may be necessary, such as for pilots flying at high altitudes. As described in Chapter 12 on the toxicology of antioxidants, vitamins continue to be the world's most widely used pharmaceutical products, based partially on the misconception that more is better.

Oxygen can produce toxicity. Estimates for some of the toxicity values for several studies are given in Table 1. This same misconception was applied to the use of oxygen in the treatment of premature human infants. Premature and young infants exposed to excess oxygen in incubators (Kinsey) were blinded (retrolental hyperplasia) by the excess free radical species. Perhaps because oxygen-induced lung lesions are nonspecific, it is possible to induce partial resistance by preexposure to 100% oxygen (Housset). In rats and humans, adaptation to acute concentrations of ozone can be achieved. Rats can adapt after 8 h to a chronic exposure of ozone but not after a 4-mo recovery period. Breathing parameters that significantly changed were breathing frequency, tidal volume, inspiratory and expiratory times, and maximum expiratory flow (Wiester). The data indicate that a reversible competitive antioxidant mechanism may be induced with chronic exposure that can be overcome. From studies in female rats, adaptation to ozone responsiveness was observed, but the results indicate smooth muscle cell function was not responsible (Szarek).

OXIDATIVE STRESS

Oxidative stress reflects the morphological consequences of a mismatch between the rate of formation of free radicals and the capacity of the cell to transform them to less toxic species. Oxidative stress can occur after an increase in overproduction of free radicals or a decrease in defense against the toxic species or both. The imbalance in free radicals may cause cellular dysfunction or death. A review of the regulation of various antioxidant enzymes after different types of oxidant stresses to the lung suggests that more than one enzyme is involved. A spectrum of possible interventional agents could be employed to provide effective strategies against lung stresses (Quinlan).

ROS may produce functional changes in lipids, proteins, and DNA. Molecular oxygen addition into polyunsaturated fatty acids initiates a chain reaction in which certain ROS radical species are formed. Resulting lipid damage can produce a gradual loss in membrane fluidity with time. The decreased fluidity reduces membrane functionality through alteration of ionic permeability.

ROS may damage proteins. Metal cations may bind to ligands on proteins, and the altered oxidative state may activate or inhibit proteins. Oxidative damage to proteins may contribute to the proteolytic activity. ROS can also damage DNA and RNA. Purines and pyrimidines can be oxidized. The oxidized bases increase with age and cause increased misread protein. Certain radicals may modify sugar phosphates and react with the backbone of DNA to produce strand breaks.

Table 1. Oxygen toxicity (RTECS)

Route	Species	Dose	Duration	Effects
Inhalation	Human	100 pph	14 h	TCLo Cough
Inhalation Reproduction	Human (female)	12 pph	10 Days wk 26–39 of Pregnancy	Cardiovascular System
Inhalation Reproduction	Rat	10 pph	12 h Day 22 of pregnancy	TCLo Respiratory system
Inhalation Reproduction	Rat	10 pph	9 h Day 22 of pregnancy	TCLo Effects on newborn
Inhalation Reproduction	Mouse	10 pph	24 h Day 8 of pregnancy	TCLo Skin and skin appendages
EPA GENETOX Program, 1988	v79 Cell Culture	Positive	N/A	Gene mutations

Note. RTECS, Registry of Toxic Effects of Chemical Substances. pph, parts per hundred.

OXYGEN INJURY

Irritant properties of oxygen have been observed for many years, especially at high partial pressures. Before the actions of free radicals were completely understood, a great deal of work was accomplished on the basic toxicity mechanisms of oxygen (Haugaard).

Injury and mortality after exposure to 100% oxygen can be diminished by surfactants that may operate by mechanisms other than those responsible for surface tension effects. It has been shown that hyperoxia arrests proliferation of an immortalized type 2 cell line (SV4OT-T2) and the expression of several growth-related genes, normally induced near the G1/S, and the boundary was altered with a block of translation of their mRNA. There are also effects on insulin-like growth factor (IGF) and transforming growth factor-beta-1 (TGF-beta-1). TGF-beta-1 was induced by O_2 exposure. Studies suggest a regulatory link between components of the IGF system and TGF-beta-1 systems in hyperoxic control of alveolar epithelial cell proliferation (Cazals). Workers exposed to coal mine dust were compared to controls, and their alveolar macrophages were compared for a number of components. In the exposed workers, tumor necrosis factor (TNF) and interleukin-6 were released in significant amounts from alveolar macrophages. Bronchioalveolar lavage fluid showed a large influx of mononuclear phagocytes, with an increased spontaneous production of oxidants, fibronectin, neutrophil chemotactic factor, and also interleukin-6 and TNF-alpha. This cytokine release was associated with increased cytokine messenger RNA expression in the lungs of coal miners (Vanhee).

Multiple different treatments have been examined using many models in an attempt to reduce lung inflammation as a result of exposure to oxygen. One such study employed a synthetic surfactant (Exosurf) that contains a mixture of Tyloxapol and cetyl alcohol. It appears that this combination in rats may increase the mean survival, increase the wet to dry lung weight ratios, and decrease thiobarbituric acid-reactive products, which were used as a measure of mortality, hyperoxic injury, and oxidized tissue products, respectively (Ghio). These results are in contrast to the observation that under other model conditions large doses of the antioxidant ascorbic acid did not attenuate lung oxidant stress (Demling). The lack of effect when vitamin C was applied directly on the lung is consistent with other studies employing ozone in vitamin C-deficient guinea pigs (Kodavanti). Hall and her co-workers, as described elsewhere in this volume, suggest that vitamin C may act at other organs rather than directly at the lung.

OZONE INJURY

The response to ozone exposure has been related to inflammation of the epithelium (Harkema, plopper). Inflammation can but does not always cause airway hyperresponsiveness, suggesting multiple causes of inflammation. Local muscarinic-reactive C-fibers play a part to the ozone effects in the isolated guinea pig lung, which suggests afferents may contribute to the toxicity (Joad). However, increased response of ozone to inhaled bronchoconstrictors occurs in both asthmatic and nonasthmatic patients (Kreit, J.). Ozone is a relatively insoluble oxidant capable of reducing lung function from the alveolar side in the range of 0.1–0.2 ppm. Higher doses of ozone can cause death (Table 2) due to lung edema. The ACGIH and the NIOSH threshold limit value (TLV, ceiling) for ozone is 0.1 ppm. The OSHA permissible exposure limit (PEL) for an 8-h time-weighted average (TWA) is 0.1 ppm (0.2 mg/m^3).

There is inconclusive or no evidence of mutagenicity from ozone found in the U.S. Environmental Protection Agency (EPA) in vitro and in vivo studies. However, there is some evidence in the mouse of carcinogenesis by inhalation, but none in the rat (NTP-TR-440, 94).

Because of concern for the public, and a perceived need to regulate industry, the Environmental Protection Agency is proposing to amend the Clean Air Act with respect to ozone. A stratospheric Ozone Information Hotline exists (1-800-296-1996) for further information. A hotline for toxicology information also exists from NIEHS (919-541-3802) and ASTDR. Notifications of Proposed Rules for ozone are found in the *Federal Register* (Federal register).

The proposed rules to be promulgated in June 1997 are based on the long-time recognition in both clinical and epidemiological research that ozone affects public health. The revised standards proposed would provide protection for children and other at risk populations against a wide range of ozone-induced health effects. These health effects include decreased lung function

Table 2. Ozone toxicity (RTECS)

Study	Species	Dose	Duration	Effects
Inhalation	Human	50 ppm	30 min	LCLo Headache, Blood pressure (BP) Depression, Dyspnea
Inhalation	Human	80 ppb	6 h	TCLo Cough Respiratory Depression
Inhalation	Human	100 ppb	1 min	TCLo Dermatitis
Inhalation	Human	1860 ppb	75 min	TCLo Lacrimation, Pulse rate Decreases, BP decrease, Cough
Inhalation	Human	1 ppm	—	TCLo Cough, Dyspnea
Inhalation	Human	600 ppb	2 h	TCLo Cough, Dyspnea
Inhalation	Rat	4800 ppb	4 h	LC50 Acute Pulmonary edema, hemorrhage
Inhalation	Mouse	12,600 ppb	3 h	LC50 Somnolence, Dyspnea
Inhalation	Rabbit	36 ppm	3 h	LC50 Somnolence, Dyspnea
Inhalation	Guinea pig	24,800 ppb	3 h	LC50
Inhalation	Hamster	10,500 ppb	4 h	LC50
Inhalation, Reproductive	Rat	1,040 ppb	24 h Day 6–9 pregnancy	TCLo Preimplant mortality, Fetotoxicity, Musculoskeletal abnormality
Inhalation, Reproductive	Rat	1500 ppb	24 h Day 17–20 pregnancy	TCLo Behavioral
Inhalation Multiple dose	Rat	950 ppb	8 h 90 days	TCLo
Inhalation Multiple dose	Rat	1600 μg/m^3	24 h 7 days	TCLo Lung weight changes Biochemical: oxidoreductases change
Inhalation Multiple dose	Rat	800 μg/m^3	24 h 3 days	TCLo Biochemical: dehydrogenases, oxidoreductases, proteins
Inhalation Multiple dose	Rat	700 ppb	2 h 28 days	TCLo Respiratory depression, Weight loss, decreased weight gain
Inhalation Multiple dose	Rat	200 ppb	24 h 12 wk	Biochemical: cytochrome oxidase, oxidoreductase
Inhalation Multiple dose	Rat	750 ppb	24 h 14 days	TCLo Weight loss, Decreased weight gain
Inhalation Multiple dose	Mouse	400 ppb	24 h 14 days	TCLo Lung weight change, Thymus weight change, Humoral immune response decreased
Inhalation Multiple dose	Mouse	1 ppm	24 h 4 wk	TCLo Thymus weight change, Humoral immune response, decreased, Weight loss, Decreased weight gain
Inhalation Multiple dose	Mouse	608 μg/m^3	26 wk	TCLo Spleen weight change
Inhalation Multiple dose	Monkey	1500 ppb	8 h 21 days	TCLo Respiratory changes
Inhalation Multiple dose	Rabbit	100 ppb	2 h 13 days	TCLo Respiratory changes
Inhalation Multiple dose	Guinea pig	2 ppm	2 days	TCLo Tracheal and bronchial changes, Lung weight changes, Weight loss, Decreased weight gain
Inhalation Multiple dose	Guinea pig	800 μg/m^3	24 h 3 days	TCLo Lung weight change, Dehydrogenases and proteins

Table 2.   Ozone toxicity (RTECS)—Continued

Study	Species	Dose	Duration	Effects
Inhalation	Quail	3 mg/m^3	24 h	TCLo
Multiple dose			7 days	Trachea and bronchial changes
				Dehydrogenases
Inhalation	Mouse	5 ppm	2 h	TCLo
Carcinogenicity			75 days	Neoplastic lung
Inhalation	Mouse	608 μg/m^3	24 wk	TC
Carcinogenicity				Equivocal tumorigenesis (lung)
Inhalation	Mouse	1 ppm	6 h	TC
Carcinogenicity			2 yr	Neoplastic lungs

(primarily in children active outdoors), increased respiratory symptoms (particularly in highly sensitive individuals), hospital admission and emergency room visits for respiratory causes (among children and adults with preexisting respiratory disease such as asthma), inflammation of the lung, and possible long-term damage to the lungs. This Federal Register notice summarized the human health effects associated with ambient ozone level exposures and is based on integrative information from human clinical, epidemiological and animal toxicological studies as presented in the Criteria Document and Staff Paper.

The effects of short term (1–3 h) and prolonged (6–8 h) exposures are summarized as follows. In healthy and asthmatic exercising adults, transient variable decrements in lung function occurred at ozone levels of 0.12 ppm or greater for 1–3 h of exposure, and at 0.08 ppm or greater for 6.6 h of exposure. Cough, throat irritation, chest pain on deep inspiration, nausea, and shortness of breath were also experienced during ozone exposure and were more severe and persistent with increasing duration and concentrations.

Exposure to ozone increases airway responsiveness to bronchoconstriction from drugs such as histamine and methacholine, which may resolve after 24 h (which is slower than for pulmonary function). In addition, the hyperresponsiveness (two- to fourfold) is usually resolved within 24 h. In asthmatic subjects the bronchial responsiveness was increased up to 100-fold.

Ozone exposure can also increase susceptibility to respiratory infection and related respiratory dysfunction. This has been reported in a large number of laboratory animal studies with generally consistent results. In a study of human subjects performing moderate exercise exposed to 0.08 ppm ozone for 6.6 h there was a decrease in alveolar macrophage function. None of the studies, however, conclusively demonstrates that human susceptibility to respiratory infection is increased by exposure to ozone.

The weight of evidence from all of these studies suggests that acute ozone exposures can impair the host-defense capability of both humans and animals, possibly by depressing alveolar macrophage function and by decreasing mucociliary clearance of inhaled particulates and microogranisms.

Studies have also consistently shown a relationship between elevated ambient ozone levels in the summertime and increased incidence of emergency room visits and hospital admissions, especially in individuals with preexisting respiratory diseases such as asthma and chronic obstructive pulmonary disease.

Respiratory inflammation that can be induced by a single exposure of humans to ozone or even by several exposures over the course of a season could be resolved entirely. Repeated acute inflammation could develop into a chronic inflammatory state, and continued inflammation could alter the structure and function of other pulmonary tissue, leading to fibrosis. In potentially vulnerable populations, such as children and older individuals, inflammation could interfere with the body's host-defense response to inhaled particles and microorganisms. Additionally, inflammation might amplify the lungs responses to allergens or toxins.

Laboratory animal studies have shown that exposure for 8 h or less can result in cell damage, inflammation, and increased leakage of proteins from blood into the alveolar air spaces. More severe effects or lower concentrations were required in studies conducted at night when the animals were active or when they were coexposed to carbon dioxide to increase their ventilation rates.

Most of the effects following acute and prolonged exposures to ozone described thus far begin to resolve within 24 h in most individuals if the exposures are not repeated.

Of greater public health concern is the potential for chronic respiratory damage resulting from repeated ozone exposures that occur over a season or a lifetime. Epidemiological studies, however, have only suggested a small but consistent decrement in lung function among inhabitants of more highly polluted communities.

Toxicological studies in animals demonstrated ozone lesions in the centriacinar regions in the lung. These lesions were similar in all species tested (monkeys, rats, mice) and appear to be concentration, time, and exposure pattern dependent.

Although changes included cell and tissue damage, postexposure damage was mainly reversible. After 78 wk of exposure, rats showed increased expiratory resistance indicative of central airway narrowing, reduced tidal volume, and reduced respiratory rate. Reduced lung volume and decreased compliance indicative of restrictive lung disease were reported in a similar study in rats. A multicenter chronic study demonstrated some of the complex interrelationships among the structural, functional, and biochemical effects in rats exposed to ozone for 20 mo. Lung biochemistry and structure were affected at 0.5 ppm and 1.0 ppm, but not at 0.12 ppm. No effects on pulmonary function were observed at any exposure level.

The laboratory animal studies to determine genotoxicity and carcinogenicity following ozone exposure have been inconclusive, particularly at the lower ozone concentrations (Federal Register).

Statistically significant changes in red blood cells (RBCs) and sera can be observed following a single short-term exposure to 0.5 ppm ozone in young adult human males for 2.75 h (Buckley). Ozone increases RBC glucose-6-phosphate dehydrogenase and lactate dehydrogenase and lowers reduced glutathione and acetylcholinesterase. The mechanisms of oxidants differ but depend on concentration and length of exposure (Chitano et al).

Ozone impairs the ability of macrophages to phagocytize particles in both humans and animals (bovine) in vivo (Mosbach). It was shown that cytokines as well as tumor necrosis factor-alpha were involved. Dipalmitoyl lecithin, the major lipid of natural surfactant, significantly reduced ozone release of macrophage chemotactic factor. Aldehydes such as hexanal, heptanal, and nonanal can be detected in the bronchioalveolar lavage fluid of rats exposed to ozone. These aldehydes may provide useful biomarkers of lipid ozonation products that may be seen following inflammation due to ozone (Pryor). Lactation appears to change the sensitivity of female rats for ozone to induce pulmonary inflammation (Gunnison). Schuller-Levis and her co-workers ( ) suggested that taurine acts to prevent oxidant damage by formation of chlorotaurine. The enzyme thought to be responsible is myeloperoxidase. These observations of the antioxidant effects of taurine are discussed in Chapter 10 of this volume with the antioxidant effects of taurine.

Rats chronically exposed to ozone appear to show thickening of the bronchiolar epithelial cells, which are composed of differentiated ciliated and Clara cells. These findings are consistent with the suggestions that ozone administered acutely on the endothelial side (i.e., intravenously, iv) may not pose the same danger to the organism (Stockstill). Reviews of more recent studies suggest that there is little to implicate ozone directly as a pulmonary carcinogen (Witschi). Nor does it appear to cause adverse cytogenetic effects in humans or in certain strains of mice (Victorin), despite its potent oxidant effects.

Ozone, a highly reactive compound, is known to inactivate bacteria, yeasts, molds, and viruses. Since it is also effective in fluids, it has been used for years to purify water for human consumption. Its essentiality or beneficial effects are currently being explored by several companies. It is suggested that the physical and chemical bases of ozone therapy are due to an ozone peroxide (Viebahn).

A sterilization technology has been described for blood and blood products utilizing ozone-induced oxidative stress (Lea, 1996). This technology is based on a cellular infusion device (CID) for the efficient delivery of sufficient ozone to whole blood or blood products to inactivate viruses with minimal damage to blood components (Smith et al, 1996a). This technology might be useful in the treatment of decubital ulceration or in sterilizing human immunoglobulin products. Human immunoglobulin (Ig) may be a key product for the treatment of patients whose immune system is compromised. The transfusion of human immunoglobulin that may be contaminated with viruses or bacteria can cause serious disease and/or death in these patients (Lee et al., 1996).

In 1994, the U.S. Food and Drug Administration mandated that all manufacturers of immunoglobulin products incorporate viral inactivation into their manufacturing processes. Ozone inactivation of enveloped and nonenveloped viruses and bacteria may be effective to treat the blood supply. This may include inactivation of the human immunodeficiency virus (HIV) and hepatitis viruses (Lea, 1996).

The simian immunodeficiency virus (SIV) causes AIDS-like disease in monkeys and other nonhuman primates. The ability of ozone-induced oxidative stress to inactivate SIV in a dose-related manner in human blood efficiently, effectively, and without damaging the blood has been claimed by Quimby (1996). Five logs SIV per milliliter was inactivated, and the results were confirmed in tissue culture and in sensitive p27 antigen assays.

Since SIV is very similar to the HIV virus, its potential inactivation should be an important step toward eliminating the risk of transferring HIV from transfusion of contaminated blood and blood products. Both lipid- and non-lipid-coated viruses have been inactivated together. Through the development of viral load inactivation models, the process can inactivate greater than 6 logs/ml of several key viruses in intravenous immunoglobulin solutions and red blood cells. Ozone-induced inactivation of more than 6 logs/ml of bovine viral diarrhea (BVD) virus, a regulatory equivalent model for hepatitis C, was also achieved in red blood cells. In addition, infectious bovine rhinotracheitis (IBR), a large herpes type virus, and the much smaller adenovirus, which is found both in bovine species and humans, were also inactivated by ozone.

Studies have been conducted that suggest that the ozone sterilization system did not adversely affect the blood components tested. Ozone is known to increase levels of cytokines, which are non-antibody proteins that act as intercellular mediators of the immune system. Cytokines can activate the natural killer (NK) lymphocyte cells that exert cytotoxic effects on abnormal and virally infected cells. The studies demonstrated that ozone did not inhibit the natural killer cells' cytotoxic potential (Smith et al, 1996a).

Additionally, cytokines activate certain white blood cells (monocytes/macrophages) and polymorphonuclear (PMN) cells to recognize, engulf, and destroy foreign materials by phagocytosis. Studies have shown that increasing concentrations of ozone mixed in blood samples increased the phagocytic activity of monocytes and polymorphonuclear cells (Smith et al 1996b). Cytokines also influence the hematopoietic system, which is responsible for the production of the body's blood cells. Thus, the effect of ozone was studied on normal hematopoiesis (colony formation, proliferation, and multilineage cell differentiation). The primitive cell known as the pluripotent cell (stem cell CD34) is capable of differentiating into all of the different types of blood cells. Studies with these stem cells indicated that ozone did not alter stem-cell viability and its differentiation into the different cell types examined (Smith et al, 1996c).

These efforts and the positive preliminary results encourage the pursuit of ozone as a possible therapeutic agent against the spread of AIDS, as well as other diseases from contaminated blood and blood products and other infections.

COMPARISON OF MODELS

Dosimetry models for estimating and comparing the regional proximal alveolar region doses (PAR) against other parameters were compared, and it was found that the PAR dose was the best estimation of the absorption of reactive gases such as oxygen and ozone (Overton).

REFERENCES

1. Buckley, R. D., Hackney, J. D., Clark, K., and Posin, C. 1975. Ozone and human blood. *Arch Environ. Health* 30:40–43.
2. Cazals, V., Mouhieddine, B., Maitre, B., Le Bouc, Y., Chadelat, K., Brody, J. S., and Clement, A. 1994. Insulin-like growth factors, their binding proteins, and transforming growth factor-beta 1 in oxidant-arrested lung alveolar epithelial cells. *J. Biol. Chem.* 269:14111–14117.
3. Chitano, P., Hosselet, J. J., Mapp, C. E., and Fabbri, L. M. 1995. Effect of oxidant air pollutants on the respiratory system: Insights from experimental animal research. *Eur. Respir. J.* 8:1357–1371.
4. Chiueh, C.C., Gilbert, D. L., and Colton, C. A., eds. 1994. The neurobiology of NO˙ and ˙OH. *Ann. N.Y. Acad. Sci.* 738.
5. Demling, R., Ikegami, K., Picard, L., and Lalonde, C. 1994. Administration of large doses of vitamin

C does not decrease oxidant-induced lung lipid peroxidation caused by bacterial-independent acute peritonitis. *Inflammation* 18:499–510.

6. Diplock, A. T. (1991). Antioxidant Nutrients and disease prevention: An interview. *Am. J. Clin. Nutr.* 53:1895–1935.

7. Emerit, I., Packer, L., and Auclair, C. 1988. *Antioxidants in Therapy and Preventive Medicine*, Vol. 264. New York: Plenum Press.

8. *Federal Register.* 1996. 61(213):56493–56496, Friday, November 1.

9. *Federal Register.* 1996. 61(241):65716–65750, Friday, December 13.

10. Foote, C. S., Valentine, J. S., Greenberg, A., and Leibman, J. F. 1995. *Selective Oxygen in Chemistry.* New York: Academic Press.

11. Ghio, A. J., Fracica, P. J., Young, S. L., and Piantadosi, C. A. 1994. Synthetic surfactant scavenges oxidants and protects against hyperoxic lung injury. *J. Appl. Physiol.* 77:1217–1723.

12. Gunnison, A. F., Hatch, G. E., Crissman, K., and Bowers, A. 1996. Comparative sensitivity of lactating and virgin female rats to ozone-induced pulmonary inflammation. *Inhal. Toxicol.* 8:607–623.

13. Harkema, J. R., Plopper, C. G., Hyde, D. M., St. George, J. A., Wilson, D. W., and Dungworth, D. U. 1993. Response of Maccaca bronchiolar epithelium to po ambient concentration of ozone. *Am. J. Pathol.* 143:857–866.

14. Haugaard, N. 1968. Cellular mechanisms of oxygen toxicity. *Physiol. Rev.* 48:312–362.

15. Housset, B. 1994. Free radicals and respiratory pathology. *C. R. Seances Soc. Biol. Fil.* 188:321–333.

16. Joad, J. P., Kott, K. S., and Bric, J. M. 1966. The local C-fiber contribution to ozone-induced effects on the isolated guinea pig lung. *Toxicol. Appl. Pharmacol.* 141:561–567.

17. Kinsey, V. E. 1956. Retrolental fibroplasia. *Arch. Ophthalmol.* 56:481–529.

18. Klein, C., B., Frenkel, K., and Costa, M. 1991. The role of oxidative processes in metal carcinogenesis. *Chem. Res. Toxicol.* 4:592–604.

19. Kodavanti, U. P., Hatch, G. E., Starcher, B., Girl, S. N., Winsett, D., Costa, D. L. 1995. Ozone-induced pulmonary functional, pathological and biochemical changes in normal and vitamin C-deficient guinea pigs. *Fundam. Appl. Toxicol.* 24:154–164.

20. Kreit, J. W., Gross, K. B., Moore, T. B., Lorenzen, T. J., D'Arcy, J., and Eschenbacher, W. L. 1989. Ozone-induced changes in pulmonary function and bronchial responsiveness in asthmatics. *J. Appl. Physiol.* 66:217–222.

21. Lea, P. 1996. Personal communication, LifeTech Corporation, Toronto, Ontario, Canada.

22. Lee, L., Dermott, W. J., Mason, K., and Lea, P. 1996. Effects of ozone on human IgG. *Int. Society for Free Radical Research VII Biennial Meeting*, Barcelona, Spain, October, abstr. 9.29.

23. Mosbach, M., Wiener-Schmuck, M., and Seidel, A. 1996. Influence of coexposure of ozone with quartz, latex, albumin and LPS on TNF-α and chemotactic factor release by bovine alveolar macrophages in vitro. *Inhal. Toxicol.* 8:625–638.

24. Nriagu, J. O., and Simmons, M.S., ed. 1994. *Environmental Oxidants.* New York: John Wiley & Sons.

25. Overton, J. H., Graham, R. C., Menache, M.G., Mercer, R. R., and Miller, F. J. 1996. Influence of tracheobronchial region expansion and volume on reactive gas uptake and interspecies dose extrapolations. *Inhal. Toxicol.* 8:723–745.

26. Parke, D. V. 1996. The phoenix of modern toxicology. *Fundam. Appl. Toxicol.* 34:1–4.

27. Pryor, W. A., Bermúdez, E., Cueto, R., and Squadrito, G. L. 1996. Detection of aldehydes in bronchioalveolar lavage of rats exposed to ozone. *Fundam. Appl. Toxicol.* 34:148–156.

28. Quimby, F. 1996. Personal communication. New York State College of Veterinary Medicine at Cornell University, Ithaca, NY.

29. Quinlan, T., Spivack, S., and Mossman, B. T. 1994. Regulation of antioxidant enzymes in lung after oxidant injury. *Environ. Health. Perspect.* 102(suppl. 2):79–87.

30. Schuller-Levis, G., Quinn, M. R., Wright, C., and Park, E. 1994. Taurine protects against oxidant-induced lung injury: Possible mechanism(s) of action. *Adv. Exp. Med. Biol.* 359:31–39.

31. Sies, H. 1997. *Antioxidants in Disease Mechanisms and Therapy.* Advances in Pharmacology, vol. 38. San Diego: Academic Press.

32. Simonian, N. A., and Coyle, J. T. 1996. Oxidative stress in neurodegenerative diseases. *Annu. Rev. Pharmacol. Toxicol.* 36:83–106.

33. Sindu, R. K., Mautz, W. J., Fujita, I., Wang, N.-S., and Kikkawa, Y. 1996. Effect of chronic exposure to ozone and nitric acid on cytochrome P-450 monooxygenase system of rat lung and liver. *Inhal. Toxicol.* 8:695–708.

34. Smith, L. M. J., Klein, D., Pinkney, K., and Dermott, W. J. 1996. Flow cytometric analysis of the effects of ozone on the cytolytic ability of human natural killer cells. *International Society for Analytical Cytology XVIII Congress*, Rimini, Italy, April, abstr. IH118.

35. Smith, L. M. J., Klein, D., Hurley, A. A., Cavanagh, G., and Dermott, W. J. 1996 Flow cytometric analysis of the effects of ozone on the phagocytic ability of human monocytes and PMNS. *International Society for Analytical Cytology XVIII Congress*, Rimini, Italy, April, abstr. IH117.

36. Smith, L. M. J., Klein, D., Yamada, J., and Dermott, W. J. 1996. Effects of oOzone on the hematopoietic activity of human pluripotent progenitor cells. *International Society for Analytical Cytology XVIII Congress*, Rimini, Italy, April, abstr. IH47.
37. Stockstill, B. L., Chang, L.-Y., Menache, M. G., Mellick, P. W., Mercer, R. R., and Crapo, J. D. 1995. Bronchiolarized metaplasia and interstitial fibrosis in rat lungs chronically exposed to high ambient levels of ozone. *Toxicol. Appl. Pharmacol.* 134:251–263.
38. Szarek, J. L., Stewart, N. L., Zhang, J. Z., Webb, J. A., Valentovic, M. A., and Catalano, P. 1995. Contractile responses and structure of small bronchi isolated from rats after 20 months' exposure to ozone. *Fundam. Appl. Toxicol.* 28:199–208.
39. Vanhee, D., Gosset, P., Boitelle, A., Wallaert, B., and Tonnel, A. B. 1995. Cytokines and cytokine network in silicosis and coal workers' pneumoconiosis. *Eur. Respir. J.* 8:834–842.
40. Victorin, K. Review of the genotoxicity of ozone. *Mutat. Res.* 277:221–238.
41. Viebahn, R. 1975. Physiocochemical basis of ozone therapy. *Erfahrungsheilkunde* 24:129–134.
42. Wiester, M. J., Tepper, J. S., Doerfler, D. L., and Costa, D. L. 1995. Ozone adaptation in rats after chronic exposure to a simulated urban profile of ozone. *Fundam. Appl. Toxicol.* 24:42–51.
43. Witschi, H. 1988. Ozone, nitrogen dioxide and lung cancer: a review of some recent issues and problems. *Toxicology* 48:1–20.

Chapter Fourteen

PEROXIDATION OF LIPIDS AND LIVER DAMAGE

Pablo Muriel

FROM FREE RADICALS TO LIPID PEROXIDATION

Oxygen Toxicity

Oxygen is lethal to mammals in a few days when 100% dioxygen is breathed at 1 atm, whereas survival time at 5 atm is approximately 1 h, making the survival time versus inspired oxygen tension curve quite steep (1).

Oxygen toxicity is associated with the capacity of this molecule to oxidize organic molecules and to produce free radical species according to the general reactions:

$$RH_2 + O_2 \rightarrow RH^{\bullet} + O_2^{\bullet-} + H^+ \text{ (1-electron transfer)}$$

$$RH_2 + O_2 \rightarrow R + H_2O_2 \text{ (2-electron transfer)}$$

For these reactions to occur at significant rates, transition metal catalysis is required.

Properties of Free Radicals

All molecules have electrons as their outermost components. The behavior of these electrons determines the properties of the molecule. Modern quantum-mechanical theory describes electrons as having an intrinsic tendency to spin, thereby generating an electromagnetic field, the effect of which can be canceled by a similar charge spinning in the opposite direction (2). Thus, the most stable configuration of electrons is a paired one in which both members have opposite spins. Given this requirement for pairing, any situation in which a species is generated with an unpaired electron will result in a potentially reactive entity known as a free radical. Therefore, a stable molecule contains an even number of electrons and a free radical is formed by gaining or losing one or more electrons. To have significant activity as a free radical, a molecule must have an unpaired electron and sufficient redox potential to be reactive. Free radicals can be generated in biological systems through a variety of processes. A major question in free radical biology is what they do once they have been formed (2).

Peroxidation of Lipids

Polyunsaturated lipids are essential to the entire supporting system of cells, including cell membranes, endoplasmic reticulum, and mitochondria. Disruption of their structural properties can therefore have dire consequences for cellular function. Peroxidation of lipids has traditionally

The author expresses his gratitude to Concepción Avalos for secretarial assistance and to Alfredo Padilla for preparing the figures. This work was supported in part by grant 4265-N9406 from CONACYT, México.

237

been thought to be a major effect of free radicals. Because of this, many of the assay methods to establish free radical-induced injury have measured by-products of the reaction of these molecules with lipids. However, other cellular components may be as important as, or more important than, lipids in free radical injury. Any assay method that assesses the effects of free radicals on a particular component of cellular architecture needs to be accompanied by a consideration of the importance of the type of injury being determined (2).

Interest in the natural peroxidation of unsatured fatty acids was initially generated in the study of rancidity of fats and oils. Lipid peroxidation (LPO) has subsequently been implicated as the molecular mechanism in a diverse range of cellular insults.

Free radicals have a particularly high affinity for electron-rich unsaturated covalent bonds, such as those found in polyunsaturated fatty acids (PUFAs) (3). The net result of this reaction is a free-radical-mediated abstraction of an electron from the unsaturated covalent bonds of a PUFA, thus generating a PUFA radical (L·) (Figure 1). Hydroxy, lipoxy, and lipid peroxy radicals

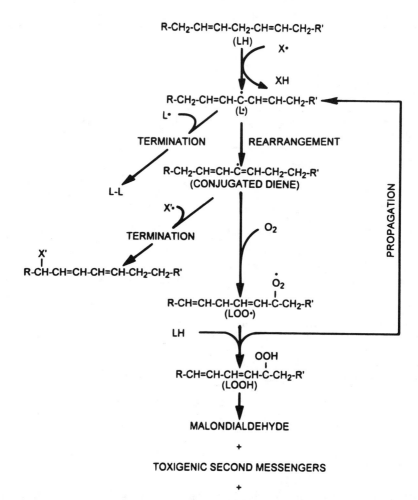

Figure 1. Lipid peroxidation (LPO). X· and X″ are free radicals, causing initiation and termination of the LPO sequence, respectively. Other abbreviations are L·, lipid radical; LOO·, lipid peroxide; and LOOH, lipid hydroperoxide (2).

have been implicated in initiating this process (3). Once formed, the lipid radical has several possible fates (4). It may rearrange to a more stable conjugated diene configuration, which enters the self-propagating LPO cascade (5). Alternatively, it may combine with another molecule, such as another lipid radical or a so-called free radical scavenger, and form a nonreactive complex. In the latter case, the dimer is formed at the expense of a cross-linkage of PUFAs within the membrane, causing a decrease in membrane fluidity (6). Because of the reactivity of PUFA radicals, the process is self-propagating. The end result is the chemical alteration of PUFAs, with the disruption of integral cellular components. Whether LPO is the major site of free radical damage to cells has been questioned (4). However, even if disruption of cellular lipids is not the final pathway to cell death, it is certainly a key route whereby additional radicals and other toxic substances are produced. Figure 1 shows the actual sequence of events in free radical attack of PUFAs. Since the process generates reactive lipid peroxides (LOO$^{\cdot}$), it is referred to as LPO.

Molecules that enter into the LPO cascade can, as shown in Figure 1, either continue to cycle causing further membrane damage, be detoxified by GSH, or generate small, diffusible, potentially toxigenic molecules (second messengers). Therefore, LPO may damage cells by its direct effect on cellular membranes, its generation of second messengers (6, 7), or its depletion of GSH, which predisposes cells to oxidative stress.

Is LPO merely the nonspecific end result of cellular damage by some other process, allowable simply by virtue of a loss of the usual homeostatic defenses against oxidative stress? Experimental data indicate that this is unlikely to be the case. For example, cells poisoned by the hepatocyte toxin thioacetamide show no LPO even when massive necrosis is demonstrable histologically (8). In fact, liver from whole rats incubated at 37°C for several hours postmortem does not shown any signs of significant LPO (8). Similarly, Wendel has shown that in mice with extensive hepatic damage secondary to experimental galactosamine/endotoxin-induced hepatitis, no LPO is detectable (9).

PEROXIDATION OF LIPIDS AND ALCOHOL-INDUCED LIVER DAMAGE

Alcohol and Liver Diseases

Alcoholic liver disease is one of the most serious medical consequences of chronic alcohol consumption. Moreover, chronic excessive alcohol use is the single most important cause of illness and death from liver disease (alcoholic hepatitis and cirrhosis) in the United States (10).

The three alcohol-induced liver conditions are fatty liver, alcoholic hepatitis, and cirrhosis. The most advanced form of alcoholic liver injury is alcoholic cirrhosis. This condition is marked by progressive development of scar tissue that chokes off blood vessels and distorts the normal architecture of the liver (11).

Role of Peroxidation of Lipids

The pathogenic importance of this peroxidative process in ethanol-induced liver damage is still a subject of controversy. The positive evidence of enhanced lipid peroxidation in the liver has only been shown when animals are fed ethanol chronically and given acute, high doses of ethanol after overnight fasting (12), or superimposed with a hypothermic condition (13). In fact, only a few studies have examined the parameters of lipid peroxidation and hepatic content of antioxidants under a chronically intoxicated state (14). In this respect the intragastric infusion model of Tsukamoto and French (15) provided an excellent opportunity to examine this question in animals with continuously sustained blood alcohol levels (BAL) at the discrete pathological stages of progressive alcohol liver disease (ALD).

Tsukamoto and Bacon (16) measured conjugated dienes in microsomal and mitochondrial lipids in rats, in which an early stage of centrilobular liver necrosis is developing, to investigate whether the induction of hepatocellular necrosis was associated with enhanced lipid peroxidation in animals continuously intoxicated with ethanol. However, they failed to show positive evidence of enhanced lipid peroxidation despite the fact that the same parameter was previously used to demonstrate enhanced lipid peroxidation in rats and baboons that were given an acute dose of ethanol after chronic ethanol feeding and overnight fasting (12). In fact, the levels of conjugated dienes in microsomal and mitochondrial lipids were even lower, and the hepatic levels of glutathione were not reduced in this model of ethanol-fed rats (14). Additionally, hepatic levels of methionine, a precursor of glutathione and α-tocopherol, one of the important antioxidants in vivo, were increased in these animals (16). These results were thus interpreted as constituting negative evidence for enhanced lipid peroxidation associated with the induction of alcoholic centrilobular liver necrosis. This conclusion corroborates that of Speisky et al. (17), who demonstrated for the first time the lack of changes in diene conjugate levels in the rat liver with ethanol-induced glutathione depletion or hepatic necrosis induced by a combination of ethanol consumption and anemia. These were in sharp contrast to the results obtained from studies on ethanol-fed baboons (12) and patients with alcoholic liver disease (18) that showed significantly elevated levels of conjugated dienes and depressed levels of glutathione in the liver. The discrepancy observed between these studies may be related to the fasting and the withdrawal state of baboons and patients at the time liver specimens were collected. In the studies of Inomata et al. (14) and Tsukamoto and Bacon (16) rats were continuously infused with ethanol and the diet until the specimens were collected. In this respect it is important to note that ethanol itself is an effective scavenger for hydroxyl radical (19) and that this free radical may be especially harmful at the time when BAL is low (19). This may explain the enhanced hepatic lipid peroxidation in baboons and patients during alcohol withdrawal (12, 18) and the reduced lipid peroxidation in the liver of the rats that were continuously intoxicated at the time of the experiment (14). Another possibility is the difference in the stage of liver pathology that could explain the differences observed. The baboons and patients appeared to have more advanced liver injury at the time when the experiments were conducted, whereas Tsukamoto and Bacon (16) examined the liver at the early stage of centrilobular necrosis. To explore this possibility, Tsukamoto et al. have recently repeated the study in their ethanol-fed rats at a time when more advanced liver injury has developed (15). The data obtained to date, indeed, indicate that the more advanced alcoholic liver injury at the time of progression to liver fibrosis was associated with an increased level of conjugated dienes in mitochondrial lipids of some of the animals. Furthermore, the increased hepatic content of methionine was no longer observable in these animals. Therefore these results support the view that lipid peroxidation in the liver is not enhanced at the initiation of hepatocyte necrosis but increased only after the substantial necrosis and consequent scar formation has taken place. The advanced stage of ALD may be associated with abnormal nutrient use, leading to the impairments in the antioxidant defense system as described in human alcoholics (20–22). Furthermore, recently it has been shown that high doses of α-tocopherol in the intragastric feeding rat model had no effect on ethanol-mediated liver damage in the rat. Also in α-tocopherol-supplemented groups, there was significant correlation between liver α-tocopherol and the extent of lipid peroxidation. Although α-tocopherol supplementation reduced hepatic peroxidation, this was not accompanied by an improvement in severity of liver injury (22).

Promotion of Lipid Peroxidation through Interactions with Cysteine, Glutathione, Vitamin E, and Iron

Whether alcohol administration in vivo results in lipid peroxidation and injury has been the subject of a long-standing debate (23–26), but more recent studies have shown evidence in favor of its occurrence in both nonhuman (27) and human (28) primates. In addition, increased ethane exhalation, an in vivo index of lipid peroxidation, was reported in alcohol abusers (29). Thus, lipid peroxidation (and associated membrane damage) is a key feature of alcoholic liver injury. It results not only from the increased oxygen radical production by the induced P-450 2E1 (30), but also from the enhanced generation of acetaldehyde, shown to be capable of causing lipid peroxidation in isolated perfused livers (31, 32). In vitro, metabolism of acetaldehyde via xanthine oxidase or aldehyde oxidase may generate free radicals, but the concentration of acetaldehyde required is much too high for this mechanism to be of significance in vivo. However, another way to promote lipid peroxidation is via GSH depletion. Binding of acetaldehyde with cysteine and/or glutathione may contribute to a depression of liver GSH (33). Rats fed ethanol chronically have significantly increased rates of GSH turnover (32). Acute ethanol administration inhibits GSH synthesis and produces an increased loss from the liver (32). GSH is selectively depleted in the mitochondria (34) and may contribute to the striking alcohol-induced alterations of that organelle. GSH offers one of the mechanism for the scavenging of toxic free radicals. Although GSH depletion *per se* may not be sufficient to cause lipid peroxidation, it is generally agreed that it may favor the peroxidation produced by other factors. GSH has been shown to spare and potentiate vitamin E (35); it is important in the protection of cells against electrophilic drug injury in general, and against reactive oxygen species in particular, especially in primates, which are more vulnerable to GSH depletion than rodents (27).

Long-term alcohol consumption was found to be associated with impaired methionine utilization and its depletion (36), as well as that of its active product, S-adenosyl-L-methionine (SAM) (37). Duce et al. (38) reported a decrease in SAM synthetase and phospholipid methyltransferase activities in cirrhotic livers. Potentially, a significant reduction of SAM may have a number of adverse effects, since it provides a source of cysteine for GSH production. Experimentally, the GSH depletion could be corrected, in part, by the administration of SAM (37, 39–41). It has been suggested that a reduction in the steady-state levels of hepatocellular GSH (in individuals who cannot support an adequate rate of synthesis) may act in synergy with other conditions that lead to hepatocellular necrosis and liver injury (42), one manifestation of which is lipid peroxidation. The latter is not only a reflection of tissue damage, but may also play a pathogenic role, for instance by promoting collagen production (43). SAM also plays a key role in the synthesis of polyamines, and it is the principal methylating agent in various vital transmethylation reactions that have been known for some time to be important to nucleic acid and protein synthesis and cell membrane function. Thus, depletion of SAM, by being detrimental to methyltransferase activity, may promote the membrane injury that has been documented in alcohol-induced liver damage (44). Moreover, SAM plays an essential role in phospholipid metabolism and the maintenance of membrane structure and function (39–41, 45).

Antioxidant protective mechanisms involve both enzymatic and nonenzymatic defense systems (46). Impairments in such defense systems have been reported in alcoholics, including alterations of ascorbic acid (47), GSH, selenium (48–50), and vitamin E (50–53). These changes could be due to direct effects of ethanol or to the malnutrition associated with alcoholism. α-tocopherol, the major antioxidant in the membrane, is viewed as the last line of defense against membrane lipid peroxidation (54–55). Bjorneboe et al. (56) reported a reduced hepatic α-tocopherol content after chronic ethanol feeding in rats receiving adequate amounts of vitamin E, as well as in the blood of alcoholics (57). Hepatic lipid peroxidation is significantly increased after chronic ethanol feeding in rats receiving a low vitamin E diet (58) (Figure 2), indicating that dietary vitamin E is an important determinant of hepatic lipid peroxidation induced by chronic ethanol feeding. The lowest hepatic α-tocopherol was found in rats receiving a combination of low vitamin E and ethanol. Hepatic α-tocopherol content was significantly reduced by both low dietary vitamin E and ethanol feeding, the latter in part because of increased conversion of α-tocopherol to α-tocopherylquinone (58) (Figure 3). In patients with cirrhosis, diminished hepatic vitamin E levels have been observed (59) (Figure 4). These deficient defense systems, coupled with increased acetaldehyde and oxygen radical generation by the ethanol-induced microsomes, may contribute to liver damage via lipid peroxidation and also via enzyme inactivation (60).

Iron overload may play a contributory role, since chronic alcohol consumption results in increased iron uptake by hepatocytes (61) and since ferric citrate-induced lipid peroxidation is accentuated in microsomes from ethanol-fed rats (61). Iron overload as well as iron deficiency in the alcoholic have been reviewed elsewhere, in conjunction with other mineral abnormalities (62).

LIPID PEROXIDATION IN OTHER TYPES OF LIVER INJURY

Paracetamol-Induced Liver Necrosis and Lipid Peroxidation

Probably the most common cause of drug-induced hepatic damage is paracetamol (acetaminophen, APAP). This drug is a relatively safe antipyretic/analgesic drug when administered at therapeutic doses but can produce hepatic centrilobular necrosis at toxic doses (63). However, the events associated with its toxicity in the liver are largely unresolved. Theories such as binding of the reactive metabolite of APAP, N-acetyl-*p*-benzoquinone imine (NAPQI), to cellular macromolecules (64, 65), lipid peroxidation (66, 67), and oxidation of critical sulfydryl groups

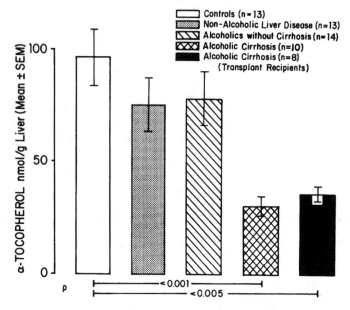

Figure 2. Effect of various liver diseases on total hepatic α-tocopherol levels. Compared with α- and β-carotene and lycopene, only the two cirrhotic groups had significantly lower α-tocopherol levels (59).

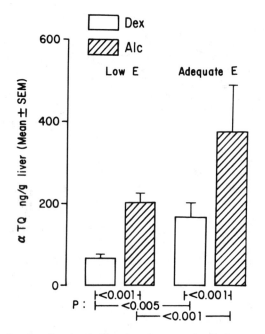

Figure 3. Effect of ethanol and/or vitamin E on hepatic α-tocopherylquinone content (αTQ). p values shown in the figure were tested by the Newman-Keuls' test (58).

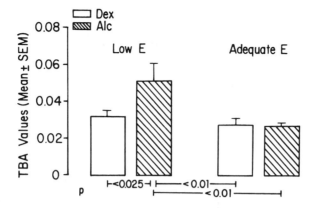

Figure 4. Effect of ethanol and/or vitamin E on hepatic lipid peroxidation. Dex, dextrin-fed controls; Alc, alcohol-fed rats; Low E, fed with low vitamin E; Adequate E, fed with adequate vitamin E. Hepatic lipid peroxidation (TBA) was significantly increased in Low E Alc rats (Newman-Keuls test). Both factors were synergistic ($p < .05$, two-way ANOVA) (58).

and alteration of calcium homeostasis (68) have been proposed. It has been demonstrated that paracetamol oxidation in the hepatocyte initiates a sequence of events that eventually leads to cell death (69). Antioxidants can inhibit these events (70, 71), suggesting that deleterious oxidative changes are involved.

Effect of Silymarin on Acetaminophen Toxicity

Silymarin is a flavonoid obtained from *Silybum marianum* (L.) Gartner, and is composed mainly of three isomers: silybinin, silydianin, and silychristin. Silymarin has been found to display a clear ability to scavenge free radicals (72).

The ability of silymarin to prevent liver damage induced by paracetamol has been studied (73). It was found that paracetamol increased the degree of lipid peroxidation while silymarin treatment prevented it (Figure 5). Furthermore, silymarin prevented the increment in serum

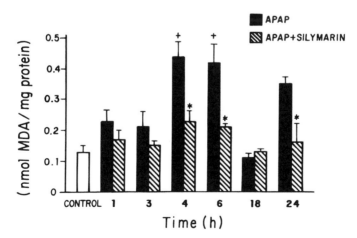

Figure 5. Effect of silymarin on the hepatic lipid peroxidation induced after 1, 3, 4, 6, 18, and 24 h of acetaminophen (APAP) intoxication. Each value represents the mean ± SEM of 10 animals in duplicate assays. Plus sign indicates means different from the control group, $p < .05$. Asterisk indicates means different from the acetaminophen-treated group, $p < .05$ (73).

markers of liver damage namely alkaline phosphatase (Figure 6), γ-glutamyl transpeptidase (γ-GT, Figure 7), and glutamic pyruvic transaminase (Figure 8), indicating an important protective effect of this compound. However, there is not enough evidence about the role of lipid peroxidation in acetaminophen toxicity. Colchicine has also shown a good hepatoprotective effect on acetaminophen-induced liver damage (74).

It appears that, in the case of an overdose of acetaminophen, the phenomena of covalent binding and lipid peroxidation may well occur simultaneously in the liver cell following depletion of GSH. Membrane deterioration (and leakage of cytosolic and membrane enzymes) due to extensive stimulation of lipid peroxidation could be an important facet of APAP toxicity (67), and the observed inhibition of lipid peroxidation by silymarin may account in part for its beneficial activity. This protective action of silymarin is probably associated with its antioxidant properties, possibly acting as a free-radical scavenger even at low levels of GSH. In fact, it has been demonstrated that silymarin is a good scavenger of superoxide and alkoxy radicals, as tested by the chemiluminescence technique (72). On the other hand, the calcium homeostasis theory (68) is not in disagreement with these data, since high levels of lipid peroxidation produce a change in the plasma membrane, which thus allows calcium accumulation (75).

Liver Damage Induced by Carbon Tetrachloride

Although carbon tetrachloride (CCl$_4$) toxicity is rarely encountered in the clinical setting, this compound provides a useful model of hepatotoxicity.

Carbon tetrachloride is a potent toxic agent able to produce different kinds of hepatic lesions. It has been used in hundreds of experimental trials to explore different aspects of the disease, using different animal species and several administration routes. Although it has been studied for more than a century, many questions still remain concerning the process of hepatocellular necrosis and subsequent regeneration or cirrhosis (76–81).

The metabolism of CCl$_4$ involves hemolytic breaking of the C$-$Cl bond, leading to the formation of free radicals that takes place in the hepatic endoplasmic reticulum through an enzyme system of electron transport from NADPH to oxygen. The free radical, CCl$_3^{\bullet}$, can bind

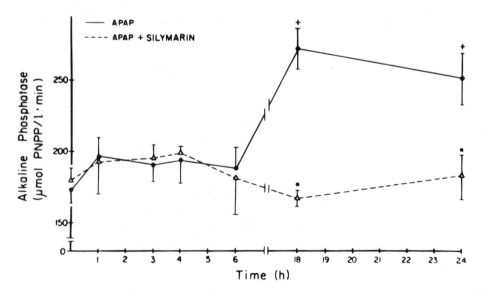

Figure 6. Time course of serum alkaline phosphatase activity after acetaminophen (APAP) intoxication. Each value represents the mean ± SEM of 10 animals in duplicate assays. Plus sign indicates means different from the control group, $p < .05$. Asterisk indicates means different from the acetaminophen-treated group, $p < .05$ (73).

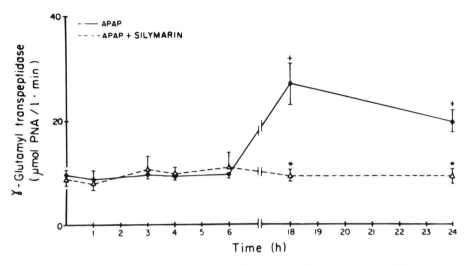

Figure 7. Time course of serum γ-glutamyl transpeptidase activity after acetaminophen (APAP) intoxication. Each value represents the mean ± SEM of 10 animals in duplicate assays. Plus sign indicates means different from the control group, $p < .05$. Asterisk indicates means different from the acetaminophen-treated group, $p < .05$ (73).

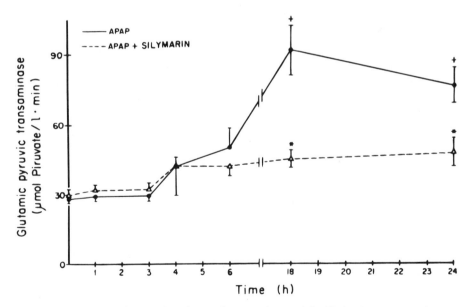

Figure 8. Time course of serum glutamic pyruvic transaminase activity. Each value represents the mean ± SEM of 10 animals in duplicate assays. Plus sign indicates means different from the control group, $p < .05$. Asterisk indicates means different from the acetaminophen-treated group, $p < .05$ (73).

covalently to a series of molecular structures, in particular to the lipids of the endoplasmic reticulum membranes. In turn, the free radicals attack the fatty acids in a peroxidation process. In tetrachloride-lesioned hepatocytes, protein synthesis is deficient; this is reflected in a decreased synthesis of hepatic lipoproteins and hence in an accumulation of neutral lipids in the liver with the consequent fatty infiltration (82).

Pathological changes in liver tissue following acute CCl_4 poisoning have been defined at the biochemical and ultrastructural level. Within 10 min, there are changes in the properties of the endoplasmic reticulum characterized by a decline in the activity of several microsomal enzymes, including Ca^{2+}-ATPase (83), glucose 6-phosphatase (84, 85), aminopyrine demethylase (84, 86), and cytochrome P-450 (84, 87), a blockage of the exit of hepatic triglycerides (88), and a disruption of protein synthesis (89). Concomitant with these functional changes, the membranous system retracts into smooth surface tubular aggregates that are considered to be denatured membrane (90). As the injury spreads within the cell, properties of the mitochondria and other cellular elements also become affected. By 12 h, central zone necrosis is evident; by 24–48 h, the necrosis is maximal (91).

Rapid, extensive lipid peroxidation of the membrane structural lipids has been proposed as the basis of CCl_4 hepatocellular toxicity (92) and has been extensively reviewed (93). A correlation between the metabolism of CCl_4, lipid peroxidation, and hepatocyte damage provides the strongest supporting evidence. Both CCl_4-induced lipid peroxidation and CCl_4 toxicity are dependent on reductive metabolism by a microsomal mixed-function oxidase with the generation of the trichloromethyl radical (92–94). $CCl_3^{\cdot}$ and $^{\cdot}CCl_3OO^{\cdot}$, a more reactive radical that arises from the reaction of $CCl_3^{\cdot}$ with oxygen, are capable of initiating lipid peroxidation by abstracting hydrogen from a polyunsaturated fatty acid (93–96).

CCl_4-enhanced lipid peroxidation, particularly of the microsomal lipids, has been observed in liver tissue homogenates (97–98), isolated hepatocytes (99, 100), and in vivo (101), and this has been associated with changes in endoplasmic reticular enzyme activity (102), in vivo fatty acid export (82, 85), and protein synthesis (89, 93). Agents that increase CCl_4 metabolism (e.g., inducers of cytochrome P-450) enhance production of malondialdehyde in vitro and increase ethane production and lethality in vivo (103). Conversely, pretreatment with SKF-525A, an inhibitor of microsomal drug metabolism, decreases both the extent of lipid peroxidation and microsomal dysfunction in vitro and toxic expression in vivo (104). Additionally, compounds such as ethylenediamine tetraacetic acid (EDTA) that inhibit lipid peroxidation also decrease the extent of microsomal cytochrome P-450 destruction in vitro (93). Propyl gallate inhibits both malondialdehyde production in rat liver homogenates in vitro and CCl_4-induced hepatic microsomal triglyceride accumulation in vivo (82). Furthermore, the free-radical scavenger silymarin (72) prevents lipid peroxidation and chronic and acute liver damage induced by CCl_4 (98, 105, 106). Thus, evidence from numerous approaches suggests that lipid peroxidation is associated with CCl_4-induced toxicity (107).

In contrast to these studies that support a primary role for lipid peroxidation in CCl_4-induced injury, other studies have raised questions about this connection. The antioxidant N,N-diphenyl-p-phenylenediamine prevents hepatocellular changes associated with CCl_4 poisoning (108), but several studies have failed to show clearly an associated decrease in microsomal lipid peroxidation (102, 108). Doses of the lipid antioxidant vitamin E, which protects against CCl_4 toxicity in vivo (109), do not affect the production of conjugated dienes in microsomal lipids (110) or changes in microsomal properties associated with lipid peroxidation (93, 111). A strict correlation does not always exist between the degree of lipid peroxidation and the toxicity of CCl_4 or related compounds. Neither 1,1-dichloroethylene, structurally very similar to CCl_4, nor $CHCl_3$, a metabolite of CCl_4 that produces liver injury similar to CCl_4, caused extensive peroxidation of microsomal lipids (112, 113). Species-specific differences in the degree of CCl_4 toxicity also do not correlate well with species-specific differences in the extent of lipid peroxidation (114). Additionally, the rapid time course of Ca^{2+} release from microsomal vesicles treated with CCl_4 does not correlate with the slower time course of lipid peroxidation (115).

Thus, while CCl_4 causes free-radical-induced oxidative damage to cells and causes lipid peroxidation, these processes may be parallel responses to the generation of free radicals. Under some conditions, lipid peroxidation may determine the extent of injury by amplifying the injury through propagation of free-radical processes, generating toxic compounds and impairing detoxification systems, but in other circumstances, even in the absence of lipid peroxidation, direct radical injury to proteins and DNA can occur. As we define the critical irreversible steps of injury and apply improved assays for lipid peroxidation, we will be able to discern the precise conditions under which lipid peroxidation contributes to the extent of injury and those under which lipid peroxidation is simply a symptom of free-radical injury (107).

Biliary Obstruction-Induced Liver Damage

The pathogenesis of hepatic injury during cholestasis is poorly understood. Recent observations suggest that oxidant or free-radical stress may play a role in cholestatic hepatic injury. A preliminary report showed that incubation of bile acids with isolated rat hepatocytes promoted oxidative modification of lipids (lipid peroxidation) concurrent with loss of hepatocyte viability and that α-tocopherol or superoxide dismutase inhibited these effects (116). In addition, elevated plasma concentrations of lipid peroxides have been reported in children with chronic cholestatic liver disease (117). Thus, there is a growing body of evidence suggesting that free radicals may be generated during cholestasis, that the diseased liver may have a diminished capacity to scavenge free radicals and that lipid peroxidation may be associated with cholestatic liver injury.

Bile duct ligation (BDL) in the rat is a useful tool for studying human cholestasis (80). It is also a good model for the evaluation of liver beneficial drugs (118–122). Thus it is important to understand its mechanism of action of toxicity (5).

Recently, Sokol et al. (123) demonstrated that peroxidative decomposition of mitochondrial lipids occurs in the chronically bile-duct-ligated rat model of cholestatic liver injury. However, in that report lipid peroxidation was studied only at day 17 after bile duct ligation (BDL) when liver damage is well established. Thus it cannot be concluded whether lipid peroxidation is the cause or the consequence of liver damage. On the basis of these considerations, we decided to study liver lipid peroxidation as well as several markers of liver damage at different times after biliary obstruction and to study two free-radical scavengers (colchicine and vitamin E) to determine the role of lipid peroxidation in this model of liver damage (124).

Figure 9 shows that total, conjugated, and unconjugated bilirubins increased after biliary obstruction. Note that total and conjugated bilirubins increased significantly ($p < .05$) as early as 1 day after ligation of the bile duct. As can be seen in Figure 10, alkaline phosphatase (AP), γ-GT, and glutamic pyruvic transaminase (GPT) serum enzyme activities increased significantly 1 day after bile duct ligation as compared with sham-operated rats. Alkaline phosphatase serum activity decreased by day 3 after surgery. However, both enzyme activities (AP and γ-GT) remained statistically different from the sham group during all the experimental period. On the other hand, the serum activity of GPT returned to normal 8 days after BDL (Figure 10). Lipoperoxidation did not increase significantly until day 3 (Figure 11), in contrast with the liver damage markers (bilirubins and serum enzyme activity) that increased after 1 day of biliary obstruction. In addition, liver glycogen content decreased to 10% of its original value at day 1 and remained low for all the experimental period (Figure 11). The sham-operated group also showed a decrease in glycogen content at days 2 and 3 but was completely recuperated by day 3.

In summary, we found that liver lipid peroxidation increases 3 days after BDL, while liver damage starts 2 days earlier (124). This indicates that peroxidation of lipids is independent of biliary obstruction and may even be a consequence rather than a cause of cell death. Moreover colchicine, a good free-radical scavenger capable of preventing liver damage induced by CCl_4, and vitamin E prevented the increase in lipid peroxidation (Figure 12) in rats subjected to BDL but were not able to protect the rats from liver injury (Table 1), thus confirming that lipid peroxidation is not an important mechanism for liver damage in experimental biliary obstruction (124). The detergent action of bile salts is more likely associated with solubilization of plasma membrane and cell death.

Naproxen-Induced Lipid Peroxidation

Nonsteroidal anti-inflammatory drugs are used over long-term periods in the treatment of patients with rheumatoid arthritis and other diseases. Chronic treatment with these drugs sometimes induces various side-effects. Naproxen, S-6-methoxy-α-methyl-2-naphthaleneacetic acid, a non-

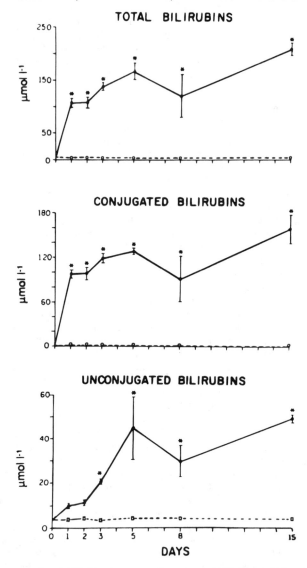

Figure 9. Time course of total bilirubins (upper panel), conjugated bilirubins (middle panel), and unconjugated bilirubins (lower panel) after biliary obstruction in the rat. Each value represents the mean ± SEM of 10 animals in duplicate assays. The dashed line connects the results obtained from control (sham-operated) animals, while the continuous line connects the results from bile-duct-ligated animals. Asterisk indicates significantly different from the control, $p < .05$ (124).

steroidal anti-inflammatory drug widely usely in cases of rheumatoid arthritis, is known to be effective and safe, but some side effects have been reported: gastrointestinal toxicity, nephrotoxicity, jaundice, and hepatotoxicity (125).

Lipid peroxidation occurred in rat liver microsomes during naproxen oxidative metabolism, although neither naproxen nor its oxidative metabolite, 6-demethylnaproxen (6-hydroxy-α-methyl-2-naphthaleneacetic acid), induced it (126). This suggested that the lipid peroxidation was caused by reactive oxygens generated from the cytochrome P-450 monooxygenase system.

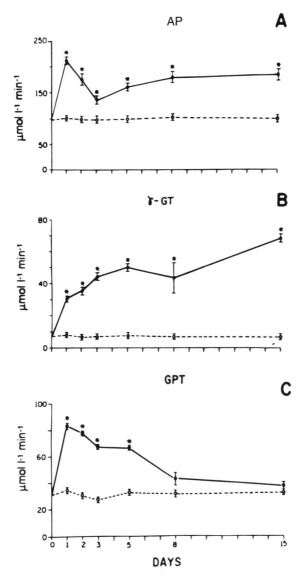

Figure 10. Time course of serum enzyme activities of (A) alkaline phosphatase, AP; (B) γ-glutamyl transpeptidase, γ-GT; and (C) glutamic pyruvic transaminase, GPT, after biliary obstruction in the rat. Each value represents the mean ± SEM of 10 animals in duplicate assays. The dashed line connects the results obtained from control (sham-operated) animals, while the continuous line connects the results from bile-duct-ligated animals. Asterisk indicates significantly different from the control, $p < .05$ (124).

On the other hand, hepatocytes have various protective devices against oxidative stress (127). If naproxen induces lipid peroxidation in hepatocytes with such defense systems, it could be a possible mechanism of the naproxen-induced hepatotoxicity.

Naproxen is administered over long periods to patients with rheumatoid arthritis. Such drugs used for chronic treatments, even if the drugs themselves are safe, may have side-effects through lipid peroxidation, because the potency of the patient's defense system against oxidative stress

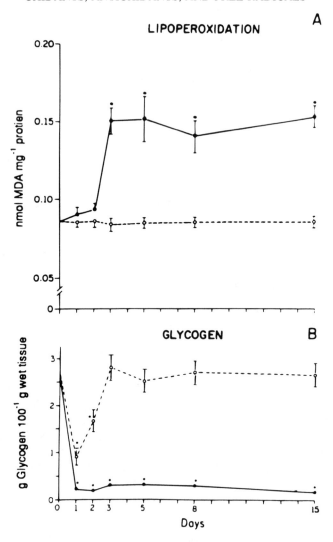

Figure 11. Time course of (A) liver lipid peroxidation and (B) liver glycogen content after biliary obstruction in the rat. Each value represents the mean ± SEM of 10 animals in duplicate assays. The dashed line connects the results obtained from control (sham-operated) animals, while the continuous line connects the results from bile-duct-ligated animals. Asterisk indicates significantly different from the control, $p < .05$ (124).

may decrease during the chronic treatments; this, in turn, may make the patient susceptible to oxidative stress due to reactive oxygen species generated from the cytochrome monooxygenase system during the metabolism of naproxen and/or other coadministration drugs.

Halothane and Sevoflurane

Halothane (2-bromo-2-chloro-1,1,1-trifluoroethane), a widely used volatile anesthetic, is toxic to the liver under certain conditions. One possible mechanism of halothane hepatotoxicity is peroxidation of microsomal lipids. Many in vivo and in vitro studies have shown the effect of lipid peroxidation in halothane-induced liver injury (128, 129). Radical metabolites are produced

LIPOPEROXIDATION

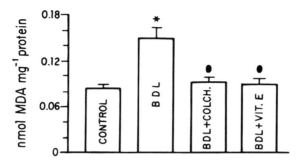

Figure 12. Effect of colchicine (COLCH.) and vitamin E (VIT. E) on liver lipid peroxidation induced by 3 days of biliary obstruction. The control group are sham-operated rats and the BDL group are bile-duct-ligated rats. Each bar represents the mean ± SEM of 10 animals in duplicate assays. Asterisk indicates significantly different from the control, $p < .05$. Bold dot indicates significantly different from the BDL group, $p < .05$ (124).

Table 1. Effect of colchicine (COLCH, 30 μg kg^{-1} po daily) and vitamin E (Vit. E, 400 IU kg^{-1} po daily) on liver damage induced by 3 days of bile-duct ligation (BDL)

Group	Bilirubins (μl L^{-1}) Total	Conjugated	Un-conjugated	Serum enzyme activities (μmol L^{-1} min^{-1}) AP	γ-GT	GPT	Glycogen content (g per 100 g wet tissue)
Control	3 ± 0.1	0.59 ± 0.98	3.3 ± 0.98	98 ± 3.4	7.0 ± 0.79	27 ± 1.99	2.8 ± 0.27
BDL	138 ± 7.3*	118.5 ± 6.6*	20.3 ± 1.5*	136 ± 5.6*	44 ± 2.4*	67 ± 2.02*	0.30 ± 0.09*
BDL + COLCH	134 ± 21.4*	105.0 ± 11.9*	18.9 ± 3.7*	149 ± 6.09*	40 ± 3.3*	66 ± 0.96*	0.38 ± 0.05*
BDL + VIT. E	141 ± 15.2*	112.0 ± 9.3*	21.4 ± 2.8*	142 ± 6.3*	42 ± 3.1*	65 ± 1.12*	0.33 ± 0.06*

Note. Results are expressed as means ± SEM ($n = 10$); asterisk indicates significantly different from control group, $p < .05$. AP, alkaline phosphatase; γ-GT, γ-glutamyl transpeptidase; GPT, glutamic pyruvic transaminase; BDL, bile-duct ligation; COLCH, colchicine; VIT. E, vitamin E.

by reductive metabolism; the production depends on cytochrome P-450 and requires NADPH (130).

Sevoflurane [fluoromethyl-2,2,2-trifluoro-1-(trifluoromethyl)ethyl ether] is a newly developed volatile inhalation anesthetic. It is a potent, nonexplosive, pleasant-smelling liquid with a vapor pressure of 200 torr at 25°C. Because of its low blood/gas partition coefficient of 0.6, sevoflurane rapidly equilibrates between the alveoli and blood (131). Due to these favorable characteristics, sevoflurane is already used commonly in Japan.

With this in mind, recently Sato et al. (131) decided to investigate whether sevoflurane causes lipid peroxidation in the liver. They found that sevoflurane potentiates microsomal lipid peroxidation in the guinea pig liver [measured as the formation of pentane, a good index of lipid peroxidation (132)]. However, they failed to obtain clear evidence of liver injury as measured by serum transaminase activity. We must be cautious in applying these results to humans (131). Further studies are required to clarify the association between sevoflurane-induced lipid peroxidation and liver injury.

CONCLUSIONS

The present viewpoint is that lipid peroxidation is not a necessary component of oxidative injury in the liver. Four general characteristics would be required if lipid peroxidation were to have a causal role in the cytotoxicity of a chemical agent or pathological condition. First, lipid peroxidation would occur in all cases of toxicity by the agent. Second, lipid peroxidation would occur in the proper sequence of events as they relate to cell death. Third, a correlation would exist between the extent of lipid peroxidation and the severity of the insult, taking into account the self-termination reactions of free radical processes and other factors that can affect such correlations. And finally, reduction or elimination of lipid peroxidation, such as by antioxidants or free radical trapping agents, would diminish the toxic expression. Methodological difficulties have precluded unequivocal determination of whether these characteristics have been met for any hepatotoxin.

Decades of study of lipid peroxidation have provided a wealth of knowledge concerning its chemistry and occurrence in biological systems, but such study has not definitively established a causal role for lipid peroxidation in pathological processes. In the absence of conclusive data, and given that the same reactive species thought to initiate lipid peroxidation also cause direct damage to proteins and DNA, it seems appropriate to view the problem of oxidative cell injury in a more general context and to focus future effort on identifying alternate molecular targets involved in oxidative injury to cells and tissues. In particular, biochemical, cellular, and molecular genetic approaches should be applied to search for target of injury in proteins and DNA. Such efforts will require an intensive input to ascertain the relative sensitivities of different proteins to various oxidative species and to identify the most reactive sites in the genome. Considerable knowledge is available on radiation damage and mutagenesis, and this information may provide a useful basis for studies of chemically induced oxidative injury. The possibility that lipid peroxidation can modulate the extent of damage to these critical sites of injury can then be addressed.

REFERENCES

1. Jamieson, D. 1989. Oxygen toxity and reactive oxygen metabolites in mammals. *Free Radic. Biol. Med.* 7:57–108.
2. Brent, J. A., and Rumack, B. H. 1993. Mechanism. Role of free radicals in toxic hepatic injury. I. Free radical biochemistry. *Clin. Toxicol.* 31:139–171.
3. Plaa, G. L., and Witschi, H. 1976. Chemicals, drugs and lipid peroxidation. *Annu. Rev. Pharmacol. Toxicol.* 16:125–141.
4. Tribble, D. L., Aw, T. Y., and Jones, D. P. 1987. The pathophysiological significance of lipid peroxidation in oxidative cell injury. *Hepatology* 7:377–387.
5. Benedetti, A., Casini, A. F., and Ferrali., M. 1977. Red cell lysis coupled to the peroxidation of liver microsomal lipids. Compartmentalization of the hemolytic system. *Res. Commun. Chem. Pathol. Pharmacol.* 17:519–528.
6. Ungemach, F. R. 1987. Pathobiochemical mechanism of hepatocellular damage following lipid peroxidation. *Chem. Phys. Lipids* 45:171–205.
7. Comporti, M. 1985. Lipid peroxidation and cellular damage in toxic liver injury. *Lab. Invest.* 53:599–623.
8. Hashimoto, S., Glende, E. A., and Recknagel, R. O. 1968. Hepatic lipid peroxidation in acute fatal human carbon tetrachloride poisoning. *N. Engl. J. Med.* 279:1082–1085.
9. Wendel, A. 1987. Measurement of in vivo lipid peroxidation and toxicological significance. *Free Radic. Biol. Med.* 3:355–358.
10. Smart, R. G., and Mann, R. E. 1991. Alcohol and the epidemiology of liver cirrhosis. *Alcohol Health Res. World* 16:217–222.
11. Grant, B. F., Dufour, M. C., and Harford, T. C. 1988. Epidemiology of alcoholic liver disease. *Semin. Liver Dis.* 8:12–25.
12. Rothschild, M. A., Oratz, M., and Schreiber, S. S. 1989. Alcohol-induced liver disease: Does nutrition play a role? *Alcohol Health Res. World* 13:229–231.
13. Kato, S., Kawase, T., Alderman, J., Inatomi, N., and Lieber, C. S. 1990. Role of xanthine oxidase in ethanol-induced lipid peroxidation in rats. *Gastroenterology* 98:203–210.
14. Inomata, T., Rao, G. A., and Tsukamoto, H. 1987. Lack of evidence for increased lipid peroxidation in ethanol-induced centrilobular necrosis of rat liver. *Liver* 7:233–239.
15. Tsukamoto, H., Gaal, K., and French, SW. 1990. Insights into the pathogenesis of alcoholic liver necrosis and fibrosis: Status report. *Hepatology* 12:599–608.
16. Tsukamoto, H., and Bacon, B. R. 1987. Status of pro and antioxidant in ethanol-induced centrilobular necrosis (abstr.). *Hepatology* 7:1078.
17. Speisky, H., Bunou, D., Orrego, H., and Israel, Y. 1985. Lack of changes in diene conjugate levels

following ethanol-induced glutathione depletion or hepatic necrosis. *Res. Commun. Chem. Pathol. Pharmacol.* 48:77–90.

18. Shaw, S., Rubin, K. P., and Lieber, C. S. 1983. Depressed hepatic glutathione and increased diene conjugates in alcoholic liver disease: Evidence of lipid peroxidation. *Dig. Dis. Sci.* 28:585–589.
19. Klein, S. M., Cohen, G., Lieber, C. S., and Cederbaum, A. I. 1983. Increased microsomal oxidation of hydroxyl radical scavenging agents and ethanol after chronic consumption of ethanol. *Arch. Biochem. Biophys.* 223:425–432.
20. Videla, L. A., and Valenzuela, A. 1982. Alcohol ingestion, liver glutathione and lipoperoxidation: metabolic interrelations and pathological implications. *Life Sci.* 31:2395–2407.
21. Tanner, A. R., Bantock, I., Hinks, L., Lloyd, B., Turner, N. R., and Wright, R. 1986. Depressed selenium and vitamin E levels in an alcoholic population: Possible relationship to hepatic injury through increased lipid peroxidation. *Dig. Dis. Sci.* 31:1307–1312.
22. Hossein, S. S. M., Meydani, M., Khettry, U., and Nanji, A. A. 1995. High-dose vitamin E supplementation has no effect on ethanol-induced pathological liver injury. *J. Pharmacol. Exp. Ther.* 273:455–460.
23. Hashimoto, S., and Recknagel, R. O. 1968. No chemical evidence of hepatic lipid peroxidation in acute ethanol toxicity. *Exp. Mol. Pathol.* 8:225–242.
24. Scheig, R., and Klatskin, G. 1969. Some effects of ethanol and carbon tetrachloride on lipoperoxidation in rat liver. *Life Sci.* 8:855–846.
25. Bunyan, J., Cawthrone, M. A., Diplock, A. T., and Green, J. 1969. Vitamin E and hepatotoxic agents. 2. Lipid peroxidation and poisoning with orotic acid, ethanol and thioacetamide in rats. *Br. J. Nutr.* 23:309–317.
26. Comporti, M., Burdino, E., and Raja, F. 1971. Fatty acids composition of mitochondrial and microsomal lipids of rat liver after acute ethanol intoxication. *Life Sci.* 10(part 11):855–866.
27. Shaw, S., Jayatilleke, E., Ross, W. A., Gordon, E. R., and Lieber, C. S. 1981. Ethanol induced lipid peroxidation: Potentiation by long-term alcohol feeding and attenuation by methionine. *J. Lab. Clin. Med.* 98:417–425.
28. Shaw, S., Rubin, K. P., and Lieber, C. S. 1983. Depressed hepatic glutathione and increased diene conjugates in alcoholic liver disease: evidence of lipid peroxidation. *Dig. Dis. Sci.* 28:585–589.
29. Lettron, P., Duchatelle, V., Berson, A., et al. 1993. Increased ethane exhalation, an in vivo index of lipid peroxidation, in alcohol-abusers. *Gut* 34:409–414.
30. Castillo, T., Koop, D. R., Kamimura, S., Triadafilopoulos, G., and Tsukamoto, H. 1992. Role of cytochrome P-450 2El in ethanol-, carbon tetrachloride- and iron-dependent microsomal lipid peroxidation. *Hepatology* 16:992–996.
31. Muller, A., and Sies, H. 1982. Role of alcohol dehydrogenase activity and the acetaldehyde in ethanol-induced ethane and pentane production by isolated perfused rat liver. *Biochem. J.* 206:153–156.
32. Morton, S., and Mitchell, M. C. 1985. Effects of chronic ethanol feeding on glutathione turnover in the rat. *Biochem. Pharmacol.* 34:1559–1563.
33. Speisky, H., MacDonald, A., Giles, G., Orrego, H., and Israel, Y. 1985. Increased loss and decreased synthesis of hepatic glutathione after acute ethanol administration. *Biochem. J.* 225:565.
34. Hirano, T., Kaplowitz, N., Tsukamoto, H., Kamimura, S., and Fernández-Checa, J. C. Hepatic mitochondrial glutathione depletion and progression of experimental alcoholic liver disease in rats. *Hepatology* 6:1423–1427.
35. Barclay, L. R. 1988. The cooperative antioxidant role of glutathione with a lipid-soluble and a water-soluble antioxidant during peroxidation of liposomes initiated in the aqueous phase and in the lipid phase. *J. Biol. Chem.* 263:16138–16142.
36. Finkelstein, J. D., Cello, J. P., and Kyle, W. E. Ethanol-induced changes in methionine metabolism in rat liver. *Biochem. Biophys. Res. Commun.* 61:475–481.
37. Lieber, C. S., Casini, A., DeCarli, L. M., Kim, C., Lowe, N., Sasaki, R., and Leo, M. A. 1990. S-Adenosyl-L-methionine attenuates alcohol-induced liver injury in the baboon. *Hepatology* 11:165–172.
38. Duce, A. M., Ortiz, P., Cabrero, C., and Mato, J. M. 1988. S-Adenosyl-L-methionine synthetase and phospholipid methyltransferase are inhibited in human cirrhosis. *Hepatology* 8:65–68.
39. Muriel, P., and Mourelle, M. 1992. Characterization of membrane fraction lipid composition and function of cirrhotic rat liver. Role of S-adenosyl-L-methionine. *J. Hepatol.* 14:16–21.
40. Muriel, P. 1993. S-Adenosyl-L-methionine prevents and reverses erythrocyte membrane alterations in cirrhosis. *J. Appl. Toxicol.* 13:179–182.
41. Muriel, P., Suárez, O.R., González, P., Zúñiga, L. 1994. Protective effect of S-adenosyl-L-methionine on liver damage induced by biliary obstruction in rats. A histological, ultrastructural and biochemical approach. *J. Hepatol.* 21:95–102.
42. Clement, B., and Guillouzo, A., eds. 1992. *Cellular and Molecular Aspects of Cirrhosis*, Vol. 216, pp. 25–37. John Libbey Eurotext Ltd.
43. Geesin, J. C., Hendricks, L. J., Falkenstein, P. A., Gordon, J. S., and Berg, R. A. 1991. Regulation of

collagen synthesis by ascorbic acid: Characterization of the role of ascorbate-stimulated lipid peroxidation. *Arch. Biochem. Biophys.* 290:127–132.

44. Yamada, S., Wilson, J. S., and Lieber, C. S. 1985. The effects of alcohol and diet on hepatic and serum gamma-glutamyltranspeptidase activities in rats. *J. Nutr.* 115:1285–1290.

45. Mato, J. M. 1986. *Progress in Protein-Lipid Interactions*, Vol. 2. New York: Elsevier.

46. Tribble, D. L., Au, T. Y., and Jones, D. P. 1987. The pathological significance of lipid peroxidation in oxidative cell injury. *Hepatology* 7:377–387.

47. Bonjour, J. P. 1979. Vitamins and alcoholism. *Int. J. Vit. Nutr. Res.* 49:434–441.

48. Korpela, H., Kumpulainen, J., Luoma, P. V., Arranto, A. J., and Sotaniemi, E . A. 1985. Decreased serum selenium in alcoholics as related to liver structure and function. *Am. J. Clin. Nutr.* 42:147–151.

49. Dworkin, B., Rosenthal, W. S., Jankowski, R. H., Gordon, G. G., and Haldea, D. 1985. Low blood selenium levels in alcoholics with and without advanced liver disease. *Dig. Dis. Sci.* 30:838–844.

50. Tanner, A. R., Bantock, Y., Hinks, L., Lloyd, B., Turner, N. R., and Wright, R. 1986. Depressed selenium and vitamin E levels in an alcoholic population. Possible relationship to hepatic injury through increased lipid peroxidation. *Dig. Dis. Sci.* 31:1307–1312.

51. Lieber, C. S. 1994. Mechanism of ethanol-drug-nutrition interactions, *Clin. Toxicol.* 32:631–681.

52. Yoshikawa, Y., Takemura, S., and Kondo, M. 1982. α-Tocopherol level in liver diseases. *Acta Vitaminol. Enzymol.* 4:311–318.

53. Bjorneboe, G. E. A., Johnsen, J., Bjorneboe, A., Morland, J., and Drevon, C. A. 1987. Effect of heavy alcohol consumption on serum concentration of fat soluble vitamins and selenium. *Alcohol Alcoholism* 1(suppl):533–537.

54. McCay, P. B. 1985. Vitamin E interaction with free radical and ascorbate *Annu. Rev. Nutr.* 5:323–340.

55. Niki, E. 1987. Interaction of ascorbate and α-tocopherol. *Ann. N.Y. Acad. Sci.* 493:186–199.

56. Bjorneboe, G. E. A., Bjorneboe, A., Hagen, B. F., Morland, J., and Drevon, C. A. 1987. Reduced hepatic α-tocopherol content after long-term administration of ethanol to rats. *Biochem. Biophys. Acta* 918:236–241.

57. Bjorneboe, G. E. A., Johnsen, J., and Bjorneboe, A. 1988. Some aspects of antioxidant status in blood from alcoholics. *Alcohol Clin. Exp. Res.* 12:806–810.

58. Kawase, T., Kato, S., and Lieber, C. S. 1989. Lipid peroxidation and antioxidant defense system in rat liver after chronic ethanol feeding. *Hepatology* 10:815–821.

59. Leo, M. A., Rosman, A., and Lieber, C. S. 1993. Differential depletion of carotenoids and tocopherol in liver disease. *Hepatology* 17:977–986.

60. Dicker, E., and Cederbaum, A. I. 1988. Increased oxygen radical-dependent inactivation of metabolic enzymes by liver microsomes after chronic ethanol consumption. *FASEB J.* 2:2901–2906.

61. Zhang, H., Loney, L. A., and Potter, B. J. 1993. Effect of chronic alcohol feeding on hepatic iron status and ferritin uptake by rat hepatocytes. *Alcohol Clin. Exp. Res.* 17:394–400.

62. Lieber, C. S. 1988. The influence of alcohol on nutritional status. *Nutr. Rev.* 46:241–245.

63. Mitchell, J. R., Jollow, D. J., Potter, W. Z., Davis, D. C., Gillette, J. R., and Brodie, B. B. 1973. Acetaminophen-induced hepatic necrosis. 1. Role of drug metabolism. *J. Pharmacol. Exp. Ther.* 187:185–194.

64. Potter, W. Z., Davis, D. C., Mitchell, J. R., Jollow, D. J., Gillette, J. R., and Brodie, B. B. 1973. Acetaminophen-induced hepatic necrosis. III. Cytochrome P-450 mediated covalent binding in vitro. *J. Pharmacol. Exp. Ther.* 187:203–209.

65. Mitchell, J. R., Jollow, D. J., Potter, W. Z., Gillette, J. R., and Brodie, B. B. 1973. Acetaminophen-induced hepatic necrosis. IV. Protective role of glutathione. *J. Pharmacol. Exp. Ther.* 187:211–217.

66. Wendal, A., and Fevertein, S. 1981. Drug induced lipid peroxidation in mice. 1. Modulation by monoxygenase activity, glutathione and selenium status *Biochem. Pharmacol.* 30:2513–2520.

67. Fairhurst, S., Barber, D. J., Clark, B., and Horton, A. A. 1982. Studies on paracetamol-induced lipid peroxidation. *Toxicology* 23:249–259.

68. Moore, M., Thor, H., Moore, G., Nelson, S., Moldeus, P., and Orrenius, S. 1985. The toxicity of acetaminophen and *N*-acetyl-*p*-benzoquinone imine in isolated hepatocytes is associated with thiol depletion and increased cytosolic Ca^{2+}. *J. Biol. Chem.* 260:13035–13040.

69. Savides, M. C., and Oehme, F. W. 1983. Acetaminophen and its toxicity. *J. Appl. Toxicol.* 3:96–111.

70. Harman, A. W. 1985. The effectiveness of antioxidants in reducing paracetamol-induced damage subsequent to paracetamol activation. *Res. Commun. Chem. Pathol. Pharmacol.* 49:215–228.

71. McLean, A. E. M., and Nuttall, L. 1978. An in vitro model of liver injury using paracetamol treatment of liver slices and prevention of injury by some antioxidants. *Biochem. Pharmacol.* 27:425–430.

72. Pascual, C., González, R., Armesto, J., and Muriel, P. 1993. Effect of silymarin and silybin on oxygen radicals. *Drug Dev. Res.* 29:73–77.

73. Muriel, P., Garciapiña, T., Pérez-Alvarez, V., and Mourelle, M. 1992. Silymarin protects against paracetamol-induced lipid peroxidation and liver damage. *J. Appl. Toxicol.* 12:439–442.

74. Muriel, P., Quintanar, M. E., and Pérez-Alvarez, V. 1993. Effect of colchicine on acetaminophen-induced liver damage. *Liver* 13:217–221.

75. Mourelle, M., and Meza, M. A. 1990. CCl_4-induced lipoperoxidation triggers a lethal defect in the liver plasma membranes. *J. Appl. Toxicol.* 10:23–27.

76. Belloti, S., Burlango, F., and Novelli, A. 1980. Rapid induction of liver cirrhosis in rats by treatment with ethanol, carbon tetrachloride and progesterone. *Boll. Soc. Ital. Biol. Sper.* 56:666–672.

77. Chatamra, K., and Proctor, E. 1981. Phenobarbitone-induced enlargement of the liver in the rat: Its relationship to carbon tetrachloride induced cirrhosis. *Br. J. Exp. Pathol.* 62:283–291.

78. Montfort, Y., and Pérez-Tamayo, R. 1978. Collagenase in experimental carbon tetrachloride cirrhosis of the liver. *Am. J. Pathol.* 92:411–418.

79. Ozeki, T., Funakoshi, K., and Iwak, K. 1985. Rapid induction of cirrhosis by administration of carbon tetrachloride plus phospholipase-D. *Br. J. Exp. Pathol.* 66:385–391.

80. Pérez-Tamayo, R. 1983. Is cirrhosis of the liver experimentally produced by CCl_4 an adequate model of human cirrhosis? *Hepatology* 3:112–120.

81. Trivedi, P., and Mowat, A. P. 1983. Carbon tetrachloride-induced hepatic fibrosis and cirrhosis in the developing rat: An experimental model of cirrhosis in childhood. *Br. J. Exp. Pathol.* 64:25.

82. Ugazio, G., and Torrielli, M. V. 1969. Effect of propyl gallate on carbon tetrachloride induced fatty liver. *Biochem. Pharmacol.* 18:2271–2274.

83. Moore, L., Davenport, G. R., and Landon, E. J. 1976. Calcium uptake of a rat liver microsomal subcellular fraction in response to in vivo administration of carbon tetrachloride. *J. Biol. Chem.* 251:1197–1201.

84. Glende, E. A., Hruszkewycz, A. M., and Recknagel, R. O. 1976. Critical role of lipid peroxidation in carbon tetrachloride-induced loss of aminopyrine demethylase, cytochrome P-450 and glucose-6-phosphatase. *Biochem. Pharmacol.* 25:2163–2170.

85. Recknagel, R. O., and Lombardi, B. 1961. Studies of biochemical changes in subcellular particles of rat liver and their relationship to a new hypothesis regarding the pathogenesis of carbon tetrachloride fat accumulation. *J. Biol. Chem.* 236:564–569.

86. Glende, E. A. 1972. On the mechanisms of carbon tetrachloride toxicity—Coincidence of loss of drug metabolizing activity with peroxidation of microsomal lipid. *Biochem. Pharmacol.* 21:2131–2138.

87. Glende, E. A. 1972. Carbon tetrachloride-induced protection against carbon tetrachloride toxicity: Role of the liver microsomal drug-metabolizing system. *Biochem. Pharmacol.* 21:1697–1702.

88. Recknagel, R. O. 1967. Carbon tetrachloride hepatotoxicity. *Pharmacol. Rev.* 19:145–208.

89. Smuckler, E. A., Iseri, O. A., and Benditt, E. P. 1962. An intracellular defect in protein synthesis induced by carbon tetrachloride. *J. Exp. Med.* 116:55–71.

90. Reynolds, E. S., and Ree, H. J. 1971. Liver parenchymal cell injury. VII. Membrane denaturation following carbon tetrachloride. *Lab. Invest.* 25:269–278.

91. Hoffman, J., Hines, M. B., and Lapan, S. 1955. Responses of the liver to injury. Effects of acute carbon tetrachloride poisoning. *AMA Arch. Pathol.* 59:429–438.

92. Recknagel, R. O., and Ghoshal, A. K. 1966. Lipoperoxidation as a vector in carbon tetrachloride hepatotoxicity. *Lab. Invest.* 15:132–146.

93. Recknagel, R. O., and Glende, E. A. 1973. Carbon tetrachloride hepatotoxicity: An example of lethal cleavage. *CRC Crit. Rev. Toxicol.* 2:263–297.

94. Castro, J. A., Díaz-Gómez, M. I., and de Ferreyra, E. C. 1972. Carbon tetrachloride effect on rat liver and adrenals related to their mixed-function oxygenase content. *Biochem. Biophys. Res. Commun.* 47:315–321.

95. Packer, J. E., Slater, T. F., and Willson, R. L. 1978. Reactions of the carbon tetrachloride-related peroxy free radical (CCl_3O_2) with amino acids: Pulse radiolysis evidence. *Life Sci.* 23:2617–2620.

96. Muriel, P., and Mourelle, M. 1992. Characterization of membrane fraction lipid composition and function of cirrhotic rat liver. Role of S-adenosyl-L-methionine. *J. Hepatol.* 14:16–21.

97. Dianzani, M. U., Baccino, F., and Comporti, C. 1966. The direct effect of carbon tetrachloride on subcellular particles. *Lab. Invest.* 15:149–156.

98. Muriel, P., and Mourelle, M. 1990. Prevention by silymarin of membrane alterations in acute liver damage. *J. Appl. Toxicol.* 10:275–279.

99. Smith, M. T., Thor, H., Hartzell, P., and Orrenius, S. 1982. The measurement of lipids peroxidation in isolated hepatocytes. *Biochem. Pharmacol.* 31:19–26.

100. McCay, P. B., Lai, E. K., Poyer, J. L., Dubose, C. M., and Janzen, E. 1984. Oxygen- and carbon-centered free radical formation during carbon tetrachloride metabolism. *J. Biol. Chem.* 259:2135–2143.

101. Comporti, M., Benedetti, A., and Casini, A. 1974. Carbon tetrachloride induced liver alterations in rats pretreated with N,N'-diphenyl-p-phenylenediamine. *Biochem. Pharmacol.* 23:421–432.

102. Ghoshal, A. K., and Recknagel, R. O. 1965. On the mechanism of carbon tetrachloride hepatotoxicity: Coincidence of loss of glucose-6-phosphatase activity with peroxidation of microsomal lipid. *Life Sci.* 4:2195–2209.

103. Rao, K. S., Glende, E. A., Jr., and Recknagel, R. O. 1970. Effect of drug pretreatment on carbon tetrachloride-induced lipid peroxidation in rat liver microsomal lipids. *Exp. Mol. Pathol.* 12:324–331.

104. Riely, C. A., Cohen, G., Lieberman, M. 1974. Ethane evolution: A new index of lipid peroxidation. *Science* 183:208–210.

105. Mourelle, M., Muriel, P., Favari, L., and Franco, T. 1989. Prevention of CCl_4 induced liver cirrhosis by silymarin. *Fundam. Clin. Pharmacol.* 3:183–191.

106. Muriel, P., and Mourelle, M. 1990. The role of membrane composition in ATPases activities of cirrhotic rats. Effect of silymarin. *J. Appl. Toxicol.* 10:281–284.

107. Tribble, D. L., Aw, T. Y., and Jones, D. P. 1987. The pathophysiological significance of lipid peroxidation in oxidative cell injury. *Hepatology* 7:377–387.

108. de Ferreyra, E. C., Castro, J. A., and Díaz-Gómez, M. I., et al. 1975. Diverse effects of antioxidants on carbon tetrachloride hepatotoxicity. *Toxicol. Appl. Pharmacol.* 32:504–512.

109. Hove, E. L. 1948. Interrelation between α-tocopherol and protein metabolism. III. The protective effect of vitamin E and certain nitrogenous compounds against CCl_4 poisoning in rats. *Arch. Biochem.* 17:467–474.

110. Benedetti, A., Ferrali, M., Chieli, E., and Comporti, M. 1974. A study of the relationships between carbon tetrachloride-induced lipid peroxidation and liver damage in rats pretreated with vitamin E. *Chem. Biol. Interact.* 9:117–134.

111. McLean, A. E. M. 1967. The effect of diet and vitamin E on liver injury due to carbon tetrachloride. *Br. J. Exp. Pathol.* 48:632–636.

112. Jaeger, R. J., Trabulus, M. J., and Murphy, S. D. 1973. Biochemicals effects of 1,1-dichloroethylene in rats: Dissociation of its hepatotoxicity from a lipoperoxidative mechanism. *Toxicol. Appl. Pharmacol.* 24:457–467.

113. Klaasen, C. D., and Plaa, G. L. 1969. Comparison of the biochemical alterations elicited in livers from rats treated with carbon tetrachloride, chloroform, 1,1,2-trichroroethane and 1,1,1,-trichloroethane. *Biochem. Pharmacol.* 18:2019–2027.

114. Díaz-Gómez, M. I., de Castro, C. R., D'Acosta, N., et al. 1975. Species differences in carbon tetrachloride-induced hepatotoxicity: The role of CCl_4 activation and of lipid peroxidation. *Toxicol. Appl. Pharmacol.* 34:102–114.

115. Waller, R. L., Glende, E. A., Jr., and Recknagel, R. O. 1983. Carbon tetrachloride and bromotrichloromethane toxicity. Dual role of covalent binding of metabolic cleavage products and lipid peroxidation in depression of microsomal calcium sequestration. *Biochem. Pharmacol.* 32:1613–1617.

116. Seto, Y., Nakashima, T., Nakajima, T., Shima, T., Sakamoto, Y., Okuno, T., and Takino, T. 1988. Involvement of oxygen radicals in bile acid-induced hepatocytes injury (abstr.). *Hepatology* 8:1452.

117. Lemonnier, F., Cresteil, D., Feueant, M., Couturier, M., Bernard, O., and Alagille, D. 1987. Plasma lipid peroxides in cholestatic children. *Acta Pediatr. Scand.* 76:928–934.

118. Muriel, P., Suárez, O. R., González, P., and Zúñiga, L. 1994. Protective effect of S-adenosyl-L-methionine on liver damage induced by biliary obstruction in rats. A histological, ultrastructural and biochemical approach. *J. Hepatol.* 21:95–102.

119. Moreno, M., and Muriel, P. 1995. Remission of liver fibrosis by interferon-alpha$_{2b}$. *Biochem. Pharmacol.* 50:515–520.

120. Rodríguez-Fragoso, L., González, M. P., and Muriel, P. 1995. Interferon-α_{2b} increases fibrolysis in fibrotic livers from bile duct ligated rats. Possible participation of the plasminogen activator. *Pharmacology*, 51:341–346.

121. Muriel, P. 1996. Alpha-interferon prevents liver collagen deposition and damage induced by prolonged bile duct obstruction in the rat. *J. Hepatol.*, 24:614–621.

122. Muriel, P. 1995. Interferon-α preserves erythrocyte and hepatocyte ATPase activities from liver damage induced by prolonged bile duct ligation in the rat. *J. Appl. Toxicol.* 15:449–553.

123. Sokol, R. J., Devereaux, M., and Rashmi, A. K. 1991. Effect of dietary lipid and vitamin E on mitochondrial lipid peroxidation and hepatic injury in the bile duct-ligated rat. *J. Lipid Res.* 32:1349–1357.

124. Muriel, P., and Suárez, O. R. 1994. Role of lipid peroxidation on biliary obstruction in the rat. *J. Appl. Toxicol.* 14:423–426.

125. Yokoyama, H., Horie, T., and Awazu, S. 1994. Oxidative stress in isolated rat hepatocytes during naproxen metabolism. *Biochem. Pharmacol.* 49:991–996.

126. Yokoyama, H., Horie, T., and Awazu, S. 1993. Lipid peroxidation in rat liver microsomes during naproxen metabolism, *Biochem. Pharmacol.* 45:1721–1724.

127. Halliwell, B. 1990. How to characterize a biological antioxidant. *Free Radic. Res. Commun.* 9:1–32.

128. Sato, N., Fujii, K., Yuge, O., and Morio, M. 1990. The association of halothane-induced lipid peroxidation with the anaerobic metabolism of halothane: An in vitro study in guinea pig liver microsomes. *Hiroshima J. Med. Sci.* 39:1–6.

129. Akita, S., Morio, M., Kawahara, M., Takeshita, T., Fujii, K., and Yamamoto, M. 1988. Halothane-

induced liver injury as a consequence of enhanced microsomal lipid peroxidation in guinea pigs. *Res. Commun. Chem. Pathol. Pharmacol.* 61:227–243.

130. Fujii, K., Morio, M., and Kikuchi, H. 1981. A possible role of cytochrome P-450 in anaerobic dehalogenation of halothane. *Biochem. Biophys. Res. Commun.* 101:1158–1163.

131. Sato, N., Fujii, K., and Yuge, O. 1994. In vivo and in vitro sevoflurane-induced lipid peroxidation in guinea-pig liver microsomes. *Pharmacol. Toxicol.* 75:366–370.

132. Sato, N., Fujii, K., and Yuge, O. 1993. Pentane formation and changes in fatty acid composition accompanying lipid peroxidation induced by carbon tetrachloride in guinea pig liver microsomes. *Toxicol. In Vitro* 7:673–675.

Chapter Fifteen

ROLE OF FREE RADICALS IN ALCOHOL-INDUCED TISSUE INJURY

Carol Colton and Sam Zakhari

The role of reactive oxygen intermediates (ROI) and reactive nitrogen intermediates (RNI) in ethanol-induced toxicity has been supported by a variety of in vivo and in vitro studies. All tissues in the body are affected by exposure to ethanol, and variations in the toxic effects may be a reflection of the ROI/RNI generating mechanisms available in that tissue as well as the level of antioxidant protection. To understand how ethanol alters the redox balance of a cell, it is important to briefly review the basic mechanisms underlying ROI and RNI generation, the types of reactions that these molecules can undergo, and the protective or reaction-terminating mechanisms that are available to prevent oxidative damage.

ROI AND RNI

The classification of a molecule as an oxyradical depends on the presence of one or more unpaired electrons associated with an oxygen atom. Molecular oxygen itself is a radical since it has two unpaired orbital electrons. However, because these electrons are in different orbits and because they spin in a parallel fashion, molecular oxygen is not highly reactive (1–3). ROI often termed reactive oxygen species or oxyradicals) are produced by the single electron reduction of O_2. For example, when O_2 accepts a single electron into its orbit, it forms the superoxide anion radical ($^{\cdot}O_2$). Further sequential single electron reduction generates hydrogen peroxide (H_2O_2), the highly reactive hydroxyl radical ($^{\cdot}OH$), and eventually water. Superoxide anion radical, H_2O_2 and the hydroxyl radical are the key ROI found in biological systems. Recently, interest has arisen in another type of reactive oxygen intermediate, those associated with nitrogen. The oxygen bound to nitrogen in a variety of molecules can undergo reduction forming $NO_x^{\cdot}$. This type of oxyradical is often termed a reactive nitrogen intermediate and is exemplified by nitric oxide (NO), the most well known of these highly reactive species (4).

Cellular Sources of ROI and RNI

All cells have the capability of generating ROI through enzymatic or nonenzymatic means, and Figure 1 provides a partial view of these mechanisms. Key ROI generating enzymes include a series of oxidoreductases such as the cytochromes involved in the transfer of electrons from O_2 to water in the mitochondria. Early in vitro studies on isolated mitochondria indicated that as much as 3–7% of the O_2 utilized in this process is leaked from the electron transport chain as superoxide anion radical or H_2O_2 (3, 5). Although it is unclear how much ROI is produced by mitochondria in intact cells, it is generally believed that this pathway serves as a major intracellular source of superoxide anion radical and hydrogen peroxide. Other heme-containing oxidases have been well described and include the membrane-bound NADPH oxidase cytochrome b_{558}

259

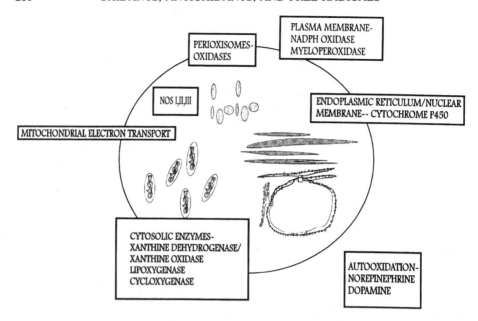

Figure 1. Sources of reactive oxygen and reactive nitrogen intermediates.

associated with phagocytic cells (6, 7) and the microsomal or mitochondrial P-450 oxidoreductases that participate in the mixed-function oxidase system (8, 9). The NADPH oxidase is commonly associated with cells of the reticular endothelial system such as neutrophils, monocytes, and tissue macrophages and is the enzyme responsible for the "respiratory burst" seen during activation of these cells (10). This enzyme generates large quantities of superoxide anion radical compared to other biological sources of superoxide anion and is used in the immune defense of the tissue (11). Cytochrome P-450 is a generic term for a large family of monooxygenases that function in the oxidation of low-molecular-weight xenobiotics (exogenous substances such as drugs or environmental chemicals). This family of enzymes is made up of heme-containing proteins that are thought to be the major source of microsomal H_2O_2 via the initial production of superoxide anion and its subsequent dismutation (8, 9, 12, 13). Other enzymes that may serve as important sources of ROI are lipoxygenase, cyclooxygenase, and xanthine oxidase. Superoxide anion is produced as a by-product in the metabolism of arachidonic acid, a membrane-derived fatty acid, by either lipooxygenase or cyclooxygenase (3, 14). Xanthine oxidase-mediated production of ROI occurs in the presence of O_2 and a product of adenosine metabolism, hypoxanthine (or xanthine). In turn, xanthine oxidase can be generated by proteolytic cleavage from a more common intracellular enzyme, xanthine dehydrogenase (1, 3, 13). Nonenzymatic sources of ROI include the autooxidation of a variety of molecules such as epinephrine.

Sources for RNI are also shown in Figure 1 and are primarily enzymatic in nature, although nonenzymatic mechanisms are known (4). There are three types of nitric oxide synthase (NOS), originally named according to their cellular location and currently termed NOS I (neuronal or nNOS), NOS II (macrophage or inducible NOS, iNOS), and NOS III (endothelial or eNOS) (4, 15). In each case, NO is generated by the five-electron reduction of arginine. This reaction is catalyzed by NOS and is dependent on the presence of specific cofactors including NADPH, FMN, FAD, calmodulin, and biopterin.

Reactions Involving ROI and RNI

The reactions of ROI with a variety of molecules have been thoroughly described (3) and are beyond the scope of this review. However, several pathways are critical to understanding the actions of ethanol and involve the reaction of ROI/RNI with transition metals such as iron or copper, the formation of lipid peroxides and alkoxyl radicals, the formation of oxidized proteins, and the modification of DNA.

Interaction with Metals

Transition metals such as iron and copper are well known to be closely involved in the production of ROI. These reactions depend on the presence of "free" metal, that is, metal that is capable of redox reactions and that can be diffusible (16, 17). The exact level of free iron or copper inside cells has not been determined and, in fact, most of this pool may be chelated to low-molecular-weight molecules such as ascorbic acid. Chelation does not prevent the metal from participating in ROI production, and the overall degree of reactivity is dependent on the type of chelating agent. A larger store of "bound" iron is available and various proteins such as ferritin, hemosiderin, and transferrin are known to sequester iron (17). Not all bound iron is nonreactive, however. For example, the iron found in certain heme-containing biological molecules such as hemoglobin undergoes redox cycling in a site-specific manner.

The production of ROI by iron-mediated processes is known generally as Fenton chemistry (Eqs. 1–3). Essentially, superoxide anion radical, which is produced by any or all of the biological processes discussed thus far, can participate in two reactions. In the first reaction (Eq. 1), superoxide anion radical acts as a reducing agent to convert Fe^{3+} to Fe^{2+} plus O_2.

$$Fe^{+++} + {}^{\cdot}O_2{}^- \rightarrow Fe^{++} + O_2 \tag{1}$$

Superoxide anion radical can also undergo a dismutation reaction where superoxide anion plus H^+ ion produces H_2O_2 (Eq. 2).

$$2\,{}^{\cdot}O_2{}^- + 2\,H^+ \rightarrow H_2O_2 + O_2 \tag{2}$$

This is a major source of H_2O_2 within the cell, and although this reaction occurs spontaneously in an aqueous environment, it can be catalyzed by the enzyme superoxide dismutase (SOD). In turn, the H_2O_2 that is generated can react with Fe^{2+} plus H^+ to generate the hydroxyl radical, Fe^{3+}, and water (Eq. 3). These reactions are frequently combined into the Haber-Weiss reaction as shown next:

$$Fe^{++} + H_2O_2 + H^+ \rightarrow Fe^{+++} + {}^{\cdot}OH + H_2O \tag{3}$$

$$H_2O_2 + {}^{\cdot}O_2{}^- \rightarrow {}^{\cdot}OH + O_2 + OH^- \tag{4}$$

Lipid Peroxidation and Protein Oxidation

One of the most common actions of ROI in cellular systems is the extraction of hydrogen from susceptible membrane lipids. This reaction can be divided into three steps: the initial production of alkoxyl or peroxyl radicals by ${}^{\cdot}OH$, the propagation of these reactions by the formation of secondary lipid radicals, and the eventual termination of the radical chain reaction by the donation of H^+ from antioxidants such as vitamin E (2). The process is termed lipid peroxidation and is associated with the oxidative degradation of the membrane. Lipids with polyunsaturated fatty acids (PUFAs) are particularly susceptible targets, and molecules with multiple double bonds are often lost during lipid peroxidation. These oxidative changes can result in changes in membrane fluidity, in membrane permeability, and in the function of membrane-bound transport systems (3).

Intracellular and membrane-bound proteins are also attacked by ROI and RNI, producing functional changes in the molecules. For example, carbonyl groups can be added to a protein during oxidation, especially in the presence of transition metals. This effect of ROI can be used as a marker of oxidative damage, and the protein carbonyl content is frequently used to provide a relative index of intensity of the damage (18). Other oxidative changes include the effect of ROI on certain amino acids containing sulfhydryl groups. Amino acids such as cysteine can be cross-linked or the SH- group can be directly oxidized. This type of modification is often associated with inactivation of the protein containing these groups (1, 3, 18). The SH- groups can also react with NO, forming S-nitrosothiol compounds. One of the most common targets of NO-mediated damage is sulfur-centered iron-containing proteins such as aconitase (4).

DNA Damage

The ability of ROI and RNI to produce long-term changes in cellular function resides, in part, with the effect of these reactive molecules on DNA. Like the events seen during lipid peroxidation, ROI can abstract H^+ from the deoxyribose sugars of the DNA molecule as well as interact with the purine and pyrimidine bases, forming single-strand breaks (19, 20). In some cases, the severity of the DNA breaks can lead to further damage because of the depletion of NAD^+ during the attempt to repair the break (3). NO also affects DNA by inactivation of the DNA repair enzyme formamidopyrimidine-DNA glycolyase (21). This protein is a zinc-finger enzyme and is an example of the type of protein inactivated by NO-mediated S-nitrosylation.

PROTECTIVE MECHANISMS

It is clear that there are multiple cellular sources of ROI and RNI and that these highly reactive molecules can produce changes in the function and structure of a cell. To protect against and mitigate the oxidative damage caused by exposure to these agents, the cell utilizes an antioxidant protection system that includes both enzymatic and nonenzymatic mechanisms. Table 1 provides a list of the enzymes involved in this process and their substrates, including SOD, catalase, and glutathione peroxidase. The levels of these enzymes vary from tissue to tissue and are under translational and transcriptional control (1, 2, 22, 23). A second class of antioxidants is represented by the presence of scavenging or chain-terminating agents such as ascorbic acid or vitamin E (2). These molecules provide H^+ to lipid radicals, and redox cycling of these molecules is necessary to regenerate their antioxidant capabilities. One of the most important intracellular antioxidants is the tripeptide glutathione (Figure 2). The SH group of glutathione is readily oxidized to -S-S- by ROI and is re-reduced by the enzyme glutathione reductase (24). Ordinarily, the reduced form of glutathione (GSH) predominates in cells such that the GSH/GSSG ratio is high (1, 24). Oxidative damage can shift this ratio. By preferentially attacking GSH, important protein SH groups are protected from the action of ROI. GSH also rapidly reacts with NO and minimizes the formation of damaging nitrosyl compounds both inside and outside the cell (25, 26, 99).

Interestingly, NO can act to protect against oxidative damage and is now considered to be part of the antioxidant protection mechanisms available to cells (21, 26, 27). This is due, in part, to the rapid reaction of NO with a variety of oxyradicals, forming less reactive molecules and thus reducing the oxidative attack. Rubbo et al. (28) have elegantly demonstrated that the level of oxidative stress is highest when the flux of superoxide anion is greater than the flux of NO and is lowest when NO flux is high. In other words, oxidative stress is directly related to superoxide anion flux and inversely related to NO flux.

ETHANOL-INDUCED ROI

Increased Production of ROI

One of the earliest observations on the role of ROI in ethanol-induced toxicity was made by Di Luzio and Hartman (29), who demonstrated that lipid peroxidation of rat liver homogenates was

Table 1. Typical antioxidants found in most cells

Antioxidant	Location	Substrate/Action
Cu/Zn SOD	Cytosol	Superoxide anion
Mn-SOD	Mitochondria	Superoxide anion
GSH peroxidase	Cytosol	Hydrogen Peroxide
Catalase	Cytosol	Hydrogen Peroxide
Glutathione reductase	Cytosol	Oxidized glutathione
Glutathione-S-transferase	Cytosol	Organic peroxides
Glutathione	Mitochondria/Cytosol/ Extracellular fluid	Protects-SH groups Scavenges oxyradicals
Ascorbic Acid	Cytosol/extracellular fluid	Scavenges oxyradicals
Vitamin E	Cell membranes	Terminates lipid radical chain reactions
Nitric Oxide	Cytosol/extracellular fluid	Promotes nitrosylation instead of oxidation

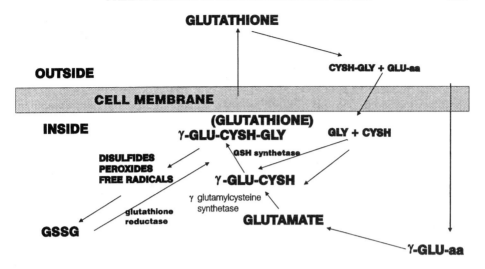

Figure 2. Antioxidant glutathione role in free-radical scavenging.

increased with ethanol treatment. A number of studies have since confirmed this finding and have indicated that the production of ROI by ethanol-treated tissues can account for the oxidative damage (13, 30–35). A recent study demonstrated that ethanol increases the production of free radicals in vivo. When acute doses of ethanol (0.2–0.6 g/kg) were given to healthy volunteers, a dose-dependent increase in isoprostanes—a free-radical—catalyzed product of arachidonic acid was detected in urine (36).

Possible Sources of Alcohol-Induced ROI

The cause of the increased ROI production has been the subject of extensive scrutiny, and several specific mechanisms have been proposed. One of the most prevalent sources of ROI during chronic treatment with ethanol is cytochrome P-4502E1 (CYP2E1). This enzyme system is induced by ethanol, is dependent on NADPH as a cofactor (9, 30, 31, 34), and is very reactive in producing $^{\cdot}O_2^-$ and H_2O_2 during NADPH oxidation (37). It is located on microsomes in most cells and catalyzes the production of acetaldehyde from ethanol. Because of its kinetic properties, cytochrome P-4502E1 produces superoxide anion radical at a high rate (37). Dai et al. (12) and others (31, 37) have demonstrated that the increased activity is maintained when the enzyme is either transiently or stably transfected into cultured cells. Although $^{\cdot}O_2^-$ is a principal product, H_2O_2 and $^{\cdot}OH$ are also formed via Fenton chemistry. Thus, the lipid peroxidation seen during chronic ethanol exposure is due, at least in part, to $^{\cdot}OH$-mediated attack on membranes. Not all ethanol-mediated lipid peroxidation appears to be dependent on ROI. Under certain conditions, cytochrome P-4502E1 can act as a peroxidase, producing $^{\cdot}OH$-independent lipid peroxidation (13, 37). In addition, other peroxidase-like enzymes in the cytochrome P-450 family are altered by exposure to ethanol including cytochrome P-4502B (13, 35, 37). In general, the activity of this multifunctional family of oxidoreductases is increased by ethanol. Other enzymes that function in a similar manner to cytochrome P-450 are affected by ethanol, including an NADH-dependent reductase found in mitochondria and oxidoreductases in nuclear or microsomal membranes (13, 31, 34). Other possible sources of ROI after ethanol administration are the cytosolic oxidases (aldehyde oxidase and xanthine oxidase) and the activation of inflammatory cells (e.g., Kupffer cells) (38). Chronic alcohol administration also increased the production of $^{\cdot}NO$, reflecting an increase in the activity of iNOS in hepatocytes, Kupffer cells, and vascular endothelial cells (39). The free radical NO could react with $^{\cdot}O_2^-$ to form peroxynitrite ($ONOO^-$), which can oxidize or nitrate lipids and proteins (40) and oxidize sulfhydryl groups (41).

The direct action of ethanol on ROI production has been clearly implicated in ethanol toxicity; however, several studies have demonstrated a role for acetaldehyde in the oxidative damage. Ethanol metabolism by alcohol dehydrogenase to acetaldehyde results in the generation of NADH, potentially shifting the NAD^+/NADH ratio (13). Changes in reducing equivalents have been proposed to favor an enhancement of the electron transport chain in mitochondria, thus increasing the "leak" of ROI from this mechanism (13). Alternatively, Mira et al. (33) have demonstrated that xanthine oxidase will metabolize acetaldehyde to acetate, resulting in the generation of superoxide anion as a by-product of this reaction. This pathway for ROI generation is supported by the fact that inhibitors of xanthine oxidase reduced ethanol-induced lipid peroxidation (35). In addition, the ethanol-mediated conversion of xanthine dehydrogenase to xanthine oxidase has been demonstrated in several tissues. The overall significance of this mechanism as a source of ROI during ethanol-induced toxicity has been challenged, however, based on the unfavorable kinetics of the reaction of xanthine oxidase with acetaldehyde (13, 35).

ROLE OF ROI IN ALCOHOL-INDUCED TISSUE INJURY

Liver Damage

Numerous mechanisms have been postulated for ethanol-induced liver damage, including changes in signal transduction, in mitochondrial function, in eicosanoid metabolism, or in redox state, as well as formation of acetaldehyde adducts, induction of hypoxia and reperfusion, induction of CYP2E1, or activation of Kupffer or Stellate cells. In addition to metabolizing many hepatotoxins, CYP2E1 oxidizes alcohol to acetaldehyde and forms reactive oxygen species and hydroxy radicals ($^{\cdot}$OH) (42, 43). The increase in these reactive oxygen species is blocked by anti-CYP2E1 immunoglobulin G (IgG) (43). Since ethanol induces CYP2E1 in vivo and in HepG2 cell line (44), interest in the role of reactive oxygen species in the genesis of liver injury has been intensified. In the rat model for alcohol-induced liver injury, the increase in microsomal lipid peroxidation and CYP2E1 levels correlated with the extent of liver damage (45), which was decreased by the oral administration of 1,2-dimethyl-3-hydroxypyrid-4-one, an iron chelator (46). Additionally, inhibition of CYP2E1 by DAS partially prevented alcohol-induced liver injury (47). CYP2E1-derived superoxide radicals increase the mobilization of iron from ferritin, which catalyses lipid peroxidation. Ferritin was much more effective in stimulating lipid peroxidation of microsomes from rats chronically fed ethanol as compared to control animals, and this stimulation was decreased by anti-CYP2E1 IgG (48). Since ferritin is the major storage form of iron within cells, increased mobilization of iron may play a role in the development of oxidative stress after ethanol treatment.

Role of 1-Hydroxyethyl Radical (HR) in Ethanol-Induced Liver Injury

The liver microsomal system oxidized ethanol in vitro to HR, a free-radical metabolite, in rat (49) and deermice (50). HR was also detected in in vivo studies (51, 52). HR-protein adducts resulted in the formation of antibodies to the HR moiety; these antibodies were detected in animals (53, 98) and in alcoholics with liver cirrhosis (54).

Albano and his colleagues (55) concluded that HR is formed from ethanol by two pathways: (1) an $^{\cdot}$OH-dependent reaction that requires iron and is inhibited by catalase, and (2) an $^{\cdot}$OH-independent pathway that involves P-450. Others have suggested that HR is formed by an oxidant derived from the interaction of $^{\cdot}O_2^-$ with transition metals such as iron, and that H_2O_2 is important for HR formation (56). Human liver microsomes also catalyzed HR formation with either NADH or NADPH as cofactor, and the addition of the chelator Chelex-100 inhibited HR formation, suggesting that iron is required to catalze this reaction (57). HR formation was nearly completely inhibited by catalase and superoxide dismutase with either NADH or NADPH as reductant. The effectiveness of NADH may be important in view of the fact that ethanol oxidation by alcohol dehydrogenase increases NADH/NAD^+ redox ratio in the liver; indeed the NADH-dependent production of ROS was increased after chronic ethanol treatment (31). Cytochrome b_5 seems to play a role in HR formation, probably due to its effects on superoxide production (57).

Macrophages

Phagocytic cells such as macrophages also contain an NADPH oxidase that is susceptible to ethanol. As mentioned previously, NADPH oxidase is a flavocytochrome (cytochrome b_{558}) and is capable of producing large quantities of ROI when stimulated with immune or inflammatory signals (6, 7, 11). Studies on alveolar macrophages and Kupffer cells have demonstrated that ethanol alters superoxide anion production in these cells (58–61). The effect is varied, however, and depends on the duration of exposure, the concentration of alcohol, and the type of agent used to induce ROI production. Both a decrease and an increase in stimulated superoxide anion production have been seen (58, 59, 61, 62). Nonstimulated (resting) levels of superoxide anion have been shown to rise within 5 h but then to return to baseline at later times (62). It is not known if this change in resting level of superoxide anion is due to the NADPH oxidase or to the activation of other enzymes affected by ethanol. Macrophage properties such as phagocytosis are also altered by ethanol and, like superoxide anion production, have been increased or decreased (60, 63), depending on the time of exposure to ethanol.

In contrast to CYP2E1 and the oxidoreductases just described, recent studies have shown that NO production by NOS, also a cytochrome P-450-like enzyme (64), is generally inhibited by exposure to ethanol. For example, acute exposure to ethanol suppressed the induction of iNOS mRNA by lipopolysaccharide (LPS) in rat alveolar macrophages (30). However, tissue specificity is evident in the action of ethanol on NO production, since an increased production of NO is found in liver perfusates from rats chronically treated with ethanol (39). Overall, the depression of NO by ethanol combined with the enhanced production of superoxide anion favors oxidative stress. This may be more critical in humans compared to the more well-studied rodent models because of the inherently poor iNOS response found in human macrophages (15, 65, 66). Thus, the ethanol-induced depression of NO seen in the lungs of humans (67) may exacerbate the damage produced by ROI.

Fetal Alcohol Syndrome

Exposure of the fetus to alcohol during early development sometimes results in a condition known as fetal alcohol syndrome or FAS, which was first described in the early 1970s. Numerous studies have advocated various mechanisms for FAS: direct inhibition of the transport of both amino acids and glucose through the placental tissue, defect in the metabolism of folic acid, hypoxia of fetal tissues (68–70), and increase in prostaglandin, among others. Diminished energy production and the formation of free radicals following hypoxia are believed to underlie cellular damage and the ensuing teratogenic effects. Recent studies have demonstrated that alcohol induces excessive cell death through the formation of free radicals in premigratory neural crest cells, which are deficient in superoxide dismutase (SOD). Amelioration of the ethanol-induced cell death and subsequent malformation is achieved in vivo by adding exogenous SOD (71). Zinc is essential for the activity of SOD, and neural crest cells are also sensitive to maternal zinc deficiency (72, 88), which is usually encountered in alcoholics.

Brain Damage

Chronic consumption of relatively large amounts of alcohol can lead to brain damage and cognitive deficits. Increasingly, ROI could contribute to the actions of ethanol in the brain. For example, Gonthier and colleagues (73) found that exposure of brain microsomes to ethanol results in the formation of hydroxyethyl radicals. Chronic ethanol administration to rodents decreased the concentration of α-tocopherol and glutathione (74) and increased the concentration of low-molecular-weight iron (75) in the brain. Iron can potentially react with H_2O_2 to generate the highly reactive hydroxyl radical ($^{\cdot}OH$). Acute exposure to ethanol has also been associated with decreases in SOD, ascorbate, and selenium (76, 77). In C6 glioma cells, exposure to ethanol for 24 h or for 10 days reduced the accumulation of nitrite in the culture supernatants of lipopolysaccharide-stimulated cells (78). Neuronal NOS is also down-regulated by ethanol such that the N-methyl-D-aspartate-stimulated activation of NOS in cultured rat cerebral cortical neurons is depressed by acute exposure to ethanol (79). These results are intriguing and may suggest a role of ROI/RNI in ethanol-induced brain damage.

MECHANISMS OF ROI-INDUCED TISSUE INJURY

Although alcohol-induced oxidant stress and the consequent tissue injury have been described, little is known about the mechanisms by which ethanol induces tissue damage. Basic research suggests several possible underlying mechanisms, including (1) increased extracellular triggering agents, (2) altered intracellular signal transduction, (3) changes in fatty acid oxygenases and eicosanoids, (4) altered nuclear transcription and gene expression, and (5) loss of antioxidant protection.

Increased Extracellular LPS

Alcohol increases lipopolysaccharides (LPS) in serum. Once cells with CD-14 interact with LPS, the intracellular signaling proceeds rapidly by activating a specific tyrosine kinase that acts on phosphoprotein (p38), the nuclear factor NF-κB, and also the mitogen-activated protein kinases MAPK1 and MAPK2 (80), which participate in intracellular signal transduction that induces gene expression (81). Protein tyrosine kinase action is also important in activating protein kinase C (PKC), a major mediator in positive regulation of the NADPH oxidase-catalyzed formation of ROI and in negative regulation of the inducible nitric oxide synthase formation of NO (82). Kupffer cell activation by lipopolysacharide leads to the synthesis and release of many cytokines (tumor necrosis factor α, interleukins 1 and 6, transforming growth factor β1), and to the induction of platelet-activating factor (PAF) and ROI that lead to oxidant stress and enhancement of tyrosine kinase signaling pathways.

Altered Intracellular Signal Transduction

Positive feedback of tyrosine kinase action with ROS provides an important point of interest in the etiology of alcohol-induced liver disease. An ROS-enhanced accumulation of tyrosine phosphates may amplify the pathway for formation of more ROI. For example, activation of PKC and NADPH oxidase was accompanied by inactivation of protein tyrosine phosphatase and a sustained protein tyrosine kinase activity with massive cellular protein tyrosine phosphorylation (83). The ROI-induced loss of protein tyrosine phosphatase (PTP) activity represents an important way in which ROI can drastically alter cell physiology independent of lipid peroxidation or cell death. Another related protein phosphatase, PP2B, is inhibited by ROI in a similar manner, leading to sustained phosphorylation and activation of phospholipase A_2, NADPH oxidase, MAPK cascade, and phosphotyrosine cinase. These tyrosine kinase signaling pathways appear to play important roles in signals participating in the development of alcoholic liver disease.

Fatty Acid Oxygenases and Eicosanoids

In addition to enhancing tyrosine phosphate or PAF signals, ROI can serve as intercellular agents to activate eicosanoid signaling mechanisms that mediate pathophysiological tissue responses (84). Cytokines trigger protein kinase signaling cascades that increase cellular levels of phosphorylated proteins and activate transcription of several important genes including cytosolic phospholipase A_2, which catalyzes the accumulation of the nonesterified fatty acid (NEFA) precursors of eicosanoids and the various "lyso-phospholipids," of which the 1-O-alkyl type can be acetylated to form PAF. PAF acetylhydrolase usually prevents the accumulation of appreciable PAF, but its rapid inactivation by peroxides permits PAF to accumulate to pathophysiological levels. PAF amplifies cellular events, increasing intracellular calcium levels, phospholipase action, eicosanoid synthesis, and the production and release of more cytokines, which, in turn, enhance the recruitment of more inflammatory cells, resulting in liver injury.

Prostaglandins, thromboxanes, and leukotrienes are formed from the accumulated NEFA by the action of cyclooxygenases and lipoxygenases; that action requires the presence of sufficient peroxide tone. Peroxides also inactivate protein tyrosine phosphatases (PTP), permitting the accumulation of active tyrosine phosphate derivatives, and inactivate IMB to permit more NF-κB to enter the nucleus and activate transcription of inflammatory genes.

Altered Nuclear Transcription and Gene Expression

The ROI formed in inflammatory/proliferative conditions may also mediate tumor necrosis factor α (TNFα) activation of NF-κB (85), transforming growth factor (TGF) β1 induction of early growth response-1 gene (86), or activate ribonucleotide reductase (87) that is essential for the DNA synthesis needed in rapid cell proliferation. The complexity of these pleiotropic actions of ROI makes it essential to understand how alcohol affects cellular signaling, and to appreciate the possibility that lipid peroxidations may be only markers rather than mediators of ROI action (89). Recent studies have shown that alcohol affects NF-κB and AP-1 transcription processes (for review see ref. 90).

Loss of Antioxidant Protection

Oxidative stress is not only due to an increased ROI but can also result from the loss of antioxidant protection mechanisms. Ethanol decreased the levels of key cellular antioxidants such as glutathione (GSH) in rat liver (91). However, other studies have reported increases, decreases, or no change in GSH content (13, 35, 88, 92–95). These findings are consistent in their inconsistency, since antioxidant levels often rise when a cell is challenged by oxidative stress and then commonly fall when the protective mechanisms become depleted or are themselves damaged (3, 23). In addition, part of the discrepancy may be due to differences in the animal and tissue studied, the nutritional status of the animal, and the concentration and duration of ethanol treatment. Overall, when compared to nonalcoholics, chronic ethanol consumption in humans is associated with a fall in liver GSH that is independent of nutritional state (35). A variety of mechanisms have been proposed to account for this fall, including the unrelenting production of ROI, the inhibition of GSH transport or production, and the direct interaction of GSH with acetaldehyde (13, 35).

Other antioxidants have also been extensively studied in both acute and chronic ethanol treatment, including SOD, catalase, glutathione peroxidase, and vitamin E (13, 35, 96, 97). Similar discrepanicies are found in the levels of these antioxidants as for GSH. Overall, the adaptive mechanisms of the cell to continued oxidative stress are critical to the degree of oxidative damage. Furthermore, many antioxidants depend on dietary factors such as selenium (for glutathione peroxidase) and zinc (for SOD) or may be dependent on dietary intake for maintenance of the proper levels, like vitamin E. Compromise in nutritional status can then be an important factor in the ROI-induced damage during ethanol toxicity (22, 96).

REFERENCES

1. Cohen, G., and Werner, P. 1993. Free radicals, oxidative stess and neurodegeneration. In *Neurodegenerative Disease*, ed. D. Calne, pp. 139–161. Philadelphia: W. B. Saunders.
2. Buettner, G. 1993. The pecking order of free radicals and antioxidants: Lipid peroxidation, α-tocopherol and ascorbate. *Arch. Biochem. Biophys.* 300:535–543.
3. Halliwell, B., and Gutteridge, J. 1989. *Free Radicals in Biology and Medicine*. Oxford: Oxford University Press.
4. Moncada, S., and Higgs, E. 1991. Endogenous nitric oxide: Physiology, pathology and clinical relevance. *Eur. J. Clin. Invest.* 21:361–374.
5. Ramasarma, T. 1982. Generation of H_2O_2 in biomembranes. *Biochim. Biophys. Acta* 694:69–93.
6. Babior, B., Kipnes, R., and Curnutte, J. 1973. The production by leukocytes of superoxide, a potential bactericidal agent. *J. Clin. Invest.* 52:741–744.
7. Segal, A., and Abo, A. 1993. The biochemical basis of the NADPH oxidase of phagocytes. *Trends Biochem. Sci.* 18:43–47.
8. Karuzina, I., and Arachakov, A. 1994. The oxidative inactivation of cytochrome P450 in monooxygenase reactions. *Free Radic. Biol. Med.* 16:73–97.
9. Bondy, S., and Naderi, S. 1994. Contribution of hepatic cytochrome P450 systems to the generation of reactive oxygen species. *Biochem. Pharmacol.* 48:155–159.
10. Babior, B. 1984. Oxidants from phagocytes: Agents of defense and destruction. *Blood* 64:959–966.
11. Nathan, C. 1987. Secretory products of macrophages. *J. Clin. Invest.* 79:319–326.
12. Dai, Y., Rashba-Step, J., and Cederbaum, A. 1993. Stable expression of human cytochrome P4502E1 in HepG2 cells: Characterization of catalytic activities and production of reactive oxygen intermediates. *Biochemistry* 32:6928–6937.

13. Nordmann, R., Ribiere, C., and Rouach, H. 1992. Implication of free radical mechanisms in ethanol-induced cellular injury. *Free Radic. Biol. Med.* 12:219–240.

14. Roy, P., Roy, S., Mitra, A., and Kulkarni, A. 1994. Superoxide generation by lipoxygenase in the presence of NADH and NADPH. *Biochim. Biophys. Acta* 1214:171–179.

15. Denis, M. 1994. Human monocytes/macrophages: NO or no NO? *J. Leukocyte Biol.* 55:682–684.

16. Aust, S., Morehouse, L., and Thomas, C. 1985. Role of metals in oxygen radical reactions. *Free Radic. Biol. Med.* 1:3–25.

17. Halliwell, B., and Gutteridge, J. 1986. Iron and free radical reactions: Two aspects of antioxidant protection. *Trends Biochem. Sci.* 11:372–376.

18. Starke-Reed, P., and Oliver, C. 1989. Protein oxidation and proteolysis during aging and oxidative stress. *Arch. Biochem. Biophys.* 275:559–567.

19. Halliwell, B. 1991. Drug antioxidant effects. *Drugs* 42:569–605.

20. Breen, A., and Murphy, J. 1995. Reaction of oxyl radicals with DNA. *Free Radic. Biol. Med.* 18:1033–1077.

21. Wink, D., and Laval, J. 1994. The Fpg protein, a DNA repair enzyme, is inhibited by the biomediator nitric oxide in vivo and in vivo. *Carcinogenesis* 15:2125–2129.

22. Draper, H., and Bettger, W. 1994. Role of nutrients in the cause and prevention of radical pathology. In *Free Radicals in Diagnostic Medicine*, ed. D. Armstrong, pp. 269–289. New York: Plenum Press.

23. Sies, H. 1991. Oxidative stress: From basic research to clinical application. *Am. J. Med.* 91(suppl. 3C):31S–38S.

24. Meister, A. 1983. Selective modification of glutathione metabolism. *Science* 220:470–477.

25. Wink, D., Cook, J., Krishna, M., Hanbauer, I., DeGraff, W., Gamson, J., and Mitchell, J. 1995a. Nitric oxide protects against alkyl peroxide-mediated cytotoxicity: Further insights into the role nitric oxide plays in oxidative stress. *Arch. Biochem. Biophys.* 319:402–407.

26. Wink, D., Cook, J., Pacelli, R., Liebmann, J., et al. 1995b. Nitric oxide (NO) protects against cellular damage by reactive oxygen species. *Toxicol. Lett.* 82–83:221–226.

27. Kanner, J., Harel, S., and Granit, R. 1992. Nitric oxide, an inhibitor of lipid oxidation by lipoxygenase, cyclcooxygenase and hemoglobin. *Lipids* 27:46–49.

28. Rubbo, H., Radi, R., Trujillo, M., Telleri, R., Kalyanaraman, B., Barnes, S., Kirk, M., and Freeman, B. 1994. Nitric oxide regulation of superoxide and peroxynitrite-dependent lipid peroxidation. *J. Biol. Chem.* 269:26066–26075.

29. Di Luzio, N., and Hartman, A. 1967. Role of lipid peroxidation in the pathogenesis of ethanol-induced fatty liver. *Fed. Proc.* 26:1436–1442.

30. Greenberg, S., Xie, J., Wang, Y., and Kolls, J. 1994. Ethanol suppresses LPS-induced mRNA for nitric oxide synthase II in alveolar macrophages in vivo and in vitro. *Alcohol* 11:539–547.

31. Kukielka, E., Dicker, E., and Cederbaum, A. 1994. Increased production of reactive oxygen species by rat liver mitochondria after chronic ethanol treatment *Arch. Biochem. Biohys.* 309:377–386.

32. Masuda, E., Kawano, S., Nagano, K., Tsuji, S., et al. 1995. Endogenous nitric oxide modulates ethanol-induced gastric mucosal injury in rats. *Gastroenterology* 108:58–64.

33. Mira, L., Maia, L., Barreira, L., and Manso, C. 1995. Evidence for free radical generation due to NADH oxidation by aldehyde oxidase during ethanol metabolism. *Arch. Biochem. Biophys.* 318:53–58.

34. Montoliu, C., Valles, S., Renau-Piqueras, J., and Guerri, C. 1994. Ethanol-induced oxygen radical formation and lipid peroxidation in rat brain: Effect of chronic alcohol consumption. *J. Neurochem.* 63:1855–1862.

35. Guerri, C., Montoliu, C., and Renau-Piqueras, J. 1994. Involvement of free radical mechanism in the toxic effects of alcohol: Implications for fetal alcohol syndrome. In *Free Radicals in Diagnostic Medicine*, ed. D. Armstrong, pp. 291–305. New York: Plenum Press.

36. Mengher, E. A., Lucey, M. R., Bensinger, S., and Fitzgerald, G. A. 1996. Alcohol-induced formation of 8-Epi PGF2: An index of oxidant stress in vivo. *Prost. Leuko. FFA* 55(suppl. 1), September

37. Ekstrom, G., and Ingelman-Sundberg, M. 1989. Rat liver microsomal NADPH-supported oxidase activity and lipid peroxidation dependent on ethanol-inducible cytochrome P-450 (P450IIE1). *Biochem. Pharmacol.* 38:1313–1319.

38. Rosser, B. G., and Gores, G. J. 1995. Liver cell necrosis: Cellular mechanism and clinical implications. *Gastroenterology* 108:252–275.

39. Wang, J., Greenberg, S., and Spitzer, J. 1995. Chronic alcohol administration stimulates nitric oxide formation in the rat liver with or without pretreatment by lipopolysaccharide. *Alcohol Clin. Exp. Res.* 19:387–393.

40. Beckman, J. S., Ye, Y. Z., Anderson, P. G., Chen, J., Accavitti, M. A., Tarpey, M. M., and White, C. R. 1994. Extensive nitration of protein tyrosines in human atherosclerosis detected by immunohistochemistry. *Biol. Chem. Hoppe-Seyler* 375:81–88.

41. Beckman, J. S., Beckman, T. W., Chen, J., Marshall, P. A., and Freeman, B. A. 1990. Apparent hydroxyl

radical production by peroxynitrite: Implications for endothelial injury from nitric oxide and superoxide. *Proc. Natl. Acad. Sci. U.S.A.* 87:1620–1624.

42. Cederbaum, A. I. 1991. Microsomal generation of reactive oxygen species and their possible role in alcohol hepatotoxicity. *Alcohol Alcoholism Suppl.* 1:291–296.
43. Rashba-Step, J., Turro, N. J., and Cederbaum, A. I. 1993. Increased NADPH and NADH-dependent production of superoxide and hydroxyl radical by microsomes after chronic ethanol treatment. *Arch. Biochem. Biophys.* 300:401–408.
44. Carroccio, A., Wu, D., and Cederbaum, A. I. 1994. Ethanol increases content and activity of human cytochrome P4502E1 in a transduced hepG2 cell line. *Biochem. Biophys. Res. Commun.* 2031:727–733.
45. French, S. W., Wong, K., Jui, L., Albano, E., Habjork, A. K., and Ingelman-Sundberg, M. 1993. Effect of ethanol on cytochrome P4502E1 (CYP 2E1), lipid peroxidation, and serum protein adduct formation in relation to liver pathology pathogenesis *Exp. Mol. Pathol.* 58:61–75.
46. Sadzadeh, S. M. H., Nanji, A. A., and Price, P. L. 1994. The oral iron chelator 1,2-dimethyl-3-hydroxy-pyrid-4-one reduces hepatic free iron, lipid peroxidation, and fat accumulation in chronically ethanol-fed rats. *J. Pharmacol. Exp. Ther.* 269:632–636.
47. Morimoto, M., Hagbjork, A. L., Wan, Y. J. Y., Fu, P. C., Clot, P., Albano, E., Ingelman-Sundberg, M., and French, S. W. 1995. Modulation of experimental alcohol-induced liver disease by cytochrome P4502E1 inhibitors. *Hepatology* 21:1610–1617.
48. Kukielka, E., and Cederbaum, A. I. 1996. Ferritin stimulation of lipid peroxidation by microsomes after chronic ethanol treatment. *Arch. Biochem. Biophys.* 3321:121–127.
49. Reinke, L. A., Moore, D. R., Hague, C. M., and McCay, P. B. 1994. Metabolism of ethanol to 1-hydroxyethyl radicals in rat liver microsomes: Comparative studies with three spin-trapping agents. *Free Radic. Res.* 21:213–222.
50. Knecht, K. T., Thurman, R. G., and Mason, R. P. 1993. Role of superoxide and trace transition metals in the production of α-hydroxyethyl radical from ethanol by microsomes from alcohol dehydrogenase-deficient deermice. *Arch Biochem Biophys.* 303:339–348.
51. Knecht, K. T., Bradford, B. U., Mason, R. P., and Thurman, R. G. 1990. In vivo formation of free radical metabolite of ethanol. *Mol. Pharmacol.* 38:26–30.
52. Moore, D. R., Reinke, L. A., and McCay, P. B. 1995. Metabolism of ethanol to 1-hydroxyethyl radicals in vivo: Detection with intravenous administration of α-(4-pyridyl-1-oxide)-*N-t*-butylnitrone. *Mol. Pharmacol.* 47:1224–1230.
53. Albano, E., Parola, H., Comoglia, A., and Dianzani, M. U. 1993. Evidence for the covalent binding of hydroxyethyl radicals to rat liver microsomal proteins. *Alcohol* 28:453–459.
54. Clot, P., Bellomo, G., Tabone, M., Arico, S., and Albano, E. 1995. Detection of antibodies against proteins modified by hydroxyethyl free radicals in patients with alcoholic cirrhosis. *Gastroenterology* 108:201–207.
55. Albano, K. E., Tomasi, A., Persson, J. O., Terelius, Y., Goria-Gatti, L., Ingelman-Sundberg, M., and Dianzani, M. U. 1991. Role of ethanol-inducible cytochrome P-450 in catalyzing free radical activation of aliphatic alcohols. *Biochem. Pharmacol.* 41:1895–1902.
56. Reinke, L. A., Rau, J. M., and McCay, P. B. 1990. Possible roles of free radicals in alcoholic tissue damage *Free Radic. Res. Commun.* 9:205–211.
57. Rao, D. N., Yang, M. X., Lasker, J. M., and Cederbaum, A. I. 1996. 1-Hydroxyethyl radical formation during NADPH and NADH dependent oxidation of ethanol by human liver microsomes. *Mol. Pharmacol.* 49(5):814–821.
58. Bautista, A., and Elliott, K. 1994. Acute ethanol intoxication regulates F-Met-Leu-Phe induced chemotaxis and superoxide release by neutrophils and Kupffer cells through modulation of the formyl peptide receptor in the rat. *Life Sci.* 54:721–730.
59. Dorio, R., and Forman, H. J. 1988. Ethanol inhibition of signal transduction in superoxide production by rat alveolar macrophages. *Ann. Clin. Lab. Sci.* 18:190–194.
60. D'Souza, N., Bagby, G., Lang, C., Deaciuc, I., and Spitzer, J. 1993. Ethanol alters the metabolic response of isolated perfused rat liver to a phagocytic stimulus. *Alcoholism Clin. Exp. Res.* 17:147–154.
61. Iwata, S., Jamieson, D., and Chance, B. 1994. Ethanol stimulates chemiluminescence from neutrophils in the liver. *Free Radic. Biol. Med.* 17:35–43.
62. Libon, C., Forestier, F., Cotte-Laffitte, J., Labarre, C., and Quero, A. 1993. Effect of acute oral administration of alcohol on superoxide anion production from mouse alveolar macrophages. *J. Leukocyte Biol.* 53:93–98.
63. Zuiable, A., Wiener, E., and Wickramasinghe, S. 1992. In vivo effects of ethanol on the phagocytic and microbial killing activities of normal human blood monocytes and monocyte derived macrophages. *Clin. Lab. Haematol.* 14:137–147.
64. White, K., and Marletta, M. 1992. Nitric oxide synthase is a cytochrome P-450 type hemoprotein. *Biochemistry* 31:6627–6631.

65. Colton, C., Wilt, S., Snell, J., Gilbert, D., Cherynshev, O., and DuBois-Dalq, M. 1995. CNS sources of reactive oxygen species. *Mol. Chem. Neuropathol.*, in press.
66. Zembala, M., Siedlar, M., Marcinkiewicz, J., and Pryjma, J. 1994. Human monocytes are stimulated for nitric oxide release in vivo by some tumor cells but not by cytokines and lipopolysaccharide. *Eur. J. Immunol.* 24:435–439.
67. Persson, M., Cederqvist, B., Wiklund, C., and Gustafsson, L. 1994. Ethanol causes decrements in airway excretion of endogenous nitric oxide in humans. *Eur. J. Pharmacol.* 270:273–278.
68. Bronsky, P. T., Johnston, M. C., and Sulik, K. K. 1986. Morphogenesis of hypoxia-induced cleft lip in CL/Fr mice. *J. Craniofac. Genet. Dev. Biol. Suppl.* 2:113–128.
69. West, J. R., Chen, W. J. A., and Pantazis, N. J. 1994. Fetal alcohol syndrome: The vulnerability of the developing brain and possible mechanisms of damage. *Metabolic Brain Dis.* 9:291–322.
70. Abel, E. L., and Hannigan, J. H. 1995. Maternal risk factors in fetal alcohol syndrome: Provocative and permissive influences. *Neurotoxicol. Teratol.* 17:445–462.
71. Kotch, L. E., Chen, S. Y., and Sulik, K. K. 1995. Ethanol-induced teratogenesis: Free radical damage as a possible mechanism. *Teratology* 52:128–136.
72. Rogers, J. M., Taubeneck, M. W., Daston, G. P., Sulik, K. K., Zucker, R. M., Elstein, K. H., Jankowski, M. M., and Keen, C. L. 1995. Zinc deficiency causes apoptosis but not cell cycle alternations in organogenesis-stage rat embryos: Effects of varying duration of deficiency. *Teratology* 52:149–159.
73. Gonthier, B., Jeunet, A., and Barret, L. 1991. Electron spin resonance study of free radicals produced from ethanol and acetaldehyde after exposure to a Fenton system or to brain and liver microsomes. *Alcohol* 8:369–375.
74. Rouache, H., Houzé, P., Orfanelli, M. T., Gentil, M., and Nordman, R. 1991. Effect of chronic ethanol intake on some anti- and pro-oxidants in rat cerebellum. *Alcohol Alcoholism* 26:257–263.
75. Rouache, H., Houzé, P., Orfanelli, M. T., Gentil, M., Bourdon, R., and Nordman, R. 1990. Effect of acute ethanol administration on the subcellular distribution of iron in rat liver and cerebellum. *Biochem. Pharmacol.* 39:1095–1100.
76. Houzé, P., Rouache, H,. Gentil, M., Orfanelli, M. T., and Nordman, R. 1991. Effect of allopurinol on the hepatic and cerebral iron, selenium, zinc and copper status following acute ethanol administration to rats. *Free Radic. Res Commun.* 12:663–668.
77. Nordmann, R., Ribiere, C., and Rouach, H. 1990. Ethanol-induced lipid peroxidation and oxidative stress in extrahepatic tissues. *Alcohol Alcoholism* 25:231–237.
78. Syapin, P. 1995. Ethanol inhibition of inducible nitric oxide synthase activity in C6 glioma cells. *Alcohol Clin. Exp. Res.* 19:262–267.
79. Chandler, L., Guzman, N., Sumners, C., and Crews, F. 1994. Magnesium and zinc potentiate ethanol inhibition of N-methyl-D-aspartate-stimulated nitric oxide synthase in cortical neurons. *J. Pharmacol. Exp. Ther.* 271:67–75.
80. Han, J., Lee, J., Tobias, P. S., and Ulevitch, R. 1993. Endotoxin induces rapid protein tyrosine phosphorylation in 70Z/3 cells expressing CD14. *J. Biol. Chem.* 268:25009–25014.
81. Avruch, J., Zhang, X., and Kyriakis, J. 1994. Raf meets Ras: Completing the framework of a signal transduction pathway. *TIBS* 19:279–283.
82. Geng, Y., Wu, Q., and Hansson, G. 1994. Protein kinase C activation inhibits cytokine-induced nitric oxide synthesis in vascular smooth muscle cells. *Biochem. Biophys. Acta* 1223:125–132.
83. Goldman, R., Ferber, E., Meller, R., and Zor, U. 1994. A role for reactive oxygen species in zymosan and B-glucan induced protein tyrosine phosphorylation and phospholipase A_2 activation in murine macrophages. *Biochem. Biophys. Acta* 1222:265–276.
84. Lands, W., and Keen, R. 199. Peroxide tone and its consequences. In *Biological Oxidation Systems*, eds. C. C. Reddy, G. A. Hamilton, and K. M. Madyastha, Vol. 2, pp. 657–665. San Diego: Academic Press.
85. Menon, S. D., Qin, S., Guy, G. R., and Tan, Y. H. 1993. Differential induction of nuclear NF-κB by protein phosphatase inhibitors in primary and transformed human cells. Requirement for both oxidation and phosphorylation in nuclear translocation. *J. Biol. Chem.* 268:26805–26812.
86. Ohba, M., Shibanuma, M., Kuroki, T., and Nose, K. 1994. Production of hydrogen peroxide by transforming growth factor-beta and its involvement in induction of egr-1 in mouse osteoblastic cells. *J. Cell. Biol.* 126:1079–1088.
87. Fontecave, M., Gerez, C., Atta, M., and Jeunet, A. 1990. High valent iron oxo intermediates might be involved during activation of ribonucleotide reductase: Single oxygen atom donors generate tyrosyl radical. *Biochem. Biophys. Res. Commun.* 168:659–664.
88. Dreosti, I., and Partick, E. 1987. Zinc, ethanol and lipid peroxidation in adult and fetal rats. *Biol. Trace Elem. Res.* 14:179–191.
89. Lands, W. 1993. Eicosanoids and health. In *Third International Conference on Nutrition in Cardiocerebrovascular Diseases*, eds. K. Lee, Y. Oike, and T. Kanazawa. *Ann. N.Y. Acad. Sci.* 676:46–59.
90. Zakhari, S., and Szabo, G. 1996. NF-κB, a prototypical cytokine-regulated transcription factor: Implications for alcohol-mediated responses. *Alcoholism Clin. Exp. Res.* 20:236A–242A.

91. Calabrese, V., Ragusa, N., and Rizza, V. 1995. Effect of pyrrolidone carboxylate (PCA) and pyridoxine on liver metabolism during chronic ethanol intake in rats. *Int. J. Tissue React.* 17:15–20.
92. Burmistrov, S. O., Kotin, A. M., and Borodkin, Y. S. 1990. Changes in activity of antioxidative enzymes and lipid peroxidation levels in brain tissue of embryos exposed prenatally to ethanol. *Journal* 1:1748–1750.
93. Devi, B., Henderson, G., Frosto, T., and Schenker, S. 1994. Effect of acute ethanol exposure on cultured fetal rat hepatocytes: Relation to mitochondria function. *Alcohol Clin. Exp. Res.* 18:1436–1442.
94. Bondy, S., and Guo, S. 1995. Regional selectivity in ethanol-induced pro-oxidant events within the brain. *Biochem. Pharmacol.* 49:69–72.
95. Reyes, E., Ott, S., and Robinson, B. 1993. Effects of in utero administration of alcohol on glutathione levels in brain and liver. *Alcohol Clin. Exp. Res.* 17:877–881.
96. Bondy, S., and Pearson, K. 1993. Ethanol-induced oxidative stress and nutritional status. *Alcohol Clin. Exp. Res.* 17:651–654.
97. Bjorneboe, A., and Bjorneboe, G. E. 1993. Antioxidant status and alcohol-related diseases. *Alcohol Alcoholism* 28:111–116.
98. Moncada, C., Torres, V., Varghese, E., Albano, E., and Israel, Y. 1994. Ethanol-derived immunoreactive species formed by free radical mechanisms. *Mol. Pharmacol.* 46:786–791.
99. Wink, D., Hanbauer, I., Laval, F., Cook, J., Krishna, M., and Mitchell, J. 1994. Nitric oxide protects against the cytotoxic effects of reactive oxygen species. In *The Neurobiology of NO and OH*, eds. C. Chiueh, D. Gilbert, and C. Colton, pp. 265–278. New York: New York Academy of Sciences.

Chapter Sixteen

OXIDANT INJURY FROM INHALED PARTICULATE MATTER

James J. McGrath

Epidemiologic studies provide compelling evidence that a relationship exists between ambient particulate matter (PM) and human mortality and morbidity. The relationship has been described in several different regions of the world for differing periods of time and appears to be most intense for PM of 10 μm or less in diameter and for mortality and morbidity in older members of the population. The purpose of this chapter is to explore the relationship between inhaled atmospheric PM and the potential for oxidative injury to the lung.

PARTICLES

Particulate matter is a nonspecific term used to describe a broad class of physically and chemically diverse substances that exist as discrete particles (liquid droplets or solids) dispersed in the ambient atmosphere over a wide range of sizes. The mass concentration of PM normally ranges from a low of 10–20 μg/m³ after a rain shower to hundreds of micrograms per cubic meter in many urban environments, and to well over 1000 μg/m³ during high-pollution episodes (1). The U.S. Environmental Protection Agency (EPA) standard for atmospheric PM is 50 μg/m³ (annual arithmetic mean) or 150 μg/m³ for 24 h (2).

Atmospheric Concentrations

Worldwide, PM concentrations in the ambient atmosphere of cities are exceedingly high. This is especially true in the megacities, which are defined as urban agglomerations with current or projected populations of 10 million or more by the year 2000 (3). These cities include urban areas in all parts of the world but they are not necessarily the most polluted cities; the megacities encompass large land areas with many people and have serious air pollution problems. In many megacities, particulate pollution commonly exceeds EPA standards and reaches several hundred micrograms per cubic meter. Megacities with especially high ambient atmospheric particulate concentrations include Bangkok, Beijing, Bombay, Cairo, Calcutta, Delhi, Jakarta, Karachi, Manila, Mexico City, Seoul, and Shanghi. These cities have ambient PM concentrations consistently above the World Health Organization guidelines, with annual average PM concentrations typically in the range of 200–600 μg/m³ and peak concentrations frequently above 1000 μg/m³.

Particle Composition

The chemical and physical properties of atmospheric PM vary greatly with time and with regions of the world and are highly dependent on meteorology. There are, however, certain particulate species that are commonly found in most urban environments. Particles originate from a variety of sources (e.g., heating, transportation, industrial activities, combustion, photochemical processes, wind-blown dust, plants) and include acid sulfate particles, diesel particles, metal particles, asbestos and silica particles, ultrafine particles, and miscellaneous other particles. The combustion of coal and heavy fuel oil yields both fine particles, from material vaporized during combustion, and coarse particles (i.e., fly ash), from noncombustible material. Particulate matter in the ambient atmosphere may range in size from <10 nm to >100 μm in diameter and is described as coarse (>2.5 μm), fine (<2.5 μm), or ultrafine (<0.1 μm). Particles less than 10 μm are capable of penetrating deeply into the lung.

The characteristic size distribution of atmospheric PM is trimodal (Figure 1) (4), consisting of fine nuclei (<0.1 μm), an accumulation range from 0.1–2.5 μm, and a coarse range (>2.5 μm). Atmospheric PM consists of primary particles introduced directly into the air and secondary particles formed by gas-to-particle conversions in the atmosphere. Particles in the nuclei range are formed from gases by condensation and as primary particles in combustion. Metals vaporized in high-temperature processes, such as smelting, which coagulate or condense without chemical reaction, form primary particles. Particles in the nuclei size range have high diffusional mobility and, consequently, collide and adhere to one another. These particles grow by coagulation with other particles and enter the second size or accumulation range. In the accumulation range, particles may grow still further by condensation of additional material or further coagulation into particles in the 0.1–2.0 μm size range; particles of this size may remain airborne for hours or days. Course particles are formed by larger particles breaking up into smaller particles.

Metals are a common component of atmospheric PM. In New York City, 26 metals have been identified in the air in greater than nanogram concentrations (5). Several transitional metals are found in urban atmospheres in the United States (6) (Table 1).

Indoor Concentrations

Particulate matter is also found in indoor environments. In general, few coarse-mode ambient particles (>2.5 μm) penetrate into closed indoor air spaces. However, a significant concentration of fine-mode particles (<2.5 μm) can enter the indoor air (e.g., >60% in certain circumstances) from outdoor sources. Particles found indoors include a high percentage of sulfates derived from outdoor air and, at times, metals and other typical ambient air fine-mode particle constituents.

EPIDEMIOLOGIC STUDIES

Early air pollution episodes in London (1952), the Meuse Valley, France (1930), and Donora, Pennsylvania (1948), clearly demonstrated the potential for high levels of particulate-based smog to produce large increases in daily mortality (7). These results have been confirmed in more recent epidemiologic studies that suggest a direct relationship between increased PM and morbidity and mortality. In both the older and more recent reports, respiratory involvement is a common association between particulate pollution and health effects.

Recent Studies

In one of the earlier studies in London, in December 1952, an air pollution episode caused 3000–4000 excess deaths; the increase in death rate was most dramatic for those 45 yr and older and occurred most frequently in those with chronic lung disease and heart disease following pollution-induced cardiorespiratory problems. Continuing study of this and other early episodes provides strong evidence that ambient PM contributes to the mortality associated with high concentrations of urban aerosols and that these aerosols are dominated by combustion products (e.g., from burning coal) and their transformation products (e.g., H_2SO_4).

In more recent studies, epidemiologic evidence has linked small but significant increases in human morbidity and mortality with PM for different cities in the world. As in the earlier studies, the elderly (>65 yr old), particularly those with preexisting cardiopulmonary disease, have distinctly higher risks than the younger age groups. Changes in daily air pollution have been associated with daily mortality rate at much lower concentrations than in the London studies in Philadelphia, Steubenville, Santa Clara, St. Louis, the Utah Valley, Detroit, and eastern Tennessee

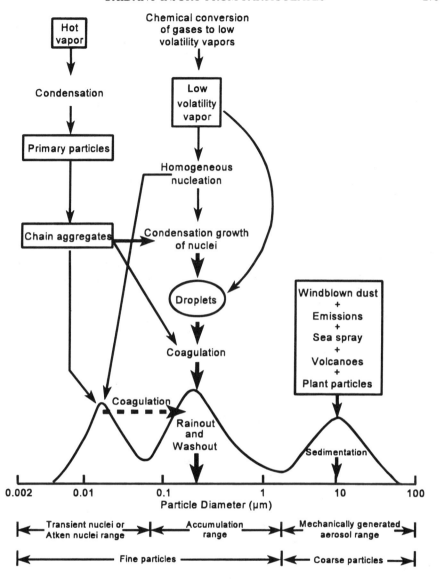

Figure 1. Schematic representation of the atmospheric aerosol showing the three major modes, sources, and removal processes for the modes and mechanisms of mass transfer between modes. From ref. 4, with permission from the IEEE.

in the 1970s and 1980s (8–11). Schwartz concluded that in Philadelphia the relative risk of dying on the high-pollution days was 1.08 and that the relative risk was highest for chronic obstructive pulmonary disease (1.25) and pneumonia (1.13) (7). Deaths also were elevated for heart disease and stroke, and respiratory factors frequently were reported as contributing causes of death. These results parallel the pattern seen in London in 1952.

PHYSIOLOGICAL RESPONSE

Between 10 and 20 m^3 of air is drawn into the body every day through the respiratory tract. The potential health effects of particles suspended in that air depend on their ability, in part, to

Table 1. Concentrations of metal associated with atmospheric particulate matter in New York City and in the United States

Metal	New York City[a]	United States[b]	Metal	New York City[a]	United States[b]
Iron	1292	130–13,800	Cadmium	6	0.2–1000
Copper	66	3–5140	Mercury	72	0.58–458
Chromium	31	2.2–124	Nickel	30	1–328
Vanadium	46	0.4–1460	Lead	1200	30–96,270

Note. Concentrations in nanograms per cubic meter.
[a] Data from Bernstein and Rahn (5).
[b] Data from Schroeder et al. (6).

penetrate the respiratory defenses and deposit, and on the body's ability to remove or neutralize them. The respiratory tract is divided into three regions: (1) the nasopharyngeal or extrathoracic region, (2) the tracheobronchial region or conducting airways, and (3) the pulmonary or gas-exchange region. Air entering the respiratory tract must pass, sequentially, through each of these anatomical zones. A sticky mucous layer that particles may come in contact with, and become attached to, coats the interior surfaces throughout the respiratory tract. In this way, a particle contacting the mucous surface can be removed from the air stream. The site and probability of a particle depositing are functions of its size, density, and shape but do not depend directly on its composition or physical state. The two most important characteristics of the aerosol are the airborne concentration and size, because these factors determine both the dose and site of deposition of the material (1).

Particle Deposition

The deposition of inhaled particles is affected by numerous factors including the anatomical and physiological characteristics of the subject. Ventilation rates and breathing patterns vary according to activity patterns and age in humans and other species. Differences in ventilation, coupled with differences in size, structure, branching patterns in the upper respiratory tract, and structure of the lower respiratory tract in both healthy and diseased subjects, cause significantly different patterns of airflow that affect particle deposition and clearance.

Particles may deposit on the mucous coating of the respiratory tract by inertial impaction, interception, sedimentation, or diffusion. Inertial impaction is a process by which inhaled particles (5–30 μm) deposit primarily in the nasopharyngeal region of the upper bronchial tree where airflow velocities and Reynolds numbers are high and the airstream changes direction abruptly due to turbulence. In this region of the respiratory tract, the larger particles often are unable to follow the moving air stream and impact the mucous-lined walls. Interception is a process by which particles of any size are removed in the airways by contacting the wall incidentally as the air stream in which they are traveling passes near an airway surface. This contact is not produced through inertia but occurs as a result of the shape, orientation, or rotation of a particle while in close proximity to the mucous surface. Sedimentation is a process by which particles larger than 0.1 μm that are suspended in a gas slowly settle out under the influence of gravity. Sedimentation is favored by slower air flows that allow time for gravitational forces to operate. Finally, diffusion is a process in which airborne particles are continuously bombarded by surrounding gas molecules (Brownian movement). This action induces a random motion in the particles that causes movement or diffusion of the particles from one region of a gas space to another. Deposition of airborne particles by diffusion is affected by particle diameter and residence time in the respiratory tract; decreasing particle diameter or increasing residence time, by breath holding, increases deposition by diffusion. Although diffusion, like sedimentation, occurs in all three respiratory zones, it is the dominant mechanism in the alveolar region.

Particle Clearance

Particulate matter deposited in the respiratory tract is cleared by mechanisms that depend on the site of deposition, on physicochemical properties of the particles, and on time elapsed since deposition. In the nasopharyngeal and tracheobronchial regions, deposited particles are cleared within hours, primarily by mucociliary transport and, presumably, the activity of airway macrophages. In this region, goblet cells produce a mucous blanket, which is continuously propelled away from the lungs by the beating cilia; PM adhering to this blanket is cleared from the airways in a matter of hours.

In the alveolar region, the primary defense against deposited PM and microorganisms is resident alveolar macrophages that, attracted to the site of particle deposition, phagocytize and transport the particles toward the mucociliary escalator (12–14). Through these mechanisms, alveolar macrophages can eliminate most particles and microorganisms from the distal airways and maintain the sterility of the alveoli. Although alveolar macrophage-mediated particle clearance is relatively slow (clearance rates are about 1% per day for the rat and 10 times higher in humans), rapid phagocytosis is important to prevent small particles from entering the interstitium (15). Once particles leave the surface and penetrate into the interstitium of the respiratory tract, their removal is slowed. In humans, although particles on the surface are cleared with biological half-times estimated to be from days to months, particles that penetrate into the interstitial spaces are cleared more slowly, with half-times ranging as high as thousands of days (16).

The precursors of alveolar macrophages are interstitial cells derived from circulating monocytes, which, in turn, arise from bone-marrow precursor cells. In the rodent, however, the alveolar macrophage can proliferate in vitro and in vivo, and in humans the alveolar macrophage can replicate in the lung (17).

The population of macrophages found throughout the respiratory tract is large and varied. Macrophages are found in the airways, the alveoli, the interstitial spaces, the intravascular spaces, and the pleura (16). In each of these areas, macrophages maintain the integrity and sterility of the respiratory tract through a complex array of functions including phagocytosis, release of reactive oxygen species (ROS) and proteolytic enzymes, and secretion of mediators for signaling. Within minutes after being inhaled, PM is found within alveolar macrophages. Within hours, many alveolar macrophages containing PM are transported out of the lung, and others can be observed in the interstitial spaces of the lung.

In addition to phagocytosis, macrophages clearly have other important roles (18). Macrophages can initiate and prolong inflammatory responses; control the activities of other cells, including neutrophils and lymphocytes; and secrete a variety of bioactive substances. More than 100 secretory products are released by macrophages. Included among the molecules released are ROS (superoxide anion, $O_2^{\cdot-}$; hydrogen peroxide, H_2O_2; hydroxyl radical, $\cdot OH$; singlet oxygen, $^1O_2$), lysosomal enzymes, interferon, components of complement, angiogenesis factor, plasminogen activator, cyclic nucleotides, prostaglandins, and granulopoietins.

The physiology of phagocytosis and the microbicidal mechanisms of alveolar macrophages, which are poorly understood, have been investigated most extensively in circulating polymorphonuclear leukocytes. Although the respiratory rate of the alveolar macrophage is greater than that of any other mammalian phagocytic cell, only a minimal increase in oxygen consumption occurs during phagocytosis. This finding is in contrast with the large "respiratory burst" characteristic of polymorphonuclear leukocytes and peritoneal macrophages (17). Moreover, although alveolar macrophages release fewer oxygen radicals in vitro than polymorphonuclear leukocytes and monocytes, the alveolar macrophage outnumbers the polymorphonuclear leukocyte by 100 times or more in the normal state.

Alveolar macrophages, as well as other phagocytic cells, kill bacteria and detoxify particles by means of ROS and lysomal enzymes and, during phagocytosis, release ROS and proteolytic enzymes into the phagosome and into the external environment (19–21). The ROS and proteolytic enzymes are essential for bacterial killing during phagocytosis. Hypochlorous acid (HOCl), generated through the H_2O_2-myeloperoxidase system, and $\cdot OH$ demonstrate the highest bactericidal activity among the oxygen-derived reactive molecules (22). Superoxide anion has trivial bactericidal activity in comparison to HOCl and $\cdot OH$, whereas H_2O_2 exhibits intermediate bactericidal activity; the activity of H_2O_2 can be increased in an iron-rich environment (23). The major mechanisms by which ROS appears to cause bacterial killing include both oxidation and decarboxylation (21).

Phagocytosis by mammalian leukocytes, which is an energy-dependent process reliant on the physical, chemical, and biological properties of the particle, has been divided into seven phases

(17, 19): (1) particle recognition, (2) reception of the message to initiate phagocytosis, (3) transmission of the message from the receptor to effector, (4) adhesion of the plasma membrane to the particle, (5) assembly of pseudopodia, (6) movement of pseudopodia to engulf the particle, and (7) fusion of the pseudopodia.

Polymorphonuclear leukocytes and alveolar macrophages activated by phagocytic stimuli, such as bacteria and other opsonized particles, respond with an oxidative burst that is characterized by the rapid consumption of molecular oxygen (O_2). This oxygen is reduced to $O_2^{\cdot-}$ by an electron donated from nicotinamide adenine dinucleotide phosphate (NADPH), in a reaction catalyzed by membrane-bound NADPH oxidase. The $O_2^{\cdot-}$ is readily dismutated to H_2O_2 and O_2 by superoxide dismutase (SOD). Hydrogen peroxide is the immediate precursor of hypochlorite, $^1O_2$, and $^\cdot OH$. These highly reactive oxygen derivatives as well as several halogen derivatives attack and modify the macromolecules of pathogens. There is strong evidence that alveolar macrophages generate H_2O_2 in the resting state and that production is augmented during phagocytosis (17). Superoxide anion production and release is augmented by membrane perturbation or phagocytosis (17). Superoxide anion, as well as SOD, is present in alveolar macrophages of several species.

The production of ROS by macrophages depends on both endogenous and exogenous factors (21). The site of the macrophage residence is important; peritoneal macrophages appear to be able to release much less $O_2^{\cdot-}$ than alveolar macrophages. The stage of the cell's maturation appears to be important because peripheral blood monocytes, as they mature in culture to alveolar macrophages, have a reduced capacity to release H_2O_2. The stage of activation of alveolar macrophages has a major effect on their production of ROS. Thus, lung macrophages activated in vivo, as observed in pulmonary sarcoidosis, release increased H_2O_2 in vitro. The mechanisms of activation are not well defined but may involve mediators such as interleukin-1 (IL-1), γ-interferon (γ-IFN), and tumor necrosis factor (TNF). Studies have demonstrated that γ-IFN can "prime" or preactivate alveolar macrophages, thereby increasing the responsiveness of the cells to other stimuli.

Although alveolar macrophages have been studied more frequently because of their accessibility through bronchoalveolar lavage (BAL) procedures, the interstitial macrophages may have greater potential to cause lung injury. Interstitial macrophages exceed alveolar macrophages in number, have an increased ability to replicate and synthesize deoxyribonucleic acid (DNA) in vitro, and are found in direct contact with pulmonary connective tissue (21). Reactive oxygen species and other inflammatory mediators released within the pulmonary interstitium might be expected to cause more damage to lung tissue than the same materials released into the alveoli.

Intravascular macrophages, which remove particles from circulating blood, have been reported in the pulmonary capillaries of sheep, calves, and goats. Normal sheep lungs have more macrophages in their blood vessels than in their alveolar spaces (24). The activity of intravascular macrophages, along with the activity of activated polymorphonuclear leukocytes attracted to the site, results in ROS and other inflammatory mediators being released, which can cause local tissue injury.

Even though much of the macrophage activity in response to inhaled PM is protective, the ROS used by macrophages and polymorphonuclear leukocytes to kill microorganisms also may cause lung damage (16). The large amounts of ROS and other inflammatory mediators released by macrophages and polymorphonuclear leukocytes during phagocytosis may have a pronounced effect on the respiratory tract, depending on their site of release. Airway macrophages, for example, involved in the ingestion of materials deposited in the small and large airways release ROS and proteolytic enzymes that can modify the barrier properties of airway epithelium (18).

Although the normal resident cell in the healthy lung is the alveolar macrophage, the polymorphonuclear leukocyte can enter the lung under certain conditions. When PM concentrations are high, or when the PM consists of toxic materials or ultrafine particles, the alveolar macrophages recruit and activate more alveolar macrophages and attract polymorphonuclear leukocytes that release more ROS and proteolytic enzymes, causing further lung injury. Within the respiratory tract, ROS may damage cell membranes and essential enzymes, or they may reduce the activity of endogenous protease inhibitors, which will increase the activity of extracellular proteases. Reactive oxygen species and other substances released into the cytoplasm may damage or kill other phagocytic cells, which, in turn, release toxic and proteolytic enzymes and substances that attract fibroblasts and elicit fibrogenic responses. Although PM stimulates alveolar macrophages to secrete ROS, continued activation may cause interstitial lung fibrosis (25). Human phagocytes that produce ROS also can cause cytogenic damage in cultured mammalian cells (26). Thus, although phagocytic cells are essential to defend the lungs, they also can injure the lung while exercising their defensive role. Intracellular levels of ROS and other free radicals can be increased by challenges such as mineral dusts and asbestos, which contain free radicals themselves (27, 28). This can cause radical-mediated tissue damage by inducing ROS release, which augments their own free radical burden; this can cause overproduction or deregulation of ROS generation (25). Once ROS production escapes from normal cellular control mechanisms, ROS have the potential to cause severe injury.

Reactive oxygen species are necessary for the normal functioning of cells, and, usually, there is little release from the cell (29); ROS that escape are readily intercepted by cellular antioxidant defenses. Extracellular release of ROS, proteases, and other agents may injure the cells and alter the extracellular matrix of the respiratory tract. Released hydrolases can be measured in the medium overlying phagocytosing macrophages (19). Ingredients that can protect macrophages against the damaging effects of ROS include vitamin E, ascorbic acid, the glutathione redox system, catalase, and SOD.

Thus, phagocytic cells through their normal function can contribute to the chemical reactions occurring in the lung. During severe challenge with inhaled PM, reactions that are occurring in the phagocytic cells can spill over into the healthy lung tissue. These reactions may be especially troubling in the injured lung where the precursors for the Haber-Weiss reaction are present and the lung is primed.

CLASSES OF PARTICLES

Although numerous types of PM are found in the ambient atmosphere, this discussion focuses on metals and ultrafine particles. Metals are considered from the perspective of initiating and participating in oxidative reactions; ultrafine particles are considered from the perspective of initiating inflammatory reactions in the lung interstitium. The phenomenon of lung overload is considered because of its importance in chronic animal inhalation studies.

Metals

Metals can occur in the atmosphere in a variety of forms (free in the ambient atmosphere, coated on particles, or complexed with other materials) and can act by multiple mechanisms to cause oxidative lung injury. They can act directly via Haber-Weiss reactions, or they can displace other metals such as iron from chemical complexes that are then free to participate in Haber-Weiss reactions (30).

Many combustion-derived metals are transition metals that are electronically stable in more than one oxidation state. As a result of this property, transition metals can catalyze the oxidative deterioration of biological macromolecules. Thus, the toxicities associated with these metals may be associated, in part, with oxidative tissue damage (30). Ionic iron, vanadium, nickel, and copper, as well as other transition metals, are common constituents of ambient aerosols and are capable of redox interaction with H_2O_2, a common weak cellular oxidant, to yield the highly toxic ˙OH. Metal ions may also enhance the production of TNFα and activate protein kinase C, as well as induce production of stress proteins.

The toxicity of each metal can vary within the lung according to its persistence, solubility, absorbability, transport, chemical reactivity, and metal complexes formed within the body. The absorption of each metal, and thus its biological availability, is influenced by its solubility in pulmonary fluids and lipids. Insoluble compounds that deposit in the airways are more likely to be cleared by airway macrophages and the mucociliary escalator. More soluble compounds will dissociate, and metal ions may be transported into pulmonary cells. Moreover, the basic mechanisms involving ROS formation are the same for the transition metal ions, and their toxicity may be due, at least in part, to oxidative tissue damage caused by ROS (30). Furthermore, the mechanisms associated with the toxicities of metal ions are similar to the effects produced by many organic redox cycling xenobiotics. In addition, studies have suggested that the ability to generate ROS by redox cycling quinones and related compounds may require metal ions. Therefore, common molecular mechanisms may be involved in the production of ROS and the toxicities of numerous xenobiotics (30).

Considering that the transition metals can catalyze the oxidative deterioration of biological macromolecules, it is plausible that inhalation of PM containing metals could cause oxidative tissue damage, including enhanced lipid peroxidation, DNA damage, and altered calcium and sulfhydryl homeostasis in the respiratory tract. The best studied of the transition metals is iron, and studies have demonstrated that iron has the ability to catalyze the formation of ROS and initiate lipid peroxidation (31–34). Guilianelli and co-workers studied the importance of iron to the toxicity of iron-containing particles using cultured tracheal epithelial cells (35). Nemalite, the most cytotoxic mineral of the three types tested, contained the most surface Fe^{2+}. Moreover, pretreatment with an iron chelator, desferrioxamine, reduced the toxic effects of nemalite.

The surface of most anthropogenically derived particles consists of a complex array of polycyclic organics, various silicates, and metal compounds (36). These materials, present as surface coatings, can make some particles more toxic than expected, based solely on particle core composition (e.g., sulfuric acid layered on the surface of a combustion particle elicits much greater lung irritancy and toxicity than does the acid alone). Functional groups on the surface of some particles can coordinate ferric iron, and such complexes can generate ·OH, causing increased lipid peroxidation. Surface complexed iron has been shown to be involved in pulmonary injury from a variety of environmental particles (37). The degree of acute inflammation following exposure to a particle appeared to correlate with the iron (specifically ferric iron) loading of the particles. Particulate matter coated with certain surface metals may be especially toxic. This may account for particles in different areas having different toxic potentials because the occurrence of trace metals varies geographically.

Garrett and co-workers (38) exposed rabbit alveolar macrophages in vitro to fly ash with and without surface coatings of various metal oxides. Cell viability and cellular adenosine triphosphate content were reduced only with the metal-coated ash particles. Amdur and co-workers exposed guinea pigs to ultrafine aerosols (<0.1 µm) of zinc oxide coated with a layer of sulfuric acid and noted that the coated particles were more toxic (39). These acid-coated aerosols are typical of primary emissions from smelters and coal combustors. Berg and co-workers measured the release of ROS from bovine alveolar macrophages stimulated with heavy metal-containing dusts <4 µm in diameter (40). Dusts, derived from waste incineration, sewage sludge incineration, an electric power station, and from two factories, incubated with alveolar macrophages caused a concentration-dependent increase in ROS release. The increase, apparent after 15 min, continued for an hour or more. The ratio of $O_2^{·-}$ and H_2O_2 secreted varied, depending on the dust, but the release of H_2O_2 correlated best, in descending order, with the content of iron, manganese, chromium, vanadium, and arsenic in the dusts. The positioning of iron first in this array is consistent with other studies examining the biological effect of iron coating the surface of particles.

Certain particles, including silica, crocidolite, kaolinite, and talc, complex considerable concentrations of ferric ion (Fe^{3+}) onto their surfaces. The potential biological importance of iron complexation was assessed by Ghio and co-workers, who examined the effects of surface Fe^{3+} on several indices of oxidative injury (41). Three varieties of silicate dusts were studied: (1) iron loaded, (2) unmodified, and (3) desferrioxamine treated. The ability of silicates to catalyze oxidant generation in an ascorbate/H_2O_2 system in vitro, to trigger respiratory burst activity and leukotriene B_4 release by alveolar macrophages, and to induce lung inflammation in the rat all increased and were proportional to the amount of Fe^{3+} complexed onto their surfaces. These results indicate that iron complexed on the surface of particles can be an important determinant in the pathogenesis of disease after PM exposure. A role for iron-induced oxidative stress in PM-related lung injury is supported further by a study demonstrating that available surface iron and the oxidizing power of mineral particles were both correlated with cytotoxicity, expression of cytokeratin-13, and formation of cross-linked envelopes of rabbit tracheal epithelial cells, and that these effects were blocked by desferrioxamine treatment (35). Nemalite, the particle containing the most iron available on its surface, was the most toxic. Ghio and Hatch examined surface components that could be responsible for the pulmonary effects of silica (42). They noted that an extracellular accumulation of surfactant following silica exposure was associated with the concentration of Fe^{3+} complexed to the surface of the particles, and that surfactant-enriched material was a target for oxidants, the production of which was catalyzed by Fe^{3+}.

The inflammatory potential of three particle types (Mount Saint Helens volcanic ash; ambient particles from Dusseldorf, Germany; and residual oil fly ash, ROFA) was evaluated following intratracheal instillation in rats (37). Iron (specifically Fe^{3+}) loading of the particles correlated with both the degree of acute inflammation [as measured by assessing polymorphonuclear leukocytes, eosinophiles, lactic dehydrogenase (LDH), and protein in lavage fluid] and nonspecific bronchial responsiveness. Although surface iron was correlated with particle acidity, H_2SO_4 instilled at comparable pH produced less inflammation compared to the particles with high surface iron. Particle toxicity increased with neutralization of the fly ash instillate and decreased when iron was removed by acid washing. These results also support the notion that ROS generation by iron present on the surface of particles may increase lung injury.

Tepper and co-workers reported that the concentration of iron (Fe^{3+}) complexed on the surface of a particle was associated with the ability of the particle to support electron transfer and to generate oxidants in vitro and to increase lung inflammation and airway hyperresponsiveness in vivo (43). Particles with or without iron complexed on the surface were instilled into the lungs of rats and evaluated for their potential to produce inflammation and airway hyperactivity. The effects of a high-iron particle (coal fly ash) before and after surface iron was removed by acid washing and the effects of an inert particle (titanium), with or without iron added to the particle surface, were evaluated. The effects of pretreating the rats with drugs to reduce iron-associated ROS formation also were studied. Although coal ash caused considerable inflammation and hyperactivity, acid washing to remove surface iron reduced the deleterious effects of the particle. However, instillation of a titanium particle coated with iron did not increase lung injury, compared to titanium alone. Pretreatment with allopurinol partially blocked lung inflammation, but desferrioxamine and an anti-neutrophil antibody were less effective. The authors concluded that the results generally support the hypothesis that ROS generation by iron on the surface of particles may exacerbate lung injury.

The response to ROFA instilled intratracheally also was compared in three rat strains (44). In general, histopathological findings in all three strains indicated severe, though transient, localized lung damage by 24 h, with near complete resolution by 1–3 wk postinstillation. Bronchoalveolar lavage fluid (BALF) protein levels increased rapidly, with the greatest increase occurring in Sprague-Dawley (SD) rats. Cell influx varied in the three strains; in all strains, alveolar macrophages increased and peaked at 72–96 h, whereas neutrophil influx rose rapidly and peaked at 24 h. The neutrophil response was greatest in the SD rats. In general, strain differences in the pulmonary response to ROFA exist, with the SD rats being most sensitive.

Dye and co-workers examined the acute injury produced in vitro by exposure to ROFA in primary cultures of rat tracheal epithelial cells. Residual oil fly ash induced cell injury as measured by release of LDH, glucose-6-phosphate dehydrogenase, glutathione (GSH) transferase, and GSH reductase. The injury from ROFA required more than 6 h to develop and obscured any effect due to concomitant exposure to ozone (45).

The possibility that preexisting lung inflammation may amplify the toxic response of the lungs to residual ROFA was explored by Costa and co-workers (37). Rats with preexisting lung inflammation, induced with lypopolysaccharide, experienced significantly greater lung injury from ROFA instilled intratracheally as measured by BAL cells and protein and LDH. Because the response to volcanic dust from Mount Saint Helens was minimal and was not amplified by preexisting inflammation, the data suggest that inflammatory lung conditions may potentiate the effects of toxic particles derived from specific emission sources.

Thus, ROS produced through chemical reactions involving iron can initiate lipid peroxidation of the cell membrane, resulting in cell death and subsequent lung injury. It is also possible that other transition metals, by virtue of their ability to redox between valence states, can generate ROS in the presence of precursor oxidants and reducing agents.

Ultrafines

Ultrafine particles, a class of particles less than 0.1 μm in diameter, appear to have great toxic potential. Although the mechanism underlying a size-related difference in toxicity is unknown, ultrafine particles may have a greater pulmonary inflammatory potential than do larger particles of the same material. It appears that ultrafine particles of low in vivo solubility and toxicity may enter the interstitium more readily than do larger particles. This results in increased contact with interstitial macrophages and the release of ROS and various mediators. Compared to larger particles, ultrafine particles have a greater particle number and a greater surface area per unit mass of inhaled particles; both of these characteristics influence toxic potential. Particle number is important because of the particle load presented to the interstitial macrophages, whereas surface area is important because chemical species on the surface may enhance ultrafine particle toxicity.

Oberdörster reported that singlet ultrafine particles, generated from thermal degradation of polytetrafluoroethylene (PTFE), with a median diameter of 26 nm were extremely toxic to rats (46). Inhaling these particles for 10–30 min resulted in acute hemorrhagic pulmonary inflammation and death. The effects were attributed solely to the ultrafine particles and not to the gas-phase component of the fumes or to reactive radicals. The calculated mass concentration of the ultrafine particles was less than 60 $\mu g/m^3$, a rather low value to cause mortality in healthy rats. Aging the fumes aggregated the ultrafine particles and decreased their toxicity significantly. On the other hand, studies with aggregated ultrafine particles have shown that inhalation of highly insoluble particles of low intrinsic toxicity, such as aggregated titanium dioxide (TiO_2) and carbon black, causes effects in animals only after repeated exposures to high concentrations (in the milligram- per-cubic-meter range) (47–51). The effects noted after prolonged exposure by inhalation to aggregated ultrafine particles included chronic lung inflammation, pulmonary fibrosis, and tumors. Oberdörster and co-workers postulated that ultrafine particles may contribute to the observed increases in mortality and morbidity associated with ambient PM air pollution (52). Lee and Seidel investigated the toxicity of pyrolysis products of perfluoropolymer fumes and demonstrated that the toxicity was associated with the small particle fraction (53). The toxicity was not observed when the fumes were filtered to remove the particles and was markedly reduced when the particles were aggregated. The authors attributed the reduced particle toxicity to a decrease in the number of toxic particles resulting from aggregation.

The potential molecular mechanisms by which ultrafine particles might produce pulmonary inflammation have been explored by Johnston and co-workers (54). They reported increases in abundance for interleukin-6 (IL-6), manganese SOD, metallothionein-I, and TNFα of 40-, 10-, 35-, and 5-fold, respectively, in lungs of rats inhaling ultrafine PTFE particles. They also observed a 55–95% increase in lavageable polymorphonuclear leukocytes 4 h posttreatment and increased messenger ribonucleic acid (mRNA) abundance expressed around all airways and interstitial regions. The authors postulated that, because IL-6 demonstrated the greatest change in mRNA abundance, IL-6 may trigger the cascade of events occurring with inhalation of ultrafine PTFE particles.

Ferin reviewed the historical concept of particles and fibers and concluded that there are no inert particles, and that even low concentrations of particles may have negative health effects (55). Ferin postulated that particles in the ultrafine range may be more toxic than larger particles, irrespective of their chemical nature, because ultrafine particles may evade the phagocytic function of alveolar macrophages to a greater extent than larger particles and, therefore, penetrate the alveolar epithelium more extensively (47,56). The result is that ultrafine particles are cleared from the lung more slowly. Extensive penetration of ultrafine particles into the interstitial space produces inflammation, damages the alveolar epithelial cells, and produces pulmonary edema.

The possibility that ambient ultrafine particles may add to the pulmonary lung burden is supported by studies of ultrafine particles in alveolar macrophages lavaged from human lungs (57). Alveolar macrophages lavaged from nonsmokers with little or no occupational exposures contained ultrafine particles. The particles were composed of a variety of elements including iron and other transition metals.

Lung Overload

The potential for inhaled particles to cause pulmonary injury has been observed in animal toxicology studies in which animals inhaled massive concentrations of poorly soluble, nonfibrous particles of low toxicity for prolonged periods. These studies show a significant depression in alveolar macrophage-mediated clearance of particles; an increased retention and accumulation of particles in the lung, accompanied by an increased access of these particles into the pulmonary interstitium; and a dose-dependent increase in pulmonary inflammatory cells. The appearance of aggregated alveolar macrophages in focal areas of the alveoli and proliferative changes in the epithelial and interstitial cells are also typical findings. The proliferative responses may be causally related to the induction of focal fibrotic lesions and the development of lung tumors in rats. The phenomenon is termed "lung" or "dust" overload.

Dust overloading occurs when the dose rate for PM is so high that it overwhelms the long-term pulmonary clearance processes (Figure 2) and is marked by a progressive reduction in particle clearance from the deep lung. The high particle lung burden impairs long-term clearance because the massive intracellular particle load prevents the alveolar macrophage from effectively phagocytizing and removing the inhaled particulate material (58, 59). As a result, particles accumulate in the alveolar spaces and ultimately move into the pulmonary interstitium. The inability of dust-laden alveolar macrophages to translocate to the mucociliary escalator is correlated with an average composite particle volume of approximately 60 μm^3/alveolar macrophage in the Fischer 344 rat. At this level, alveolar macrophage-mediated clearance ceases and agglomerated particle-laden macrophages remain in the alveolar region.

Suppression of particle transport by the alveolar macrophage leads to increased interstitial dust uptake and a prolonged inflammatory response, caused, most likely, by the persistent, possibly excessive elaboration of chemotactic and chemokinetic factors by the alveolar macrophage (59). Particle types that have been associated with lung overload include TiO_2, toner, polyvinyl chloride, carbon black, diesel soot, talc, coal dust, petroleum coke, oil shale, and volcanic fly ash (50, 60–70).

Lung overload from diesel exhaust (DE) particles has been the subject of extensive study (71). Prolonged exposure of rats to high concentrations of DE initiates a progression of cellular changes starting as early as 2 wk after exposure and leading eventually to lung tumors. The first change in the lung following exposure to DE is an increase in the number of alveolar macrophages filled with DE particles. With continued exposure, inflammation becomes evident.

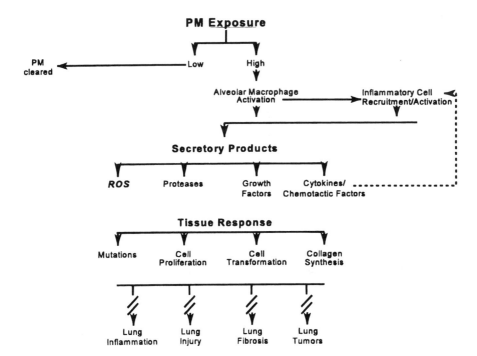

Figure 2. Potential mechanisms for PM-associated lung injury from ROS and other secretory products.

After 6–12 mo of exposure, fibrotic changes (i.e., increases in the number of fibroblasts, interstitial macrophages, and collagen fibers) become apparent. After 18–24 mo of exposure, epithelial cells undergo metaplasia, and particle-containing macrophages aggregate in the interstitium, where focal fibrosis develops; tumors begin to appear at this time. Debris from dead and disintegrated macrophages accumulates in the focal lesions, thus exposing local tissues to both the particles and the degradation products of the disintegrating macrophages.

Impaired particle clearance due to lung overload has been observed in rats, mice, hamsters, and dogs; however, the effects differ markedly among these species (i.e., mice and hamsters are less prone to developing chronic inflammation and pulmonary fibrosis, and lung tumors have been observed only in rat studies). Although it is not known why particle overload is associated with pulmonary inflammation and neoplasia in one species (rat) but not in another (hamster), it is possible that differences in the macrophage responses or the antioxidant systems may be involved (72).

The formation of ROS and the resulting oxidant injury are potential mechanisms for the overload effect. Mice have less marked fibrotic and proliferative responses to lung overload with DE particles than rats, and antioxidant (glutathione) levels increase in mice, but not in rats (64, 73). The focal point in lung overload pathophysiology is inflammation with activation of alveolar macrophages and polymorphonuclear leukocytes and release of ROS and proteases and a number of mediators, including the inflammatory cytokines and growth factors (74). Release of these mediators can lead to epithelial and interstitial cell proliferation. Increasing access of particles into the pulmonary interstitium where they can interact directly with the interstitial cells can amplify cell proliferation. Formation of DNA adducts, possibly due to the action of ROS, has been described in rats exposed to diesel soot and carbon black particles and may lead to lung tumor formation (75, 76). The potential importance of ROS for lung injury is seen in studies reporting that in vitro exposure of rat lung epithelial cells to inflammatory cells lavaged from the lungs of quartz-exposed rats caused a significantly increased mutational frequency in the *hprt* gene, indicating that mediators (e.g., ROS) released from inflammatory cells are causally involved in mutational events (77). Thus, a sequence of events leading to pulmonary overload appears to be initiated in rats inhaling high concentrations of PM. The sequence involves alveolar macrophage activation; acute inflammation and impaired clearance; further particle accumulation; chronic inflammation and focal fibrosis; epithelial cell proliferation; mutations; metaplasia; and, ultimately, lung tumors.

Although most overload studies have been conducted at extremely high concentrations, some were conducted at concentrations comparable to occupational exposure limits for "nuisance dusts" (46).

OXIDATIVE LUNG INJURY CAUSED BY REACTIVE OXYGEN SPECIES

Free radicals are defined as molecules that contain one or more unpaired electrons (78); they are highly reactive and therefore transient. Included in this broad definition are the hydrogen atom (one unpaired electron), most transition metals, and the oxygen molecule itself. Although free radicals are generated in vivo as by-products of normal metabolism, they also can be produced when an organism is exposed to ionizing radiation, to drugs capable of redox cycling, or to xenobiotics that already exist as free radicals or are converted to radical species by metabolic processes. In the respiratory tract, the cellular targets at risk from free radical damage depend on the nature of the radical and its site of generation.

Free radicals can be generated from a wide variety of sources in the respiratory tract (21). These sources include autoxidation of small molecules such as thiols, hydroquinones, and catecholamines; numerous cycling enzymes during their catalytic cycling, including xanthine oxidase; the mitochondrial electron transport system; the endoplasmic reticulum and nuclear membrane transport system; peroxisomes; and plasma membranes.

Reactive Oxygen Species Generation and Secretion

The secretory products released by phagocytic cells during phagocytosis include ROS, proteases, antiproteases, polypeptide hormones, and other bioactive mediators. Although these biologic mediators all can contribute to inflammatory processes in the lung, directly or indirectly, as reviewed by Sibille and Reynolds (21), only the ROS are considered further in this section.

Reactive oxygen species are produced and released in the respiratory tract during normal phagocytic cell activity but are usually neutralized by the antioxidant mechanisms in the tissue. In response to elevated particle concentrations deposited in the lung, however, intense phagocytic activity by macrophages and polymorphonuclear leukocytes recruited to the lung increases the generation and release of ROS and other inflammatory mediators. The released ROS can, in turn, exhaust the antioxidant defenses of the lungs. Extreme examples of this condition have been observed in animals in particle overload studies.

Ultrafine particles depositing in the lung can evade phagocytosis and enter the interstitium, presumably to be engulfed by the interstitial macrophages. This can result in ROS and other inflammatory mediators being released in the interstitium, where they can severely damage the lung parenchyma.

It is plausible that particles consisting of the transition metals, especially iron, can accelerate the generation of ROS. It is uncertain how the transition metals might contribute to inflammatory reactions in the normal lung, however, except when deposited in high concentrations or as ultrafine particles. Nonetheless, in the inflamed or injured lung, where precursors of ROS, such as $O_2^{\cdot-}$, are present, transition metals could increase the generation of ROS and other inflammatory mediators and cause further lung damage. Potential targets for free radical injury are presented in Table 2.

A transition metal, with iron being the most likely candidate, is required for the generation of $\cdot OH$, a potent ROS, from $O_2^{\cdot-}$ or H_2O_2 (31). These substances react according to the iron-catalyzed Haber-Weiss reaction:

$$Fe^{3+} + O_2^{\cdot-} \rightarrow Fe^{2+} + O_2$$

$$Fe^{2+} + H_2O_2 \rightarrow Fe^{3+} + OH^- + \cdot OH$$

In this reaction, $O_2^{\cdot-}$ reduces iron, which, in turn, reduces H_2O_2 to form $\cdot OH$. Consequently, iron and iron-carrying proteins appear to play a crucial role in promoting free-radical reactions (79). Other reductants (e.g., ascorbate) also can reduce Fe^{3+}, indicating that sources of peroxides in the presence of transition metals can generate $\cdot OH$ in the absence of $O_2^{\cdot-}$.

Similar reactions may be commonly associated with most membranous fractions in the respiratory tract, including mitochondria, microsomes, and peroxisomes (30). Thus, excessive concentrations of ROS can be generated and released by ultrafine and metal-containing particles and by particles during lung overload conditions. The ROS, and other inflammatory mediators, released in response to these conditions can react with any of the organic constituents of lung cells, cell membranes, or the lung lining fluid. The products of these reactions are themselves highly reactive. Accordingly, secondary reactions can occur that initiate chain reactions or break chemical

Table 2.   Cellular free radical attack sites

Cellular site	Biochemical effects
Macromolecules	
Proteins	Peptide chain scission, denaturation
DNA	Strand scission, base modification
Other molecules	
Unsaturated and thiol- containing amino acids	Protein denaturation, cross-linking, enzyme inhibition Organelle and cell permeability changes
Nucleic acid bases	Cell cycle changes, mutations
Carbohydrates	Cell surface receptor changes
Unsaturated lipids	Cholesterol and fatty acid oxidation Lipid cross-linking Organelle and cell permeability changes
Cofactors	Decreased availability and activity of nicotinamide and flavin- containing cofactor, ascorbate and porphyrin oxidation
Antioxidants	Depleted and regenerated

bonds, or both, with the consequent fragmentation of macromolecules (DNA, proteins, cell membrane lipoproteins, and structural or mucus glycoprotein). The ultimate result depends on the nature and concentration of the inhaled PM, but could include oxidative lung injury resulting in lung inflammation, fibrosis, and tumors.

As described earlier, phagocytic cells release ROS either into the phagosome or into the external environment in response to particulate stimulation. Excessive dust loading of the macrophages, as observed in pneumoconiosis and asbestosis, also activates the cells to release increased amounts of mediators, including ROS. In contrast to inert dusts, toxic silica particles and asbestos fibers provoke an inflammatory response in the lungs (74, 80, 81). Macrophages are the predominant inflammatory cell in chronic silicosis and asbestosis and are instrumental in the pathogenesis of lung disease induced by mineral fibers (18).

Both silica (SiO_2) and TiO_2 dusts instilled into the lungs of rats elicited dose-related increases in BALF, LDH, and total protein that were more pronounced and progressive with SiO_2 (74). Although all doses of SiO_2 elicited persistent increases in alveolar macrophage fibronectin release, TiO_2-stimulated alveolar macrophage fibronectin releases only at doses >50 mg/kg. Increased SiO_2 retention was observed for all doses, whereas TiO_2 retention was increased only at doses >50 mg/kg. In contrast to highly toxic dusts such as SiO_2, nuisance-type dusts (e.g., TiO_2) are not usually associated with adverse pulmonary responses. However, exposure to normally innocuous materials can result in chronic inflammation and fibrosis when pulmonary dust burdens are reached that overload particle clearance mechanisms. Workers with silicosis, asbestosis, or coal-worker's pneumoconiosis have alveolar macrophage populations activated to release increased levels of ROS, alveolar macrophage-derived fibroblast growth factor, and fibronectin, the last representing a glycoprotein that is chemotactic for fibroblasts (82).

Alveolar macrophages become activated at sites of tissue injury and inflammation. These activated macrophages, in turn, release large quantities of ROS and nitrogen metabolites, arachidonic acid derivatives, proteases and lysosomal enzymes, fibronectin, cytokines, and growth factors. These secreted products amplify inflammatory and immune reactions and are meant to destroy invading organisms and initiate healing. However, unregulated or persistent release of these products by activated phagocytic cells may cause extensive tissue injury. Specific substances that trigger macrophage activation include bacterial products, γ-interferon, inflammatory cytokines (IL-1, TNFα), tumor promoters, and possibly asbestos fibers (83). The response to these agents is similar and results in the synthesis of new proteins and the release of potent mediators.

Thus, phagocytes are recognized as major components of inflammatory reactions in the lung. The macrophage population is morphologically and functionally heterogeneous, includes alveolar, interstitial, intravascular, and airway macrophages, and, under normal conditions, represents the majority of phagocytes in the respiratory tract. Inhaled PM and microorganisms are cleared from the respiratory tract by the macrophages using numerous and diverse secretory products. Macrophages play an important role in inflammatory processes by releasing ROS and proteolytic enzymes, as well as other bioactive materials, and by recruiting polymorphonuclear leukocytes. The polymorphonuclear leukocytes, although practically absent from normal alveoli, when recruited in inflammatory states can outnumber macrophages and release substantial amounts of ROS and proteases that contribute to the degradation of lung matrix during inflammation.

SUMMARY AND CONCLUSIONS

Particulate matter is inhaled and deposited in the lung in virtually all environments. The quantity of material inhaled and deposited depends both on the characteristics of the aerosol and on the physiological and anatomical characteristics of the host. The physiologic response depends on the nature and dose of PM deposited in the respiratory tract. Low doses of nontoxic PM are usually phagocytosed, detoxified, and cleared from the alveoli by the resident macrophage population. Reactive oxygen species, released into the phagolysosome during these activities, may escape from the cell in small quantities but are rapidly neutralized by the antioxidant defenses of the lung. As part of the response to excessive doses of PM or toxic PM, the alveolar macrophage population increases in size and polymorphonuclear leukocytes are activated and recruited to the lung. These cells release innumerable inflammatory mediators, including ROS, that can damage the lung parenchyma (Figure 2). Moreover, if the PM is inhaled in sufficiently high concentrations and if the exposures are long enough, lung overload can develop in experimental animals, and severe lung inflammation and, ultimately, tumors can occur.

Transition metals, inhaled as PM or constituents of PM, may cause oxidant injury to the lung from ROS formation. This may be especially true in the injured lung in which precursor oxidants and reducing agents are present.

Ultrafine aerosols, which have high deposition efficiencies, large numbers per unit mass of particles, and large surface areas available for chemical reaction, are small enough to enter the lung interstitium, where they can initiate a severe inflammatory reaction in which ROS and other mediators are released.

Thus, it is clear that PM has the potential to cause oxidative lung injury, and it is reasonably well established that daily fluctuations in ambient PM are associated with mortality and morbidity. Although it is reasonable to assume that chronic oxidative lung injury could sensitize the lung to the additional stress produced by elevated PM levels, further research is essential to determine the mechanisms by which changes in ambient PM can trigger acute events such as fluctuations in daily mortality.

REFERENCES

1. Schreck, R. 1982. Respiratory airway deposition of aerosols. In *Air Pollution-Physiological Effects*, eds. J. J. McGrath and C. D. Barnes, pp. 183–221. New York: Academic Press.
2. U.S. Environmental Protection Agency. 1994. *National Air Quality Trends Report, 1993*, pp. 19–23. Washington, DC: U.S. EPA.
3. United Nations Environmental-World Health Organization. 1992. *Urban Air Pollution in Megacities of the World*, p. 230. Geneva: WHO.
4. Whitby, K. T., and Cantrell, B. 1976. Atmospheric aerosols—Characteristics and measurements. *Proc. Int. Conf. Sens. Assess.* 1976, pp. 1–6.
5. Bernstein, D. M., and Rahn, K. A. 1979. New York Summer Aerosol Study: Trace element concentration as a function of size. *Ann. N.Y. Acad. Sci.* 322:87–97.
6. Schroeder, W. H., Dobson, M., Kane, D. M., and Johnson, N. D. 1987. Toxic trace elements associated with airborne particulate matter: A review. *JAPCA* 37:1267–1285.
7. Schwartz, J. 1994. What are people dying of on high air pollution days? *Environ. Res.* 64:26–35.
8. Schwartz, J., and Dockery, D. W. 1992. Increased mortality in Philadelphia associated with daily air pollution concentrations. *Am. Rev. Respir. Dis.* 145:600–604.
9. Pope, C. A. III, Schwartz, J., and Ransom, M. R. 1992. Daily mortality and PM10 pollution in Utah valley. *Arch. Environ. Health* 47:211–217.
10. Dockery, D. W., Schwartz, J., and Spengler, J. D. 1992. Air pollution and daily mortality: Associations with particulates and acid aerosols. *Environ. Res.* 59:362–373.
11. Dockery, D. W., Pope, C. A. III, Xu, X., Spengler, J. D., Ware, H. J., Fay, M. E., Ferris, B. G., Jr., and Speizer, F. E. 1993. An association between air pollution and mortality in six U.S. cities. *N. Engl. J. Med.* 329:1753–1759.
12. Warheit, D. B., Overby, L. H., George, G., and Brody, A. R. 1988. Pulmonary macrophages are attracted to inhaled particles on alveolar surfaces. *Exp. Lung Res.* 14:51–66.
13. Oberdörster, G. 1988. Lung clearance of inhaled insoluble and soluble particles. *J. Aerosol. Med.* 1:289–330.
14. Snipes, M. B. 1989. Long-term retention and clearance of particles inhaled by mammalian species. *Toxicology* 20:175–211.
15. Bailey, M. R, Fry, R. A., and James, A. C. 1985. Long-term retention of particles in the human respiratory tract. *J. Aerosol Med.* 16:295.
16. Brain, J. D. 1985. Macrophages in the respiratory tract. In *Handbook of Physiology, Volume 1. Circulation and Nonrespiratory Functions*, eds. A. P. Fishman and A. B. Fisher, pp. 447–471. Bethesda, MD: American Physiological Society.
17. Hocking, W. G., and Golde, D. W. 1979. The pulmonary-alveolar macrophage [first of two parts]. *N. Engl. J. Med.* 301:580–587.
18. Brain, J. D. 1988. Lung macrophages: How many kinds are there? What do they do? *Am. Rev. Respir. Dis.* 137:507–509.
19. Green, G. M., Jakob, G. J., Low, R. B., and Davis, G. S. 1977. Defense mechanisms of the respiratory membrane. *Am. Rev. Respir. Dis.* 115:479–513.
20. Johnston, R. B., Godzik, C. A., and Cohn, Z. A. 1978. Increased superoxide anion production by immunologically activated and chemically elicited macrophages. *J. Exp. Med.* 148:115–127.

21. Sibille, Y., and Reynolds, H. Y. 1990. Macrophages and polymorphonuclear neutrophils in lung defense and injury *Am. Rev. Respir. Dis.* 141:471–501.
22. Klebanoff, S. J. 1968. Myeloperoxidase-halide-hydrogen peroxide antibacterial system. *J. Bacteriol.* 95:2131–2138.
23. Quie, P. G., White, J. G., Holmes, B., and Good, R. A. 1967. In vitro bactericidal capacity of human polymorphonuclear leukocytes. Diminished activity in chronic granulomatous disease in childhood. *J. Clin. Invest.* 46:668–679.
24. Warner, A. E., Barry, B. E., and Brain, J. D. 1986. Pulmonary intravascular macrophages in sheep: Morphology and function of the mononuclear phagocyte system. *Lab. Invest.* 55:276–288.
25. Doelman, C. J. A., Leurs, R., Oosterom, W. C., and Bast, A. 1989. Mineral dust exposure and free radical-mediated lung damage. *Exp. Lung Res.* 16:41–55.
26. Weitberg, A. B., Weitzman, S. A., Destrempes, M., Latt, S. A., and Stossel, T. P. 1983. Stimulated human phagocytes produce cytogenetic changes in cultured mammalian cells. *N. Engl. J. Med.* 308:25–29.
27. Kandaswami, C., Morin, G., and Sirois, P. 1988. Lipid peroxidation in rat alveolar macrophages exposed to chrysotile fibers. *Toxicol. In Vitro* 2:117–120.
28. Vallyathan, V., Shi, X., Dalal, N. S., Irr, W., and Castranova, V. 1988. Generation of free radicals from freshly fractured silica dust. *Annu. Rev. Respir. Dis.* 18:1213–1219.
29. Klein, C. B., Frenkel, K., and Costa, M. 1991. The role of oxidative processes in metal carcinogenesis. *Chem. Res. Toxicol.* 4:592–604.
30. Stohs, S. J., and Bagchi, D. 1995. Oxidative mechanisms in the toxicity of metal ions. *Free Radic. Biol. Med.* 18:321–336.
31. Aust, S. D. 1989. Metal ions, oxygen radicals and tissue damage. *Bibl. Nutr. Dieta* 43:266–277.
32. Minotti, G., and Aust, S. 1987. The role of iron in the initiation of lipid peroxidation. *Chem. Phys. Lipids* 44:191–208.
33. Imlay, J. A., Chin, S. M., and Linn, S. 1988. Toxic DNA damage by hydrogen peroxide through the Fenton reaction in vivo and in vitro. *Science* 240:640–642.
34. Halliwell, B., and Gutteridge, J. M. C. 1986. Iron and free radical reactions: Two aspects of antioxidant protection. *Trends Biochem. Sci.* 11:372–375.
35. Guilianelli, C., Baeza-Squiban, A., Boisvieux-Ulrich, E., Houcine, O., Zalma, R., Guennou, C., Pezerat, H., and Marano, F. 1993. Effect of mineral particles containing iron on primary cultures of rabbit tracheal epithelial cells: Possible implication of oxidative stress. *Environ. Health Perspect.* 101:436–442.
36. Costa, D. L., Lehmann, J. R, Smith, S., and Dreher, K. L. 1995. Amplification of particle toxicity to the lung by pre-existent inflammation. Presented at HERL Open House, U.S. Environmental Protection Agency, Research Triangle Park, NC, June 7, abstr. 36.
37. Costa, D. L., Tepper, J. S., Lehmann, J. R., Winsett, D. W., and Ghio, A. J. 1994. Surface complexed iron, lung inflammation and hyperactivity. *Colloquium on Particulate Air Pollution and Human Mortality and Morbidity,* Irvine, CA, January.
38. Garrett, N. E., Campbell, J. A., Stack, H. F., Waters, M. D., and Lewtas, J. 1981. The utilization of the rabbit alveolar macrophage and Chinese hamster ovary cell for evaluation for the toxicity of particulate materials. II. Particles from coal-related processes. *Environ. Res.* 24:366–376.
39. Amdur, M. O., and Chen, L. C. 1989. Furnace-generated acid aerosols: Speciation and pulmonary effects. *Environ. Health Perspect.* 79:147–150.
40. Berg, I., Schluter, T., and Gercken, G. 1993. Increase of bovine alveolar macrophage superoxide anion and hydrogen peroxide release by dusts of different origin. *J. Toxicol. Environ. Health* 39:341–354.
41. Ghio, A. J., Kennedy, T. P., Whorton, A. R., Crumbliss, A. L., Hatch, G. E., and Hoidal, J. R. 1992. Role of surface complexed iron in oxidant generation and lung inflammation induced by silicates. *Am. J. Physiol.* 263:L511–L518.
42. Ghio, A., and Hatch, G. E. 1993. Lavage phospholipid concentration after silica instillation in the rat is associated with complexed (Fe^{3+}) on the dust surface. *Am. J. Cell Mol. Biol.* 8:403–407.
43. Tepper, J. S., Lehmann, J. R., Winsett, D. W., Costa, D. L., and Ghio, A. J. 1994. The role of surface-complexed iron in the development of acute lung inflammation and airway hyperresponsiveness. *Am. Rev. Respir. Dis.* 149(4, pt. 2): A839.
44. Jaskot, R. H., Costa, D. L., Kodavanti, U. P., Lehmann, J. R., Winsett, D., and Dreher, K. L. 1995. Comparison of lung inflammation and airway reactivity in three strains of rats exposed to residual oil fly ash particles. Presented at HERL Open House, U.S. Environmental Protection Agency, Research Triangle Park, NC, June 7, abstr. 43.
45. Dye, J. A., Richards, J. R., and Dreher, K. L. 1995. Injury of rat tracheal epithelial cultures by exposure to ozone and/or residual oil fly ash. Presented at HERL Open House, U.S. Environmental Protection Agency, Research Triangle Park, NC, June 7, abstr. 37.
46. Oberdörster, G. 1995. Lung particle overload: Implications for occupational exposures to particles. *Regul. Toxicol. Pharmacol.* 27:123–135.

47. Ferin, J., Oberdörster, G., and Penney, D. P. 1992. Pulmonary retention of ultrafine and fine particles in rats. *Am. J. Respir. Cell Mol. Biol.* 6:535–542.
48. Oberdörster, G., Ferin, J., Gelein, R., Soderholm, S. C., and Finkelstein, J. 1992. Role of the alveolar macrophage in lung injury: Studies with ultrafine particles. *Environ. Health Perspect.* 97:193–199.
49. Heinrich, U. 1994. Carcinogenic effects of solid particles. In *Toxic and Carcinogenic Effects of Solid Particles in the Respiratory Tract*, eds. U. Mohr, D. L. Dungworth, J. L. Mauderly, and G. Oberdörster, pp. 57–73. Washington, DC: ILSI Monographs, ILSI Press.
50. Mauderly, J. L., Snipes, M. B., Barr, E. B., Belinsky, S. A., Bond, J. A., Brooks, A. L., Chang, I.-Y., Cheng, Y. S., Gillett, N. A., Griffith, W. G., Henderson, R. F., Mitchell, C. E., Nikula, K. J., and Thomassen, D. G. 1994. *Pulmonary Toxicity of Inhaled Diesel Exhaust and Carbon Black in Chronically Exposed Rats. Part I: Neoplastic and Nonneoplastic Lung Lesions.* Cambridge, MA, Health Effects Institute. Research report 68.
51. Nikula, K. J., Snipes, M. B., Barr, E. B., Griffith, W. C., Henderson, R. F., and Mauderly, J. L. 1994. Influence of particle-associated organic compounds on the carcinogenicity of diesel exhaust. In *Toxic and Carcinogenic Effects of Solid Particles in the Respiratory Tract*, eds. U. Mohr, D. L. Dungworth, J. L. Mauderly, and G. Oberdörster, pp. 565–568. Washington, DC: ILSI Monographs, ILSI Press.
52. Oberdörster, G., Ferin, J., Gelein, R., Mercer, P., Corson, N., and Godleski, J. 1995. Low-level ambient air particulate levels and acute mortality/morbidity: Studies with ultrafine Teflon particles. *Respir. Crit. Care Med.* 151:A66.
53. Lee, K. P., and Seidel, W. C. 1991. Pulmonary response of rats exposed to polytetrafluoroethylene and tetrafluoroethylene hexafluoropropylene copolymer fume and isolated particles. *Inhal. Toxicol.* 3:237–264.
54. Johnston, C., Finkelstein, J., Gelein, R., Baggs, R., Mercer, P., Corson, N., Nguyen, K., and Oberdörster, G. 1995. Early alterations in the mRNA abundance of IL-6, MnSOD, MT and TNF associated with ultrafine Teflon particle exposure. *Am. J. Crit. Care Med.* 151:A66.
55. Ferin, J. 1994. Pulmonary retention and clearance of particles. *Toxicol. Lett.* 72:121–125.
56. Takenaka, S., Dornhöfer-Takenaka, and Muhle, H. 1986. Alveolar distribution of fly ash and of TiO_2 after long-term inhalation by Wistar rats. *J. Aerosol Med.* 17:361–364.
57. Godleski, J. J., Hatch, V., Hauser, R., Christiani, D., Gazula, G., and Sioutas, C. 1995. Ultrafine particles in lung macrophages of healthy people. *Respir. Crit. Care Med.* 151:A264.
58. Morrow, P. E. 1988. Possible mechanisms to explain dust overloading of the lungs. *Fundam. Appl. Toxicol.* 10:369–384.
59. Morrow, P. E. 1992. Dust overloading of the lungs: Update and appraisal. *Toxicol. Appl. Pharmacol.* 113:1–12.
60. Martin, J. C., Daniel, H., and Le Bouffant, L. 1975. Short- and long-term experimental study of the toxicity of coal-mine dust and of some of its constituents. In *Inhaled Particles IV* (Proceedings of an international symposium organized by the British Occupational Hygiene Society, September 22–26, 1975), Part I, eds. W. H. Walton and B. McGovern, pp. 361–371. Edinburgh: Pergamon Press.
61. MacFarland, H. N., Coate, W. B., Disbennett, D. B., and Ackerman, L. J. 1982. Long-term inhalation studies with raw and processed shale dusts. *Ann. Occup. Hyg.* 26:213–226.
62. Lee, K. P., Trochimowicz, H. J., and Reinhardt, C. F. 1985. Pulmonary responses of rats exposed to titanium dioxide (TiO_2) by inhalation for two years. *Toxicol. Appl. Pharmacol.* 79:179–192.
63. Wehner, A. P., Dagle, G. E., Clark, M. L., and Buschbom, R. L. 1986. Lung change in rats following inhalation exposure to volcanic ash for two years. *Environ. Res.* 40:499–517.
64. Heinrich, U., Muhle, H., Tanaka, S., Ernst, H., Fuhst, R., Mohr, U., Pott, F., and Stober, W. 1986. Chronic effects on the respiratory tract of hamsters, mice, and rats after long-term inhalation of high concentrations of filtered and unfiltered diesel engine emissions. *J. Appl. Toxicol.* 6:383–395.
65. Heinrich, U., Fuhst, R., and Mohr, U. 1992. Tierexperimentelle inhalationsstudien zur frage der tumorinduzierenden wirkung von dieselmotorabgasen und zwei test stäuben. In *Auswirkungen von Dieselmotorabgasen auf die Gesundheit*, pp. 21–30. Munich: GSF.
66. Mauderly, J. L., Jones, R. K., Griffith, W. C., Henderson, R. F., and McClellan, R. O. 1987. Diesel exhaust is a pulmonary carcinogen in rats exposed chronically by inhalation. *Fundam. Appl. Toxicol.* 9:208–221.
67. Klonne, D. R., Burns, J. M., Halder, C. A., Holdsworth, C. E., and Ulrich, C. E. 1987. Two-year inhalation toxicity study of petroleum coke in rats and monkeys. *Am. J. Ind. Med.* 11:375–389.
68. Muhle, H., Creutzenberg, O., Bellmann, B., Heinrich, U., and Mermelstein, R. 1990. Dust overloading of lungs: investigations of various materials, species differences and irreversibility of effects. *J. Aerosol Med.* 3(suppl. 1):S111–S128.
69. National Toxicology Program. 1993. Toxicology and Carcinogenesis Studies of Talc in F344/N Rats and B6C3F₁ Mice. *Tech. Rep.* Ser. No. 421. NIH Publication No. 93–315.
70. Heinrich, U., Fuhst, R., Rittinghausen, S., Creutzenberg, O., Bellmann, B., Koch, W., and Levsen, K. 1995. Chronic inhalation exposure of Wistar rats and two different strains of mice to diesel engine exhaust, carbon black, and titanium dioxide. *Inhal. Toxicol.* 7:533–556.

71. Health Effects Institute. 1995. Diesel Exhaust: A Critical Analysis of Emissions, Exposure and Health Effects, p. 294.
72. Oberdörster, G. 1994. Extrapolation of results from animal inhalation studies with particles to humans? In *Toxic and Carcinogenic Effects of Solid Particles in the Respiratory Tract*, eds. U. Mohr, D. L. Dungworth, J. L. Mauderly, and G. Oberdörster, pp. 335–353. Washington, DC: ILSI Monographs, ILSI Press.
73. Henderson, R. F., Pickrell, J. A., Jones, R. K., Sun, J. D., Benson, J. M., Mauderly, J. M., and McClellan, R. 1988. Response of rodents to inhaled diluted diesel exhaust: Biochemical and cytological changes in bronchoalveolar lavage fluid in the lung tissue. *Fundam. Appl. Toxicol.* 11:546–567.
74. Driscoll, K. E., Maurer, J. K., Lindenschmidt, R. C., Romberger, D., Rennard, S. I., and Crosby, L. 1990. Respiratory tract responses to dust: Relationships between dust burden, lung injury, alveolar macrophage fibronectin release, and the development of pulmonary fibrosis. *Toxicol. Appl. Pharmacol.* 106:88–101.
75. Bond, J. A., Johnson, N. F, Snipes, M. B., and Mauderly, J. L. 1990. DNA adduct formation in rat alveolar type II cells: Cells potentially at risk for inhaled diesel exhaust. *Environ. Mol. Mutagen.* 16:64–69.
76. Wolff, R. K., Bond, J. A., Henderson, R. F., Harkema, J. R., and Mauderly, J. L. 1990. Pulmonary inflammation and DNA adducts in rats inhaling diesel exhaust or carbon black. *Inhal. Toxicol.* 2:241–254.
77. Driscoll, K. E., Carter, J. M., Howard, B. W., and Hassenbein, D. G. 1994. Mutagenesis in rat lung epithelial cells after in vivo silica exposure or ex vivo exposure to inflammatory cells. *Am. J. Respir. Crit. Care Med.* 149:A553.
78. Halliwell, B., and Gutteridge, J. M. C. 1984. Oxygen toxicity, oxygen radicals, transition metals and disease. *Biochem. J.* 219:1–14.
79. Halliwell, B. 1987. Oxidants and human disease: Some new concepts. *FASEB J.* 1:358–364.
80. Beck, B. D., Brain, J. D., and Bohannon, D. E. 1982. An in vivo hamster bioassay to assess the toxicity of particulates for the lung. *Toxicol. Appl. Pharmacol.* 66:9–29.
81. Warheit, D. B., Hansen, J. F., and Hartsky, M. A. 1991. Physiological and pathophysiological pulmonary responses to inhaled nuisance-like or fibrogenic dusts. *Anat. Rec.* 231:107.
82. Rom, W. N., Bitterman, P. B., Rennard, S. I., Cantin, A., and Crystal, R. G. 1987. Characterization of the lower respiratory tract inflammation of nonsmoking individuals with interstitial lung disease associated with chronic inhalation of inorganic dusts. *Am. Rev. Resp. Dis.* 135:1429–1434.
83. Branchaud, R. M., Garant, L. J., and Kane, A. B. 1993. Pathogenesis of mesothelial reactions to asbestos fibers: Monocyte recruitment and macrophage activation. *Pathobiology* 61:154.

Chapter Seventeen

EDEMAGENIC GASES CAUSE LUNG TOXICITY BY GENERATING REACTIVE INTERMEDIATE SPECIES

Carmen M. Arroyo and Jill R. Keeler

PHOSGENE (OCCl$_2$)

Phosgene, an acyl halide, is a toxic gas. Some of its reactions include acylation, chlorination, decarboxylation, and dehydration (1). The exact toxic mechanisms of phosgene remain unclear. However, because of its diversity of reactions and because phosgene is so reactive, it is possible that some of its toxicological mechanisms may be free radical mediated. Electron paramagnetic resonance (EPR)/spin trapping techniques have successfully been applied to determine and identify free radical intermediates in toxicology (2). Spin trapping allows one to determine if short-lived free radicals are involved as reaction intermediates by scavenging the reactive radical to produce more stable nitroxide radicals (3).

The reactions between phosgene and spin trapping agents must be understood prior to applying EPR/spin trapping techniques to investigate possible free radical mechanisms in the interaction of phosgene with biological systems. In this section we describe the reaction of phosgene with various nitrone spin traps.

Materials and Methods

N-t-Butyl-α-phenylnitrone (PBN), α-(4-pyridyl-1-oxide)-N-t-butylnitrone (POBN), 5,5-dimethyl-1-pyrroline-N-oxide (DMPO), and 3,3,5,5-tetramethylpyrroline-N-oxide (M$_4$PO) were obtained from Sigma Chemical Co., St. Louis, MO.

PBN (35.2 mg) was dissolved in 20 µl dimethyl sulfoxide (DMSO, Sigma), vortexed for 5 min, then 1 ml of doubly distilled, doubly deionized water was added. The PBN solution was again mixed for 2 min until the solution was homogenized. POBN (60 mg) was dissolved in 1 ml of doubly distilled, doubly deionized H$_2$O. DMPO and M$_4$PO were purified using a modified method described by Buettner and Oberley (4). Activated charcoal was successively added to a 10% aqueous solution of the nitrone compounds until complete homogeneity was obtained and the mixture was stirred for 2 min at room temperature. The charcoal suspension was filtered and the filtrate was monitored by EPR at a high receiver gain (1.25×10^5) for nitrone impurities. All spin trap preparation was performed in a dim room.

Each nitrone solution was transferred to a Pyrex glass tube (12 × 85 mm) containing a side arm with ground glass connection joined to an EPR quartz flat cell (60 × 0.30 × 16.9 mm). The Pyrex tube was sealed with a rubber stopper. The sealed solutions were exposed to 45 ppm (182 mg/m^3) of phosgene (OCCl$_2$, Matheson Gas Products, East Rutherford, NJ). [^{13}C]Phosgene, 99%, 1.1 M solution (4 ml) in benzene, was obtained from Cambridge Isotope Laboratories,

Woburn, MA. The glass apparatus was inverted, and the EPR spectrum was recorded and analyzed with a Varian E-109, X-band spectrometer at 100 kHz magnetic field modulation. Each experiment was performed in triplicate.

Results and Discussion

Phosgene reacted with POBN to generate two overlapping spin adduct EPR spectra (Figure 1). One consisted of a triplet of doublets with hyperfine coupling constants $a_N = 1.58$ mT and $a_H = 0.26$ mT, obtained by computer simulation of the spin adduct EPR spectrum. The second consisted of a 1:2:2:1 quartet with hyperfine coupling constants $a_N = a_H = 1.41$ mT. This spin adduct corresponds to the t-butylaminoxyl radical, the reduction product of the nitroso spin trap 2-methyl-2-nitrosopropane (MNP) (5). This result indicates that the t-butylaminoxyl portion of the POBN has split off, subsequently adding a proton. Since these spin adducts are observed simultaneously and their formation does not depend on the decomposition of one to form the other, their production occurs independently by two different mechanisms.

Consistent with the EPR results (Figure 1), two possible dissociation mechanisms for phosgene are conceivable. One mechanism (Eq. 1) would involve a heterolytic cleavage and the other (Eq. 2) a homolytic cleavage.

$$Cl_2CO \rightarrow ClCO^+ + Cl^- \tag{1}$$

$$Cl_2CO \rightarrow ClC^{\cdot}O + Cl^{\cdot} \tag{2}$$

Addition of the carbamoyl monochloride radical (Eq. 2) to POBN, in a direct spin trapping type reaction, would yield a spin adduct containing a single β hydrogen to interact with the nitroxide electron yielding a triplet of doublets EPR spectrum. On the other hand, the electron-deficient $ClCO^+$ (Eq. 1) could destabilize the carbon-nitrogen double bond through an electron transfer process. This could conceivably destabilize POBN, leading to the scission of the t-butylaminoxyl portion, which would subsequently add a proton, yielding the 1:2:2:1 quartet EPR spectrum.

PBN in the presence of phosgene also yields overlapping spin adduct EPR spectra (Figure 2). One consists of a 1:2:2:1 quartet similar to the spectrum obtained in the reaction of POBN with phosgene. This quartet can be computer simulated using hyperfine coupling constants $a_N = a_H = 1.45$ mT and is also attributed to the t-butylaminoxyl radical. However, differing from POBN, PBN in the presence of phosgene yields two triplets of doublet spin adduct EPR spectra.

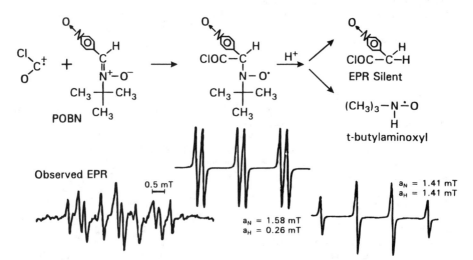

Figure 1. Reaction of POBN with the carbamoyl monochloride radical of phosgene. The observed EPR spectrum was recorded using the following conditions: modulation amplitude 0.1 mT; receiver gain 1 × 10^5; microwave power 20 mW. The computerized spectrum is the summation of the two different components by a factor of 0.5 using a linewidth of 0.1 mT. From ref. 53, with permission.

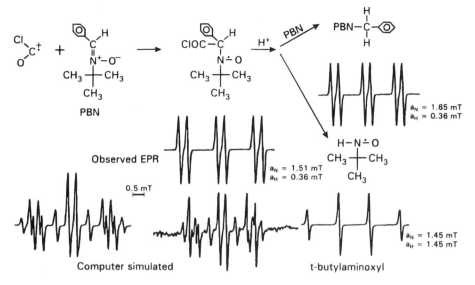

Figure 2. Mechanistic representation of the reaction of carbamoyl monochloride radical with the nitrone PBN. The first-derivative EPR spectrum observed was obtained under the following conditions: modulation amplitude 0.08 mT; microwave power 20 mW; receiver gain 1.6×10^5. The computerized spectrum is the addition of the three different components by a factor of 0.33 using a linewidth of 0.109 mT. From ref. 53, with permission.

One has hyperfine coupling constants $a_N = 1.51$ mT and $a_H = 0.36$ mT. The other has hyperfine coupling constants $a_N = 1.65$ mT and $a_H = 0.36$ mT. This latter triplet of doublets ($a_N = 1.65$ mT) is more persistent and becomes more prominent in time, suggesting that its formation depends on the disappearance of the former ($a_N = 1.51$ mT). Because of the electron-withdrawing properties of the carbamoyl monochloride group, a smaller interaction between the unpaired nitroxide electron and the nitrogen nucleus would be expected. This smaller interaction would yield a lower nitrogen hyperfine coupling constant. Therefore, the spin adduct with $a_N = 1.51$ mT is attributed to the addition of the carbamoyl monochloride to PBN. Furthermore, several RCOCl type compounds are known to react, losing their chloride (6). Among these are aromatic substitution reactions. Loss of the chloride from the carbamoyl chloride–PBN spin adduct would lower the electron-withdrawing properties of the carbamoyl moiety. This would permit a larger interaction of the unpaired nitroxide electron with the nitrogen nucleus and increase the nitrogen hyperfine coupling constant. The results suggest that the PBN spin adduct with $a_N = 1.65$ mT is the product of a reaction between the carbamoyl monochloride adduct and another species in the solution, possibly PBN. It is conceivable that PBN could, in the absence of a catalyst, undergo an aromatic substitution type reaction at the position *para* to the nitrone functional group. The additional triplet of doublets was not observed in the POBN/phosgene reaction mixture. This suggests that the carbamoyl monochloride adduct either is (1) less reactive; (2) reactive, forming a product that yields an EPR spectrum indistinguishable from the observed triplet of doublets (Figure 1); or (3) highly reactive and that the observed triplet of doublet (Figure 1) corresponds to the product of a carbamoyl monochloride adduct reaction. Compared to PBN, the POBN adduct nitrogen hyperfine coupling constant, $a_N = 1.58$ mT, and the aromatic structural difference between POBN and PBN suggest that the first alternative is more likely. Differing from PBN, POBN has an aminoxide-type functional group in the position *para* to the nitrone functional group; therefore, a possible aromatic substitution-type reaction would have to occur at the *ortho* or *meta* positions, which would be less favorable.

The addition of phosgene to a DMPO solution generated two spin adducts: a triplet of doublets with hyperfine structure $a_N = 1.598$ mT and $a_H = 2.273$ mT, in addition to the typical DMPO-

H adduct (7) with hyperfine coupling constants a_N = 1.66 mT and $a_{H(2)}$ = 2.254 mT. The DMPO-H contribution to the overall EPR spectrum is less than 5%; however, since this adduct is short-lived ($t_{1/2}$ < 1 min) its contribution to the overall EPR spectrum may not accurately reflect the actual amount formed. Two possibilities exist for the formation of the DMPO adduct yielding a triplet of doublets EPR spectrum. The first adduct is the direct trapping of a carbamoyl monochloride radical. The second possibility is a scission at the nitrone as occurred with POBN and PBN. Similar to POBN and PBN, this scission in DMPO would yield a nitroxide with an α-carbon, C(5), containing two methyl groups that would not produce additional hyperfine structure. However, the α-hydrogen would split each EPR line of the primary nitroxide triplet into doublets, yielding an EPR spectrum consisting of a triplet of doublets. The formation of DMPO-H would argue against scission at the nitrone. Instead, its formation is consistent with an initial shifting of electron density from the double bond toward the nitrone carbon caused by an electron-deficient carbamoyl monochloride cation (Eq. 2). The nitrone carbon could then become rapidly protonated prior to scission occurring, thus stabilizing the integrity of the DMPO molecule and yielding the DMPO-H adduct. POBN and PBN are structurally very different from DMPO, and it is possible that this rapid protonation is not favored prior to scission, yielding the t-butylaminoxyl radical.

An alternate mechanism for DMPO-H formation is the ejection, by one of the reactants, of an electron into the medium forming a hydrated electron (e_{aq}^-). Hydrated electrons react with DMPO followed by rapid protonation yielding DMPO-H. Hydrated electrons also react rapidly with nitrous oxide (Eqs. 3 and 4), yielding hydroxyl radicals (˙OH) (8).

$$e_{aq}^- + N_2O \rightarrow N_2 + O^- \tag{3}$$

$$O^- + H \rightarrow \text{˙OH} \tag{4}$$

The addition of phosgene to N_2O-saturated solutions of DMPO or M_4PO (a DMPO analog) did not yield DMPO-OH nor M_4PO-OH EPR spectra, indicating that no e_{aq}^- were produced.

Although the EPR triplet of doublets observed for POBN, PBN, and DMPO is consistent with the formation of a carbamoyl monochloride adduct, the homolytic dissociation of phosgene (Eq. 2) generating a carbamoyl monochloride radical remains to be established. ^{13}C-Labeled phosgene can be used to address the nature of this spin adduct. Use of an aprotic solvent would address the question regarding homolytic dissociation of phosgene. A heterolytic dissociation of phosgene (Eq. 1) into ionic species would appear unlikely in a nonpolar aprotic solvent. Therefore, the experiments just described were repeated in benzene using ^{13}C-labeled phosgene. The choice of benzene as a solvent was twofold: (1) It is an aprotic solvent and would minimize a possible heterolytic dissociation of phosgene as in Eq. 1; (2) if this type of dissociation were to occur, benzene is known to undergo aromatic substitution by reactive species such as $COCl^+$. Thus, any $COCl^+$ formed would be preferentially removed from the solution by the solvent due to the much larger concentration of benzene than the spin trap concentration.

The addition of ^{13}C-labeled phosgene to a PBN solution in benzene generates a 1:1:2:2:2:2:1:1 octet spin adduct EPR spectrum (Figure 3). Computer simulation of this spectrum (Figure 3) yields hyperfine coupling constants a_N = 1.25 mT, a_H = 0.624 mT, and a_{13C} = 1.25 mT. This result indicates the direct addition of the phosgene carbon to PBN and supports the formation of a carbamoyl monochloride radical. No evidence is observed in the EPR spectrum of a scission at the PBN nitrone group as occurred in aqueous solutions, supporting the argument that no highly reactive electron deficient species such as $COCl^+$ was produced to interact with PBN. The octet EPR spectrum transforms in time, yielding the triplet EPR spectrum shown in Figure 3 superimposed over some residual peaks corresponding to the original octet. The formation of this pronounced triplet suggests a rearrangement of carbamoyl monochloride in which the β hydrogen is lost.

In summary, the experiments described show the possible heterolytic and homolytic dissociation of phosgene into highly reactive species. These species may be partially responsible for the pathogenesis of acute lung injury from inhalation of phosgene.

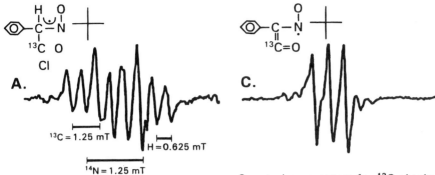

Spectral parameters for ^{13}C obtained
from the observed spin adducts

Computer simulated

Spin adduct	Nuclear spin	HFCC(mT)
PBN	0.5	1.250
M₄PO	0.5	0.700
DMPO	0.5	0.665

HFCC, hyperfine coupling constant.

Figure 3. Spin adduct formation of ^{13}COCl$_2$. (A) EPR spectrum of PBN-^{13}COCl adduct in benzene. Receiver gain 1.25×10^5; modulation amplitude 0.1 mT. (B) Computer simulation of spectrum A using the following parameters $a_N = 1.24$ mT; $a_H = 0.625$ mT and $a_{13C} = 1.250$ mT using a linewidth of 0.25 mT. (C) Final spin adduct detected in the reaction of PBN and ^{13}C-phosgene. Insert: Spectral parameters for ^{13}C-phosgene obtained from observed nitrone spin adducts. From ref. 53, with permission.

BIS(TRIFLUOROMETHYL)DISULFIDE (TFD)

Bis(trifluoromethyl)disulfide (TFD) is an industrial fumigant, and inhalation of this organofluorine compound causes varying degrees of pulmonary edema (9). Currently, novel organophosphorus compounds containing one or two different *S*-trifluoromethyl functionalities have been synthesized in the reaction between the tricoordinate phosphorus ester and TFD (10). The toxic effects of this halogenated disulfide are important from the standpoint of human health and industrial safety. Furthermore, the mechanism(s) responsible for the primary lesion of pulmonary damage has not yet been explored. This study was undertaken as part of an attempt to develop testing procedures for possible future toxicological tests on by-products of TFD. Therefore, we have used electron paramagnetic resonance (EPR)/spin trapping techniques to examine free-radical formation by TFD under different experimental conditions.

Spin trapping is a technique in which short-lived free radicals can be accumulated by an addition reaction to a spin trap to generate a long-lived free radical product, a spin adduct, which can be detected by EPR (reviews in refs. 11 and 12). We report here the metal-dependent oxidation of TFD generating thiyl and reactive oxygen species detected by three different spin trap agents, 3,3,5,5-tetramethylpyrroline-*N*-oxide (M₄PO), 5,5-dimethylpyrroline-*N*-oxide (DMPO) and 2-methyl-2-nitrosopropane (MNP). In addition, we discuss the metal-catalyzed oxidation of this halogenated disulfide with the formation of highly reactive and damaging free radicals. The metal-catalyzed oxidation of toxic by-products was repressed by being strongly complexed to zinc, which is catalytically inert.

Free-radical intermediates have been inferred from product studies, demonstrated by the use of chemical probes and identified by EPR/spin trapping techniques. Intervention studies show that an antioxidant compound, composed of zinc gluconate and vitamin E, competed with radical-adducts formation. The recording of such type of competition reactions introduces another analytical application of the EPR/spin trapping techniques.

Materials and Methods

The spin trap agents DMPO and M_4PO were obtained from Aldrich (Milwaukee, WI) and were purified according to the method of Buettner and Oberley (4) by repeatedly treating the solution with activated charcoal until all free-radical impurities were eliminated (as verified by EPR). DMPO concentration was measured spectrophotometrically ($\lambda = 227$ nm, $\epsilon = 8 \times 10^3$ M^{-1} cm^{-1}) (13). The concentration of the purified aqueous solution of M_4PO was determined using $\epsilon_{228} = 1.01 \times 10^4$ $M^{-1}cm^{-1}$ (16). DMPO (25 mM) and M^4PO (10 mM) solutions were prepared using double-deionized, double-distilled water. Each nitrone solution was transferred to a Pyrex glass tube (12 × 85 mm) containing a side arm with ground glass connection joined to an EPR quartz flat cell (60 × 0.30 × 16.9 mm). The Pyrex tube was sealed with a rubber stopper. The sealed solutions were exposed to TFD.

TFD was obtained from SCM Corporation (Gainsville, FL) at a purity of ~96%. It is an amber-colored odorous gas with a boiling point of 38°C (14). Liquid TFD was distilled into a crimped sealed vial. Gaseous TFD from the vial headspace was diluted to 50 ppm in nitrogen. Gas samples were analyzed on a Hewlett-Packard 5880A gas chromatograph (GC), and TFD (ppm) concentrations in diluted samples were determined by GC output and dilution ratio.

The nitroso compound MNP was obtained from Sigma Chemical Company (St. Louis, MO). MNP (2 mg) was dissolved in 1 ml benzene and vortexed for 5 min. The reason for the choice of benzene as a solvent was twofold: (1) MNP is known to be more soluble in nonpolar solvents than in water and (2) MNP is volatile; therefore, it is more probable that a sufficient amount of MNP to trap thiyl radicals would remain in benzene as compared to water during N_2 gas bubbling. The experiments in benzene were carried out using the experimental design described for the nitrones.

Superoxide dismutase-polyethylene glycol (PEG-SOD) from bovine erythrocytes, catalase from bovine liver (EC 1.11.1.6), mannitol, and DETAPAC were purchased from Sigma and ethanol from Fisher Company, Philadelphia. ZE caps™ (soft gel capsules containing 75 mg zinc gluconate; vitamin E 200 IU) were obtained from Everett Laboratories, Inc. (East Orange, NJ).

Unless otherwise stated, the instrumental conditions were as follows: field intensity 335.0 mT; modulation width 0.1 mT and microwave power 10 mW. The EPR spectra were obtained using a 10.5 × 60 × 0.3 mm quartz flat cell (Wilmad Glass Company, Inc. Buena, NJ) connected to an U-tube arm as previous described (49). EPR measurements were made at room temperature. Laboratory illumination was dimmed throughout the study to prevent photolytic degradation of trap agents.

Results

Air-saturated solutions of M_4PO (10 mM) produced, after the addition of TFD (~40–62 ppm), the EPR spectrum given in Figure 4A. The spectrum is the result of two overlapping spin adduct radicals. A triplet spectrum observed, in the lines labeled A with $a_N = 1.81 \pm 0.02$ mT, represents a way in which spin trapping artifacts may arise, that of direct reduction of a nitrone spin trap to a nitroxide free radical. The reduction to the spin adduct has been proposed as a decay process (15). The other spin adduct (B) is the result of a primary nitrogen triplet that has split into doublets due to the interaction of the unpaired nitroxide electron with one β-hydrogen. The hyperfine coupling constants, $a_N = 1.529 \pm 0.02$ mT and $a^\beta_H = 1.685 \pm 0.02$ mT, are consistent with the hydroxyl spin adduct, M_4PO-OH (16).

Oxygen was an absolute requirement for the observation of the spin-trapped radical (B). When the solution was thoroughly bubbled with nitrogen (~20 min) before the addition of TFD, only a triplet signal (spin adduct A) was detected as indicated in Figure 4B. In the presence of 1 mM diethylenetriaminepentaacetic acid (DETAPAC), a heavy-metal chelating agent, no signal was observed from the N_2-bubbled solution (data not shown). This indicated the necessity for both oxygen and metal ions to be present. The observations suggest redox metal ion catalysis of TFD oxidation.

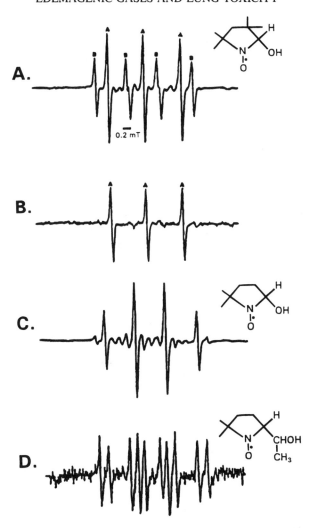

Figure 4. EPR spectra observed after the addition of TFD to different solutions of M₄PO and DMPO. (A) Air-saturated solution of M_4PO (10 mM). (B) Nitrogen-saturated solution of M_4PO (10 mM). (C) Aerobic solution of DMPO (25 mM). (D) Aerobic solution of DMPO (25 mM) and ethanol (2 M). EPR spectrometer conditions for (A), (B), and (C): scan time 480 s, time constant 0.5 s, and modulation amplitude 0.05 mT; (D) scan time 240 s, time constant 0.250 s, and modulation amplitude 0.1 mT. From ref. 60, with permission.

The production of M_4PO-OH (signal B, Figure 4A) is not conclusive evidence for the presence of ˙OH or similar species. The formation of M_4PO-OH could originate from the direct addition of ˙OH to M_4PO. However, it could also originate from the presence of superoxide radical anion ($O_2^{-˙}$). $O_2^{-˙}$ rapidly dismutates, forming H_2O_2, which in the presence of trace metals yields ˙OH. Alternatively, and similar to DMPO, M_4PO-O_2^{-} adduct could rapidly decompose, forming M_4PO-OH. DMPO is often preferred for hydroxyl radical detection because (1) DMPO-OH yields a distinctive 1:2:2:1 EPR quartet; and (2) DMPO has well-defined and well-characterized EPR signals for the spin adducts originating from radical scavengers after the reaction of the scavengers with the primary radicals formed. Aerobic solutions of DMPO (25 mM), after exposure to TFD, generated the EPR spectrum given in Figure 4C. The signal has a 1:2:2:1 pattern with separation of 1.49 $\pm$ 0.02 mT. Tangible hyperfine lines are observed in addition to the DMPO-OH signal. This additional component was dependent upon the concentration of TFD. The EPR spectral parameters for this component are a_N = 1.53 mT and a_H = 1.72 mT. These parameters are possibly due to a DMPO adduct of a thiyl radical (17,18).

To provide additional insight into the generation of the observed spin adducts, similar spin trapping experiments were performed in the presence of ethanol. Ethanol is known to react with hydroxyl radicals, yielding the hydroxyethyl radicals, which are easily trapped by DMPO. The addition of ethanol to aerobic solutions of DMPO, followed by TFD exposure, generated an EPR spectrum that corresponds to the hydroxyethyl adduct, with a_N = 1.57 $\pm$ 0.02 mT and a^β_H = 2.24 $\pm$ 0.02 mT (Figure 4D) (19):

$$HO^\bullet + CH_3CH_2OH \rightarrow H_2O + CH_3C^\bullet HOH \text{ (hydroxyethyl radical)}$$

The formation of DMPO-OH was progressively inhibited (Figure 5) when solutions of DMPO containing different concentrations of ethanol were exposed to TFD. Concomitantly, the second component grows to be more pronounced as indicated by arrows in Figure 5C. Mannitol (1 M) also reduced the DMPO-OH EPR signal (data not shown). As illustrated in Figure 5, formation of DMPO adducts was suppressed to a great extent by ethanol but not totally abolished even at high concentration. Since the reaction rates for DMPO and ethanol are similar, it is expected that at high ethanol concentrations the formation of the hydroxyethyl radical will predominate and the DMPO-OH EPR signal will be completely abolished. Therefore, because the formation of M_4PO-OH and DMPO-OH adducts is dependent on the presence of oxygen in the solution and the DMPO-OH adduct is not completely eliminated at high ethanol concentrations, the results suggest that the formation of M_4PO-OH and DMPO-OH does not occur via direct formation of $^\bullet OH$ in the reaction mixtures. $^\bullet OH$ may be formed indirectly from $O_2^{-\bullet}$ and the M_4PO-OH/DMPO-OH may also be formed, in part, from the decomposition of M_4PO-O_2^- and DMPO-O_2^-. This explains the formation of the hydroxyethyl adducts of DMPO in addition to the persistence of the M_4PO-OH and DMPO-OH adducts at high ethanol concentrations. In support of the production of $O_2^{-\bullet}$ in the reaction mixtures are: (1) that the observations of hydroxyl and hydroxyethyl adducts of M_4PO and DMPO are dependent on the presence of oxygen in the reaction mixtures and (2) the additional weak EPR signals observed in Figure 5D. These signals can be computer simulated using hyperfine coupling constants a_N = 1.430 mT, a^β_H = 1.165 mT, a^γ_H = 0.120 mT for the DMPO adduct, which correspond to those previously observed for DMPO-O_2^- (4).

In an attempt to clarify the generation of M_4PO-OH and DMPO-OH spin adducts, the possible formation of $O_2^{-\bullet}$ was investigated. Hypothetically, $O_2^{-\bullet}$ might be formed by the addition of e^-_{aq} or by a direct electron transfer to molecular oxygen. $^\bullet OH$ can be produced from $O_2^{-\bullet}$ by trace amount of iron salt via the following mechanisms (Fenton-type reaction):

$$2O_2^{-\bullet} + 2H_2O \rightarrow H_2O_2 + 2HO^- + O_2$$

$$O_2^{-\bullet} + Fe^{+3} \rightarrow O_2 + Fe^{+2}$$

$$Fe^{+2} + H_2O_2 \rightarrow {}^\bullet OH + {}^-OH + Fe^{+3}.$$

Alternatively, the spin adduct DMPO-O_2^- could decompose into DMPO-OH or by another minor side reaction, releasing hydroxyl radicals, which are subsequently spin trapped, giving DMPO-OH (20). Since hydroxyl radical scavengers did not completely abolish the observed DMPO adducts, suggesting the presence of $O_2^{-\bullet}$, the effect of PEG-SOD on the rate of DMPO adducts formation by TFD was explored. Superoxide dismutase catalyzes the conversion of $O_2^{-\bullet}$ to hydrogen peroxide (H_2O_2) and removes $O_2^{-\bullet}$ from the reaction mixtures, by causing its dismutation to O_2 plus H_2O_2. The heme enzyme catalase converts H_2O_2 to water. Figure 6 displays the effect of PEG-SOD, catalase, and PEG-SOD/catalase on the EPR signal of DMPO adducts. These experiments showed that individually PEG-SOD and catalase (to a greater extent) significantly diminish the production of the reactive oxygen species (ROS) DMPO adducts, and that the combination of these enzymes totally inhibits their formation. No changes were observed when PEG-SOD and/or catalase were independently added after adduct formation (data not shown). These results indicate the involvement of $O_2^{-\bullet}$ as the primary oxygen radical species in this system.

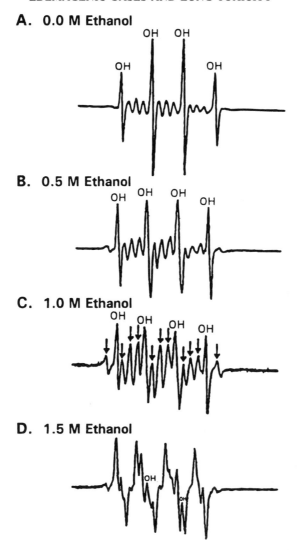

A. 0.0 M Ethanol

B. 0.5 M Ethanol

C. 1.0 M Ethanol

D. 1.5 M Ethanol

Figure 5. EPR spectra following TFD exposure of solutions containing DMPO (25 m*M*) and varying concentrations of ethanol. From ref. 60, with permission.

It is of interest to determine the rate of production of these free radicals in the presence of iron (III) or copper (II) salts, and the metal ion chelator, DETAPAC, because the overall reduction and oxidation mechanism of organic disulfides, monosulfides, and thiols is known to be influenced by metals (21). For this purpose, several solutions, each containing equal concentrations of TFD ($\approx$53–60 ppm) and equal concentrations of DMPO (25 m*M*), were monitored for various lengths of time and their EPR spectra recorded. Figure 7 shows the EPR signal intensity of the DMPO-OH low-field EPR line plotted against the postexposure time in the presence of various salts. An amount of ferric (Fe^{+3}) salt of 0.3 m*M* was sufficient to increase by an order of magnitude the intensity of the DMPO-OH low-field EPR line. The same amount of copper salt slightly increased the intensity of the DMPO-OH low-field peak compared to the control. The metal ion chelator, DETAPAC, inhibited radical formation.

Two EPR signals were detected (Figure 8) when aprotic solutions containing MNP (10 m*M*,

A. DMPO + TFD (49 ppm)

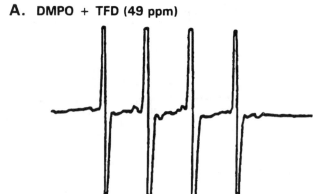

B. DMPO/PEG-SOD + TFD (49 ppm)

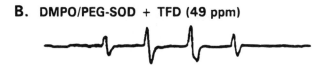

C. DMPO/Catalase + TFD (49 ppm)

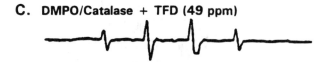

D. DMPO/PEG-SOD/Catalase + TFD (49 ppm)

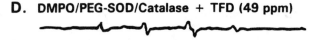

Figure 6. Effect of ROS scavengers on EPR signal of DMPO adducts. (A) DMPO (25 mM) exposed to 49 ppm of TFD. (B) DMPO (25 mM) in the presence of PEG-SOD (500 units/ml) exposed to TFD (49 ppm). (C) DMPO (25 mM) plus catalase (5 units/mg) exposed to TFD (49 ppm). (D) DMPO (25 mM) containing PEG-SOD (500 units/ml) and catalase (5 units/mg) exposed to TFD (49 ppm). EPR spectrometer conditions: receiver gain 5×10^4; modulation amplitute 0.1 mT; scan time 480 s. From ref. 60, with permission.

pH 7) were exposed to TFD in a sealed EPR apparatus. The stronger signal (I) in Figure 8A consisted of three sharp lines ($a_N = 1.53$ mT) and was easily identified as di-*tert*-butyl nitroxide (DTBN, I), a decomposition product of MNP trapped by MNP (22).

$$(CH_3)_3C-N^{\cdot}-O$$
$$|$$
$$C(CH_3)_3$$
$$(I)$$

The other EPR signal (II) in Figure 5A consisted of three lines with a different nitrogen hyperfine coupling constant $a_N = 1.39$ mT and a g value of 2.005. This spectrum was difficult to analyze because the intensity of the EPR peaks changed during postexposure time. An attempt was made to identify it by reviewing reported EPR parameters of detected spin adducts. Chignell et al. (23) have reported that when benzylsulfonamide ($C_6H_5CH_2SO_2NH_2$) was ultraviolet (UV) irradiated in

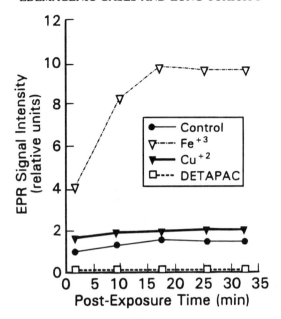

Figure 7. Comparison of the low-field DMPO adducts by EPR signal intensity of TFD alone (control) and TFD in the presence of iron(III), copper(II), and DETAPAC. Abscissa = time in minutes, ordinate = relative units. ● Control, aerobic solutions of DMPO and TFD. ∇ FeCl$_3$ about 3/10,000 mol. ▼ CuCl$_2$ about 3/10,000 mol □ DETAPAC 1 mM. From ref. 60, with permission.

the presence of MNP, a three-line spectrum was obtained (a_N = 1.39 mT; g = 2.0055) in addition to the DTBN EPR signal. This radical was assigned the structure of

$$(CH_3)_3C-N-O^\bullet$$
$$\mid$$
$$SO_2NH_2$$

a sulfonyl spin adduct of MNP. The spectrum of MNP-cysteine sulfonyl radical and MNP-sulfur trioxide radical adduct produced in the cysteine sulfinic acid/hematin/H$_2$O$_2$ system (24) appears similar to the one obtained in Figure 8. The EPR parameters, reported in this study of TFD, have a strong resemblance to the one of a sulfur-centered adduct; therefore, the structure of spin adduct (**II**) has been assigned as

$$(CH_3)_3C-N-O^\bullet$$
$$\mid$$
$$SO_2CF_3$$
$$(II)$$

TFD-derived sulfoxide could result from the known reaction of thiyl radicals with oxygen (25).

$$RS^\bullet + O_2 \rightarrow RSOO^\bullet$$

The computer-simulated spectrum of the superimposed components **I** and **II** using the splitting constants in Table 1 is shown in Figure 8B. Furthermore, excess concentrations of MNP (100 mM) in benzene yielded the well-characterized EPR spectrum (Figure 8C) of two nitroxide radicals (RN$^\bullet$-O) interacting in close vicinity (26, 27). The five broad EPR lines detected (Figure 8C) became narrower under anaerobic conditions as displayed in Figure 8D.

Besides trapping short-lived reactive species, EPR/spin trapping could be used as an analytical

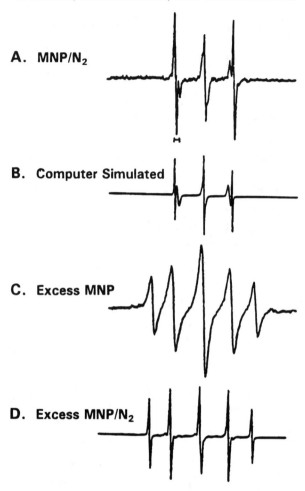

Figure 8.   EPR spectra of aprotic solutions of MNP (pH 7) exposed to 53 ppm TFD. (A) Anaerobic solution of MNP (10 mM). (B) Simulated spectrum of radical I and II using the splitting constant in Table 1. (C) Excess concentration of MNP (100 mM). (D) N$_2$-saturated MNP solution (100 mM). Modulation amplitude 0.05 mT, time constant 0.5 s, scan time 480 s, and receiver gain 5 × 10^4. From ref. 60, with permission.

Table 1.   Splitting constants of nitroxide radicals

			Splitting constant* (mT)	
		g-value	a$_N$	line width
(CH$_3$)$_3$C$-$N$^{\cdot}-$O | C(CH$_3$)$_3$	(I)	2.006	1.543	0.05
(CH$_3$)$_3$C$-$N$^{\cdot}-$O | SO$_2$CF$_2$	(II)	2.005	1.390	0.15

*Determined by computer simulation of EPR spectra.
Values are accurate to within ±0.002 mT.

tool for intervention studies. Competition reactions between therapeutic potential agent and free radical reaction could be monitored. Reactions between a high-potency zinc plus antioxidant, ZE cap™ (Figure 9), and TFD-produced radical adducts were investigated by EPR as a demostrative example. To determine whether ZE cap is an efficient free-radical scavanger of radicals produced by the metal catalyzed oxidation of TFD, several aprotic solutions containing MNP (0.01 M) exposed to TFD and varying concentrations of ZE cap™ were monitored. Figure 10 shows the EPR signal obtained from the adducts of MNP under different experimental conditions. The EPR peaks (indicated by an asterisk) at the low and high magnetic field correspond to a butoxyl butyl nitroxide radical (a_N = 2.66 mT) of MNP previosly reported (22). These results clearly indicate that there is an inverse relationship between the ZE cap™ concentration and the MNP-adduct's EPR signal intensity. The insert of Figure 10 shows the decrease of MNP adducts' middle-field EPR peak intensity versus postexposure time when aprotic solutions containing different ZE cap™ concentrations and MNP (0.01 M) were exposed to TFD. In this case the center-field MNP adduct EPR line was measured immediately after exposure to TFD. These results indicate that ZE cap™ effectively competes with MNP for free radicals produced by TFD.

Discussion

The results obtained in this investigation clearly implicate the production of thiyl and reactive oxygen species during TFD oxidation. The rate of production of these free radicals was found to be increased in the presence of iron(III) and copper(II). Furthermore, the metal ion chelator DETAPAC and the ROS scavengers ethanol, mannitol, and PEG-SOD/catalase were found to inhibit free-radical production. Moreover, there exists an oxygen dependence in this oxidation process. ZE caps™ inhibited the formation of the produced reactive free radicals.

A mechanism consistent with these observations is (28):

Initiation reaction:

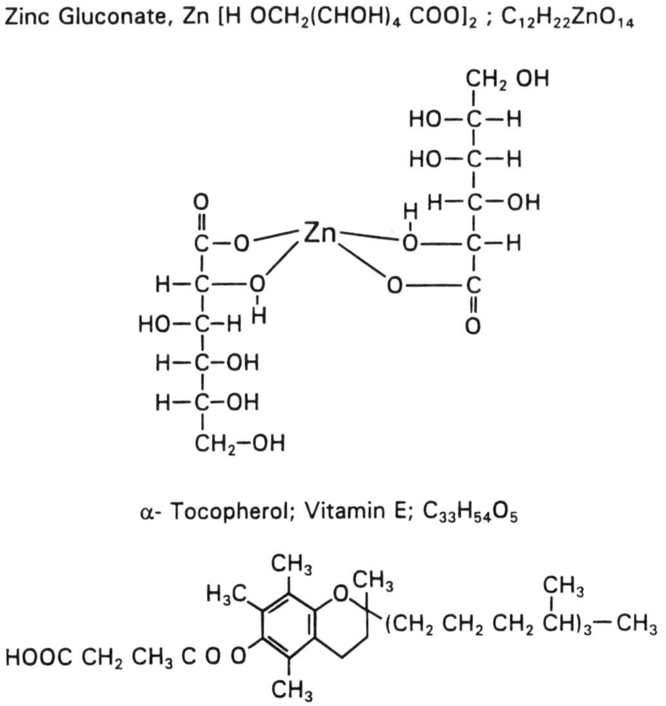

Zinc Gluconate, Zn [H OCH$_2$(CHOH)$_4$ COO]$_2$; C$_{12}$H$_{22}$ZnO$_{14}$

α- Tocopherol; Vitamin E; C$_{33}$H$_{54}$O$_5$

Figure 9. Chemical formula and structure of zinc gluconate and α-tocopherol. From ref 60, with permission.

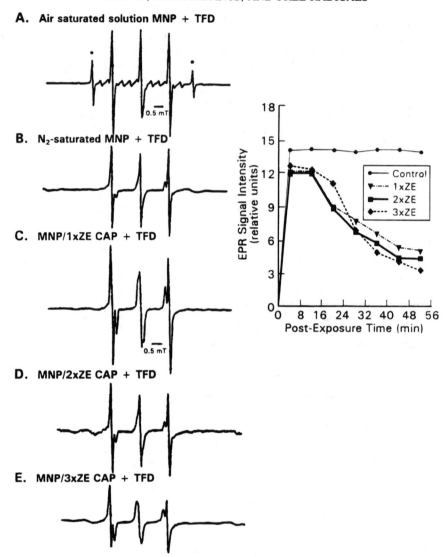

Figure 10. Effect of ZE caps™ on the EPR signals generated from aprotic solutions of MNP exposed to TFD (52 ppm). Administration of 75 mg of zinc gluconate and 200 IU vitamin E in a commercial form. (A) Air-saturated solution of MNP plus TFD. (B) N_2-saturated solution of MNP plus TFD. (C) MNP solution/ $1 \times$ ZE cap™ (1×10^{-3} M) plus TFD. (D) MNP solution/$2 \times$ ZE cap™ (2×10^{-3} M) plus TFD. (E) MNP solution/$3 \times$ ZE cap™ (3×10^{-3} M) plus TFD. Insert: Radical production monitored by EPR signal intensity as a function of post-exposure time. EPR spectrometer conditions: receiver gain 5×10^4; modulation amplitute 0.1 mT; scan time 240 s. From ref. 60, with permission.

$$CF_3SSCF_3 + M^{n+} \rightarrow 2CF_3S^{\bullet} + M^{(n-1)+} + H^+ \tag{5}$$

Propagation reaction:

$$2CF_3S^{\bullet} + O_2 \xrightarrow{H+} CF_3SO_2^{\bullet} \rightarrow 2CF_3SOOH + CF_3S^{\bullet}$$
$$\downarrow O_2 \tag{6}$$
$$\rightarrow\rightarrow CF_3SOOOH \rightarrow O_2^{-\bullet} + CF_3SOH$$

$$CF_3SOOH \rightarrow \quad CF_3SO^{\bullet} \quad + \ ^{\bullet}OH$$
$$\downarrow H^+ \tag{7}$$
$$CF_3SOH$$

$$DMPO + \ ^{\bullet}OH \rightarrow DMPO-OH \tag{8}$$

$$M_4PO + \ ^{\bullet}OH \rightarrow M_4PO-OH \tag{9}$$

$$DMPO + CF_3S^{\bullet} \rightarrow DMPO-SCF_3 \tag{10}$$

$$MNP + CF_3S^{\bullet}O_2 \rightarrow MNP-SO_2CF_3 \tag{11}$$

The free radical chain reactions that are initiated by the oxidation of TFD (as seen by reactions 8, 9, 10, and 11 of this study) could quite easily be propagated in a cellular environment with its antioxidant capacity compromised. The occurrence of such reactions in vivo in the vicinity of vital molecules may be seriously damaging. Thus it appears possible that hydroxyl and thiyl free radicals generated during the metal-catalyzed oxidation of TFD could be responsible for the observed pathophysiological effects. In addition, the high reactivity of the sulfur and hydroxyl radicals makes them prime suspects in the initiation of other free-radical processes such as lipid peroxidation. Cells injured by primary radical damage, or by any other mechanism, begin to malfunction and cannot maintain their tight compartmentalization of ions, not only of Ca^{2+} (29), but also of ions of such transition metals as iron and copper (30). This release of ions from their normal locations (31) can cause more widespread free radical reactions, including lipid peroxidation, facilitated also by impaired functioning of antioxidants (32).

If the pulmonary edema observed upon exposure of TFD is caused by radical formation, a good therapeutic approach is to block that source. Chain-breaking antioxidants, such as α-tocopherol (vitamin E), have been found to have protective effects in many human diseases (30). The reactivity of vitamin E with free radicals is considered to be its major biochemical function. Vitamin E inhibits lipid peroxidation in membranes by scavenging peroxy and alkoxy radicals, thus breaking the chain reaction. Another approach is to bind transition metal ions in ways that will stop them from partipating in free radical reactions (33). Mathew and Walker (21) reported that zinc ions inhibited the consumption of oxygen by the iron thiol system. Such a mechanism could equally apply to decompartmentalized iron and oxygen, in which case zinc inhibits the formation of the damaging reactive oxygen species (Scheme I).

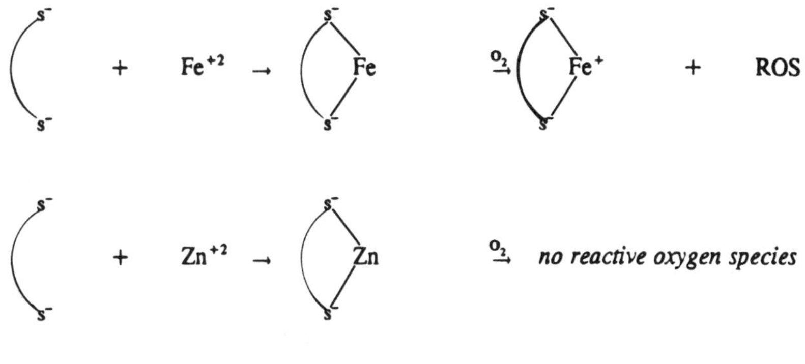

Scheme I

Therefore, in terms of antioxidant therapy, it might be better to block the source of excessive radical formation, that is, to remove H_2O_2, or bind malplaced metal ions.

Isolated rabbit lung preparations using two oxygen radical-generating enzyme systems resulted in lung injury characterized by increased capillary permeability and edema. The injury was inhibited by hydrogen peroxide or hydroxyl scavengers (catalase and dimethyl thiourea, a potent ˙OH scavenger) but not by superoxide dismutase (34). These results demonstrate that chemically generated reactive oxygen species cause acute lung injury in the isolated salt-perfused rabbit lung preparation and suggest that hydrogen peroxide or its ˙OH-derived product caused the injury. These data sustain the possibility of the involvement of toxic reactive oxygen species in the pathophysiologic mechanism of TFD.

Calcium stimulates, and zinc inhibits, a wide variety of cell types (35–42). Furthermore, Zn^{2+} has been shown to have an antioxidant role(s) in defined chemical systems (43). Two mechanisms have been elucidated: the protection of sulfhydryl groups against oxidation, and the inhibition of the production of reactive oxygen species by transition metals (43). Although this study reports in vitro EPR/spin trapping data, the authors suspect that the mechanism(s) of zinc action have important implications for therapeutic protection against TFD-induced radical reactions.

In summary, the known chemical reactivity of radicals makes them prime suspects in the pulmonary pathological findings associated with exposure to TFD (14). Thus it appears that some of the free radicals generated during the scission of the sulfur-sulfur bond of TFD might be responsible for increased permeability and the profound pulmonary edema caused by this organofluorine disulfide compound.

PERFLUOROISOBUTYLENE (PFIB)

The chemistry of perfluoroisobutylene (PFIB) has been studied since 1953 (44). This fluoro-olefin is of interest because of its high electrophilic properties. PFIB is produced by thermal decomposition of polytetrafluoroethylene (e.g., Teflon) and reacts with practically all known nucleophiles (45, 46). Overheating of insulating material generates fumes that may pose a serious health hazard to the respiratory tract in humans, resulting in so-called "polymer fume fever" with symptoms ranging from slight irritation to severe pulmonary edema (46). The high electrophilicity of PFIB is a result of the strong electron-attracting effect of the fluorine atoms of the CF_3 groups and the possibity of the vinylic fluorine atoms to conjugate with the C=C bond. PFIB is a colorless, highly toxic gas, approximately 10 times as toxic to the lungs as phosgene (46). It is possible that it exerts its effects by the depletion of intracellular nucleophiles (47). Reactive nucleophiles occurring in vivo include amines, thiols, and alcohols.

When PFIB reacts with charged nucleophiles (Nu^-), a vinylic fluorine atom may be replaced. This process can be accompanied by bonding of the nucleophile to PFIB. In a number of cases, allylic substitution products are formed because of elimination of a fluoride anion from one of the CF_3 groups. The ratio of vinylic and allylic substitution products is determined by the nature of the entering nucleophile and by the reaction conditions:

$$
(CF_3)_2C{=}CF_2 \xrightarrow{Nu^-}
\begin{array}{c}
CF_3 \\
| \\
F_2C{-}C{-}CFNu \\
| \ominus \ | \\
F \quad F \\
\textit{carbanion}
\end{array}
$$

$$\xrightarrow[\ ]{-F^-} (CF_3)_2C{=}CFNu \xrightarrow{Nu^-} (CF_3)_2C{=}C(Nu)_2$$
$$\xrightarrow[-F^-]{} F_2C{=}\underset{CF_3}{C{-}}CF_2Nu \xrightarrow[-F^-]{Nu^-} NuCF{=}\underset{CF_3}{C}CF_2Nu$$

In addition, the fluorine could be replaced by an addition-elimination mechanism, such that the nucleophile adds to the π-bond to form a relatively stable carbanion (48).

However, because of its diversity of reactions and because PFIB is so reactive, it is possible that some of its toxicological mechanisms may be free radical mediated. Electron paramagnetic resonance (EPR)/spin trapping techniques have been successfully applied to determine if short-lived free radicals are involved as reaction intermediates by scavenging the reactive radical to produce more stable nitroxide radicals. Therefore, first the reaction of PFIB and spin trapping agents must be understood prior to applying EPR/spin trapping techniques to investigate possible free-radical mechanisms in the interaction of PFIB with biological systems. In this section we describe the reaction of PFIB with nitrone and nitroso spin traps.

Experimental Procedures

Chemicals

PFIB was purchased from Flura Corporation (Newport, TN). 4-POBN, PBN, and DBNBS were obtained from Sigma Chemical Co. (St. Louis, MO). MNP was obtained from Aldrich Chemical Co. (Milwaukee, WI). Dried benzene was acquired from Fisher Scientific (Pittsburgh, PA). Nitrogen gas was obtained from Matheson Co. (Secaucus, NJ). All chemicals were commercial samples of high purity and used as supplied.

Aqueous solutions (0.1–0.2 mM) of 4-POBN and DBNBS were prepared and transferred to an EPR/flat cell glass apparatus tube as previously described (49). Concentration-dependent studies were performed with PFIB in which the aqueous solutions of 4-POBN or DBNBS were reacted with 15 μl [24.1 ppm; 1 ppm = 8.2 mg/m^3 (3)], 30 μl (48 ppm), and 60 μl (96.4 ppm) of PFIB. The PFIB concentration of 15 μl generated a weak EPR signal with poor resolution, making the characterization and identification of the signal difficult. A clear EPR spectrum with defined signal intensity and better resolution was generated using 30 μl. Therefore, 30 μl was injected into the trap solutions contained in the EPR sealed-glass apparatus using gas-tight syringes from Supelco, Inc. (Bellefonte, PA). Anaerobic conditions were achieved by bubbling N_2 for approximately 15 min through the aqueous solutions at a constant rate (0.1 ml/min).

Aprotic solutions (0.1–0.2 mM) of PBN and MNP were also prepared and purged with N_2 for 10 min to remove oxygen, which causes EPR line broadening. These solutions, contained in the EPR sealed-glass apparatus, were exposed to 48 ppm PFIB and were analyzed by EPR.

EPR Spectra Measurements

EPR spectra were measured at room temperature using a Varian E-109 Century series X-band EPR spectrometer equipped with a TM_{110} microwave cavity and an E-102 microwave bridge. Spectrometer conditions are given in the figure legends.

Data Analysis

EPR spectra were stored on a COMPAQ Deskpro 386S computer. The hyperfine coupling constants were measured either with a spectrum simulation program written by P. Kuppusamy or directly from the spectrum. Each experiment was performed five times.

Results

PFIB reacts rapidly with nitrone and nitroso spin traps, forming various spin adducts (Figures 11–14). These results suggest that in solution it is not the PFIB itself that reacts directly with the spin traps, but instead, the spin adducts originate from spin trap reactions with highly reactive intermediates formed in the decomposition of the dissolved PFIB. It is known that PFIB decomposes in aqueous solutions to form fluorophosgene (F_2CO), and furthermore, fluorophosgene decomposes, ultimately leading to the formation the carbon dioxide radical anion ($CO_2^{\cdot-}$) and hydrogen fluoride (Eq. 1) (50–52):

$$F_2CO \rightarrow CO_2^{\cdot-} + 2HF \tag{12}$$

Similar to phosgene (Cl_2CO) (53), two initial dissociation mechanisms for fluorophosgene are conceivable. These mechanisms involve the heterolytic cleavage (Eq. 13) and/or the homolytic cleavage (Eq. 14) of a fluorine-carbon bond.

$$F_2CO \rightarrow FCO^+ + F^- \tag{13}$$

$$F_2CO \rightarrow FC^{\cdot}O + F^{\cdot} \tag{14}$$

In order to identify reactive intermediates formed in the decomposition of PFIB, PFIB was dissolved in solutions containing nitrone or nitroso spin traps. The reactions were carried out under protic (aqueous) and aprotic conditions.

Reactions with Nitrone Spin Traps

Addition of PFIB to an aerated aqueous solution containing the water soluble nitrone spin trap 4-POBN yields the EPR spectrum shown in Figure 11. This EPR spectrum consists of three overlapping spin adduct EPR spectra. One spin adduct yields an EPR spectrum consisting of a triplet of doublets indicated by the open circles. This spin adduct can be computer simulated using hyperfine coupling constants $a_N = 1.550$ mT and $a_H^\beta = 0.3$ mT, which correspond to the previously characterized POBN-CO_2^- spin adduct (5,50). The formation of this spin adduct is also consistent with the dissociation products of PFIB (Eq. 1). The second spin adduct identified in Figure 11A consists of a 1:2:2:1 quartet with hyperfine coupling conststants $a_N = a_H^\beta = 1.45$ mT and corresponds to the t-butylaminoxyl radical (H-N'(O)-Bu^t) resulting from the hydrolysis or decomposition of 4-POBN (5). The third spin adduct yields an EPR spectrum consisting of a triplet of doublets computer simulated using hyperfine coupling constants $a_N = 1.525$ mT and $a_H^\beta = 0.250$ mT. This spin adduct has nitrogen and β-hydrogen couplings similar to those reported for the reaction of the phosgene-derived carbonyl chlorides and 4-POBN ($a_N = 1.58$ mT, $a_H^\beta = 0.26$ mT) (53). Therefore, its formation is attributed to the reaction (in a similar fashion as occurs with phosgene) of one of the carbonyl monofluoride-type species (Eqs. 2 and 3) with 4-POBN. Addition of a carbonyl fluoride to 4-POBN produces a spin adduct containing a β-hydrogen to interact with the nitroxide electron yielding a spin adduct EPR spectrum consisting of a triplet of doublets. The smaller nitrogen hyperfine coupling ($a_N = 1.525$ mT) for the 4-POBN-COF adduct when compared to the 4-POBN-COCl adduct ($a_N = 1.58$ mT) from phosgene is attributed to the larger electron-withdrawing capability of the more electronegative fluorine. Figure 11B shows the overall computer simulation that matches the experimental EPR spectrum in Figure 11A. This simulated EPR spectrum was obtained using a linewidth of 0.16 mT and after added the three described spin adducts EPR simulation. Figure 11C shows the

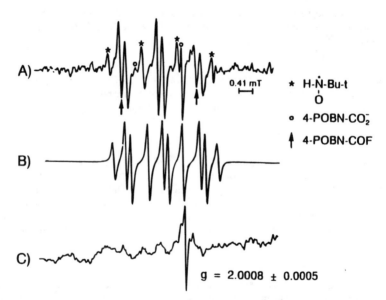

Figure 11.   Proposed mechanism for the reaction of PFIB with O-nucleophiles under aerobic conditions. (A) POBN adducts; the spectrometer conditions were 10 mW microwave power, 0.05 mT modulation amplitude, 2 s time constant, 2.5×10^5 receiver gain, and 0.27 mT/min scan rate. (B) Computer simulation that best represented the experimental one. (C) CO_2· radical. Spectrometer conditions: 10 mW microwave power, 0.1 mT modulation amplitude, 1 s time constant, 1.6×10^5 receiver gain, 1.25 mT/min scan rate.

EPR spectrum obtained in the absence of the 4-POBN spin trap corresponding to the $CO_2^{\cdot -}$ radical anion ($g = 2.0008 \pm 0.0005$). This EPR spectrum was obtained using a closed EPR glass apparatus described earlier in the experimental procedures and its g value is consistent with previously reported g values for $CO_2^{\cdot -}$ (54).

The decomposition of PFIB depends on the presence of oxygen and/or protons in the reaction medium. For instance, when PFIB is dissolved in a deaerated (N_2 saturated) aprotic (benzene) solution containing the nitrone spin trap PBN, it generates an adduct that yields the EPR spectrum shown in Figure 12A. PBN differs from 4-POBN in that it is less water soluble and does not contain the aminoxide group in the aromatic *para* position to the t-butylnitrone functional group. As shown in Figure 12B, the spin adduct EPR spectrum (Figure 12A) can be computer simulated (0.08 mT linewidth) as a triplet of doublets ($a_N = 1.10$ mT, $a_H^{fl} = 0.25$ mT) with an additional γ-coupling $a^\gamma = 0.05$ mT. Two possible mechanisms that cannot be differentiated by the experimental results are conceivable to explain the EPR spectrum in Figure 12A. One involves the defluorination of PFIB to form a carbon centered radical on PFIB and a fluorine radical (Eq. 15).

$$\begin{array}{c} CF_3 \\ \diagdown \\ \diagup \quad C = CF_2 \\ CF_3 \end{array} \quad \longrightarrow \quad \begin{array}{c} CF_3 \\ \diagdown \\ \diagup \quad C = \overset{\cdot}{C} - F \quad + \quad F^{\cdot} \\ CF_3 \end{array} \qquad (15)$$

Another mechanism would be the direct addition of the PFIB to the PBN nitrone carbon and simultaneous defluorination of PFIB forming a fluoride radical (Eq. 16). The carbon-centered PFIB radical (Eq. 15) will react with PBN in a direct spin trapping type reaction to yield an identical adduct as shown in Eq. 16.

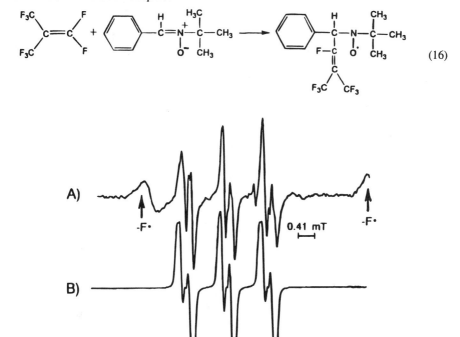

Figure 12. Proposed mechanism for the reaction of PFIB with nitrone PBN. (A) EPR signal of the detected PBN adducts (B) Computer simulation that best fit the experimental one (A). Spectrometer conditions: 10 mW microwave power, 0.1 mT modulation amplitude, 2 s time constant, 1.25×10^5 receiver gain, and 0.27 mT/min scan rate.

Fluorine has a nuclear spin $I = \frac{1}{2}$. Therefore, the interaction of a single fluorine nucleus with the unpaired nitroxide electron would cause each line in the triplet of doublets to be further split into doublets yielding the overall triplet of quartets observed in Figure 12A. The broad outer peaks in the EPR spectrum (Figure 12A) can be explained as the PBN trapping of fluorine radicals yielding a spin adduct with a hyperfine coupling for the β-fluorine of 5.1 mT (55). This observation also supports the mechanisms described in Eqs. 15 and 16.

Reactions with Nitroso Spin Traps

Dissolving PFIB in an aerated aqueous solution containing the nitroso spin trap DBNBS (56) yields the spin adduct EPR spectrum shown in Figure 13A. This EPR spectrum consists of a triplet of triplets with each triplet arranged in a 1:2:1 pattern indicating that the unpaired nitroxide electron is interacting with two equivalent nuclei with a nuclear spin, $I = \frac{1}{2}$. The interaction of the unpaired nitroxide electron with two equivalent β fluorine nuclei ($I_F = \frac{1}{2}$) would yield an EPR spectrum in which each line in the primary triplet is further split into a 1:2:1 triplet. Therefore, it is possible that a reactive intermediate is formed during the decomposition of PFIB and prior to the formation of fluorophosgene. Equation 17 shows a possible mechanism that would explain the observed EPR results in Figure 13. Addition of the intermediate $F_2C^\cdot OO^-$ to the DBNBS nitroso nitrogen would yield a spin adduct containing two β-fluorines. Such an intermediate could rapidly dehydrate, leaving a F_2COH functional group attached to the DBNBS nitroso nitrogen. In support of this type of adduct is the lack of γ-hydrogen splittings usually observed in DBNBS adducts. The electron-withdrawing capability of the fluorines would minimize the delocalization of the unpaired nitroxide onto the aromatic ring and thus minimizing the interaction of the unpaired nitroxide electron with the nuclei of the aromatic protons. Figure 13B shows the computer simulation that matches the EPR spectrum in Figure 13A. This simulation was obtained using hyperfine coupling constants, $a_N = 1.365$ mT and $a_{F(2)}{}^\beta = 1.035$ mT, and a linewidth of 0.1 mT.

Addition of PFIB to an aerated aprotic (benzene) solution containing the nitroso spin trap MNP yields a spin adduct EPR spectrum consisting of a 1:2:3:2:1 quintet (Figure 14). This EPR spectrum could also originate from the $F_2C^\cdot OO^-$ species (Eq. 17). In this case there are no solvent protons required for the dissociation of the PFIB. However, it is possible that the carbanion

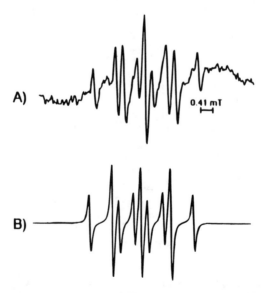

A)   0.41 mT

B)

Figure 13. Formation of DBNBS adduct when PFIB reacted with DBNBS. EPR signal of the observed DBNBS adduct (A) including the computer simulation (B) Spectrometer conditions: 10 mW microwave power, 0.1 mT modulation amplitude, 1 s time constant, 1.25×10^5 receiver gain, and 1.25 mT/min scan rate.

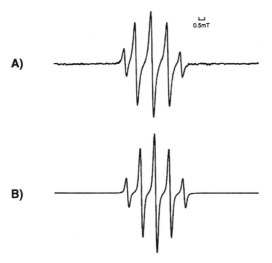

Figure 14. Nitration of PFIB by the nitroso spin trap MNP. Spectrometer conditions: 10 mW microwave power, 0.1 mT modulation amplitude, 1 s time constant, 1.25×10^5 receiver gain, and 1.25 mT/min scan rate.

reacts with solvent molecules, subsequently releasing the $F_2C^\cdot OO^-$ species, which reacts with MNP directly, adding to the nitroso nitrogen. The EPR pattern in Figure 14A suggests that the primary nitrogen and β-fluorine couplings are similar if not identical. Therefore, the spectrum can be computer simulated using hyperfine coupling constants, $a_N = 2.755$ mT $a^\beta_{F(2)} = 2.765$ mT (Figure 14B).

$$(17)$$

$$(18)$$

These results suggest that PFIB decomposes when dissolved, forming various reactive intermediates initiated by the attack of dissolved oxygen to form $F_2C^\cdot OO^-$. In aqueous environments this intermediate rapidly forms fluorophosgene, which then decomposes to ultimately yield the carbon dioxide radical anion and hydrogen fluoride.

Discussion

The ability of PFIB to enter into reactions with diverse nucleophiles distinguishes it from other highly hydrophobic gases. PFIB is an extremely powerful electrophile whose role as a toxicant is modulated by lung thiol levels (57). The cellular thiol nucleophilic protectants glutathione and cysteine have different major roles, which include the function of glutathione as a cofactor in transport and transferase reactions, in protein synthesis, and in detoxication of reactive intermediates formed intracellularly. Initial studies on the mechanism of pulmonary injury by PFIB (57, 58) showed that the amounts of nonprotein thiol and glutathione in lung were reduced by 30% and 49%, respectively, in animals exposed to PFIB. Pretreatment with cysteine esters protected against toxicity and raised cysteine levels by 100%. Furthermore, the authors concluded that the role of glutathione as a cofactor in transferase and peroxidation reactions may not be important for protection against PFIB.

This study provides evidence that PFIB is reactive toward nucleophilic reagents to yield substitution and addition radical byproducts. Furthermore, PFIB undergoes nucleophilic reactions typical of fluoro-olefin, such as addition, vinylic, and allylic substitution. In the addition reactions, the nucleophilic catalysis acquires special significance, since it provides a possible mechanism for its toxicity. The toxicity of PFIB may be correlated with its susceptibility to nucleophilic attack (52) and the generation of reactive intermediate species. A striking correlation exits between fluoro-olefin and their toxicological properties: The toxicity of a halogenated olefin is directly proportional to the reactivity of that olefin to nucleophiles (59). Therefore, it is probable that raising the overall levels of nucleophiles (thiols) increases the level of protection by neutralizing the incoming PFIB before it can damage cellular constituents. Further research is required to demonstrate the scavenging role of lung nucleophiles in reducing the toxicity of PFIB.

REFERENCES

1. Babad, H., and Zeiler, A. G. 1973. The chemistry of phosgene. *Chem. Rev.* 73:75–91.
2. McCay, P. B. 1987. Review: Application of EPR spectroscopy in toxicology. *Arch. Toxicol.* 60:133–137.
3. Janzen, E. G. 1971. Spin trapping. *Acc. Chem. Res.* 4:31–40.
4. Buettner, G. R., and Oberley, I. W. 1978. Consideration in spin trapping of superoxide and hydroxyl radicals in aqueous solutions using 5,5-dimethyl-1-pyrroline-N-oxide. *Biochem. Biophys. Res. Commun.* 83:69–74.
5. Buettner, G. R. 1987. Spin trapping: ESR parameter of spin adducts. *Free Radic. Biol. Med.* 3:259–303.
6. Kim, J. K., and Caserio, M. C. 1981. Acyl-transfer reactions in the gas phase. The question of tetrahedral intermediates. *J. Am. Chem. Soc.* 103:2124–2127.
7. Carmichael, A. J., Arroyo, C. M., and Cockerham, L. G. 1988. Reaction of disodium chromaglycate with hydrated electrons. *Free Radic. Biol. Med.* 4:215–218.
8. Halliwell, B., and Gutteridge, J. M. C. 1989. *Free Radical Biology and Medicine*, 2nd ed., p. 25. Oxford: Clarenden Press.
9. Nold, J. B., Petrali, J. P., and Moore, D. H. 1991. Progressive pulmonary pathology of two organofluorine compounds. *Inhal. Toxicol.* 3:123–137.
10. Lopusinski, A. 1989. Chemistry of S-trifluoromethyl organophosphorothioates and their structural analogs, a convenient synthesis of organophosphorus fluoridates. *Phosphorus Sulfur Silicon* 45:137–143.
11. Mottley, C., and Mason, R. P. 1989. Nitroxide radical adducts in biology: Chemistry, applications, and pitfalls. In *Biological Magnetic Resonance*, eds. L. J. Berliner and J. Reuben, Vol. 8, pp. 489–546. New York: Plenum Press.
12. Janzen, E. G. 1980. A critical review of spin trapping in biological systems. In *Free Radicals in Biology*, ed. W. A. Pryor, Vol. IV, pp. 115–154. New York: Academic Press.
13. Kalyanaraman, B., Felix, C. C., and Sealy, R. C. 1982. Photoionization of melanin precursors: An electron spin resonance investigation using the spin trap 5,5-dimethyl-1-pyrroline-1-oxide. *Photochem. Photobiol.* 36:5–12.
14. Moore, D. H., and Wall, H. G. 1991. The effects of exercise following exposure to bis(trifluoromethyl)disulfide. *Drug Chem. Toxicol.* 14:343–352.
15. Giottta, G. J., and Wang, H. H. 1972. Reduction of nitroxide free radicals by biological materials. *Biochem. Biophys. Res. Commun.* 46:1576–1580.
16. Janzen, E. G., Shetty, R. V., and Kunanec, S. M. 1981. Spin trapping chemistry of 3,3,5,5-tetramethylpyrroline-N-oxide: An improved cyclic spin trap. *Can. J. Chem.* 59:756–758.
17. Davies, M. J., Forni, L. G., and Shuter, S. L. 1987. Electron spin resonance and pulse radiolysis studies on the spin trapping of sulphur-centered radicals. *Chem. Biol. Interact.* 61:177–188.

18. Buettner, G. R. 1985. Thiyl free radical production with hematoporphyrin derivative, cysteine and light: A spin trapping dtudy. *FEBS Lett.* 177:295–299.

19. Searle, A. J., and Tomasi, A. 1982. Hydroxyl free radical production in iron-cysteine solutions and protection by zinc. *J. Inorg. Biochem.* 17:161–166.

20. Carmichael, A., Mossoba, M. M., Riesz, P., and Rosenthal, I. 1985. Food dye-sensitized photoreaction in aqueous media. *Photobiochem. Photobiophys.* 10:13–21.

21. Mathews, A. P., and Walker, S. 1909. The action of metals and strong salt solutions on the spontaneous oxidation of cystein. *J. Biol. Chem.* 6:299–312.

22. Perkins, M. J., Ward, P., and Horsfield, A. 1970. A probe for homolytic reactions in solution. Part III. Radicals by hydrogen abstraction. *J. Chem. Soc. B* 395–400.

23. Chignell, C. F., Kalyanarama, B., Mason R. P., and Sik R., H. 1980. Spectroscopy studies of cutaneous photosensitizing agents-I. Spin trapping of photolysis products from sulfinamide, 4-aminobenzoic acid and related compounds. *Photochem. Photobiol.* 32:563–571.

24. Harman, L. S., Mottley, C. and Mason, R. P. 1984. Free radical metabolites of L-cysteine oxidation. *J. Biol. Chem.* 259:5606–5611.

25. Schafer, K., Bonifacic, M., Bahnemann, D., and Asmus, K.-D. 1978. Addition of oxygen to organic sulfur radicals. *J. Phys. Chem.* 82:2777–2780.

26. Swartz, H. M., Bolton, J. R., and Borg, D. C. 1972. *Biological Applications of Electron Spin Resonance,* p. 532. New York: Wiley-Interscience, John Wiley & Sons.

27. Joshi, A. and Yang, G. C. 1981. Spin Trapping of radicals generated in the UV photolysis of alkyl disulfides. *J. Org. Chem.* 46:3736–3738.

28. Parker, A. J., and Kharasch, N. 1959. The scission of the sulfur-sulfur bond. *Chem. Rev.* 59:583–628.

29. Orrenius, S. 1985. Oxidative stress studied in intact mammalian cells. *Philos. Trans. R. Soc. B* 311:673–677.

30. Halliwell, B., and Gutteridge, J. M. C. 1984. Lipid peroxidation, oxygen radicals, cell damage and antioxidant therapy. *Lancet* 1:1396–1398.

31. Albert, A. 1985. *Selective Toxicity.* New York: Chapman and Hall.

32. Halliwell, B., and Gutteridge, J. M. C. 1988. Free radicals and antioxidant protection: Mechanisms and significance in toxicology and disease. *Hum. Toxicol.* 7:7–13.

33. Halliwell, B., and Gutteridge, J. M. C. 1985. The importance of free radicals and catalytic metal ions in human disease. *Mol. Aspects Med.* 8:89–93.

34. Tate, R. M., Van Benthuysen, K. M., Shasby, D. M., McMurtry, I. F., and Repine, J. E. 1982. Oxygen-radical-mediated permeability edema and vasoconstrition in isolated perfused rabbit lungs. *Am. Rev. Respir. Dis.* 126:802–806.

35. Nugteren, D. H., Beerthuis, R. K., and Van Dorp, D. A. 1966. The enzymic conversion of *all-cis*-8,11,14-eicosatrienoic acid into prostaglandin E$_1$. *Recl. Trav. Chim. Pays-Bas Belg.* 85:405–419.

36. Chvapil, M., Ryan, J. N., and Brada, Z. 1972. Effects of selected chelating agents and metals on the stability of liver lysosomes. *Biochem. Pharmacol.* 21:1097–1105.

37. Chvapil, M. 1973. New aspects in the biological role of zinc: A stabilizer of macromolecules and biological membranes. *Life Sci.* 13:1041–1049.

38. Chvapil, M., Ryan, J. N., Elias, S. L., and Peng, Y. M. 1973. Protective effect on zinc on carbon tetrachloride-induced liver injury in rats. *Exp. Mol. Pathol.* 19:186–196.

39. Willson, R. L. 1977. In *Iron Metabolism,* CIBA Foundation Symposium 51 (new series), pp. 331–354. Amsterdam: Excerpta Medica.

40. Cho, C. H., Dai, S,. and Ogle, C. W. 1977. The effect of zinc on anaphylaxis *in vivo* in the guinea-pig. *Br. J. Pharmacol.* 60:607–608.

41. Marone, G., Findlay, S. R., and Lichtenstein, L. M. 1981. Modulation of histamine release from human basophils *in vitro* by physiological concentrations of zinc. *J. Pharmacol. Exp. Ther.* 217:292–298.

42. Mousli, M., Gies, J.-P., Bertrand, C., Pelen, F., Bronner, C., and Landry, Y. 1990. The sensitivity to Zn^{+2} discriminates between typical and atypical mast cells. *Agents Actions* 30:102–105.

43. Bray, T. M., and Bettger, W. J. 1990. The physiological role of Zn as an antioxidant. *Free Radic. Biol. Med.* 8:281–291.

44. England, D. C. and Krespan, C. G. 1966. Fluoroketenes. I. Bis(trifluoromethyl)ketene and its reactions with fluoride ion. *J. Am. Chem. Soc.* 88:5582–5587.

45. Smith, L. W., Gardner, R. J., and Kennedy, G. L. 1982. Short-term inhalation toxicity of perfluoroisobutylene. *Drug Chem. Toxicol.* 5:295–303.

46. Oberdorster, G., Ferin, J., Gelein, R., Finkelstein, J., and Baggs, R. 1994. Effects of PTFE fumes in the respiratory tract: A particle effect? *Aerospace Medical Association 65th Annual Scientific Meeting* 538:A52.

47. Lailey, A. F., Hill, L. Lawston, I. W., Stanton, D., and Upshall, D. G. 1991. Protection by cysteine esters against chemically induced pulmonary oedema. *Biochem. Pharmacol.* 42:PS47–54.

48. Tedder, J. M., and Walton, J. C. 1980. The importance of polarity and steric effects in determining the rate and orientation of free radical addition to olefins. *Tetrahedron* 36:701–707.

49. Arroyo, C. M., and Kohno, M. 1991. Difficulties encontered in the detection of niric oxide (NO) by spin trapping techniques. A cautionary note. *Free Radical Res. Commun.* 14:145–155.

50. Riesz, P., Berdahl, D., and Christman, C. L. 1985. Free radical generation by ultrasound in aqueous and nonaqueous solutions. *Environ. Health Perspect.* 64:233–252.

51. Fawcett, F. S., Tullock, C. W., and Coffman, D. D. 1962. The chemistry of carbonyl fluoride. I. The fluorination of organic compounds. *J. Am. Chem. Soc.* 84:4275–4284.

52. Zeifman, Y. B., Ter-Gabrielyan, Y. G., Gambaryan, N. P., and Knunyants, I. L. 1984. The chemistry of perfluoroisobutylene. *Uspekhi Khim.* 53:431–461.

53. Arroyo, C. M., Feliciano, F., Kolb, D. L., Keeler, J. R., Millette, S. R., and Stotts, R. R. 1993. Autoionization reaction of phosgene ($OCCl_2$) studied by electron paramagnetic resonance/spin trapping techniques. *J. Biochem. Toxicol.* 8:107–110.

54. Ovenall, D. W., and Whiffen, D. H. 1961. Electron spin resonance and structure of the CO_2^- radical ion. *Mol. Phys.* 4:135–144.

55. Haire, L. D., Krygsman, P. H., Janzen, E. G., and Oehler, U. M. 1988. Correlation of radical structure with EPR spin adduct parameters: Utility of the proton, carbon-13, and nitrogen-14 hyperfine splitting constant of aminoxyl adducts of PBN-nitronyl-^{13}C for three-parameters scatter plots. *J. Org. Chem.* 53:4535–4542.

56. Kaur, H., Leung, K. H. W., and Perkins, M. J. 1981. A water-soluble, nitroso-aromatic spin-trap. *J. Chem. Soc. Chem. Commun.* 1981:142–143.

57. Lailey, A. F., Leadbeater, L., Maidment, M. P., and Upshall, D. G. 1989. The mechanism of chemically-induced pulmonary oedema. *Proc. 3rd Int. Symp. Protection Against Chemical Warfare Agents*, 11–16 June, Sweden, pp.153–161.

58. Makulova, I. D. 1965. Clinical aspect of acute poisoning with perfluoroisobutylene. *Gig. Tr. Prof. Zabol.* 9:20–23.

59. Cook, E. W., and Pierce, J. S. 1973. Toxicology of fluoro-olefins. *Nature* 242:337–338.

60. Arroyo, C. M., Kirby, S. D., Werrlein, R. J., McCarthy, R. L., Moran, T. S. and Keeler, J. R. 1994. Reactive oxygen species produced in metal-catalyzedoxidation of bis(trifluoromethyl) disulfide and protection by ZE.™ *J. Toxicol. and Environ. Health.* 41:329–344.

Chapter Eighteen

Lipid Peroxidation and Antioxidant Depletion Induced by Blast Overpressure

Nabil M. Elsayed

INTRODUCTION

Detonation of explosives or firing of large-caliber weapons in the military, and accidental (occupational) or intentional (terrorist) explosions, as well as a variety of high-energy impulse noise that exceeds safe threshold limits in civilian situations, can cause injury or even death (1–5). The common feature in all these events is that they produce a region of instantaneous rise in atmospheric pressure that is termed blast overpressure (BOP). This rise in pressure is usually followed by exponential decay back to baseline ambient level. The rise and fall in pressure occur within a period of micro- to milliseconds until steady-state equilibrium takes place. Figure 1 shows a graphic representation of a simple or ideal blast wave occurring in an open field with no reflections from walls or other structures. Such a blast wave can be described physically by a typical Friedlander waveform, in which the mathematical derivation of the incident wave is the integral of pressure changes over time (P/dt) for the area under the peak (impulse) of both positive (above ambient), and negative (below ambient) phases. On the other hand, if the explosion occurs in an enclosure or a large-caliber weapon is fired from or into an enclosure, the blast wave generated will be reflected from the surfaces to exhibit a complex

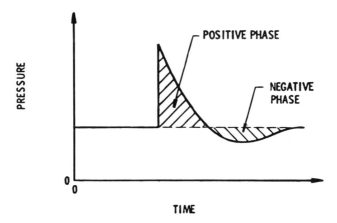

Figure 1. Pressure-time histogram of an ideal blast wave in a free-field environment. Impulse is the integral of pressure over time, P/dt. Positive and negative phases denote the pressure above or below ambient atmospheric pressure, respectively. From ref. 3, with permission.

waveform that requires different mathematical treatment to describe it (Figure 2). The impulse as well as the damage to both soft tissues of human or animal bodies and solid structures from complex blast waves is usually greater than that caused by simple waves. Although the magnitude of the blast peak overpressure weighs heavily on the biological response, other factors contribute to the overall effect of each BOP exposure. For example, there is the total impulse value (positive + negative changes in pressure), duration of the overpressure exposure, orientation of the body to the incident blast wave, etc. Initially, BOP waves will travel radially from the blast source at a velocity greater than the velocity of sound in air (6) until they impact a body or a structure, and then the velocity changes (increase or decrease depending on the structure's composition).

Exposure to blast can cause auditory and/or nonauditory damage. The latter can lead to death with no apparent external signs of injury (1–5). In general, blast-induced injury can be divided into three broad categories: (1) primary blast injury, in which the blast wave travels through air or water, impacting the body to cause internal damage, sometimes with no visible external signs of injury; (2) secondary blast injury, in which the blast wave propels objects or fragments to impact the body and cause the injury; and (3) tertiary blast injury, in which the incoming blast wave would displace the body and cause the injury by impacting it against solid objects (3).

Biologically, exposure to BOP results in injury particularly to the gas-filled organs: ears, lungs, and gastrointestinal tract. Among such organs, the ears are the most susceptible to damage, but they are also the easiest to protect using simple protective devices like ear plugs, ear muffs, etc. The lungs, however, are equally very sensitive, but are much more difficult to protect, and BOP-induced lung injury can be lethal. Upon impact of the incident blast wave on the thorax, the pressure wave propagates within the thorax, and gross displacement of the thoracic wall leads to compression/decompression of intrathorathic structures (6). In previous studies with sheep, we have shown histologically that exposure to BOP cause pulmonary edema and hemorrhage that result in alveolar flooding (Figure 3), and consequently, impairment of the gas exchange functions of the lung as illustrated by decreased arterial O_2 tension (P_aO_2); meanwhile, CO_2 tension (P_aCO_2) usually remains unchanged (Figure 4). Cooper et al. (6) examined rat lung ultrastructure after exposure to BOP in the laboratory using a blast-wave generator in which compressed air was applied to an aluminum diaphragm that disrupted discharging a blast wave. They found considerable damage and tearing to interalveolar septa with capillary rupture and intra-alveolar hemorrhage, and "ballooning" of the endothelium into the capillary lumen. The

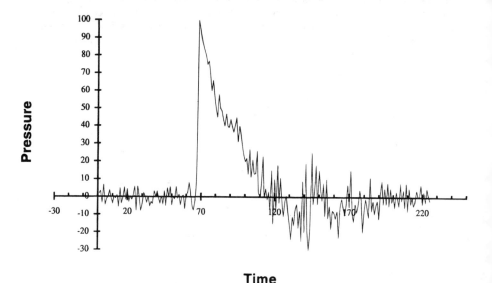

Figure 2. Pressure-time histogram of a blast wave within an enclosure showing reflections from walls and other objects.

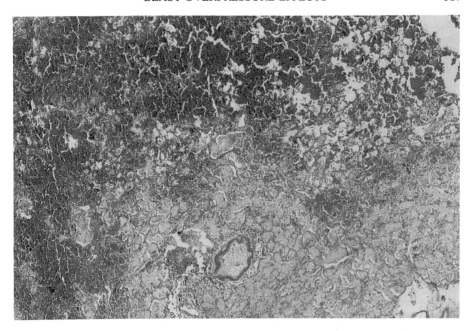

Figure 3. Section from lung of a sheep exposed to a single blast causing level 4-equivalent injury. Lower right area, alveolar edema; upper left area, alveolar hemorrhage. From ref. 50, with permission.

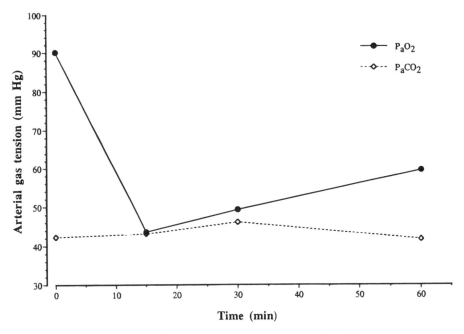

Figure 4. Arterial oxygen tension (P_aO_2) and alveolar carbon dioxide tension (P_aCO_2) measured in arterial sheep blood after exposure to a blast wave of approximately 150 kPa generated in the laboratory using compressed air-driven shock tube. (K. T. Dodd, unpublished observations).

injury was progressive for the first 24 h. Recently, we have reported that exposure to BOP can also induce lipid peroxidation in plasma or lung tissue of sheep, rabbits, and rats (7–10). In addition, a study from the People's Republic of China observed that lipid peroxidation occurs in cochlea of guinea pig ears exposed to BOP (11).

Historically, experimental studies of blast-induced injury date back to the Balkan Wars. For example, in 1914, a Swiss experimental animal researcher, Franchino Rusca, observed that three soldiers who had been killed by a bursting grenade had no external injuries. He then exposed rabbits to blast from exploding 100 g of dynamite and found that they all died similarly without external signs of injury (12). The death was found later to result from pulmonary embolism. During World War I, many shell-shock casualties exhibited psychophysiological symptoms after prolonged exposure to heavy artillery, leading to the belief that blast affect primarily the nervous system (3). In 1918–1919, an American researcher, David Hooker, began conducting blast-injury studies at Sandy Hook Proving Ground, New Jersey, using dogs, rabbits, cats, and frogs exposed to muzzle blast from mortars and naval cannons (13). Blast research continued and intensified during World War II, and by the Germans, British, and Swedes, led the investigations of the effects of nonnuclear blast, with the Germans pioneering quantitative experimentation. American research programs that led the way on the effects of exposure to nuclear blast contributed also to nonnuclear blast research and added significantly to our knowledge. For a more detailed historical overview the reader is referred to the 1991 study of Phillips and Richmond (3). It is worth noting that a large body of research has been produced by the former Soviet Union (14–16), People's Republic of China (11, 17), and East European Countries, particularly Serbia in former Yugoslavia (18). However, most of their research is classified and therefore is not available through open literature.

Despite the continued research and characterization of the different aspects of blast-induced damage, the biochemical mechanism(s) of injury is not yet understood. However, since 1993, a series of preliminary studies from our laboratory at Walter Reed Army Institute of Research suggested that exposure to BOP may be associated with free radical generation resulting in the formation of lipid peroxidation products in sheep and rabbit plasma (7–9). More recently, we found that concomitant with lipid peroxidation, antioxidant contents are depleted, and calcium transport is perturbed in rat lung tissue after BOP exposure (10).

EXPERIMENTAL

We used for the present study adult male Sprague-Dawley rats, certified virus free (CVF), weighing 350–375 g. After a few days of acclimatization upon arrival, the rats were randomly divided into four groups. All rats were deeply anesthetized before exposure to avoid pain, using sodium pentobarbital (60 mg/kg body weight), then exposed either to blast overpressure at 1 of 3 peak pressures of approximately 60 kPa (low), 90 kPa (medium), or 140 kPa (high). The fourth, a control group, was handled similarly except that the rats were not exposed to blast. Immediately after exposure, the rats were euthanized, and the lungs were perfused in situ with ice-cold saline until blanched to flush out the blood, then extracted free of connective tissue and airways, blotted dry on gauze, and weighed. Lung tissue was then homogenized on ice in phosphate buffer, pH 7.4, using a Polytron homogenizer (Brinckman Instruments Co., Westburg, New York). Aliquots of lung homogenate were used for determination of total protein by the method of Lowry et al. (19), and several antioxidant contents, namely, α-tocopherol, ascorbate, and glutathione, and total sulfhydryls were determined using established methods as described previously in detail (10). Lipid peroxidation was assessed by measuring both conjugated dienes in the lipids extracted according to Folch et al. (20), and malondialdehyde formation by the thiobarbituric acid method according to Beuge and Aust (21).

Blast exposures were simulated in our laboratory using a compressed air-driven shock tube illustrated in Figure 5. The shock tube is a 533 cm (17.5 ft) long, 30 cm (12 in) diameter, horizontally mounted, circular steel tube divided into a 76-cm (2.5-ft) compression chamber separated from a 450-cm (15-ft) expansion chamber by one or more Mylar (polyester) sheets (Du Pont Co., Wilmington, DE). The peak pressure obtained varies depending upon the number and thickness of the Mylar sheets placed in the compression chamber. To measure the peak pressure produced, we placed at the end of the expansion chamber piezoresistive pressure

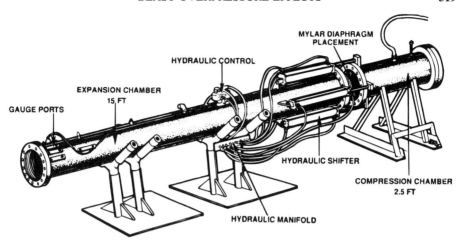

Figure 5. Schematic drawing of the compressed air-driven shock tube used to simulate blast overpressure at the Department of Respiratory Research, Walter Reed Army Institute of Research, Washington, DC.

transducers (low-impedance quartz pressure transducers) specifically designed for pressure-time (impulse) measurements (Piezotronics, Inc., Depew, NY). The pressure-time graphic representation produced is specific for each blast and is considered its specific signature. Other transducers are usually placed at different locations in the open field or enclosure to provide complete evaluation of the incident and reflected BOP waves. The shock tube has been shown as a useful tool to simulate BOP in the laboratory for many years (22, 23). To illustrate the similarities between simulated and actual explosions, a comparison of the signatures of two blasts, one simulated in the laboratory in the course of this study and the other representing detonation of defined charge (i.e., defined weight and distance from the transducer) of dynamite, is shown in Figure 6.

RESULTS AND DISCUSSION

In general, the blast peak pressures produced were selected to be nonlethal; thus all rats survived the exposures. In lung tissue homogenate, total protein content did not change in any of the exposed groups relative to unexposed control. On the other hand, water-soluble antioxidants (ascorbate, glutathione, total sulfhydryls) decreased significantly in a peak pressure-dependent manner. Similarly, the lipid-soluble α-tocopherol content declined significantly. Concomitant with such a decrease in antioxidant contents, lipid peroxidation measured as conjugated diene in lipid extract and as total thiobarbituric acid-reactive substances (TBARS) in whole lung tissue homogenate increased significantly with increased blast peak pressure. Because the units of each measurement differed too greatly from each other to be included in a single figure, the percent change from unexposed control were plotted against blast peak pressure and presented in Figure 7.

Lipid peroxidation or oxidative rancification (24) represents one form of tissue damage (25–27) associated with disease states and drug-induced toxicity that proceeds by free radical-initiated chain reactions. Measurement of malondialdehyde indirectly by the thiobarbituric acid method is one of the most used assays to estimate lipid peroxidation, despite many problems and nonspecificity associated with the assay (28, 29). Measurements of hydrocarbons such as ethane and pentane are also popular methods for assessment of lipid peroxidation, particularly in vivo, although they introduce different kind of problems (29, 30). Reactions 1–7 show the formation of hydrocarbon gases from polyunsaturated fatty acids, and reactions 8–13 illustrate the mechanism of malondialdehyde formation.

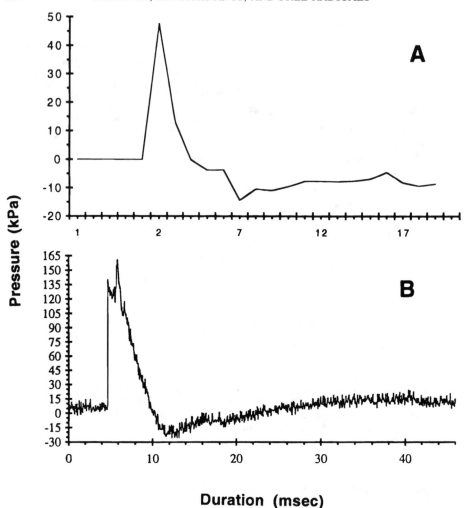

Figure 6.   Pressure-time histograms (blast signatures) of (A) 0.364 kg TNT detonated at 10.2 m from the pressure transducer and of (B) a peak pressure of 136 kPa simulated in the laboratory using the shock tube. The pressure transducer was placed immediately at the nozzle of the shock tube.

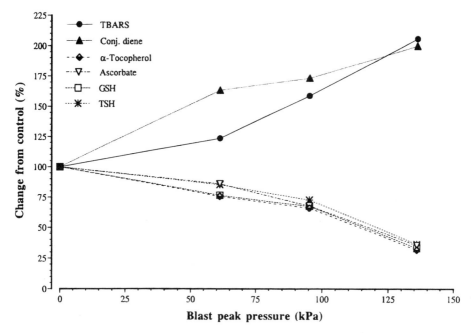

Figure 7. Effects of increasing blast peak pressure on lipid peroxidation assessed by measuring conjugated diene in extracted lipids and thiobarbituric acid-reactive substances (TBARS) in whole lung homogenate, and the lung endogenous antioxidants: ascorbate, glutathione, total sulfhydryls, and α-tocopherol represented as percent change from control. Control values, expressed as mean ± SE, n = 5, were: TBARS, 0.17 ± 0.025 nmol/mg protein; conjugated dienes as the ratio of absorbance E_{232}/E_{215} nm, 0.30 ± 0.035; α-tocopherol, 0.41 ± 0.04; ascorbate, 46.9 ± 3.2 nmol/mg protein; GSH, 18.7 ± 4.5; TSH, 21 ± 3.2 nmol/mg protein

Since oxygen (O_2) metabolism results in production of reactive oxygen species in all living organisms, it was implicated in lipid peroxidation (31). Direct or indirect radical attack on membrane lipids by gases (hyperoxia, ozone, and nitrogen dioxide), chemicals (carbon tetrachloride and paraquat), radiation, or, as suggested by the results of these studies, blast overpressure can induce lipid peroxidation. Despite extensive studies over many years, the reactive species that initiate lipid peroxidation in biological systems have not been identified (32). Several mechanisms have been proposed for the induction of lipid peroxidation.

Superoxide anion radical ($O_2^{-\cdot}$), a major product of O_2 metabolism, is one of the species that has been suggested to play a major role in initiation of lipid peroxidation (31, 33). However, this proposal faces many problems that were reviewed by Kappus (32). First, it is impossible for $O_2^{-\cdot}$, which is the predominant species in the aqueous solution at physiological pH 7.4, to abstract a hydrogen atom from a methylene group of an unsaturated fatty acid. Only the undissociated perhydroxy radical ($HO_2^{\cdot}$) that is present at lower pH, such as in phagocytic vacuoles, is sufficiently reactive to initiate lipid peroxidation (34, 35). Second, formed $O_2^{-\cdot}$ radical would dismutate either spontaneously or be catalyzed by superoxide dismutase, giving rise to singlet oxygen ($^1O_2$) and hydrogen peroxide (H_2O_2) (33). However, it is improbable that the former would initiate lipid peroxidation, and the latter is unable to oxidize unsaturated fatty acids. Third, the hydroxyl radical ($^{\cdot}OH$) formed in a Haber-Weiss reaction, although capable of initiating lipid peroxidation, requires traces of transition metals (36–38), as shown in reactions 14 and 15. Copper and other metals can catalyze the reaction as shown in reactions 16 and 17 (39, 40).

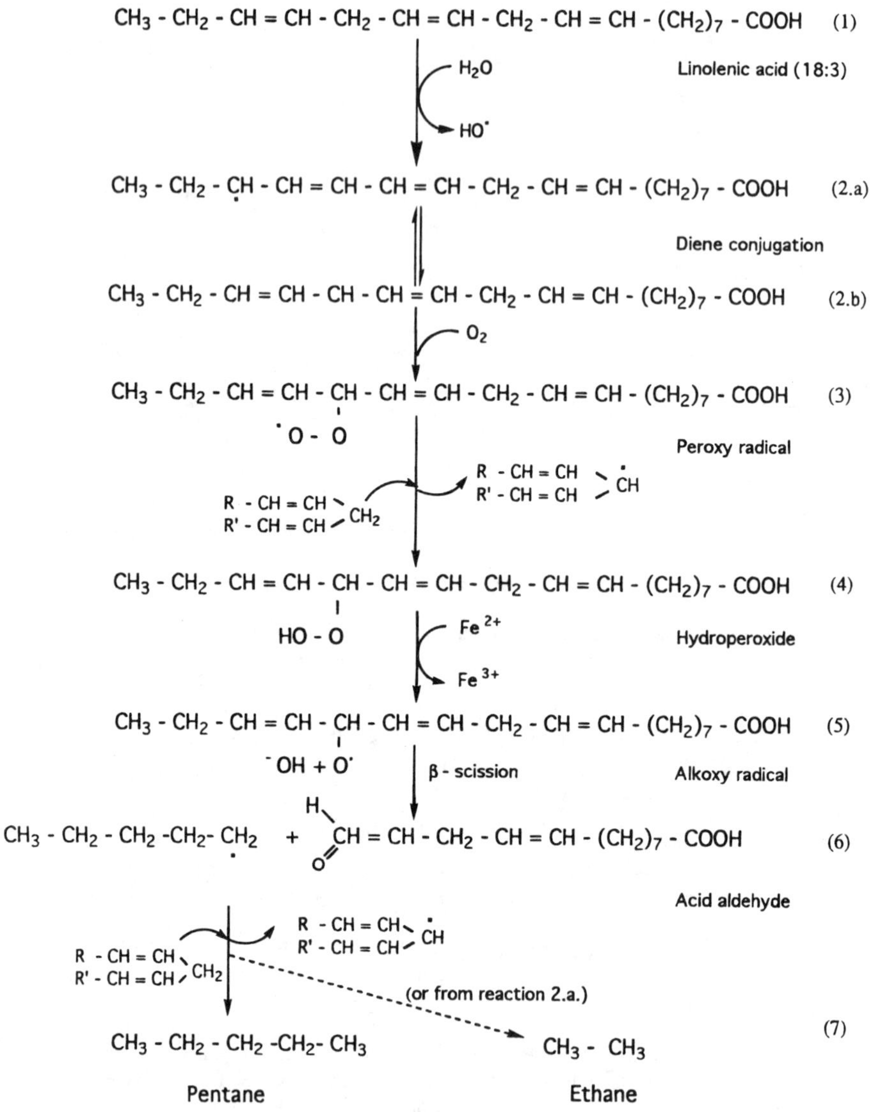

$$CH_3 - CH_2 - CH = CH - CH_2 - CH = CH - CH_2 - CH = CH - (CH_2)_7 - COOH \quad (1)$$

H_2O

$HO^\cdot$

Linolenic acid (18:3)

$$CH_3 - CH_2 - \overset{\cdot}{C}H - CH = CH - CH = CH - CH_2 - CH = CH - (CH_2)_7 - COOH \quad (2.a)$$

Diene conjugation

$$CH_3 - CH_2 - CH = CH - CH - CH = CH - CH_2 - CH = CH - (CH_2)_7 - COOH \quad (2.b)$$

O_2

$$CH_3 - CH_2 - CH = CH - CH - CH = CH - CH_2 - CH = CH - (CH_2)_7 - COOH \quad (3)$$

$^\cdot O - O$

Peroxy radical

$R - CH = CH$
$R' - CH = CH$ CH_2

$R - CH = CH$
$R' - CH = CH$ $\overset{\cdot}{C}H$

$$CH_3 - CH_2 - CH = CH - CH - CH = CH - CH_2 - CH = CH - (CH_2)_7 - COOH \quad (4)$$

$HO - O$

Fe^{2+}

Fe^{3+}

Hydroperoxide

$$CH_3 - CH_2 - CH = CH - CH - CH = CH - CH_2 - CH = CH - (CH_2)_7 - COOH \quad (5)$$

$^-OH + O^\cdot$

β - scission

Alkoxy radical

$$CH_3 - CH_2 - CH_2 - CH_2 - \overset{\cdot}{C}H_2 \quad + \quad \overset{H}{\underset{O}{\diagdown}}CH = CH - CH_2 - CH = CH - (CH_2)_7 - COOH \quad (6)$$

Acid aldehyde

$R - CH = CH$
$R' - CH = CH$ CH_2

$R - CH = CH$
$R' - CH = CH$ $\overset{\cdot}{C}H$

(or from reaction 2.a.)

$$CH_3 - CH_2 - CH_2 - CH_2 - CH_3 \qquad\qquad CH_3 - CH_3 \quad (7)$$

Pentane   Ethane

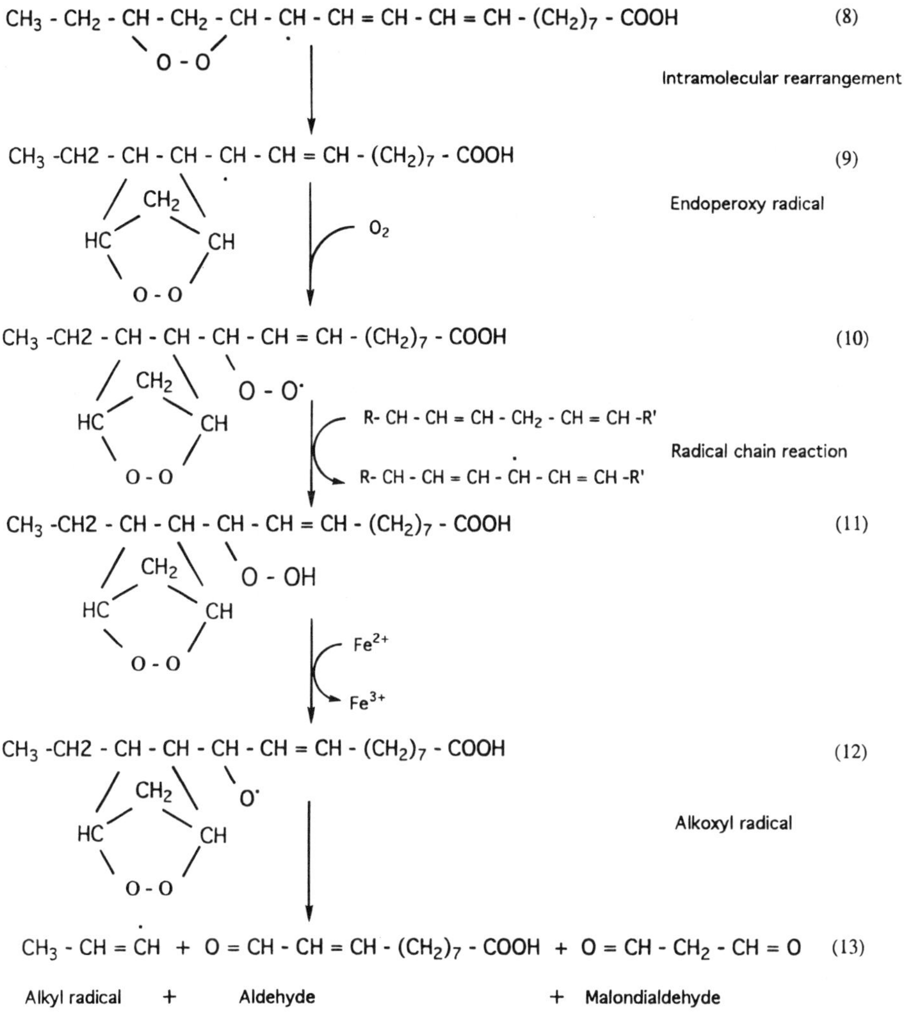

$$CH_3 - CH_2 - CH - CH_2 - CH - CH - CH = CH - CH = CH - (CH_2)_7 - COOH \qquad (8)$$

(with O-O across the CH-CH_2, and radical dot)

Intramolecular rearrangement

$$CH_3 - CH2 - CH - CH - CH - CH = CH - (CH_2)_7 - COOH \qquad (9)$$

(CH_2 ring structure: HC and CH with O-O bridge, radical dot)

$- O_2$

Endoperoxy radical

$$CH_3 - CH2 - CH - CH - CH - CH = CH - (CH_2)_7 - COOH \qquad (10)$$

(ring structure with O-O· group)

$R- CH - CH = CH - CH_2 - CH = CH -R'$

$R- CH - CH = CH - CH - CH = CH -R'$

Radical chain reaction

$$CH_3 - CH2 - CH - CH - CH - CH = CH - (CH_2)_7 - COOH \qquad (11)$$

(ring structure with O - OH group)

Fe^{2+}

Fe^{3+}

$$CH_3 - CH2 - CH - CH - CH - CH = CH - (CH_2)_7 - COOH \qquad (12)$$

(ring structure with O· group)

Alkoxyl radical

$$CH_3 - CH = CH \ + \ O = CH - CH = CH - (CH_2)_7 - COOH \ + \ O = CH - CH_2 - CH = O \qquad (13)$$

Alkyl radical   +   Aldehyde   +   Malondialdehyde

$$O_2^{-\bullet} + Fe^{3+} + \rightarrow O_2 + Fe^{2+} \qquad (14)$$

$$H_2O_2 + Fe^{2+} + H^+ \rightarrow {}^{\bullet}OH + Fe^{3+} + H_2O \qquad (15)$$

$$O_2^{-\bullet} + Cu^{2+} + \rightarrow O_2 + Cu^+ \qquad (16)$$

$$H_2O_2 + Cu^+ \rightarrow {}^{\bullet}OH + Cu^{2+} \qquad (17)$$

These reactions will occur only if iron or copper ions are present at the $O_2^{-\bullet}$ production site. However, the yield of ${}^{\bullet}OH$ from such systems is relatively low, but can be enhanced in the presence of metal chelators such as ethylenediamine tetraacetic acid (EDTA) (38, 41), possibly due to a faster reduction of the Fe^{3+} chelates compared to free Fe^{3+}. Fourth, if a semiquinone such as ubiquinone reacts with H_2O_2 formed in one of the preceding reactions, ${}^{\bullet}OH$ radical would be formed (42–44). Fifth, lipid peroxidation can be formed also by a ferrous-oxygen complex,

$$Fe^{2+} + O_2 \rightarrow Fe^{2+}O_2 \leftrightarrow Fe^{3+}O_2^{-\bullet} \qquad (18)$$

$$Fe^{3+}O_2^{-\bullet} + AH \rightarrow Fe^{3+}O_2^{2-\bullet} + A^{\bullet} + H^+ \qquad (19)$$

The product of this reaction, the perferryl ion, may be formed if the iron is complexed with a chelator (32, 36, 41). It is such a reaction or a modification of it involving oxoferryl hemoglobin that we believe could explain (at least in part) blast-induced lipid peroxidation, as we suggested in recent preliminary reports (45, 46). The question of how endogenous reducing agents such as ascorbate and glutathione participate in metal reduction in cells can be answered, at least in part, by the significant depletion of antioxidant content after blast. Another mechanism is the paradoxical pro-oxidant function of antioxidants that we are suggesting (47–49) occurs depending upon the conditions of microenvironment they exist in.

SUMMARY

Figure 8 summarizes a proposed sequence of events that possibly occurs in the lung as a result of a nonlethal exposure to BOP. The immediate outcome of the exposure is rupture of capillaries and alveolar septa. Consequently, pulmonary edema and hemorrhage occur, and red blood cells rupture, loosing their integrity and releasing hemoglobin and oxyhemoglobin. In the presence of endogenous hydroperoxides, peroxides, and antioxidants, hemoglobin and oxyhemoglobin will be oxidized to methemoglobin, and/or to oxoferryl hemoglobin. These reactions and accumulation of their products can lead to oxidative stress, causing antioxidant depletion and lipid peroxidation. We have found similar reactions with myoglobin, suggesting that such free-radical-mediated reactions may not be limited to the lung, and may occur also in the muscles. In the lung, if these events are severe enough death may occur; otherwise, pulmonary edema and hemorrhage can cause respiratory insufficiency, hypoxemia, or other manifestations of organ

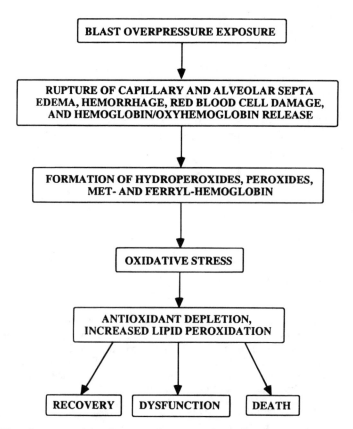

Figure 8.   Flow chart summarizing the proposed sequence of events that occur in the lung in response to a single blast exposure.

dysfunction. If the exposure is not too severe, the lung will recover from blast-induced injury, and the lesions will be resolved.

ACKNOWLEDGMENTS

I would like to thank Drs. Nikolai Gorbunov, and Valerian Kagan for critical comments and useful suggestions, and Jennifer Morris, Joseph Viduya, and Peter Grant for their technical assistance.

DISCLAIMER

The opinions and assertions expressed in this report are the personal views of the author and do not reflect the views of the Department of the Army or Department of Defense. The experimental studies described adhered to the National Institutes of Health "Guide for the Care and Use of Laboratory Animals."

REFERENCES

1. Benzinger, T. 1950. Physiological effects of blast in air and water. In *German Aviation Medicine, World War II*, Vol. 2, pp. 1225–1259. Washington, DC: U.S. Government Printing Office.
2. Clemedson, C.-J. 1956. Blast injury. *Physiol. Rev.* 36:336–354.
3. Phillips, Y. Y. III, and Richmond, D. R. 1991. Primary blast injury and basic research: A brief history. In *Textbook of Military Medicine, Part I, Vol. 5, Conventional Warfare. Ballistic, Blast, and Burn Injuries*, eds. R. F. Bellamy and R. Zajtchuk, pp. 221–240. Washington, DC: Office of the Surgeon General, Department of The Army.
4. Roberto, M., Hamernik, R. P., and Turrentine, G. A. 1989. Damage of the auditory system associated with acute blast trauma. *Ann. Oto. Rhinol. Laryngol.* 98:23–34.
5. Hamernik, R. P., Patterson, J. H., Turrentine, G. A., and Ahroon, W. A. 1989. The quantitative relation between sensory cell loss and hearing thresholds. *Hearing Res.* 38:199–212.
6. Brown, R. F. R., Cooper, G. J., and Maynard, R. L. 1993. The ultrastructure of rat lung following acute primary blast injury. *Int. J. Exp. Pathol.* 74:152–162.
7. Elsayed, N. M., Dodd, K. T., Fitzpatrick, T. M., and Morris, J. R. 1993. Does high energy impulse noise result in free radical-mediated injury? *Proc. Int. Conf. Critical Aspects of Free Radicals in Chemistry, Biochemistry, and Medicine*, Vienna, Austria, p. 66.
8. Elsayed, N. M., Dodd, K. T., Morris, J. R., and Ghosal, A. 1993. Biochemical changes in sheep plasma after exposure to high energy impulse noise. *Toxicologist* 13:336 (abstr.)
9. Elsayed, N. M., Dodd, K. T., and Fitzpatrick, T. M. 1994. Biochemical alterations associated with pulmonary contusion injury resulting from exposure to air blast. *Am. J. Respir. Crit. Care Med.* 149:A1029 (abstr.).
10. Elsayed, N. M., Tyurina, Y. Y., Tyurin, V. A., Menshikova, E. V., Kisin, E. R., and Kagan, V. E. 1996. Antioxidant depletion, lipid peroxidation, and impairment of calcium transport induced by air blast overpressure in rat lungs. *Exp. Lung Res.* 22:179–200.
11. Liu, Z. 1992. Experimental study on the mechanism of free radical in bast trauma induced hearing loss. *Chin. J. Otorhinolaryngol.* 27:24–61.
12. Rusca, F. 1915. Experimentelle untersuchungen über traumatische druckwirkung der explosionen. *Dtsch. Z. Chir.* 132:315–375.
13. Hooker, D. R. 1924. Physiological effects of air concussion. *Am. J. Physiol.* 67:219–274.
14. Deneiga, V. G. 1966. Some information on the biophysics of pneumatic contusion. *Biofizika* 11:371–374 (in Russian).
15. Chesnokov, P. T., and Kholodny, A. Y. 1970. Pathomorphological shifts in the organism caused by explosive blast waves. *Voen Med. Zh.* 8:33–36 (in Russian).
16. Shereshevskiy, G. M. 1979. Contusions. *Voen Med. Zh.* 11:22–25.
17. Wang, Z. G. 1979. Research on blast injury in China. *Chuang. Shang. Tsa. Chih.* 6:222–228.
18. Cernak, I., Savic, J., Zunic, G., Ivanoyic, I., Radosevic, P., Malicevic, Z., and Davidovic, L. 1994. The involvement of CNS in the general response of organism to the pulmonary blast injury. *Proc. 7th Int. Symp. Weapons Truamatology and Wound Ballistics*, St. Petersburg, Russia, p. 92.
19. Lowry, O. H., Rosenbrough, N. J., Farr, A. L., and Randall, R. J. 1951. Protein measurement with the Folin phenol reagent. *J. Biol. Chem.* 193:265–275.
20. Folch, J., Lees, M., and Sloan-Stanley, G. H. 1957. A simple methods for isolation and purification of total lipids from animal tissues. *J. Biol. Chem.* 226:497–509.
21. Buege, J. A., and Aust, S. D. 1978. Microsomal lipid peroxidation. *Methods Enzymol.* 52:302–310.

22. Cassen, B., Curtis, L., and Kistler, K. 1950. Initial studies of the effect of laboratory produced air blast on animals. *J. Aviat. Med.* 21:38–47.

23. Celander, H., Clemedson, C., Erickson, U., and Hultman, H. 1955. The use of compressed-air-operated shock tube for physiological blast research. *Acta Physiol. Scand.* 32:6–13.

24. Sinnhubeer, R. O., and Yu, T. C. 1958. Characterization of the red pigment formed in the 2-thiobarbituric acid determination of oxidative rancidity. *Food Res.* 23:626–633.

25. Sevanian, A., and Hochstein, P. 1985. Mechanisms and consequences of lipid peroxidation in biological systems. *Annu. Rev. Nutr.* 5:365–390.

26. Dianzani, M. U., and Ugazio, G. 1978. Lipid peroxidation. In *Biochemical Mechanism of Liver Injury*, ed. T. F. Slater, pp. 45–76. San Diego: Academic Press.

27. Halliwell, B., and Gutteridge, J. M. C. 1989. Lipid peroxidation: A radical chain reaction. In *Free Radicals in Biology and Medicine*, 2nd ed., pp. 188–276. Oxford: Clarendon Press.

28. Janero, D. R. 1990. Malondialdehyde and thiobarbituric acid-reactivity as diagnostic indices of lipid peroxidation and peroxidative tissue injury. *Free Radic. Biol. Med.* 9:515–540.

29. Tappel, A. L. 1980. Measurement of and protection from in vivo lipid peroxidation. In *Free Radicals in Biology*, ed. W. A. Pryor, Vol. IV, pp. 1–47. San Diego: Academic Press.

30. Cohen, G. 1979. Lipid peroxidation: Detection in·vivo and in vitro through the formation of saturated hydrocarbon gases. In *Oxygen Free Radicals and Tissue Damage*, Ciba Foundation Symposium 65, pp. 177–185. Amsterdam: Excerpta Medica.

31. Halliwell, B., and Gutteridge, J. M. C. 1989. *Free Radicals in Biology and Medicine*, 2d ed., pp. 1–85. Oxford: Clarendon Press.

32. Kappus, H. 1985. Lipid peroxidation: Mechanisms, analysis, enzymology, and biological relevance. In *Oxidative Stress*, ed. H. Sies, pp. 273–310. San Diego: Academic Press.

33. Fridovich, I. 1983. Superoxide radical: An endogenous toxicant. *Annu. Rev. Pharmacol. Toxicol.* 23:239–257.

34. Bielski, B. H. J., Cabelli, D. E., and Arudi, R. L. 1985. Reactivity of hydroxyl/superoxide radicals in aqueous solutions. *J. Phys. Chem. Ref. Data* 14:1041–1100.

35. Gebicki, J. M., and Bielski, B. H. J. 1981. Comparison of the capacities of the perhydroxyl and superoxide radicals to initiate chain oxidation of linoleic acid. *J. Am. Chem. Soc.* 103:7020–7022.

36. Aust, S. D., and Svingen, B. A. 1982. The role of iron in enzymatic lipid peroxidation. In *Free Radicals in Biology*, ed. W. A. Pryor, Vol. V, pp. 1–28. San Diego: Academic Press.

37. Gutteridge, J. M. C. 1982. The role of superoxide and hydroxyl radicals in phospholipid peroxidation catalyzed by iron salts. *FEBS Lett.* 150:454–458.

38. Ursini, F., Maiorino, M., Hochstein, P., and Erstner, L. 1989. Microsomal lipid peroxidation: Mechanisms of initiation. The role of iron and iron chelates. *Free Radic. Biol. Med.* 6:31–36.

39. Gutteridge, J. M. C., and Wilkins, S. 1982. Copper-dependent hydroxyl radical damage to ascorbic acid. *FEBS Lett.* 137:327–330.

40. Rowley, D. A., and Halliwell, B. 1983. Superoxide-dependent and ascorbate-dependent formation of hydroxyl radicals in the presence of copper salts: A physiologically significant reaction? *Arch. Biochem. Biophys.* 225:279–284.

41. Floyd, R. A. 1983. Direct demonstration that ferrous iron complexes of di- and triphosphate nucleotides catalyze hydroxyl free radical formation from hydrogen peroxide. *Arch. Biochem. Biophys.* 225:263–270.

42. Winterbourn, C. C. 1981. Cytochrome c reduction by semiquinone radicals can be indirectly inhibited by superoxide dismutase. *Arch. Biochem. Biophys.* 209:159–167.

43. Nohl, H., and Jordan, W. 1983. OH·—centered by adriamycin semiquinone and H_2O_2: An explanation for the cardiotoxicity of anthracycline antibiotics. *Biochem. Biophys. Res. Commun.* 114:197–205.

44. Bates, D. A., and Winterbourn, C. C. 1982. Deoxyribose breakdown by adriamycin semiquinone and H_2O_2: Evidence for hydroxyl radical participation. *FEBS Lett.* 145:137–142.

45. Gorbounov, N., Osipov, A., Day, B., Kagan, V., and Elsayed, N. 1995. Nitric oxide protects from oxidative stress by reducing ferryl myoglobin and ferryl hemoglobin radicals. *FASEB J.* 9:A892 (abstr.).

46. Gorbounov, N., Osipov, A. N., Day, B. W., Zayas-Rivera, B., Kagan, V. E., and Elsayed, N. M. 1995. Antioxidant function of nitric oxide against oxoferryl hemoprotein-induced oxidations in the presence of organic hydroperoxides. Int. Congr. Toxicology VII. *Int. Toxicol.* 7:48-PF-2 (abstr.).

47. Gorbounov, N. V., Osipov, A. N., Day, B. W., Zyas-Rivera, B., Kagan, V. E., and Elsayed, N. M. 1995. Reduction of ferrylmyoglobin and ferrylhemoglobin by nitric oxide: A protective mechanism against ferryl hemoprotein-induced oxidations. *Biochemistry* 34:6689–6699.

48. Gorbounov, N. V, Osipov, A. N., Sweetland, M. A., Day, B. W., Elsayed, N. M., and Kagan, V. E. 1996. NO-redox paradox: Direct oxidation of α-tocopherol and α-tocopherol mediated oxidation of ascorbate. *Biochem. Biophys. Res. Commun.* 219:835–841.

49. Elsayed, N. M., Gorbounov, N. V., Kisin, E. R., Kagan, V. E., and Kozlov, A. V. 1996. Prooxidant action of water soluble antioxidants and their role in blast overpressure-induced oxidative stress. *Fundam. Appl. Toxicol.* 30:313 (abstr.).

50. Sharpnack et al. 1991. In *Textbook of Military Medicine, Conventional Warfare: Ballistic, Blast, and Burn Injuries*, pp. 271–294. Washington, DC: Office of the Surgeon General.

Chapter Nineteen

FOOD ANTIOXIDANTS: THEIR DUAL ROLE IN CARCINOGENESIS

Saura C. Sahu and Sidney Green

INTRODUCTION

Oxidation of lipid components of food is a major problem for food manufacturers. Often synthetic antioxidants such as butylated hydroxyanisole (BHA) and butylated hydroxytoluene (BHT) are used to minimize this problem. Thus, as food additives, the antioxidants play an important role in the safety of foods. However, the demonstration that the synthetic antioxidant BHA induces tumors in rat forestomach (1) has drawn attention to the fact that such phenolic compounds may be double-edged biological swords. Because of this safety concern, the replacement of synthetic antioxidants by natural antioxidants is widely advocated (2, 3). However, like the synthetic antioxidants, the natural antioxidants may possess certain properties that could lead to questions regarding their safe use; some of them demonstrate pro-oxidant properties leading to oxidative DNA and protein damage (3).

Human beings are simultaneously exposed to a great variety of food antioxidants, both synthetic and natural. Amounts of antioxidants ingested orally via the diet are increasing. Besides, in addition to the single antioxidants derived from natural sources, the use of their mixtures is also being advocated. However, the antioxidants, synthetic or natural, may be beneficial individually, but in combination they may produce significant antagonistic or synergistic effects. Such synergistic actions of chemicals on development of tumors in a wide variety of target organs have been reported (4). For example, enhancement of the BHA-induced carcinomas in rat forestomach by ascorbate (5) and of the estradiol-induced tumors in hamster kidney by quercetin (6) has been observed. Although most chemical exposures in the real world involve mixtures rather than single agents, the scientific database for these mixtures has been generated almost entirely from studies on individual agents; very little information is available on interaction of chemicals in experimental carcinogenesis. Therefore, there is a need for evaluating the toxicity of complex mixtures.

Naturally occurring plant phenols are present extensively in human diet. They are present in fruits, vegetables, cereals, coffee, and tea. They are also present in human food as food additives. They include, but are not limited to, flavonoids and anthraquinones (7), catechols (8), caffeic acid (9), catechins (10), gallates (11), and chlorogenic acid (12). The human diet includes milligrams to perhaps a gram of the natural plant polyphenols (8). Grapes contain as much as 0.5 mM ellagic acid (13). Propyl gallate, a synthetic food additive, is ingested at a level of about 3.9 mg/person/day (14). The estimated daily intake of flavonoids in the average American diet is about 1 g (7).

The present review attempts to discuss a dual role of the food antioxidants in mutagenesis and carcinogenesis, and the possible involvement of oxygen free radicals in producing these effects.

327

DIET, OXYGEN, AND CANCER

Diet modulates toxicity, including carcinogenesis. A complex but close relationship appears to exist among diet, oxygen, and cancer. Epidemiological studies indicate that dietary antioxidant deficiency is an important factor in cancer risk (15). A number of natural mutagens and carcinogens are activated when reactive oxygen species are produced (16). Ionizing radiation and certain carcinogens known to produce reactive oxygen species induce tumors (16, 17). Some antioxidants demonstrate antimutagenic and anticarcinogenic activity (18, 19).

The range of intra- and extracellular antioxidant defenses should be adequate to protect against oxidative damage. However, over production of reactive oxygen species by dietary and environmental factors can overwhelm the antioxidant defenses and induce oxidative stress, which can cause oxidative damage to cellular macromolecules and lead possibly to cancer. The polyunsaturated membrane lipids are prime targets of active oxygen. Cellular proteins, particularly the sulfur-containing proteins, are susceptible to oxidative damage. Lipid peroxidation products have been shown to induce protein degradation, cross-linking, and amino acid modifications (20). Nucleic acids are also major target molecules for endogenous active oxygen. The relatively long life of hydrogen peroxide in most cellular environments enables it to diffuse appreciable distances before reaction (21). It can cross the nuclear membrane and enter into the nucleus. In the presence of transition metals such as iron and copper, bound to DNA, it will generate the highly reactive hydroxyl radicals at the metal binding site by the Fenton and/or Haber-Weiss reaction. The hydroxyl radicals react with deoxyribose and allylic positions in the base residues of DNA, producing site-specific strand breaks and modified bases (22). Formation of 8- hydroxy-deoxyguanosine (8-OH-dG) in DNA is a reliable marker of the oxidative DNA damage (23, 24). Measurement of 8-OH-dG in human urine indicates that the level of endogenous oxidative DNA damage in humans can be significantly high depending on the dietary and environmental conditions (25). From the measurement of the products of DNA oxidation excreted in urine it has been calculated that in normal humans an average of 10^4 "oxidative hits" on DNA occur per day for each cell in the body (26). This finding indicates that DNA is a major target molecule of oxidant attack in vivo.

Genetic damage may result either from direct attack of the hydroxyl radicals on DNA or indirectly, through initiation of lipid peroxidation. Lipid peroxidation can induce oxidative DNA damage. Oxygen free radicals may trigger the peroxidation of nuclear membrane lipids with formation of peroxyl radicals that can react with DNA; the altered DNA in a proliferating tissue may initiate the carcinogenic process. The proximity of nuclear DNA to the nuclear membrane may facilitate the interaction of DNA with the peroxyl radicals and other reactive intermediates formed during membrane lipid peroxidation. It is possible that the hydroxyl radicals could initiate peroxidation of lipids in the nuclear membrane via metal-catalyzed Haber-Weiss reaction; the levels of lipid peroxides and intermediate radicals may then be amplified by a chain reaction. Consequently, the reactive species oxidize nuclear DNA when the lipid peroxidation reaction takes place in the nuclear membrane. Since the nuclear membrane regulates the transport of mRNA into the cytoplasm and aids the process of nuclear division, peroxidation of nuclear membrane lipids and proteins, in addition to aiding oxidative nuclear DNA damage, may disrupt these critical cellular functions. Thus, oxygen radicals may play an important role in mutagenesis and carcinogenesis via peroxidation of the nuclear membrane lipids.

Malonaldehyde, a major product of lipid peroxidation, reacts with DNA, producing the adduct 1-N^6-ethenoadenine (27). The normal adult human liver accumulates substantial amounts of oxidative DNA modifications (malonaldehyde-deoxyguanosine adducts) resulting from endogenous lipid peroxidation (28).

The unrepaired and/or misrepaired oxidative DNA damage, prior to cell replication, can induce gene mutations and putatively "initiate" the exposed cell(s), converting it to an irreversibly altered preneoplastic lesion, which may undergo subsequent "promotion" and "progression" to potential malignancy. Such natural oxidative DNA damage may be responsible for spontaneous tumors (16, 17), and tumors induced by nongenotoxic carcinogens (29) such as dioxin (30) and peroxisome proliferators (31).

ANTIOXIDANTS AS ANTICARCINOGENS

Some of the antioxidants present in food exhibit anticarcinogenic activities, as follows.

Butylated hydroxyanisole (BHA) is a synthetic antioxidant that is used as a food additive. It is a non-mutagen in the Ames test. It inhibits rat hepatocarcinogenesis induced by diethylnitroso-amine (32) and 3,2'-dimethyl-4-aminobiphenyl (33) as well as rat lung carcinogenesis induced by 2,2'-dihydroxy-di-n-propylnitrosoamine (18).

Butylated hydroxytoluene (BHT) is another synthetic antioxidant and is widely used as a food additive. It inhibits carcinogenicity in various organs, including liver (34, 35), intestine (36), colon (37), breast (38), lung (39), and skin (40), in rodents induced by a variety of carcinogens. BHT at a dietary level of 100–1000 ppm inhibits liver carcinogenesis induced by concurrent feeding of 2-acetyl-aminofluorene (50 ppm) for 76 wk without any adverse effects (41).

Vitamin C is a water-soluble natural antioxidant present abundantly in fruits and vegetables. It is a scavenger of singlet oxygen, superoxide, and hydroxyl radicals (42). It inhibits the endogenous formation of N-nitroso compounds (43), and is therefore a scavenger of nitrite in the stomach, where nitrosoamines can be produced from nitrites and amines. Ascorbate inhibits mutagenesis of N-methyl-N'-nitro-N nitrosoguanidine by a free radical mechanism (44). Ascorbic acid given simultaneously with estradiol dipropionate (EP) inhibits the promoting activity of EP on the uterine sarcomas in mice induced by 1,2-dimethylhydrazine (45). Epidemiological studies indicate a negative correlation between the consumption of vitamin C and risk of cancer at various sites (46).

Vitamin E is the major lipid-soluble antioxidant found in cell membranes, where it protects against lipid peroxidation. Its major dietary sources are vegetable oils, margarine, and shortening. It is an inhibitor of oxidation-related diseases (47). It inhibits mouse skin (40), mouse colon (48), and hamster buccal pouch (49) carcinogenesis. It prevents the early events during hepatocar-cinogenesis, the induction of phenotypically altered foci in rat liver initiated with diethylnitroso-amine (50).

Carotenoids, which include beta-carotene, are metabolized to vitamin A and exhibit antioxidant properties. The intake of dietary vitamin A or beta-carotene is inversely related to the incidence of lung cancer (51). The natural vitamin A inhibits hepatocarcinogenesis induced by 2-acetyl-aminofluorene (52): mammary carcinogenesis induced by N-methyl-N-nitrosourea (53) and 7,12-dimethyl-benz[a]anthracene (54) in rats.

Ellagic acid, a plant phenolic antioxidant present in a variety of fruits and nuts, inhibits the mutagenicity of benzo[a]pyrene 7,8-diol-9,10-epoxide-2 in strain TA 100 of *Salmonella typhimurium* and in Chinese hamster V79 cells (55). Feeding rats with ellagic acid supplemental diet for 3 wk results in a significant protection against the methylation of O^6-guanine in esophageal DNA by N-nitrosobenzylmethylamine (56). Ellagic acid at a dietary concentration of 0.4 g/kg produces a significant decrease in esophageal tumors induced by N-nitrosobenzylmethylamine after 20 wk (57). Conflicting results have been reported on the effects of ellagic acid on carcinogenesis induced by benzo[a]pyrene (58–60) and 3-methylcholanthrene (61). Flavonoids, which are widely present in edible plants, have been reported to be antioxidants (62) and anticarcinogens (63). They protect against induction of micronuclei in whole body gamma-ray irradiated mice (64). This radioprotective effect of flavonoids may be attributed to their ability to scavenge hydroxyl radicals. Several plant phenols including quercetin demonstrate antiprolifer-ative properties in cells in culture. Quercetin inhibits the proliferation of the L-1210 and P-388 leukemia and Ehrlich ascites tumor cells (65), NK/Ly tumor cells (66), OVCA 433 cells (67), HeLa and Taji lymphoma cancer cells (68), and squamous carcinoma cells (69) in culture. Apigenin inhibits the growth of mammary carcinoma cells (70).

Some flavonoids have shown anticarcinogenic activity in rodents. Quercetin inhibits rat mam-mary (71), colon (72), and skin tumors (73). Apigenin inhibits skin papillomas initiated by DMBA and promoted by TPA in female SENCAR mice (74). Therefore, the use of quercetin as a cancer preventive agent has been suggested (8, 63).

Curcumin, the major yellow pigment in turmeric, is an antioxidant (75, 76). It inhibits activities of cyclo-oxygenase and lipoxygenase in mouse epidermis (77). It inhibits the formation of 8-hydroxydeoxyguanosine in cultured mouse fibroblasts (78). It is an antimutagen in in vitro (79), and an anticarcinogen in vivo (80). It inhibits the tumor promotion induced by phorbol 12-

myristate on mouse skin (81). Curcumin inhibits protein kinase C activity induced by 12-*O*-tetradecanoyl phorbol 13-acetate in NIH 3T3 cells (82). Curcumin fed at the level of 0.03% in the diet for 4 wk significantly reduced the formation of DNA-benzo[a]pyrene adducts (83). Administration of 2% curcumin in the diet inhibits azoxymethanol-induced colonic neoplasia in mice (84). Topical administration of 3 or 10 μmol curcumin 5 min prior to the application of 20 nmol benzo[a]pyrene or 2 nmol 7,12-dimethylbenz[a]anthracene once a week for 10 wk followed a week later by promotion with 15 nmol 12-*O*-tetradecanoylphorbol-13-acetate twice weekly for 21 wk significantly decreased the number of skin tumors per mouse (85).

Tannic acid, which is present in tea, fruits, and vegetables, is an antioxidant. It has been documented to have protective effects against chemically induced carcinogenesis in rodents. It inhibits rat colon carcinogenesis initiated by 2,2′-dihydroxy-di-*n*-propylnitrosoamine (86), mouse lung and forestomach carcinogenesis induced by benzo[a]pyrene (87), and mouse skin carcinogenesis initiated by benzo[a]pyrene or *N*-methylnitrosourea (88).

Tea is rich in tannins, flavonoids, and other polyphenols, the compounds that have been shown to inhibit mutagenicity and carcinogenicity. Tea extracts exhibit antioxidant properties (89) and inhibit in vitro mutagenesis (90). In experimental animals, tea extracts inhibit chemically induced skin, esophageal, stomach, colon, and mammary tumors (90).

Phytic acid, which is present in cereals, nuts, oil seeds, and legumes, is a natural antioxidant (91). Sodium phytate at a level of 1% in the drinking water inhibits azoxymethane- and 1,2-dimethylhydrazine-induced rat and mouse colon carcinogenesis, respectively, when given simultaneously with or after carcinogen exposure (92, 93). It inhibits cell proliferation of rat colonic epithelial cells (94), but shows no inhibitory effect on colon carcinogenesis in rats (86).

ANTIOXIDANTS AS CARCINOGENS

Some of the food antioxidants that have demonstrated carcinogenic potential are described next.

Butylated hydroxyanisole (BHA), which has been shown to be an anticarcinogen, is also a rodent carcinogen. The main target organ is the forestomach, but the bladder is also affected (95). Continuous exposure of rats to BHA at a dose of 2% in the diet for 104 wk results in forestomach carcinoma (96, 97). Also, BHA promotes forestomach carcinogenesis in rats initiated with *N*-methyl-*N*′-nitro-*N*-nitroguanidine (9, 98), *N*-methylnitrosourea (99, 100), or *N*-dibutylnitrosoamine (101) at a much lower dose than required for carcinogenicity. For example, a dose of 0.6% BHA is sufficient for effective promotion of forestomach carcinogenesis in rats initiated with *N*-methyl-*N*′-nitro-*N*-nitrosoguanidine (9). BHA induces marked cell proliferation in both rat (102) and hamster (103) forestomach epithelium. It is a strong promoter in rat forestomach carcinogenesis initiated by *N*-methyl-*N*′-nitro-*N*-nitrosoguanidine (98). No DNA adduct was detected by the ^{32}P-postlabeling method in the forestomach of rats given either a single or repeated oral administration of BHA for 5 days (104).

Butylated hydroxytoluene (BHT), an antioxidant and anticarcinogen, is an effective tumor promoter in mouse lung (105), rat liver (106), urinary bladder (107), and gastrointestinal tract (36).

Vitamin C causes DNA single-strand breaks in Chinese hamster ovary cells (108). Ascorbic acid oxidizes deoxyguanosine to 8-hydroxydeoxyguanosine in vitro (109), which causes miscoding in the DNA template (24) and induces mutations in vivo (110). Hydrogen peroxide and hydroxyl radicals appear to be involved in the pro-oxidant activities of ascorbate, which is stimulated by iron or copper ions (44). Sodium ascorbate in the presence of iron or copper ions induces oxidative damage to human red blood cells in vitro, producing changes in cell morphology and membrane proteins, and significant increase in methemoglobin (111). Sodium ascorbate at 5% level in the diet promotes urinary bladder cancer in F344 rats initiated with *N*-butyl-*N*-(4-hydroxybutyl)nitrosoamine (112). However, a comparative study using F344 and Lewis rats has shown that the animal strain and diet strongly influence this promoting activity of sodium ascorbate (113). Also, ascorbic acid amplifies the promoting effect of potassium carbonate on the bladder carcinogenesis in rats initiated by *N*-butyl-*N*-(4-hydroxy-butyl)nitrosoamine (114). Some of the carotenoids possess anticarcinogenic properties. Beta-carotene, a precursor of vitamin A, is a lipid soluble antioxidant and protects against a variety of tumorigenesis including lung cancer (51). However, a recent epidemiological study (115), in which a total of 29,133 male smokers aged 50–69 yr from southwestern Finland participated for a period of 5–8 yr, has concluded that dietary supplementation with beta-carotene or alpha-tocopherol (vitamin E) does not reduce the incidence of lung cancer. In fact, the study reports that beta-carotene increases the incidence of lung cancer among heavy smokers. The incidence of lung cancer was 18% higher among smokers who took beta-carotene than among those who did not. These results have raised doubts about the beneficial health claims about the antioxidants.

Leo et al. (116) have shown that high vitamin A intake potentiated by ethanol results in hepatotoxicity in rats. Rats given diets with normal vitamin A and ethanol produced the expected proliferation of smooth endoplasmic reticulum. However rats fed a fivefold increased level of vitamin A and ethanol showed much more striking lesions in the liver.

Most of the flavonoids that show anticarcinogenic properties are mutagenic in the Ames test under aerobic conditions, with quercetin being the most potent (117). Oxygen is essential for their mutagenicity. The mutagenicity of flavonoids appears to be dependent upon the absence of excision repair capability, as well as the presence of the pkM101 plasmid, which enhances error-prone repair of DNA lesions; elimination of either or both of these factors results in loss of their mutagenicity (118).

Several flavonoids including quercetin enhance the mutagenicity of 2-acetylaminofluorene when tested under aerobic conditions in the *Salmonella typhimurium* mutagenicity assay (119) and appear to exhibit similar enhancing influence on other carcinogens. In an in vivo host-mediated bacterial mutation assay, 2-amino-3,5-dimethyl(4,5-*f*)imidazoquinoline (MeIQ) and 3-amino-1-methyl-5*H*-pyridol(4,3-*b*)indole (Trp-P-2) exerted more potent mutagenic effects in mice fed either quercetin or rutin than in mice fed control diet (120).

Quercetin induces recombinational mutations in the BMT-11 mouse fibrosarcoma cells and FM3A mouse mammary tumor cells in culture (121), similar to the type of mutations found in human cancer cells (122). Quercetin, a potent mutagen in the Ames test (117), is a controversial carcinogen in rodents. Carcinogenicity studies in a number of experimental animals have yielded conflicting results, and its carcinogenic potential remains unresolved. In the original study, Pamukcu et al. (123) reported a very high frequency of intestinal tumors (80%) and a lesser frequency of bladder tumors (20%) in Norwegian rats fed a diet containing 0.1% quercetin for approximately 1 yr. Later Erturk et al. (124) reported hepatoma induction in Sprague-Dawley and Fisher-344 rats fed a diet containing 2% quercetin or rutin for 2 yr. However, some other studies have failed to find carcinogenic effects of quercetin (125). In a 2-yr feeding study by the National Toxicology Program (NTP), quercetin showed "some evidence of carcinogenic activity" (126, 127) in male F344 rat kidney.

It has been suggested that flavonoids play an important role in the induction of gastric cancer in humans (128). Epidemiological studies (129) that have associated a high incidence of gastric cancer in France with high consumption of alcohol containing flavonoids (128) appear to support this hypothesis.

The data available in the literature at the present time do not permit definite conclusions about the human risk due to dietary flavonoid intake. Of the hundreds of flavonoids present in the human diet, only a few have been tested for mutagenicity or carcinogenicity. Even for quercetin, which has been tested for mutagenicity and carcinogenicity, it is not possible to draw definite conclusions because of the discrepancies in the results of carcinogenesis studies.

Besides mutagenicity and carcinogenicity, the toxicity of flavonoids to animal and human cells has also been demonstrated. The flavonoids quercetin, myricetin, and kaempferol are toxic to isolated guinea pig enterocytes, producing cell damage (130). Several flavonoids including quercetin are potent inhibitors of cellular DNA and RNA polymerases (131), suggesting their potential toxic effects on normal cells. They have been shown to be cytotoxic to several types of human cancer cells in culture (68, 132).

Tannic acid, which provides protection against certain types of cancer, induces liver tumors when administered subcutaneously to mice at a dose of 0.25 ml once a week for 12 wk and maintained without any treatment thereafter for 1 yr (133). Repeated subcutaneous doses of 150–200 mg tannic acid/kg body weight administered every 5 days for 200 days induces hepatocellular adenoma in more than half of the rats surviving 100 days or more (134). However, tannic acid showed no carcinogenic potential in rats given in drinking water a daily intake of 131–290 mg/kg body weight (135).

Phytic acid, which demonstrates anticarcinogenic properties, also significantly enhances the development of rat urinary bladder cancer initiated by N-butyl-N-(4-hydroxybutyl)nitrosoamine (136).

Caffeic acid induces hyperplasia in the rat forestomach epithelium in a short period of 4 wk (102) and squamous-cell carcinomas in the long term (137) when given at a dose of 2% in the diet. It also promotes forestomach carcinogenesis at a dietary dose of 1% in rats initiated with 7,12-dimethylbenz[a]anthracene (138).

Catechol at a dietary dose of 0.8% strongly enhances the rat forestomach and glandular stomach carcinogenesis initiated by N-methyl-N'-nitro-N-nitrosoguanidine (138). Continuous treatment with catechol for 104 wk induces hyperplasia in the forestomach and adenocarcinomas in the glandular stomach (137, 139).

INTERACTIONS OF ANTIOXIDANTS

In addition to the direct quenching of reactive oxygen species, the dietary antioxidants can interact with each other, producing beneficial or adverse health effects. For example, BHA in combination with sodium ascorbate produces more severe forestomach carcinogenesis in rats compared to BHA alone (5). Vitamin A (54) and BHT (39) each have significant anticarcinogenic activity in the rat mammary gland, but an enhanced inhibition of mammary tumors is achieved by their combined administration (140). However, this increased anticarcinogenic activity is accompanied by the potentiation of vitamin A hepatotoxicity by BHT (141). Dietary supplementation with vitamin E or selenium alone is not effective in preventing a chemically induced mammary tumor. However, supplementation with both micronutrients prevents tumor development (142).

The carcinogenic doses of individual food antioxidants used in the studies just described may be high compared to the possible human intake levels; however, it is conceivable that the mixtures of these compounds, even at very low noneffective individual doses, may exert carcinogenic activity in combination. This possibility deserves further investigation.

ANTIOXIDANTS AS PRO-OXIDANTS

Certain natural antioxidants in human food are rich sources of oxygen free radicals. They can produce reactive oxygen species by their autoxidation and subsequent redox cycling, and thus can act as pro-oxidants. For examples, quinones and hydroquinones, which are widespread in the human food supply, are excellent sources of oxygen free radicals because they generate oxygen radicals during their redox cycling at physiological pH (143). Hydroquinone and catechol generate hydrogen peroxide and superoxide radicals (144). Polyphenolic pyrogallol produces superoxide radicals (145). In the presence of metal ions, the hydrogen peroxide and superoxide radicals thus produced can generate the highly reactive hydroxyl radicals. Then the hydroxyl radicals can react with DNA bases and induce mutation. It has been reported that the hydroquinones increase the mutagenic activity of pyrolysis products of amino acids and proteins under aerobic conditions (146). This mutagenic activity may be due to the active oxygen formed during the reaction of hydroquinones with amino acids.

BHA is considered an "epigenetic" carcinogen. It does not react with DNA producing DNA-BHA adducts as detected by the ^{32}P-postlabeling technique (104). On the other hand, BHA stimulates production of superoxide in rat forestomach (147). The active oxygen such as superoxide can lead to oxidative DNA damage (17). It has been demonstrated that BHA enhances the cryosatile-induced oxidative stress in the mouse lung (148).

Flavonoids are generally considered antioxidants because they prevent oxidation of polyunsaturated fatty acids (149), scavenge superoxide (62, 150) and hydroxyl radicals (151), and inhibit the activities of lipoxygenases and cyclooxygenases (152). However, they also exhibit pro-oxidant activities. Several polyphenolic flavonoids including quercetin act as oxidants by generating reactive oxygen species in the presence of metal ions such as iron or copper (153–155).

Caffeic acid is not mutagenic in the bacterial test systems (156), but induces forestomach carcinoma in rats (137). Transition metals enhance its clastogenic activity (156), suggesting involvement of active oxygen species. In vitro studies on the caffeic acid-induced DNA damage show that it undergoes autoxidation in the presence of transition metals, producing hydroxyl radicals that oxidize DNA (157).

Glutathione (GSH) is a natural antioxidant and plays a major role in the detoxification of carcinogens. Glutathione (GSH), gamma-glutamyltransferase (GGT), glutathione S-transferase (GST), and glutathione peroxidase take part in the cellular detoxification mechanisms and protect against oxidative damage. High levels of GGT and GST in enzyme-altered foci are early markers of hepatocarcinogenesis in rodents (158). However, GSH is mutagenic to the bacterial (159) and mammalian (160) systems in vitro. The antioxidant GSH has also been shown to induce lipid peroxidation catalyzed by GGT (161). Thus, GSH can produce reactive oxygen species and act as a pro-oxidant. Such an oxidative environment rich in active oxygen species may be generated in the GGT-positive preneoplastic liver foci. The reaction of active oxygen and/or products of lipid peroxidation with DNA in these preneoplastic foci may lead to hepatocarcinogenesis (162).

Flavonoids such as quercetin and myricetin induce concentration-dependent DNA strand breaks concurrent with lipid peroxidation in isolated rat liver nuclei (154, 155). Myricetin and quercetin produce hydrogen peroxide and superoxide radical by their autoxidation when treated with beef heart mitochondria (163). Polyphenolic 2-hydroxyemodine, a direct-acting mutagen like quercetin, generates reactive oxygen species and induces DNA strand breaks (164). Gossypol induces oxidative DNA damage in human lymphocytes (165). Polyphenolic compounds including quercetin react with DNA producing the oxidative base product 8-hydroxydeoxyguanosine (166).

FREE RADICAL MECHANISMS OF ANTIOXIDANT CARCINOGENICITY

The mechanism of mutagenicity and potential carcinogenicity of food antioxidants has not been elucidated. However, most of them are phenolic compounds, which are susceptible to autoxidation. The reactive oxygen species produced by these compounds (143, 145) may be due to their autoxidation and subsequent redox cycling. The facts that oxygen is essential for the mutagenicity of phenolic compounds such as flavonoids, catechol, and hydroquinone (167), and that the mutagenicity of quercetin is modulated by the presence of exogenous antioxidants (168), suggests the involvement of active oxygen.

One possible mechanism of mutagenicity and carcinogenicity of phenolic antioxidants may be their ability to produce active oxygen, which may react with DNA, yielding oxidative DNA damage. The other possibility is that the oxidized antioxidant may react with DNA, producing DNA-antioxidant adducts. Formation of such an adduct for quercetin has been suggested (118). It has been demonstrated that the aerobic mutagenicity of quercetin in *Salmonella typhimurium* is strongly dependent on the absence of excision repair. This strong dependence on the presence of an excision-repair deficiency in the test organism suggests the presence of a bulky covalent DNA adduct (118).

ANTIOXIDANTS AS DOUBLE-EDGED BIOLOGICAL SWORDS

Food antioxidants are thought to be beneficial because they prevent oxidation of food components. Some of these antioxidants show anticarcinogenic activity (19). Their beneficial action has led to their increasing consumption and to their suggested use in cancer therapy and prevention. For example, use of quercetin (63), vitamin C, and vitamin E (169) as therapeutic cancer-preventing drugs has been suggested. However, recent observations indicate that these antioxidants also have toxic and cancer-enhancing potential because of their pro-oxidant properties (169). They can produce reactive oxygen species by their autoxidation and subsequent redox cycling. It has been demonstrated that the cellular antioxidant glutathione (170) acts as a pro-oxidant and forms superoxide-dependent hydroxyl radicals in the presence of iron (171). The pro-oxidant properties of phenolic antioxidants may be due to such superoxide-dependent hydroxyl radicals.

The mechanisms of the pro- and antioxidant properties of phenolic antioxidants are not known. However, their hydroxyl groups would undergo autoxidation to their quinoid structures followed by redox cycling, which would generate reactive oxygen species. This can account for their pro-oxidant behavior. On the other hand, they would also react with oxygen free radicals, produced by other compounds, providing them with radical-trapping properties. This would account for their antioxidant behavior. Thus, these molecules have the potential of acting as both pro-

and antioxidants depending on the redox state of their biological environment. In the cellular environment, these two opposing effects may be competitive, and thus these compounds can play a dual role in mutagenesis and carcinogenesis.

SUMMARY

Food antioxidants are generally considered safe because of their well-known property of preventing lipid peroxidation. Also, some of them show anticarcinogenic activity. However, they can also act as pro-oxidants and induce oxidative DNA damage, which may lead to mutagenesis and carcinogenesis. Thus they can have a dual role in carcinogenicity. Their potential to act both as pro- and antioxidants is dependent upon the redox state of their biological environment. In the cellular environment, these two opposing effects may be competitive, and thus they may act as double-edged biological swords and may play a dual role in mutagenesis and carcinogenesis. In addition, their interactions with other dietary components such as sulfur compounds and trace metals can influence their redox environment, and thus they may play an important role in their beneficial or harmful properties.

REFERENCES

1. Ito, N., Fukushima, S., Hagiwara, A., Shibata, M., and Ogiso, T. 1983. Carcinogenicity of BHA in rats. *J. Natl. Cancer Inst.* 70:343–352.
2. Stich, H. 1991. The beneficial and hazardous effects of simple phenolic compounds. *Mutat. Res.* 95:119–128.
3. Aruma, O. 1994. Nutrition and health aspects of free radicals and antioxidants. *Food Chem. Toxicol.* 32:671–683.
4. Arcos, J., Woo, J., and Lai, D. 1988. Database on binary combination effects of chemical carcinogens. *Environ. Carcinogen. Rev.* 6:1–150.
5. Shibata, M., Hirose, M., Kagawa, M., Boonyaphiphat, P., and Ito, N. 1993. Enhancing effects of concomitant ascorbic acid administration on BHA-induced forestomach carcinogenesia in rats. *Carcinogenesis* 14:275–280.
6. Zhu, B., and Liehr J. 1994. Quercetin increases the severity of estradiol-induced tumorigenesis in hamster kidney. *Toxicol. Appl. Pharmacol.* 125:149–158.
7. Brown, J. P. 1980. A review of the genetic effects of naturally occurring flavonoids, anthraquinones and related compounds. *Mutat. Res.* 75:243–277.
8. Stich, H., and Rosin, M. 1984. Naturally occurring phenolics as antimutagenic and anticarcinogenic agents. *Adv. Exp. Med. Biol.* 177:1–29.
9. Williams, G. M. 1986. Epigenetic promoting effects of BHA. *Food Chem. Toxicol.* 24:1163–1166.
10. Ruch, R. J., Chen, S. J., and Klaunig, J. E. 1989. Prevention of cytotoxicity by antioxidant catechins isolated from Chinese green tea. *Carcinogenesis* 10:1003–1008.
11. Heijden, C., Janssen, P., and Strik, J. 1986. Toxicology of gallates: A review. *Food Chem. Toxicol.* 24:1067–1070.
12. Van Buren, J., De Vos, L., and Pilnik, W. 1973. Measurement of chlorogenic acid and flavonol glycosides in apple juice. *J. Food Sci.* 38:656–658.
13. Dhingra, B. S., and Davis, A. 1988. Determination of free ellagic acid by reversed-phase high-performance liquid chromatography. *J. Chromatogr.* 447:284–286.
14. U.S. National Research Council. 1981. *Health Effects of Nitrate, Nitrite and N-Nitroso Compounds,* Part 1. pp. 6–20 Washington, DC: National Academy Press.
15. U.S. National Academy of Sciences. 1982. *Diet, nutrition and cancer.* Washington, DC: National Academy Press.
16. Ames, B. N. 1983. Dietary carcinogens and anticarcinogens. *Science* 221:1256–1264.
17. Cerutti, P. A. 1985. Pro-oxidant states and tumor promotion. *Science* 227:375–381.
18. Hartman, P., and Shankel, D. 1990. Antimutagens and anticarcinogens. *Environ. Mol. Mutagen.* 15:145–182.
19. Namiki, M. 1990. Antioxidants/antimutagens in food. *Crit. Rev. Food. Sci. Nutr.* 29:273–300.
20. Hunt, J. V., and Dean, R. T. 1989. Free radical mediated degradation of proteins. *Biochem. Biophys. Res. Commun.* 162:1976–1084.
21. Pryor, W. A. 1986. Oxy-radicals and related species: Their formation, lifetimes and reactions. *Annu. Rev. Physiol.* 48:657–667.
22. Pryor, W. A. 1988. Why is the hydroxyl radical the only radical that commonly adds to DNA? *Free Radic. Biol. Med.* 4:219–223.

23. Kasai, H., Crain, P. F., Kuchino, Y., Nishimura, A., Ootsuyama, A., and Tanooka, H. 1986. Formation of 8-hydroxyguanine moiety in cellular DNA by agents producing oxygen radicals and evidence for its repair. *Carcinogenesis* 7:1849–1851.

24. Kuchino, Y., Mori, F., Kasai, H., Inoue, H., Iwai, S., Miura, K., Ohtsuka, E., and Nishimura, S. 1987. Misreading of DNA templates containing 8-hydroxydeoxyguanosine at the modified base and at adjacent residues. *Nature* 327:77–79.

25. Ames, B. N. 1989. Endogenous oxidative DNA damage, aging and cancer. *Free Radic. Res. Commun.* 7:121–128.

26. Ames, B. N., Shigenaga, M., and Hagen, T. M. 1993. Oxidants, antioxidants and the degenerative diseases of aging. *Proc. Natl. Acad. Sci. U.S.A.* 90:7915–7922.

27. Miller, J. 1994. Recent studies on metabolic activation of chemical carcinogens. *Cancer Res. (Suppl.)* 54:1879S–1881S.

28. Chaudhury, A., Nokubo, M., Blair, I., and Marnett, L. 1994. Detection of endogenous malonaldehyde-deoxyguanine adducts in human liver. *Science* 265:1580–1582.

29. Roe, F. 1989. Nongenotoxic carcinogens. *Mutagenesis* 4:407–411.

30. Stohs, S. 1990. Oxidative stress induced by dioxin. *Free Radic. Biol. Med.* 9:79–90.

31. Moody, D., Reddy, J., Lake, B., Popp, J., and Reese, D. 1991. Peroxisome proliferators and nongenotoxic carcinogenesis. *Fundam. Appl. Toxicol.* 16:233–248.

32. Ito, N., Tsuda, H., Tatematsu, M., Tagawa, Y., Kagawa, M., and Asamoto, M. 1988. Enhancing effects of various hepatocarcinogens on induction of preneoplastic foci in rats. *Carcinogenesis* 9:387–394.

33. Ito, N., Hirose, M., Shibata, M., Tanaka, H., and Shirai, T. 1989. Modifying effects of simultaneous treatment with BHA on rat tumor induction by 3,2'-dimethyl-4-aminobiphenyl- and N-methylnitrosourea. *Carcinogenesis* 10:2255–2259.

34. Williams, G., Maeura, Y., and Weisberger, J. 1983. Simultaneous inhibition of liver carcinogenicity and enhancement of bladder carcinogenicity of N-2-fluorenylacetamide by BHT. *Cancer Lett.* 19:55–60.

35. Williams, G., Tanaka, T., and Maeura, Y. 1986. Dose-related inhibition of aflatoxin B_1-induced hepatocarcinogenesis by phenolic antioxidants BHA and BHT. *Carcinogenesis* 7:1043–1050.

36. Weisberger, E., Evarts, R., and Wenk, M. 1977. Inhibitory effect of BHT on intestinal carcinogenesis in rat by azoxymethane. *Food Cosmet. Toxicol.* 15:139–141.

37. Reddy, B., Maeura, Y., and Weisberger, J. 1983. Effects of various levels of dietary BHA on methylazoxymethanol acetate-induced colon carcinogenesis in mice. *J. Natl. Cancer Inst.* 71:1299–1305.

38. Cohen, L., Polansky, M., Furuyama, K., Reddy, M., Berke, B., and Weisberger, J. 1984. Inhibition of chemically induced mammary carcinogenesis in rats by BHT. *J. Natl. Cancer Inst.* 72:165–174.

39. McCormick, D., Major, N., and Moon, R. 1984. Inhibition of DMBA-induced rat mammary carcinogenesis by concomitant or postcarcinogen antioxidant exposure. *Cancer Res.* 44:2858–2863.

40. Slaga, T., and Bracken, W. 1977. The effects of antioxidants on skin tumor initiation. *Cancer Res.* 37:1631–1635.

41. Williams, G., Tanaka, T., Maruyama, H., Maeura, Y., Weisberger, J., and Zang, E. 1991. Modulation by BHT of liver and bladder carcinogenesis induced by 2-acetylaminofluorene. *Cancer Res.* 51:6224–6230.

42. Anderson, R., and Lukey, P. 1987. A biological role for ascorbate in the selective neutralization of extracellular phagocyte-derived oxidants. *Ann. N.Y. Acad. Sci.* 498:229–246.

43. Mackerness, C., Leach, S., Thompson, M., and Hill, M. 1989. The inhibition of bacterially mediated N-nitrosation by vitamin C. *Carcinogenesis* 10:397–399.

44. Norkus, E., and Kuenzig, W. 1985. Studies on the antimutagenic activity of ascorbic acid. *Carcinogenesis* 6:1593–1598.

45. Turusov, V., Trukhanova, L., and Parfenov, Y. 1991. Modifying effects of ascorbic acid on the promoting stage of uterine sarcomogenesis induced in mice by dimethylhydrazine and estradiol-dipropionate. *Cancer Lett.* 56:29–35.

46. Chen, L. H., Boissonneault, G. A., and Glauert, H. P. 1988. Vitamin C, vitamin E and cancer (review). *Anticancer Res.* 8:739–748.

47. Bieri, J., Corash, L., and Hubbard, V. 1983. Medical uses of vitamin E. *N. Engl. J. Med.* 308:1063–1071.

48. Cook, M., and McNamara, P. 1980. Effect of dietary vitamin E on dimethylhydrazine-induced colon tumors in mice. *Cancer Res.* 40:1329–1331.

49. Shklar, G. 1982. Oral mucosal carcinogenesis in hamsters inhibited by vitamin E. *J. Natl. Cancer Inst.* 68:791–797.

50. Ura, H., Denda, A., Yokose, Y., Tsutsumi, M., and Konishi, Y. 1987. Effects of vitamin E on the induction of enzyme altered foci in rat liver. *Carcinogenesis* 8:1595–1600.

51. Shekelle, R., Liu, S., Lepper, M., Maliza, C., and Rassob, A. 1981. Dietary vitamin A and risk of cancer in Western Electric study. *Lancet* 2:1185–1189.

52. Sarkar, A., Mukherjee, B., and Chatterjee, M. 1994. Inhibitory effects of beta-carotene on chronic AAF-induced hepatocarcinogenesis. *Carcinogenesis* 15:1055–1060.

53. Moon, R., Grubbs, C., Sporn, M., and Goodman, D. 1977. Retinyl acetate inhibits mammary carcinogenesis induced by N-methyl-N-nitrosourea. *Nature* 267:620–621.
54. McCormick, D., Burns, F., and Albert, R. 1981. Inhibition of benzo(a)pyrene-induced mammary carcinogenesis by retinyl acetate. *J. Natl. Cancer Inst.* 66:559–564.
55. Wood, A., Huang, M., Chang, R., and Conney, A. 1982. Inhibition of mutagenicity of bay-region diolepoxide of polycyclic aromatic hydrocarbons by natural plant phenols. *Proc. Natl. Acad. Sci. U.S.A.* 79:5513–5517.
56. Barch, D., and Fox, C. 1988. Selective inhibition of methylbenzylnitrosamine-induced formation of esophageal O^6-methylguanine by dietary ellagic acid in rats. *Cancer Res.* 48:7088–7092.
57. Mandal, S., and Stoner, G. 1990. Inhibition of N-nitrosobenzylmethylamine-induced esophageal tumorigenesis in rats by ellagic acid. *Carcinogenesis* 11:55–61.
58. Chang, R., Huang, M., Wood, C., and Conney, A. 1985. Effects of ellagic acid and flavonoids on tumorigenicity of benzo(a)pyrene on mouse skin and in newborn mouse. *Carcinogenesis* 6:1127–1133.
59. Lesca, P. 1983. Protective effects of ellagic acid and other plant phenols on benzo(a)pyrene-induced neoplasia in mice. *Carcinogenesis* 4:1651–1653.
60. Smart, R., Huang, M., Chang, R., Wood, A., and Conney, A. 1986. Effects of ellagic acid on the formation of benzo(a)pyrene-derived DNA adducts in mice. *Carcinogenesis* 7:1669–1675.
61. Mukhtar, H., Das, M., and Bickers, D. 1986. Inhibition of methylcholanthrene-induced skin tumorigenesis in mice by chronic feeding of ellagic acid in drinking water. *Cancer Res.* 46:2262–2265.
62. Robak, J., and Gryglewski, R. 1988. Flavonoids are scavengers of superoxide anions. *Biochem. Pharmacol.* 37:837–841.
63. Ip, C., and Ganther, H. E. 1991. Combination of blocking agents and suppressing agents in cancer prevention. *Carcinogenesis* 12:365–367.
64. Shimoi, K., Masuda, S., Esaki, S., and Kinae, N. 1994. Radioprotective effects of flavonoids in gamma-ray irradiated mice. *Carcinogenesis* 15:2669–2672.
65. Suolinna, E., Buchsbaum, R., and Racker, E. 1975. The effect of flavonoids on aerobic glycolysis and growth of tumor cells. *Cancer Res.* 35:1865–1872.
66. Molnar, J., Beladi, I., Domonkos, K., Foldeak, S., Bola, K., and Veckenstedt, A. 1981. Antitumor activity of flavonoids on NK/Ly ascites tumor cells. *Neoplasma* 28:11–18.
67. Scambia, G., Ranellettti, F., Panici, P., Pianelli, M., Bonano, G., Rumi, C., Larocca, L., and Mancuso, S. 1990. Inhibitory effects of quercetin on OVCA 433 cells. *Br. J. Cancer* 62:942–946.
68. Ramanathan, R., Tan, C., and Das, N. P. 1992. Cytotoxic effects of plant polyphenols and fat-soluble vitamins on malignant human cultured cells. *Cancer Lett.* 62:217–224.
69. Kandaswami, C., Perkins, E., Soloniuk, D., Drzewiecki, G., and Middleton, E. 1991. Antiproliferative effects of citrus flavonoids on human squamous cell carcinoma in vitro. *Cancer Lett.* 56:147–152.
70. Hirano, T., Oka, K., and Akiba, M. 1989. Antiproliferative effects of synthetic and naturally occurring flavonoids on tumor cells of human breast carcinoma cell line ZR-75-1. *Res. Commun. Chem. Pathol. Pharmacol.* 64:69–78.
71. Verma, A. K., Johnson, J. A., Gould, M. N., and Tanner, M. A. 1988. Inhibition of 7,12-dimethylbenz(a)anthracene- and N-nitrosomethylurea-induced rat mammary cancer by dietary flavonol quercetin. *Cancer Res.* 48:5754–5758.
72. Deschner, E. E., Ruperto, J., Wong, G., and Newmark, H. 1991. Quercetin and rutin as inhibitors of azoxymethanol-induced colonic neoplasia. *Carcinogenesis* 12:1193–1196.
73. Kato, R., Nakadate, T., Yamamoto, S., and Sugimura, T. 1983. Inhibition of TPA-induced tumor promotion by quercetin: Possible involvement of lipoxygenase inhibition. *Carcinogenesis* 4:1301–1305.
74. Wei, H., Tye, L., Bresnick, E., and Birt, D. 1990. Inhibitory effect of apigenin, a plant flavonoid, on epidermal ornithine decarboxylase and skin tumor promotion in mice. *Cancer Res.* 50:499–502.
75. Toda, S., Miyase, T., Arichi, H., Tanizawa, H., and Takino, Y. 1985. Natural antioxidants. *Chem. Pharmacol. Bull.* 33:1725–1728.
76. Srinivas, L., and Shalini, V. 1991. DNA damage: Protection by turmeric. *Free Radic. Biol. Med.* 11:277–283.
77. Huang, M., Lysz, T., Ferraro, T., Abidi, T., Laskin, J., and Conney, A. 1991. Inhibitory effects of curcumin on lypoxygenase and cyclooxygenase activities in mouse skin. *Cancer Res.* 51:813–819.
78. Shih, C., and Lin. J. 1993. Inhibition of 8-hydroxydeoxyguanosine formation by curcumin. *Carcinogenesis* 14:709–712.
79. Nagabhusan, M., Amonkar, A., and Bhide, S. 1987. Antimutagenicity of curcumin. *Food Chem. Toxicol.* 25:545–547.
80. Kuttan, R., Bhanumathy, P., Nirmala, K., and George, M. 1985. Potential anticancer activity of turmeric. *Cancer Lett.* 29:197–202.
81. Huang, M., Smart, R., Wong, C., and Conney, A. 1988. Inhibitory effect of curcumin and caffeic acid on tumor promotion in mouse skin. *Cancer Res.* 48:5941–5946.

82. Liu, J., Lin, S., and Lin, J. 1993. Inhibitory effects of curcumin on protein kinase C. *Carcinogenesis* 14:857–861.
83. Mukundan, M., Chako, M., Annapurna, V., and Krishnaswamy, K. 1993. Effects of turmeric and curcumin on BP-DNA adducts. *Carcinogenesis* 14:493–496.
84. Huang, M., Deschner, E., Newmark, H., Wang, Z., Ferraro, T., and Conney, A. 1992. Effects of dietary curcumin on azoxymethane-induced colonic epithelial cell proliferation. *Cancer Lett.* 64:117–121.
85. Huang, M., Wang, Z., Georgiadis, C., Laskin, J., and Conney, A. 1992. Inhibitory effects of curcumin on tumor initiation by benz(a)pyrene and dimethylbenz(a)anthracene. *Carcinogenesis* 13:2183–2186.
86. Hirose, M., Ozaki, K., Takaba, K., Fukushima, S., Shirai, T., and Ito, N. 1991. Modifying effects of naturally occuring antioxidants in rats. *Carcinogenesis* 12:1917–1921.
87. Athar, M., Khan, W., and Mukhtar, H. 1989. Effect of dietary tannic acid on epidermal, lung and forestomach tumorigenicity in mice. *Cancer Res.* 49:5784–5788.
88. Mukhtar, H., Das, M., Khan, W., Wang, Z., Bik, D., and Bickers, D. 1988. Exceptional activity of tannic acid in protecting against DMBA-induced skin tumorigenesis in mice. *Cancer Res.* 48:2361–2365.
89. Ho, C., Chen, O., Shi, H., Zhang, K., and Rosen, R. 1992. Antioxidant effects of polyphenolic extracts of Chinese tea. *Prevent. Med.* 21:520–525.
90. Stich, H. 1992. Teas and tea components as inhibitors of carcinogen formation in model systems. *Prevent. Med.* 21:377–384.
91. Graf, E., and Eaton, J. 1990. Antioxidant functions of phytic acid. *Free Radic. Biol. Med.* 8:61–69.
92. Shamsuddin, A., Elsayed, A., and Ullah, A. 1988. Suppression of large intestinal cancer in rats by inositol hexaphosphate. *Carcinogenesis* 9:577–580.
93. Shamsuddin, A., Ullah, A., and Chakravarthy, A. 1989. Inositol and inositol hexaphosphate suppresses cell proliferation and tumor formation in mice. *Carcinogenesis* 10:1461–1463.
94. Nielsen, B., Thompson, L., and Bird, R. 1987. Effects of phytic acid on colonic epithelial cell proliferation. *Cancer Lett.* 37:317–325.
95. Nera, E., Iverson, F., Lok, E., Armstrong, C., Karpinsky, K., and Clayson, D. 1988. A carcinogenesis reversibility study of the effects of BHA on the forestomach and urinary bladder in male rats. *Toxicology* 53:251–268.
96. Ito, N., Fukushima, S., Tamano, S., Hirose, M., and Hagiwara, A. 1986. Dose response in BHA-induction of forestomach carcinogenesis in rats. *J. Natl. Cancer Inst.* 77:1261–1265.
97. Masui, T., Hirose, M., Imaida, K., Fukushima, S., Tamano, S., and Ito, N. 1986. Sequential changes of the forestomach of rats, hamsters and mice treated with BHA. *Jpn. J. Cancer Res.* 77:1083–1090.
98. Hirose, M., Kagawa, K., Ogawa, K., Yamamoto, A., and Ito, N. 1989. Antagonistic effects of diethylmaleate on promotion of forestomach carcinogenesis by BHA in rats. *Carcinogenesis* 10:2223–2226.
99. Tsuda, H., Sakata, T., and Musui, T. 1984. Modifying effects of BHA on induction of neoplastic lesions in rat liver and kidney. *Carcinogenesis* 5:525–531.
100. Imaida, K., Fukushima, S., Shirai, T., Masui, T., and Ito, N. 1984. Promoting activities of BHA, BHT and ascorbate on forestomach and urinary bladder carcinogenesis in rats. *Gann* 75:769–775.
101. Imaida, K., Fukushima, S., Inoue, K., Hirose, M., and Ito, N. 1988. Modifying effects of concomitant treatment with BHA and BHT on liver, forestomach and urinary bladder carcinogenesis in male rats. *Cancer Lett.* 43:167–172.
102. Hirose, M., Masuda, A., Imaida, K., Kagawa, M., Tsuda, H., and Ito, N. 1987. Catechol strongly enhances rat stomach carcinogenesis. *Jpn. J. Cancer Res.* 78:317–321.
103. Hirose, M., Masuda, A., Hasegawa, R., Wada, S., and Ito, N. 1990. Regression of BHA-induced hyperplasia in forestomach of hamsters. *Carcinogenesis* 11:239–244.
104. Saito, K., Nakagawa, S., Yoshitake, A., Hirose, M., and Ito, N. 1989. DNA-adduct formation in the forestomach of rats treated with 3-*tert*-butyl-4-hydroxyanisole. *Cancer Lett.* 48:189–195.
105. Witschi, H., and Morse, C. 1983. Enhanced lung tumor formation in mice by dietary BHT. *J. Natl. Cancer Inst.* 71:859–866.
106. Peraino, C., Fry, R., Staffeldt, E., and Christopher, P. 1977. Enhancing effects of phenobarbitone and BHT on AAF-induced hepatic tumorigenesis in rats. *Food Cosmet. Toxicol.* 15:93–96.
107. Imaida, K., Fukushima, S., Shirai, T., Ohtani, M., Nakanishi, K., and Ito, N. 1983. Promoting activities of BHA and BHT on two-stage urinary bladder carcinogenesis in the liver of rats. *Carcinogenesis* 4:895–899.
108. Gulati, D., Witt, K., Anderson, B., Zeiger, E., and Shelby, M. 1989. Chromosome aberration and sister chromatid exchange in Chinese hamster ovary cells. *Environ. Mol. Mutagen.* 13:133–193.
109. Kasai, H., and Nishimura, S. 1984. Hydroxylation of deoxyguanosine at the C-8 position by ascorbic acid. *Nucleic Acids Res.* 12:2137–2145.
110. Kobayashi, S., Yoshida, K., Ueda, K., Sakai, H., and Komano, T. 1988. Induction of mutations by 8-hydroxydeoxyguanosine. *Nucleic Acids Res.* 19:29–32.
111. Shinar, E., Rachmilewitz, E., Shiefter, A., Rahamim, E., and Saltman, P. 1989. Oxidative damage to

human red cells induced by copper and iron complexes in presence of ascorbate. *Biochim. Biophys. Acta* 1014:66–72.

112. Fukushima, S., Imaida, K., Sakata, T., Okamura, T., Shibata, M., and Ito, N. 1983. Promoting effects of sodium ascorbate on two-stage urinary bladder carcinogenesis in rats. *Cancer Res.* 43:4454–4457.

113. Mori, S., Kurata, Y., Takeuchi, Y., Toyama, M., Makino, S., and Fukushima, S. 1987. Influence of strain and diet on promoting effects of sodium ascorbate in two-stage urinary bladder carcinogenesis in rats. *Cancer Res.* 47:3492–3495.

114. Fukushima, S., Kurata, Y., Hasegawa, R., Asamoto, M., Shibata, M., and Tamano, S. 1991. Ascorbic acid amplification of bladder carcinogenesis promotion by K_2CO_3. *Cancer Res.* 51:2548–2551.

115. Heinonen, O., and Albanes, D. 1994. The effect of Vitamin E and beta-carotene on the incidence of lung cancer and other cancers in male smokers. *N. Engl. J. Med.* 330:1029–1035.

116. Leo, M., Arai, M., Sato, M., and Liebers, C. 1982. Hepatotoxicity of vitamin A and ethanol in rats. *Gastroenterology* 82:194–205.

117. Nagao, M., Morita, N., Yahagi, T., Shimizu, M., Kuroyanagi, K., Fukuoka, M., Yoshihira, K., Natori, S., Fujino, T., and Sugimura, T. 1981. Mutagenicities of 61 flavonoids an 11 related compounds. *Environ. Mutagen.* 3:401–419.

118. MacGregor, J. T., and Wilson, R. E. 1988. Flavone mutagenicity in *Salmonella typhimurium*: Dependence on the pKM101 plasmid and excision-repair deficiency. *Environ. Mol. Mutagen.* 11:315–322.

119. Ogawa, S., Hirayama, M., Tokuda, K., and Fukui, S. 1987. Enhancement of mutagenicity of 2-acetylaminofluorene by flavonoids and the structural requirements. *Mutat. Res.* 190:107–112.

120. Aldrick, A. J., Lake, B. G., and Rowland, I. R. 1989. Modification of in vivo heterocyclic amine genotoxicity by dietary flavonoids. *Mutagenesis* 4:365–370.

121. Suzuki, S., Takada, T., Sugawara, Y., Muto, T., and Kominami, R. 1991. Quercetin induces recombinational mutations in cultured cells. *Jpn. J. Cancer Res.* 82:1061–1064.

122. Thein, S., Jeffrey, A., Gori, H., Cotter, F., Flint, J., O'Corner, N., Wealtherall, D., and Wainscoat, J. 1987. Detection of somatic changes in human cancer DNA by DNA finger print analysis. *Br. J. Cancer* 55:353–356.

123. Pamukcu, A. M., Yalciner, S., Hatcher, J. F., and Bryan, G. T. 1980. Quercetin, a rat intestinal and bladder carcinogen, present in bracken fern. *Cancer Res.* 40:3468–3472.

124. Erturk, E., Hatcher, J. F., Nunoya, T., Pamukcu, A. M., and Bryan, G. T. 1984. Hepatic tumors in Sprague-Dawley and Fisher-344 female rats exposed chronically to quercetin or its glycoside rutin. *Cancer Res.* 25:95 (abstr.).

125. Ito, N., Hagiwara, A., Tamano, S., Kagawa, M., Shibata, M., Kurata, Y., and Fukusima, S. 1989. Lack of carcinogenicity of quercetin in F344/DuCrj rats. *Jpn. J. Cancer Res.* 80:317–325.

126. National Toxicology Program. 1991. NTP Technical Report (No. 409) on the Toxicity and Carcinogenesis Studies of Quercetin in F344/N Rats. NIH Publication No. 91-3140. Research Triangle Park, NC: U.S. Department of Health and Human Services, Public Health Service, National Toxicology Program.

127. Dunnick, J. K., and Hailey, J. R. 1992. Toxicity and carcinogenicity studies of quercetin, a natural component of foods. *Fundam. Appl. Toxicol.* 19:423–431.

128. Bull, P., Yanez, L., and Nervi, F. 1987. Mutagenic substances in red and white wine in Chile: A high risk for gastric cancer. *Mutat. Res.* 187:113–141.

129. Hoey, J., Montvernay, C., and Lambert, R. 1981. Wine and tobacco: Risk factors for gastric cancer in France. *Am. J. Epidemiol.* 113:668–674.

130. Canada, A. T., Watkins, W. D., and Nguyen, T. D. 1989. The toxicity of flavonoids to guinea pig enterocytes. *Toxicol. Appl. Pharmacol.* 99:357–361.

131. Ono, K., and Nakane, H. 1990. Mechanisms of inhibition of various cellular DNA and RNA polymerases by several flavonoids. *J. Biochem.* 108:609–613.

132. Li, S. Y., Teh, B. S., Seow, W. K., Li, F., and Thong, Y. H. 1990. Effects of plant flavonoids on cancer cells in vitro. *Cancer Lett.* 53:175–181.

133. Kirby, K. 1960. Induction of tumors by tannin extracts. *Br. J. Cancer* 14:147–150.

134. Korpassy, B. 1961. Tannins as hepatic carcinogens. *Prog. Exp. Tumor Res.* 2:245–290.

135. Onodera, H., Kitaura, K., Mitsumori, K., Yoshida, J., Takahashi, M., and Hayashi, Y. 1994. Study on the carcinogenicity of tannic acid in rats. *Food Chem. Toxicol.* 32:1101–1106.

136. Takaba, K., Hirose, M., Ogawa, K., Hakoi, K., and Fukushima, S. 1994. Modification of N-butyl-N-(4-hydroxybutyl)nitrosamine-initiated urinary bladder carcinogenesis in rats by phytic acid. *Food Chem. Toxicol.* 32:499–503.

137. Hirose, M., Fukushima, S., Shirai, T., Hasegawa, R., Kato, T., Tanaka, H., Asakawa, E., and Ito, N. 1990. Stomach carcinogenicity of caffeic acid and catechol in rats. *Jpn. J. Cancer Res.* 81:207–212.

138. Hirose, M., Fukushima, S., Kurata, Y., Tsuda, H., Tatematsu, M., and Ito, N. 1988. Modification of N-methyl-N'-nitro-N-nitrosoguanidine-induced forestomach and glandular stomach carcinogenesis by phenolic antioxidants in rats. *Cancer Res.* 48:5310–5315.

139. Hirose, M., Fukushima, S., Tanaka, H., and Ito, N. 1993. Carcinogenicity of catechol in rats and mice. *Carcinogenesis* 14:525–529.

140. McCormick, D., May, C., Thomas, C., and Detrisac, C. 1986. Anticarcinogenic and hepatotoxic interactions between retinyl acetate and BHT in rats. *Cancer Res.* 46:5264–5269.

141. McCormick, D., Hultin, T., and Detrisac, C. 1987. Potentiation of vitamin A hepatotoxicity by BHT. *Toxicol. Appl. Pharmacol.* 90:1–9.

142. Harvath, P. M., and Ip, C. 1983. Synergistic effects of vitamin E and selenium on the chemoprevention of mammary carcinogenesis in rats. *Cancer Res.* 43:5335–5341.

143. Wilson, R. L. 1990. Quinones, semiquinone free radicals and one-electron transfer reactions. *Free Radic. Res. Commun.* 8:201–217.

144. Nakayama, T., Church, D. F., and Pryor, W. A. 1988. Quantitative analysis of the hydrogen peroxide formed in aqueous cigarette tar extract. *Free Radic. Biol. Med.* 7:9–15.

145. Marklund, S., and Marklund, G., 1974. Involvement of superoxide anion radical in autoxidation of pyrogallol. *Eur. J. Biochem.* 47:469–474.

146. Yoshida, D., Matsumoto, T., Okamoto, H., Mizusaki, S., Kushi, A., and Fukuhara, Y. 1986. Formation of mutagens by heating foods and model systems. *Environ. Health Perspect.* 67:55–58.

147. Kahl, R., Weinke, S., and Kappus, H. 1989. Production of reactive oxygen species by butylated hydroxyanisole. *Toxicology* 59:179–194.

148. Ahmad, I., Krishnamurthi, K., Arif, J., Asquin, M., Mahmood, N., Athar, M., and Rahman, Q. 1995. Augmentation of cryosotile-induced oxidative stress by BHA in mice lungs. *Food Chem. Toxicol.* 33:209–215.

149. Torel, J., Cillard, J., and Cillard, P. 1986. Antioxidant activity of flavonoids and reactivity with peroxy radicals. *Phytochemistry* 25:383–385.

150. Yuting, C., Rongliang, Z., Zhongjian, J., and Yong, J. 1990. Flavonoids as superoxide scavengers and antioxidants. *Free Radic. Biol. Med.* 9:19–21.

151. Husain, S. R., Cillard, J., and Cillard, P. 1987. Hydroxyl radical scavenging activity of flavonoids. *Phytochemistry* 26:2489–2491.

152. Nakadate, T., Yamamoto, S., Aizu, E., and Kato, R. 1984. Effects of flavonoids on TPA-caused tumor promotion in relation to lipoxygenase inhibition by these compounds. *Gann* 75:214–222.

153. Rahman, A., Shahabuddin, M., Hadi, S., Parish, J., and Ainley, K. 1989. Strand scission in DNA induced by quercetin and Cu(II): Role of Cu(I) and oxygen free radicals. *Carcinogenesis* 10:1833–1839.

154. Sahu, S., and Washington, M. 1991. Quercetin-induced lipid peroxidation and DNA damage in isolated rat liver nuclei. *Cancer Lett.* 58:75–79.

155. Sahu, S., and Gray, G. 1993. Interaction of flavonoids and trace metals: Myricetin-induced DNA damage and lipid peroxidation. *Cancer Lett.* 70:73–79.

156. Stich, H., Rosin, M., Wu, C., and Powrie, W. 1981. A comparative genotoxicity study of chlorogenic acid. *Mutat. Res.* 90:201–212.

157. Inoue, S., Ito, K., Yamamoto, K., and Kawanishi, S. 1992. Caffeic acid causes metal-dependent damage to cellular DNA. *Carcinogenesis* 13:1497–1502.

158. Pitot, H. 1990. Altered hepatic foci: Their role in murine hepatocarcinogenesis. *Annu. Rev. Pharmacol. Toxicol.* 30:465–500.

159. Ross, D., Moldeus, P., Sies, H., and Smith, M. 1986. Mechanism and relevance of glutathione mutagenicity. *Mutat. Res.* 175:127–131.

160. Thust, R. 1988. Mechanism of cytogenetic genotoxicity of exogenous glutathione in V-79 cells: Implication of hydrogen peroxide and oxidative chromosome damage. *Cell Biol. Toxicol.* 4:241–257.

161. Stark, A., Zeiger, E., and Pagano, D. 1993. Glutathione metabolism by gamma-glutamyltranspeptidase leads to lipid peroxidation: Relevance to hepatocarcinogenesis. *Carcinogenesis* 14:183–189.

162. Stark, A., Russell, J., Langenbach, R., Pagano, D., Zeiger, E., and Huberman, E. 1994. Localization of oxidative damage by a glutathione-gamma-glutamyltranspeptidase system in preneoplastic lesions of livers from carcinogen treated rats. *Carcinogenesis* 15:343–348.

163. Hodnick, W. F., Kung, F. S., Roettger, W. J., Bohmont, C. W., and Pardini, R. S. 1986. Inhibition of mitochondrial respiration and production of toxic oxygen radicals by flavonoids. *Biochem. Pharmacol.* 35:2345–2357.

164. Kodama, M., Kamioka, Y., Nakayama, T., Nagata, C., Morooka, N., and Ueno, Y. 1987. Generation of free radical and hydrogen peroxide from 2-hydroxyemodin, a direct-acting mutagen, and DNA strand breaks by active oxygen. *Toxicology Lett.* 37:149–156.

165. Chen, Y., Sten, M., Nordenskjold, M., Lambert, B., Matlin, S. A., and Zhou, R. H. 1986. The effect of gossypol on the frequency of DNA-strand breaks in human leukocytes in vitro. *Mutat. Res.* 164:71–78.

166. Leanderson, P., and Tagesson, C. 1990. Cigarette-smoke-induced DNA damage: Role of hydroquinone and catechol in the formation of the oxidative DNA-adduct, 8-hydroxydeoxyguanosine. *Chem. Biol. Interact.* 75:71–81.

167. Dean, B. J. 1985. Recent findings of the genetic toxicology of benzene, toluene, xylene and phenols. *Mutat. Res.* 154:153–181.

168. Ochiai, M., Nagao, M., Wakabayashi, K., and Sugimura, T. 1984. Superoxide dismutase acts as an enhancing factor for quercetin mutagenesis in rat-liver cytosol by preventing its decomposition. *Mutat. Res.* 129:19–24.

169. Kahl, R. 1986. The dual role of antioxidants in the modification of chemical carcinogenesis. *J. Environ. Sci. Health* C4:47–92.

170. Ketterer, B. 1988. Protective role of glutathione and glutathione transferases in mutagenesis and carcinogenesis. *Mutat. Res.* 202:343–361.

171. Rowley, D. A., and Halliwell, B. 1982. Superoxide dependent formation of hydroxyl radicals in the presence of thiol compounds. *FEBS Lett.* 138:33–36.

Chapter Twenty

ANTICARCINOGENIC EFFECTS OF SYNTHETIC PHENOLIC ANTIOXIDANTS

Gary M. Williams and Michael J. Iatropoulos

INTRODUCTION

The anticarcinogenic properties of the synthetic phenolic antioxidants butylated hydroxyanisole (BHA) and butylated hydroxytoluene (BHT) were first reported by Wattenberg (1) in a study showing that at 5000 ppm in the diet both phenolics inhibited polycyclic aromatic hydrocarbon-induced forestomach tumors in mice and mammary tumors in rats. Subsequently, inhibitory effects were shown against a variety of chemical carcinogens in mice and rats (Table 1). The effective concentrations range from 125 to 6600 ppm. At high concentrations in the diet, these agents induce phase II conjugating enzyme systems (2) and this detoxification has been proposed to be the basis for their anticarcinogenicity (3). However, the phenolic antioxidants inhibit the effects of agents such as nitrosamines that are not effectively detoxified by conjugation (4). Moreover, they have been demonstrated to be anticarcinogenic at levels as low as 100–125 ppm (5,6), which is well below levels that induce conjugation enzymes (7). Accordingly, we have postulated that these two antioxidants act as free radical trapping agents at low levels to inhibit carcinogenicity (4,6).

This chapter reviews four studies from this laboratory reporting results on anticarcinogenic effects of BHA and BHT with emphasis on low-dose effects that are relevent to cancer prophylaxsis at levels acceptable for human intake (8).

Table 1. Inhibition of chemical-induced carcinogenesis by phenolic antioxidants

Carcinogen	Species	Organ	Antioxidant (ppm)	Reference
2-Acetylaminofluorene	Rat	Liver, mammary gland	BHT, 6600	31
Azoxymethane	Rat	Colon	BHT, 6600	32
Aflatoxin B$_1$	Rat	Liver	BHT, 1000	21
			BHA, 1000	6
			BHA, 125	27
Benzo[a]pyrene	Mouse	Forestomach	BHA, 5000	1
			BHT, 5000	
7,12-Dimethylbenz[a]-anthracene	Rat	Mammary gland	BHT, 300	33
Methylazoxymethanol acetate	Mouse	Colon, lung	BHA, 300	34
N-Nitrosodimethylamine	Mouse	Lung	BHA, 5000	35

MATERIALS AND METHODS

Experiment I

Williams et al. (9) and Maeura et al. (10) reported on the effect of BHT on hepatocarcinogenicity of 2-acetylaminofluorene (AAF). AAF was obtained from Aldrich Chemical Co. (Milwaukee, WI), and butylated hydroxytoluene (BHT) was obtained from Sigma chemical Co. (St. Louis, MO). Powdered NIH-07 and semisynthetic AIN-76A diets were used as basal diets. AAF was added to the NIH-07 diet at a concentration of 200 ppm, and BHT at concentrations of 300, 1000, 3000, or 6000 ppm. In other groups, 6000 ppm BHT, 200 ppm AAF with 6000 ppm BHT, and 200 ppm AAF were added to the AIN-76A diet. All compounds were given for 24 wk. The cumulative AAF dose is calculated to be 1680 mg/kg. Five hundred and thirty-one 11-wk-old male Fischer 344 (F344) rats were used in this experiment.

Experiment II

Williams et al. (11) reported on the effect of BHT on hepatocarcinogenicity of a low dose of AAF. AAF and BHT obtained from the same companies were prepared under similar conditions and added to the NIH-07 diet at 50 ppm for AAF and 100, 300, 1000, or 6000 ppm for BHT. In other groups, 50 ppm AAF and 50 ppm AAF with 100 ppm BHT were added to the AIN-76A diet. All compounds were administered for up to 76 wk. The cumulative AAF dose is calculated to be 1330 mg/kg. Interim sacrifices were performed at 12, 25, 36, and 48 wk and a terminal at 76 wk. A total of 351 F344 male rats, 11 wk of age, was used in this experiment.

Experiment III

Williams et al. (12) reported on the effect of BHA and BHT on hepatocarcinogenicity of aflatoxin B_1 (AFB$_1$). AFB$_1$ was obtained from Aldrich Chemical Co.; BHT and BHA were obtained from Sigma Chemical Co. AFB$_1$ was dissolved in dimethyl sulfoxide (DMSO) and given by gavage 3 times a week for 20 wk at 25 mg/kg. BHT or BHA were mixed in NIH-07 diet and were given at 1000 or 6000 ppm starting 1 wk before and continuing 1 wk after AFB$_1$ exposure. Administration of BHT or BHA was discontinued at wk 22 and all groups were maintained on the basal diet for another 23 wk. The cumulative AFB$_1$ dose was 1500 mg/kg. Interim sacrifices were performed at 12 and 24 wk and the experiment was terminated at 45 wk. A total of 267 F344 male rats was utilized.

Experiment IV

Williams and Iatropoulos (6) reported on the effect of BHA and BHT on hepatocarcinogenicity of a low dose of AFB$_1$. AFB$_1$ was obtained from Sigma Chemical Co., butylated hydroxyanisole (BHA) was obtained from Aldrich Chemical Co., and BHT from Fluka Chemie AG, Buchs, Switzerland. AFB$_1$ dosing solutions were prepared 3 times weekly and were delivered by gavage for 40 wk following 2 wk of BHA or BHT administration, at 5 μg/kg body weight (bw). BHA and BHT were dissolved in corn oil, mixed weekly with the AIN-76A diet at a concentration of 5, 25, or 125 ppm, and were given 2 wk prior to AFB$_1$ administration. The cumulative AFB$_1$ dose was 600 μg/kg. BHA and BHT were administered for 42 and AFB for 40 wk. Three hundred and twenty-four 6-wk-old male F344 rats were used in this experiment.

Morphology

In all experiments, standardized slices from each liver lobe were taken and processed as described previously (13). Experiment I was terminated at 25 wk. In experiment II, 3 or 4 randomly selected rats were killed at wk 12, 24, 36, and 48 to determine the incidence of altered hepatocellular foci. At 76 wk, the experiment was terminated. In experiment III, interim sacrifices were performed at 12 and 24 wk and the final sacrifice at 45 wk. In all three studies, the livers were subjected to histochemical reaction for gamma-glutamyltransferase (GGT) using the method of Rutenberg et al. (14). For morphometric analysis of GGT-reacted liver slides a Videoplan image analyzer (Carl Zeiss, Inc.) was used. The numbers of GGT$^+$ hepatocellular altered foci (HAF) and the whole liver areas of the sections were recorded and the numbers of foci per square centimeter calculated. In addition, the occurrence of hepatocellular adenomas and carcinomas was also recorded. In experiment IV, in order to demonstrate inhibition of AFB-induced carcinogenesis by concomitant feeding of BHA or BHT, the development of HAF, in the group with AFB$_1$ alone (group 2), was monitored by killing 4 rats at 16-, 24-, and 32- wk intervals. The marker used for was the placental form of glutathione S-transferase (GST-P) (15). The number of GST-P$^+$ clusters of more than three cells over the entire sampled and scanned liver area (in square centimeters) was calculated and expressed as positive foci number per square centimeter. The same image analyzer described earlier was used.

Statistical Analysis

In experiment I, the between-group differences multiple comparisons using the Bonferroni criteria were used to assess whether or not a dose effect exists among the five different levels. Differences in tumor incidence were analyzed using the Kruskal-Wallis test.

In experiments II and III, the means were compared using an one-tailed Student's t-test. The trends in liver foci over time were tested using linear regression analysis.

In experiment IV, mean lesion multiplicity and mean total lesion area were analyzed using factional analysis of variance (ANOVA). Scheffe's multiple comparison test was also used to test for significantly different pairs within each significant factor (16). Dose-response on lesion multiplicity was also examined and compared among groups using Armitage's test.

Only statistically significant differences are reported.

RESULTS AND DISCUSSION

In a series of studies, the dose-response characteristics of inhibition by BHA and BHT of chemical-induced liver cancer in male F344 rats have been examined.

Experiment I

BHT at 300, 1000, 3000, or 6000 ppm was fed concurrently with the potent rat liver carcinogen AAF at 200 ppm (9,10). AAF induced a substantial number of preneoplastic GGT$^+$ HAF by 6 wk, and this increased to 66.1/cm^2 by 25 wk. Feeding of BHT alone did not induce foci or neoplasms by 25 wk. Administration of BHT together with AAF reduced the development of foci throughout the entire experimental period in a dose-dependent manner. BHT also produced a reduction in the incidence of rats with liver tumors from 100% in group 7 (given AAF alone), to 56% in group 11, (given AAF plus 6000 ppm of BHT). The total numbers of tumors per rat in all groups were significantly reduced. Moreover, the reduction in tumors in each BHT group was dose related.

In the group fed the AIN-76 (semisynthetic) diet (group 13) containing 200 ppm AAF, the number of HAF at 6 wk was increased and was comparable to that in the group given 200 ppm AAF in NIH-07 diet (group 7). After wk 8, all animals of group 13 died. In the group fed the AIN-76 diet (group 12) containing 200 ppm AAF and 6000 ppm BHT, the number of foci and incidence of neoplasms were greater than those in the group given the same substances in NIH-07 diet (group 11), although BHT still had an effect.

Thus, BHT produced in the liver a dose-dependent reduction of the number of foci, tumor incidence, and number of neoplasms per animal.

In this study, no bladder neoplasms occurred in rats given either AAF or BHT alone, but groups fed both AAF and BHT developed bladder neoplasms in proportion to the dose of BHT. The effect of BHT in inhibiting liver neoplasia and enhancing bladder cancer could be explained by BHT causing an alteration of the metabolism of AAF in the liver, resulting in less activation and formation of greater amounts of metabolites excreted in the urine, as reported by Grantham et al. (17). Such metabolites could be further processed in the urine, yielding carcinogenic

Table 2.   Effect of simultaneously administered butylated hydroxytoluene (BHT) on induction of liver γ-glutamyltransferase (GGT)-positive altered foci in rats by a low dose of 2-acetylaminofluorene (AAF) fed for 76 wk

Group identification	Number of GGT+ foci/cm² of liver[a]				
	12 wk[b]	24 wk	36 wk	48 wk	76 wk
1. Control	0	0	0	0.1	0.9
2. 50[c] ppm AAF + 100 ppm BHT	1.2	1.2	4.3	4.8[d]	9.6
3. 50 ppm AAF	1.5[e]	5.1[e]	7.0[e]	7.4[e]	11.5[e]

[a] Mean, derived from 3 or 4 rats/group at each period.
[b] Weeks of treatment.
[c] Concentration of chemicals in diet.
[d] Significantly less than group 3 at $p < .05$.
[e] Significantly greater than group 1 at $p < .05$.

products. In addition, BHT at high dose levels may exert an additional promoting effect on bladder carcinogenesis (18).

Experiment II

A second experiment was performed, under similar conditions, to determine whether at a lower exposure to AAF, closer to environmental levels of carcinogens, BHT would be inhibitory at lower levels without the complication of an increase in bladder cancer (11). Feeding of AAF alone at 50 ppm produced a progressive increase in the incidence of GGT+ HAF (Table 2), up to 11.5/cm² by 76 wk, which was less than that induced by 200 ppm AAF at 25 wk. Concurrent feeding of BHT together with AAF reduced HAF at all time points in a dose-related manner (Figure 1). In control rats (group 1) or those fed 6000 ppm BHT (group 10), no foci were evident up to 36 wk, but small numbers were present at 48 and 76 wk. All doses of BHT reduced the incidence of AAF-induced foci, although at 100 ppm the reduction was significant at 48 wk (Table 2).

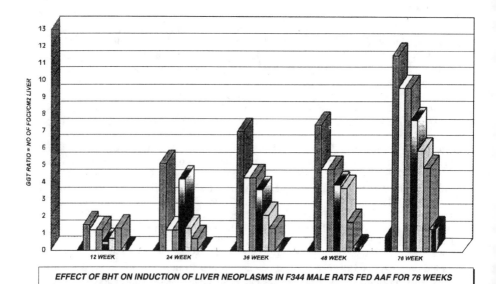

Figure 1.   Effect of BHT on induction of GGT-positive hepatocellular altered foci in F344 rats fed AAF for 76 wk.

Table 3. Effect of simultaneously administered butylated hydroxytoluene (BHT) on induction of neoplasms in rats by a low dose of 2-acetylaminofluorene (AAF) fed for 76 wk

Group identification	Percent multiplicity of adenomas	Percent multiplicity of carcinomas	Percent multiplicity of both neoplasms
1. Control	0	0	0
2. 50[a] ppm AAF + 100 ppm BHT	14.7[b]	1.6[c]	16.3
3. 50 ppm AAF	19.2[d]	4.8[d]	25.0[d]

[a] Concentration of chemicals in diet.
[b] Mean.
[c] Significantly less than group 3 at $p < .05$.
[d] significantly greater than group 1 at $p < .05$.

Liver neoplasms in rats were evident first at 48 wk; that is, 1 in 4 rats in group 2 fed AAF had a small adenoma. At 76 wk, all rats in the AAF-exposed groups displayed multiple liver neoplasms. BHT did not reduce the incidence (percentage) of rats with adenomas, which was 100% in all groups, although at all 5 doses it produced a dose-related reduction in the multiplicity of adenomas and carcinomas, including at 100 ppm (Table 3, Figure 2).

In group 8, fed AAF in the AIN-76A diet, the increase in the incidence of GGT+ foci was similar to group 2, but not as marked. When administered in AIN-76A diet with AAF, BHT at 100 ppm reduced the incidence of AAF-induced foci at 48 and 76 wk in this diet also. In the groups (8 and 9) maintained on AIN-76A diet, the incidence and multiplicity of liver tumors were similar to those in the groups given the same compounds in NIH-07 diet (groups 2 and 3). BHT at 100 ppm reduced the multiplicity of neoplasms.

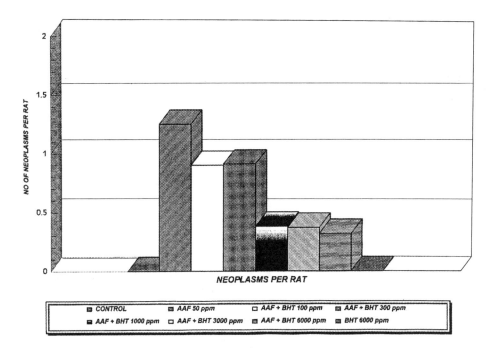

Figure 2. Effect of BHA or BHT on induction of liver neoplasms in F344 male rats fed AAF for 76 wk.

With respect to the bladder, only feeding of 6000 ppm BHT together with AAF resulted in an increase in the incidence and multiplicity of bladder neoplasms, as previously. BHT at 3000 ppm plus AAF increased nodular hyperplasia of the bladder. Lower concentrations were not associated with an increase in bladder pathology. These results indicate that the inhibition by BHT of liver neoplasia, resulting from low-level carcinogen exposure, may be achieved at doses that do not produce adverse effects.

The finding of inhibition of liver carcinogenesis by BHT at 100 ppm is particularly noteworthy. It is not known whether this dose is capable of modulating AAF biotransformation as has been reported for 6600 ppm (17), but that seems unlikely since we have shown that the activity of the phase II enzyme glutathione S-transferase was not induced by concentrations below 3000 ppm (7). Alternatively, the inhibition may have resulted from trapping of reactive AAF metabolites by BHT. Support for this possibility comes from several studies. Chipman and Davies (19) examined the genotoxicity in cultured hepatocytes of AAF and its reactive metabolite, N-hydroxy-AAF, using DNA repair to monitor DNA damage. Addition of BHT to the cultures reduced the DNA repair elicited by both chemicals. The protective effect against AAF could have resulted from inhibition of enzymatic activation, that is, N-hydroxylation, but since BHT was also protective against the N-hydroxy derivative, reduction of bioactivation alone could not be the entire explanation. Also, Richer et al. (20) reported that addition of BHA or BHT to microsome preparations reduced the binding of [^{3}H-] AAF and [^{3}H]-N-OH-AAF to calf thymus DNA. They also showed that addition of the antioxidants to cultured rat hepatocytes inhibited [^{3}H]-AAF binding to DNA, corroborating the finding of Chipman and Davies (19). Accordingly, the antioxidant protection likely stemmed from scavenging of free radicals from reactive AAF products. Mikuni et al. (21) have shown that this occurs for free radicals derived from the alkylating agent N-methyl-N'-nitro-N-nitrosoguanidine.

Experiment III

To determine whether phenolic antioxidants can inhibit the carcinogenicity of a human carcinogen without shifting the organotropism, as occurred with AAF, the protective effect against aflatoxin B_1 (AFB$_1$) was studied (12). Administration of AFB$_1$ (group 2) resulted in induction of GGT$^+$ HAF by 12 wk. At 24 wk, the incidence of foci was further increased to 11.5/cm^2 in group 2, but no liver neoplasms were evident. In rats of groups fed BHT or BHA during AFB$_1$ administration, a dose-related reduction in the number of foci occurred at all times. At the end of the experiment (45 wk), administration of AFB$_1$ alone for 20 wk induced liver neoplasms in 63% of rats with a multiplicity of 1.4 neoplasms/rat. Feeding BHT or BHA together with AFB$_1$ reduced the liver neoplasms incidences to only 0–13% and multiplicities to 0–0.2/rat at the end of the study. In group 4 (AFB$_1$ + BHT 6000), rats developed no hepatocellular neoplasms.

Neoplasms in other organs were rare and were not affected by antioxidant treatment, except for a possible reduction of colon cancer. Thus, BHA and BHT inhibited the hepatocarcinogenicity of concurrently administered AFB$_1$ without shifting the organotropism, in contrast to what was found with AAF at high concentrations of BHT.

Lotlikar et al. (22) have provided evidence that induction of glutathione S-transferase activity by BHA reduces AFB$_1$ binding to DNA. Accordingly, as with AAF, it is important to establish whether antioxidants inhibit AFB$_1$ carcinogenicity at concentrations unlikely to produce enzyme induction.

Experiment IV

To further examine the potential of BHA and BHT to inhibit AFB$_1$ carcinogenicity, a study was performed in which lower concentrations, 5, 25, and 125 ppm, of antioxidants were fed together with a dose of AFB$_1$ that was reduced to 5 μg/kg per day 3 times per week for 40 wk for a cumulative dose of 0.6 mg/kg (6). With AFB$_1$ alone, morphometric analysis of GST-P$^+$ HAF revealed a low incidence at 16 wk. The multiplicity of foci increased progressively up to 12.9/cm^2 at 42 wk (Table 4).

At 42 wk (final sacrifice), the mean number of HAF per square centimeter was very low in the controls and was substantially increased in rats given only AFB$_1$ (Table 4). The values of

Table 4. Effect of simultaneously administered BHA or BHT on reduction of induced GST-P-positive hepatocellular altered foci[a] in rats by a low dose of aflatoxin B_1 given for 40 wk

Group identification	GST-P$^+$ HAF/cm^2
1. Vehicle (DMSO) control	0.75
2. AFB$_1$[b]	12.90
3. BHA 5 ppm + AFB$_1$	11.30
4. BHA 25 ppm + AFB$_1$	11.09
5. BHA 125 ppm + AFB$_1$	7.72[c]
6. BHT 5 ppm + AFB$_1$	11.13
7. BHT 25 ppm + AFB$_1$	11.40
8. BHT 125 ppm + AFB$_1$	9.35[c]

[a] Foci of >3 cells/cm^2 of liver.
[b] At 5 μg/kg body weight, 3 × wk for 40 wk.
[c] Statistically significant compared to group 2 at $p < .01$.

group 5 given 125 ppm BHA and 8 given 125 ppm BHT together with AFB$_1$ were significantly reduced in comparison to group 2 (Figure 3). The lower concentrations of antioxidants did not reduce the induction of foci by AFB$_1$.

Thus, the antioxidants were effective at about the same low concentration as protected against AAF hepatocarcinogenicity.

CONCLUSIONS

As described here, substantial documentation shows that BHA and BHT are effective in inhibiting chemically induced liver carcinogenesis at low doses when the exposure to carcinogen is low. The mechanism(s) for the inhibition is not fully clear and, possibly, may involve several actions, including modulation of biotransformation and trapping of carcinogen radical species. Moreover, the mechanism may differ for different types of carcinogen. The crucial question, however, is whether antioxidants can be practically used in human cancer prevention (8). In prophylaxis,

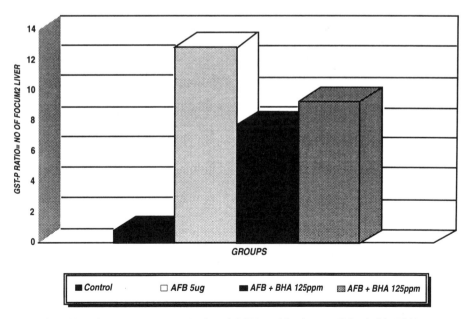

Figure 3. Effect of BHA or BHT on induction of GST-P-positive hepatocellular foci in F344 male rats by AFB given orally for 40 wk.

only a low risk of adverse effects is acceptable. Both BHA and BHT are approved food additives that have been used for at least 30 years and for which either temporary or permanent acceptable daily intakes have been established. Nevertheless, the finding that BHA and BHT have exhibited carcinogenic or neoplasm promotion activities in rodent models (18) is of concern. These effects, however, occurred at high doses of 1000 ppm or greater, and in the case of BHA, carcinogenicity was limited to the nonglandular portion of the rodent stomach (forestomach), a tissue that does not exist in humans. The phenolic antioxidants are not DNA reactive (23), and it seems clear that their high-dose effects are related to cellular toxicity, including exertion of a promoting action (24). Such epigenetic effects for diverse agents have thresholds and have not occurred in humans at low exposures (25,26). Accordingly, these effects do not represent a human hazard at the low doses of antioxidants that might be considered for cancer prophylaxis (27).

Strong support for the potential of phenolic antioxidants as anticancer agents in humans comes from several observations: (a) Ethoxyquin (28), BHT, and BHA (6,12), have been shown to be potent inhibitors of the carcinogenicity of a human carcinogen, AFB_1; and (b) BHA and BHT have been shown to be inhibitors at concentrations as low as 100–125 ppm. To be effective inhibitors of human cancer, antioxidants would have to be present continuously in order to be available to counteract intermittent carcinogen exposures. The finding that antioxidants can be inhibitory at low doses opens this possibility. Moreover, ingestion of antioxidants can be expected to confer other benefits (see ref. 28). Future research should examine their efficacy at even lower doses (>50 ppm) against the very low levels of food-borne carcinogens to which humans are exposed, such as polycyclic aromatic hydrocarbons, heterocyclic amines, mycotoxins, and nitrosamines. Such investigations would be difficult with neoplasms as the endpoint, but could be done economically using hepatocellular foci as the measured parameter, based upon the studies detailed earlier showing that reduction of antecedent foci is predictive of a reduction of subsequent neoplasms.

The current weight of evidence favors the potential benefits of phenolic antioxidants. It is intriguing that since the introduction of phenolic antioxidants as food additives in around 1960, the only cancer that has increased substantially is lung cancer, which is due to cigarette smoking, whereas, for unknown reasons, cancers of the stomach and liver have decreased. Explanations could be offered as to how antioxidants could have contributed to these decreases. There is evidence that adequate intake of certain natural antioxidants, including vitamin C, beta-carotene, vitamin A, vitamin E, and selenium, is associated with reduced risk for some cancers (29,30). It may be speculated that low levels of synthetic exogenous antioxidants could complement the actions of natural antioxidants. Thus, although significant uncertainties remain, synthetic antioxidants are attractive candidates for one approach to the prevention of human cancer.

ACKNOWLEDGMENTS

The authors thank N. Rivera for preparing the manuscript.

REFERENCES

1. Wattenberg, L. W. 1972. Inhibition of carcinogenic and toxic effects of polycyclic hydrocarbons by phenolic antioxidants and ethoxyquin. *J, Natl. Cancer Inst.* 48:1425–1430.
2. Talalay, P., and Benson, A. M. 1982. Elevation of quinone reductase activity by anticarcinogenic antioxidants. *Adv. Enzyme Regul.* 20:28–30.
3. Wattenberg, L. W. 1985. Chemoprevention of cancer. *Cancer Res.* 45:1–8.
4. Williams, G. M. 1993. Inhibition of chemical-induced experimental cancer by synthetic phenolic antioxidants. In *Antioxidants: Chemical, Physiological, Nutritional and Toxicological Aspects, eds.* G. M. Williams, H. Sies, G. T. Baker III, J. W. Erdman, Jr., and C. J. Henry, pp. 303–308. Princeton, NJ: Princeton Scientific Press.
5. Tanaka, T., Maruyama, H., Maeura, Y., Weisburger, J. H., and Williams, G. M. 1991. Modulation by butylated hydroxytoluene of liver and bladder carcinogenesis induced by chronic low level exposure to 2-acetylaminofluorene. *Cancer Res.* 51:6224–5230.
6. Williams, G. M. and Iatropoulos, M. J. 1996. Inhibition of the hepatocarcinogenicity of aflatoxin B_1 in rats by low levels of the phenolic antioxidants butylated hydroxyanisole and butylated hydroxytoluene. *Cancer Lett.* 104:49–53.

7. Furukawa, K., Maeura, Y., Furukawa, N. T. and Williams, G. M. 1984. Induction by butylated hydroxytoluene of rat liver gamma-glutamyltranspeptidase activity in comparison to expression in carcinogen-induced altered lesions. *Chem. Biol. Interact.* 48:43–58.

8. Williams, G. M. 1994. Interventive prophylaxis of liver cancer. *Eur. J. Cancer Prev.* 3:89–99.

9. Williams, G. M., Maeura, Y., and Weisburger, J. H. 1983. Simultaneous inhibition of liver carcinogenicity and enhancement of bladder carcinogenicity of N-2-fluorenylacetamide by butylated hydroxytoluene. *Cancer Lett.* 19:50–60.

10. Maeura, Y., Weisburger, J. H., and Williams, G. M. 1984. Dose-dependent reduction of N-2-fluorenylacetamide-induced liver cancer and enhancement of bladder cancer in rats by butylated hydroxytoluene. *Cancer Res.* 44:1604–1610.

11. Williams, G. M., Tanaka, T, Maruyama, H., Maeura, Y., Weisburger, J. H., and Zang, E. 1991. Modulation by butylated hydroxytoluene of liver and bladder carcinogenesis induced by chronic low level exposure to 2-acetylaminofluorene. *Cancer Res.* 51:6224–6230.

12. Williams, G. M., Tanaka, T., and Maeura, Y. 1986. Dose-dependent inhibition of aflatoxin B_1 induced hepatocarcinogenesis by the phenolic antioxidants, butylated hydroxyanisole and butylated hydroxytoluene. *Carcinogenesis* 7:1043–1050.

13. Williams, G. M., and Watanabe, K. 1978. Quantitative kinetics of development of N-2-fluorenylacetamide-induced, altered (hyperplastic) hepatocellular foci resistant to iron accumulation and of their reversion of persistence following removal of carcinogen. *J. Natl. Cancer Inst.* 61:113–121.

14. Rutenberg, A. M., Kim, H., Fishbein, J., Hanker, J. S., Wasserkrug, H. C., and Seligman, R. 1968. Histochemical and ultrastructural demonstration of gamma-glutamyl transpeptidase activity. *J. Histochem. Cytochem.* 17:517–526.

15. Shimizu, A., Nakamura, Y., Harada, M., Ono, T., Sato K., Inone, T., and Kawisawa, M. 1989. Positive foci of glutathione S-transferase placental form in the liver of rats given furfural by oral administration. *Jpn. J. Cancer Res.* 80:608–611.

16. Scheffe, H. 1959. *The Analysis of Variance.* Wiley, New York.

17. Grantham, P. H., Weisburger, J. H., and Weisburger, E. K. 1973. Effect of the antioxidant butylated hydroxytoluene (BHT) on the metabolism of the carcinogens N-2-fluorenylacetamide. *Food Cosmet. Toxicol.* 11:209–217.

18. Ito, N., Fukushima, S., and Tsuda, H. 1985. Carcinogenicity and modification of the carcinogenic response by BHA, BHT and other antioxidants. *CRC Crit. Rev. Toxicol.* 15:109–150.

19. Chipman, J. K., and Davies, J. E. 1988. Reduction of 2-acetylaminofluorene-induced unscheduled DNA synthesis in human and rat hepatocytes by butylated hydroxytoluene. *Mutat. Res.* 207:193–198.

20. Richer, N., Marion, M., and Denzeau, F. 1989. Inhibition of binding of 2-acetylaminofluorene to DNA by butylated hydroxytoluene and butylated hydroxyanisole in vitro. *Cancer Lett.* 47:211–216.

21. Mikuni, T., Tasuta, M., and Kamachi, M. 1987. Scavanging effect of butylated hydroxytoluene on the production of free radicals by the reaction of hydrogen peroxide with N-ethyl-N′-nitrosoguanidine. *J. Natl. Cancer Inst.* 79:281–283.

22. Lotlikar, P. D., Clearfield, M. S., and Thee, E. C. 1984. Effect of butylated hydroxyanisole on in vivo and in vitro hepatic aflatoxin B_1-DNA binding in rats. *Cancer Lett.* 24:241–250.

23. Williams, G. M., McQueen, C. A., and Tong, C. 1990. Butylated hydroxyanisole and butylated hydroxytoluene I. Genetic toxicology studies. *Food Chem. Toxicol.* 28:793–798.

24. Williams, G. M. 1986. Epigenetic promoting effects of butylated hydroxyanisole. *Food Chem. Toxicol.* 24:1163–1166.

25. Williams, G. M. 1987. Definition of a human cancer hazard. In *Nongenotoxic Mechanisms in Carcinogenesis.* Banbury Report 25, pp. 367–380. Cold Spring Harbor, NY: Cold Spring Harbor Laboratory.

26. Williams, G. M., Karbe, E., Fenner-Crisp, P. A., Iatropoulos, M. J., and Weisburger, J. H. 1996. Risk assessment of carcinogenesis in food with special consideration of non-genotoxic carcinogens. Scientific arguments for use of risk assessment and for changing the Delaney Clause specifically. *Exp. Toxicol. Pathol.* 48:209–215.

27. Whysner, J., and Williams, G.M. 1996. Butylated hydroxyanisole mechanistic data and risk assessment: Conditional species-specific cytotoxicity, enhanced cell proliferation and tumor promotion. *Pharmacol. Ther.* 71:137–151.

28. Cabral, J. R. P., and Neal, G. E. 1983. The inhibitory effects of ethoxyquin on the carcinogenic action of aflatoxin B_1 in rats. *Cancer Lett.* 19:125–132.

29. Birt, D. F. 1986. Update on the effects of vitamins A, C, and E and selenium on carcinogenesis. *Proc. Soc. Exp. Biol. Med.* 183:311–320.

30. Rogers, A. E., and Longnecker, M. P. 1988. Dietary and nutritional influences on cancer: A review of epidemiologic and experimental data. *Lab. Invest.* 59:729–759.

31. Ulland, B. J., Weisburger, J. H., Yamamoto, R., and Weisburger, E. K. 1973. Antioxidants and carcinogenesis: Butylated hydroxytoluene, but not diphenyl-p-phenylenediamine, inhibits cancer induction by N-2-fluorenylacetamide in rats. *Food Cosmet. Toxicol.* 11:199–207.

32. Weisburger, E. K., Evarts, R. P., and Wenk, M. L. 1977. Inhibitory effect of butylated hydroxytoluene (BHT) on intestinal carcinogenesis in rats by azoxymethane. *Food Cosmet. Toxicol.* 15:139–141.
33. Cohen, L. A., Polansky, M., Furuya, K., Reddy, M., Berke, B., and Weisburger, J. H. 1984. Inhibition of chemically induced mammary carcinogenesis in rats by short-term exposure to butylated hydroxytoluene (BHT): Interrelationships among BHT concentration, carcinogen dose, and diet. *J. Natl. Cancer Inst.* 72:165–172.
34. Reddy, S. B., Maeura, Y., and Weisburger, J. H. 1983. Effect of various levels of dietary butylated hydroxyanisole on methylazoxymethanol acetate-induced colon carcinogenesis in CF1 mice. *J. Natl. Cancer Inst.* 71:1299–1305.
35. Chung, F. L., Wang, M., Carmella, S. G., and Hecht, S. S. 1986. Effects of butylated hydroxyanisole on the tumorigenicity and metabolism of *N*-nitrosodimethylamine and *N*-nitrosopyrrolidine in A/J mice. *Cancer Res.* 46:165–168.

INDEX

CIDEP (chemically induced dynamic electron polarization), 4
CIDNP (chemically induced dynamic neutron polarization), 4
Cigarette smoke, free-radical-generating pollutants in, 105
Cirrhosis, 238
CLAS (Cholesterol Lowering Atherosclerosis Study), 114
Clean Air Act, 229
Cobalamin monitoring, by homocysteine measurement, 54
Cobalt, as catalyst, 143–144
Coenzyme Q$_{10}$ (ubiquinone; CoQ)
  antioxidant effects, 15
  as chain-breaking antioxidant, 48
  functions, 52
  measurement, 52
  mechanism of action, 96–97
  nomenclature system, 10–11
Cold-induced stress, glutathione response to, 177
Collagen formation, reduction of, 199
Colorectal cancer risk, antioxidants and, 84, 85
Column-switching technique, for 80H2DG measurement, 44
Copper
  as catalyst, 143–144
  characteristics, 218
  deficiency, 218
  dietary intake, 218
  interaction, with ROS, 261
  LDL oxidation and, 116
  in particulate matter, 276
  preloading, postischemic cardiac oxidative injury and, 147
  sources, dietary, 208
  toxicity, 218
Copper-phenanthroline complex, 178
Coronary artery disease
  PUFA and, 102–104
  risk factors, 102, 113
  vitamin E and, 82, 102–104
Cortisol, vitamin C and, 167–168
Coulometric array method
  for flavonoids, 57, 58
  for polyphenols, 57, 58
Curcumin, anticarcinogenic activity, 329–330
Cu, Zn-superoxide dismutase, 126, 127, 131
Cyanide poisoning, 5
Cysteine
  in GSH synthesis, 173
  intracellular, 175
Cytochrome P-450, 260
Cytochrome P-4502E1, in alcohol-induced injury, 263, 264
Cytokines, ozone effect on, 232, 233

D

Decomposition, of radicals, 3–4
Deferoxamine, postischemic reperfusion injury and, 148–149

Dehydroascorbic acid (DHA), 182
Dehydrogenations, 2–3
Deoxyguanosine, aromatic hydroxylation, 31
Deoxyhemoglobin, 38
Deoxyribose assay, for hydroxyl free radical, 28–29
DETAPAC (diethylenetriaminepentaacetic acid), 297, 305
DHA (dehydroascorbic acid), 182
DHA (docosohexaenoic acid), 101
DHBAs (2,5-dihydroxybenzoic acids), 31–33
Diabetes, vitamin E and, 106
Diesel exhaust particles, lung overload from, 283–284
Diet. See also Antioxidants, in food
  carcinogenesis and, 328
Diethylenetriaminepentaacetic acid (DETAPAC), 297, 305
2,5-Dihydroxybenzoic acids (DHBAs), 31–33
3,4-Dihydroxyphenylalanine, 30
Dimerization, 3
2,4-Dinitrophenylhydrazine (DNPH), 46, 47
Dioxygen (molecular oxygen), 5
Dipalmitoyl lecithin, 232
Disproportionation, 3
Di-*tert*-butyl nitroxide (DTBN), 300
Dityrosine, 46
DMPO, 297–298
DMPO-OH, 298
DNA adducts
  measurement, 41–43
  mutagenicity, 41
  production, 41
DNA damage, radical-induced
  lipid peroxidation and, 328
  markers, 328
  measurement, 40–44
  RNS and, 262
  ROS and, 262
  strand breaks, 101
DNPH (2,4-dinitrophenylhydrazine), 46, 47
Docosohexaenoic acid (DHA), 101
Doxorubicin. See Adriamycin
Drugs. See also specific drugs
  toxicities, free-radical-mediated, 136
DTBN (di-*tert*-butyl nitroxide), 300
Dust particles, lung overload from, 283

E

EBPs (vitamin E binding proteins), 107
EC-superoxide dismutase, 127
Eicosanoids, in alcohol-induced tissue injury, 266
Eicosapentaenoic acid (EPA), 101
Electron paramagnetic resonance spectroscopy (EPR), 3–4
Electron shuttling, 181
Electron spin resonance (ESR)
  measurement, of hydroxyl free radical, 29
  spin trap approach
    description of, 29
    for phosgene, 291–295

V

Vanadium, in particulate matter, 276
Very-low density lipoprotein (VLDL), 107
Viruses, ozone-induced inactivation, 232–233
Vitamin A (retinol)
  characteristics, 126–127, 209
  functions, 55, 209
  isoforms
    A_1, 8–9
    A_2, 8–9
  mutagenicity, 214
  prostate cancer and, 211–212
  tetatogenicity
    in animals, 213–214
    in humans, 210–211
  toxicity
    acute animal, 212
    acute human, 209–210
    animal reproductive studies, 213–214
    chronic human, 210–212
    repeat dose animal, 212–213
  vitamin E antagonism and, 97
Vitamin C (ascorbic acid)
  adrenocortical function and, 167
  anticarcinogenic activity, 83, 105, 329
  antioxidant activity, 16, 50
  carcinogenicity, 330–331
  characteristics, 126, 214
  colorectal cancer and, 86
  coprotection with vitamin E and GSH, 181–182
  coronary artery disease progression and, 82
  deficiency, 182
    pituitary-adrenal axis and, 167, 168
    wound healing and, 214–215
  diabetes and, 106
  functions, 50, 96
  insulin release and, 168
  lung oxidative stress and, 229
  measurement, 50
  oxidation, 6–7
  protection, against ozone exposure, 105–106
  redox activity, 13–14
  response, to exercise, 129
  sources, dietary, 208
  supplementation
    for cardiovascular disease prevention, 115
    coronary heart disease progression and, 114–115
    for immune disorders, 167
  toxicity
    in animals, 215–216
    in humans, 215
    LD50 values, 216
Vitamin E binding proteins (EBPs), 107
Vitamin E (α-tocopherol), 16
  absorption, 79, 107
  air pollution damage and, 87–88
  anticancer activity, 98–99, 329
  antioxidant effects, 14–15
  biological activity, 95–96
  biopotency, 9–10

Vitamin E (α-tocopheròl) (*Continued*)
  cancel risk and, 83–86, 104–105
  cardiovascular disease and, 81–82
  cataracts and, 86–87
  cell senescence and, 99
  characteristics, 51, 126, 216
  chemical formula, 303
  colorectal cancer and, 86
  coprotection with vitamin C and GSH, 181–182
  coronary artery disease progression, vitamin C intake and, 82
  deficiency, 80, 216–217
    arteriosclerosis and, 102
    cancer induction and, 104
    genetic predisposition for, 98
    symptoms of, 98, 99
  derivatives, 79
  diabetes and, 106
  dietary consumption
    coronary heart disease and, 102–104
    requirements for, 54, 97
  distribution, 107
  formation, 54
  functions, 79, 80, 98, 217
    anticancer, 104–105
    cell membrane stabilization, 99
  hepatic lipid peroxidation and, 241, 242
  in human tissues, 79
  interactions, with protein kinase C, 98
  LDL content, 120
  in low density lipoprotein, 103
  measurement, 52, 53
  membrane-confined, regeneration by ascorbic acid, 50
  microsomal fractions, 97
  mitochondrial fractions, 96–97
  oxidation products, 96
  polyunsaturated fats and, 95
  prevention of lipid peroxidation, 51–52
  protection, against lung injury, 105
  reducing activity, 96
  regeneration, 49
  requirements, 80–81
  research, future, 107–108
  response, to exercise, 129–130
  side chain, 96
  side effects/safety, 88
  smoking and, 87–88
  sources, 80
    dietary, 97–98, 208
  structure, 95, 303
  supplementation
    cancel risk and, 86
    for cardiovascular disease prevention, 115
    coronary heart disease progression and, 114–115
    coronary heart disease risk and, 102
    exercise and, 129–130
    LDL oxidation lag phase and, 117–118
    for LDL oxidation resistance, 104
    lung disease risk and, 105
  synergistic interactions, 96